Electromechanical Systems and Devices

Electromechanical Systems and Devices

Sergey E. Lyshevski

CRC Press
Taylor & Francis Group
Boca Raton London New York

CRC Press is an imprint of the
Taylor & Francis Group, an **informa** business

CRC Press
Taylor & Francis Group
6000 Broken Sound Parkway NW, Suite 300
Boca Raton, FL 33487-2742

© 2008 by Taylor & Francis Group, LLC
CRC Press is an imprint of Taylor & Francis Group, an Informa business

No claim to original U.S. Government works
Printed in the United States of America on acid-free paper
10 9 8 7 6 5 4 3 2 1

International Standard Book Number-13: 978-1-4200-6972-3 (Hardcover)

Library of Congress Cataloging-in-Publication Data

Lyshevski, Sergey Edward.
    Electromechanical systems and devices / Sergey E. Lyshevski. -- 1st ed.
       p. cm.
    Includes bibliographical references and index.
    ISBN 978-1-4200-6972-3 (alk. paper)
    1. Power electronics. 2. Electromechanical devices. I. Title.

TK6271.A93 1988
621.31'7--dc22   2007040400

**Visit the Taylor & Francis Web site at
http://www.taylorandfrancis.com**

**and the CRC Press Web site at
http://www.crcpress.com**

080096573

Dedication

*Dedicated to my family with a deep appreciation and
admiration of their love, devotion, and support*

Contents

Preface

We encounter electromechanical systems in daily activities. Because of the enormous impact of electromechanical systems, which has been signified in recent years, as well as enabling actuation/sensing/control/fabrication technologies, this book is written. The overall goal is to introduce and coherently cover electromechanical systems and their components. In particular, we will examine motion devices (actuators, motors, transducers, sensors, and others), power electronics, controllers, and so on. The major emphasis is focused on high-performance electromechanical systems addressing analysis, design, and implementation issues. Different electromechanical systems are widely used as electric drives and servosystems. A variety of enabling electromechanical systems and devices are covered. Recent trends in engineering have increased the emphasis on integrated analysis, design, and control. The scope of electromechanical systems has been expanded, and systems integrate actuators, sensors, power electronics, integrated circuits (ICs), microprocessors, digital signal processors (DSPs), and so on. Even though the basic fundamentals have been developed, some urgent areas were downgraded or less emphasized. This book aims to ensure descriptive features, extend the results to the modern hardware-software developments, utilize enabling solutions, place the integrated system perspectives in favor of a sound engineering, as well as focus on the unified studies.

This book facilitates comprehensive studies and covers the design aspects of electromechanical systems with high-performance motion devices. We combine traditional engineering topics and subjects with the latest technologies and developments in order to stimulate new advances in design of state-of-the-art systems. The major objective of this book is to provide a deep understanding of the engineering underpinnings of integrated technologies. The modern picture of electromechanics, energy conversion, electric machines, and electromechanical motion devices is provided.

The demand for an educational book in electromechanical systems and devices far exceeds what was previously expected by the academy, industry, and the engineering community. Although excellent textbooks in electric machines, power electronics, ICs, microcontrollers, and DSPs were published, and outstanding books about control are available, the need for a comprehensive study of electromechanical systems is evident. There is a lack of books that comprehensively cover and examine high-performance electromechanical systems. This book targets the frontiers of electromechanical engineering and science examining basic theory, emerging technologies, advanced software, and enabling hardware. The author is disturbed by the recent increase in the number of students whose good programming skills and sound theoretical background are matched with their complete inability to solve the simplest engineering problems. The major aims of this book are to demonstrate the application of cornerstone fundamentals in analysis and design of electromechanical systems, cover emerging software and hardware, develop and introduce the rigorous theory, and help the reader to develop strong problem-solving skills. This book offers an in-depth presentation and contemporary coverage facilitating the developments of problem-solving skills. This book is readable, comprehendible, and accessible to students and engineers because it develops a thorough understanding of integrated perspectives, and, by means of practical worked-out examples, documents how to use the results. Engineers and students who master this book will know what they are doing, why they are doing it, and how to do it. To avoid possible difficulties, the material is presented in sufficient details. In particular, basic results (needed to fully understand, appreciate, and apply the knowledge) are covered for use by those whose background and expertise could be deficient to some extent. Step-by-step, this book guides the reader through various aspects, for example, from rigorous theory to advanced applications, from coherent design to systems analysis, and so on.

Analysis and optimization are very important in designing advanced systems. Competition has prompted cost and product cycle reductions. To accelerate analysis and design, ensure productivity and creativity, integrate advanced control algorithms, attain rapid prototyping, generate C codes, and visualize the results, MATLAB® (with embedded Simulink®, Real-Time Workshop®, Control, Optimization, Signal Processing, Symbolic Math, and other application-specific environments and toolboxes) is used. The book demonstrates the MATLAB capabilities and helps the reader to master this viable environment, studies important practical examples, helps increase designer productivity by showing how to use the advanced software, and so on. MATLAB offers a rich set of capabilities to efficiently solve a variety of complex design and analysis problems. The reader can easily modify the studied application-specific problems and utilize the reported MATLAB files for particular systems. The examples covered in this book consist of a wide spectrum of practical systems and devices. Users can easily apply these results, modify the findings, and develop new MATLAB files and Simulink block diagrams. For various enterprise-wide practical examples, efficient analysis and design methods to solve practical problems are demonstrated. The documented results provide the solutions for various simulation, analysis, control, and optimization tasks as applied to electromechanical systems and devices.

In line with the specifically related topics philosophy and flexibility of course content, this book is adaptable to a wide variety of courses fulfilling overall and specific objectives as well as goals. *Electromechanical Systems and Devices* forms a versatile and modular basis for the following courses:

- Energy Conversion
- Electric Machines
- Electromechanics
- Electromechanical Systems
- Mechatronics

The above-listed courses are offered as undergraduate and graduate courses by many electrical and mechanical engineering departments around the world. This book is balanced and works nicely for one- and two-semester courses because the material is arranged to guarantee the greatest degree of flexibility in the choice of topics. Different course settings are possible at the undergraduate and graduate levels.

The revision of this book depends on students, engineers, scholars, and faculty who would be kind enough to provide their suggestions. This valuable communication, especially concerning potential deficiencies and shortcomings, will be greatly appreciated. The author welcomes any comments and corrections to ensure the highest degree of quality and accuracy. Your comments and suggestions are very welcome. It was a great pleasure to work on this book, and I hope readers will enjoy and will like this book.

Sergey Edward Lyshevski
E-mail: Sergey.Lyshevski@mail.rit.edu
Web site: www.rit/~seleee

Acknowledgments

Many people contributed to this book. I would like to express my sincere acknowledgments and gratitude to many colleagues and peers who have provided valuable and deeply treasured feedback. Thanks go to students who have taken courses I taught; their feedback was very helpful. Many examples reported in the book to some degree reflect the research performed under numerous grants and contracts performed for the U.S. federal government, laboratories, high-technology companies, and other agencies. Those opportunities have been very useful and beneficial. It gives me a great pleasure to acknowledge the help I received from many people in the preparation of this book. The outstanding CRC Press team, especially Nora Konopka (Acquisitions Editor, Electrical Engineering), Jessica Vakili (Project Coordinator) and Gail Renard (Project Editor), tremendously helped and assisted me.

Sincere gratitude goes to MathWorks, Inc. for supplying the MATLAB® environment (MathWorks, Inc., 24 Prime Park Way, Natick, MA 01760-15000 http://www.mathworks.com.) Many thanks to all of you.

About the Author

Sergey Edward Lyshevski was born in Kiev, Ukraine. He received his M.S. (1980) and Ph.D. (1987) degrees from Kiev Polytechnic Institute, both in electrical engineering. From 1980 to 1993, Dr. Lyshevski held faculty positions at the Department of Electrical Engineering at Kiev Polytechnic Institute and the Academy of Sciences of Ukraine. From 1989 to 1993, he was the microelectronic and electromechanical systems division head at the Academy of Sciences of Ukraine. From 1993 to 2002, he was with Purdue School of Engineering as an associate professor of electrical and computer engineering. In 2002, Dr. Lyshevski joined Rochester Institute of Technology as a professor of electrical engineering. Dr. Lyshevski serves as a full professor faculty fellow at the U.S. Air Force Research Laboratories and Naval Warfare Centers.

Dr. Lyshevski is the author of 14 books. He is the author or coauthor of more than 300 journal articles, handbook chapters, and regular conference papers. His current research and teaching activities include the areas of electromechanical and electronic systems, micro- and nanoengineering, intelligent large-scale systems, molecular processing, systems informatics, and biomimetics. Dr. Lyshevski has made significant contributions in the devising, design, application, verification, and implementation of various advanced aerospace, electromechanical, and naval systems. He has made more than 30 invited presentations nationally and internationally.

1

Introduction to Electromechanical Systems

Every day, one utilizes and largely depends on thousands of electromechanical systems. More than 99.9% of electricity is produced by electric machines (generators) that convert one form of energy (nuclear, hydro, solar, thermal, wind, and other) into electric energy. By increasing the efficiency of power generators by 1%, one will reduce the oil and coal consumption by millions of barrels and tons per day. Synchronous generators, which induce the voltage in power systems, are covered in this book. Conventional synchronous generators are utilized in power plants, whereas permanent-magnet synchronous generators are widely used in auxiliary power units, low/medium-power alternative energy modules, and so on. Our major emphasis is focused on high-performance electromechanical systems and motion devices for electric drives and servosystems. For example, in "high-end" applications such as computer and camera hard drives, two electromechanical systems (drive and servo) are utilized, as reported in Figure 1.1. Without those electromechanical systems and actuators, one would not be able to assess the hard drive memory. In cars, there are hundreds of electromechanical systems, from the starter/alternator to various solenoids, fans, microphones, speakers, and even a traction electric drive in hybrid cars.

A phenomenal growth in electromechanical systems has been accomplished as a result of:

1. Raising industrial/societal needs and growing market;
2. Affordability and overall superiority of electromechanical systems as compared to any other (hydraulic, pneumatic, and other) drives and servos;
3. Rapid advances in actuators, sensors, power electronics, integrated circuits (ICs), microprocessors, and digital signal processors (DSPs) hardware;
4. Matured cost-effective fabrication technologies.

Fundamental and applied developments in actuators, electric machines, sensors, and power electronics notably contributed and motivated current

FIGURE 1.1
Two hard drives. In the hard drive on the left, to displace the pointer, the rotational motion of the stepper motor (at the top left) is converted to the translational motion (displacement) by using a gear (mechanical kinematics). The direct-drive limited-angle axial topology actuator is illustrated in the hard drive on the right. Permanent-magnet synchronous motors (at the center of the hard drives), controlled by power electronics, rotate the disk. As documented, the phase windings are on the stator (stationary machine member), whereas the radial segmented permanent magnets are on the rotor.

societal progress, welfare, and technological advances. There is a need to further expand meaningful developments by:

1. Devising, advancing and integrating leading-edge actuation and sensing paradigms;
2. Enhancing/devising device physics of electromechanical motion devices, thereby ensuring enabling performance and capabilities;
3. Advancing hardware and developing novel software;
4. Developing and implementing advanced fabrication technologies;
5. Applying enabling energy conversion concepts, and so on.

The concurrent design provides the end-users with the needed coherency integrating subsystems and components within high-performance electromechanical systems. With the stringent requirements on electromechanical systems performance and capabilities, the designer applies the advanced concepts in analysis, design, and optimization. This introductory chapter discusses the electromechanical systems and their components reporting introductory and conceptual features.

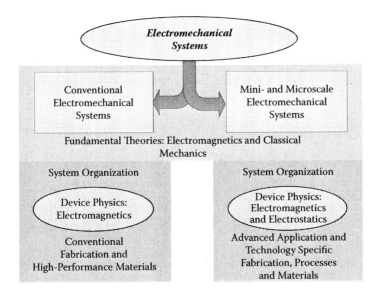

FIGURE 1.2
Electromechanical systems.

In general, electromechanical systems are classified as:

- Conventional electromechanical systems,
- Mini-/micro-scale electromechanical systems.

The operating principles and basic foundations of conventional and mini-/micro-scale electromechanical systems are identical or similar. The overall device physics and analysis are based on the electromagnetics and classical mechanics. However, the device physics specifics (electromagnetic phenomena and effect utilized, for example, electromagnetics versus electrostatics), system organization, and fabrication technologies (including processes and materials used) can be profoundly different. Figure 1.2 illustrates those features.

The structural and organizational complexity of electromechanical systems has been increased drastically as a result of hardware advancements and stringent performance requirements imposed. To meet the rising demands on the system complexity and performance, novel solutions and design concepts have been introduced and applied. In addition to the coherent choice of system components (subsystems, modules, devices, etc.), there are various issues that must be addressed in view of the constantly evolving nature of integrated developments in design, analysis, optimization, complexity, diagnostics, packaging, and so on. The *optimum-performance* systems can be designed only by applying the advanced hardware and software. Integrated multidisciplinary features approach quickly. As documented in Figure 1.3,

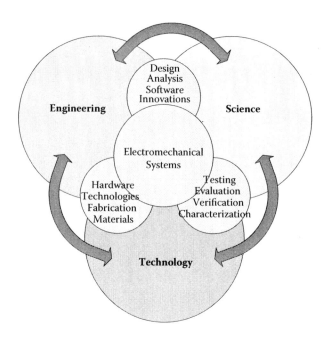

FIGURE 1.3
Integration of engineering, science, and technology.

integration of engineering (electrical, mechanical, and computer), science and technologies are taking place.

One of the most challenging problems in the electromechanical systems design is the development and integration of advanced hardware components (actuators, sensors, power electronics, integrated circuits [ICs], microcontrollers, digital signal processors [DSPs], and others), device/system-level optimization, and software developments (environments, tools, and computation algorithms to perform CAD, control, sensing, data acquisition, simulation, visualization, virtual prototyping, and evaluation). Attempts to design high-performance electromechanical systems guaranteeing the integrated design can be pursued through analysis of complex patterns and paradigms of enabling devices, systems, and technologies. For example, permanent-magnet-centered actuation and sensing solutions are an enabling paradigm and technologies, whereas digital electronics may ensure overall advantages as compared to the analog ICs. Recent trends in engineering have increased the emphasis on integrated design and analysis of advanced systems. The design process is evolutionary in nature. It starts with a given set of requirements and specifications. High-level functional design can be performed first in order to produce specification at the component level. Using the advanced components, the initial design is performed, and the electromechanical system performance is studied against the requirements. If requirements and specifications are not met, the designer revisits and refines the system organization and optimizes the design devising

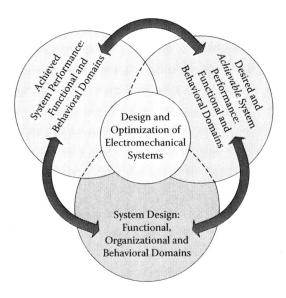

FIGURE 1.4
Interactive design flow.

and evaluating other alternative solutions. At each level of the design hierarchy, the system performance in the behavioral domain is used to evaluate and refine (if needed) the structural design and solutions applied. Each level of the design hierarchy corresponds to a particular abstraction level and has the specified set of activities and design tools, which support the design at this level. Different concepts are applied and used to design actuators and ICs as a result of different device physics, operational principles, behavior, physical properties, and performance criteria imposed for these components. The level of hierarchy must be defined, and there is no need to study the behavior of thousands of transistors in each IC because electromechanical systems might integrate hundreds of ICs components. The end-to-end behavior of ICs is usually evaluated because ICs are assumed to be optimized. The design flow is illustrated in Figure 1.4. The fundamental and technological limits and constraints are imposed and exist there. Only through sound structural and behavioral designs can an optimal performance be accomplished. The required electromechanical system performance cannot exceed the best (*achievable*) performance, which is a result of the fundamental and technological limits. Distinct performance estimates, metrics, and measures are used integrating different specifications on the efficiency, robustness, redundancy, power/torque density, accuracy, and other requirements tailored to specific hardware solutions. The so-called *systems design* should be applied with a great care. To apply a high-abstraction-level methodology, one should have a great deal of expertise on the available hardware, possible solutions, performance analysis, capabilities assessment, and so on. The blind application of the *systems design*, without the use of fundamentals, usually results in catastrophic failure.

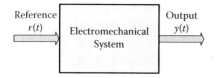

FIGURE 1.5

Electromechanical system with input reference $r(t)$ and output $y(t)$.

The electromechanical systems performance and capabilities are measured using many criteria, for example, functionality, efficiency, stability, robustness, sensitivity, transient behavior, accuracy, disturbance attenuation, noise immunity, thermodynamics, and so on. The specifications depend on the requirements imposed in the full operating envelope. For example, examining the system dynamics, the designer can analyze and optimize the input-output transient dynamics. In particular, denoting the reference (command) and output variables as $r(t)$ and $y(t)$, the tracking error $e(t) = r(t) - y(t)$ is minimized, and behavior for $y(t)$ and $e(t)$ is optimized by using different performance criteria. For example, to optimize the output system dynamics, one can minimize the tracking error and settling time ensuring robustness. The electromechanical system with the reference $r(t)$ and output $y(t)$ is documented in Figure 1.5. Transient responses are optimized designing closed-loop systems with control laws by applying the performance functionals that include time and tracking error. For example, the minimizing functionals can be given as

$$J = \min_e \int_0^\infty |e|\, dt, \qquad J = \min_e \int_0^\infty e^2 dt, \qquad J = \min_{t,e} \int_0^\infty t\, |e|\, dt,$$

and so on. The state and control variables (x and u), which significantly affect the system performance (stability, efficiency, sensitivity, etc.), also should be integrated in the design as discussed in Example 7.1 and covered in this book.

Among the most important criteria imposed, one can emphasize efficiency, accuracy, and stability in the full operating envelope. A wide variety of other requirements are usually imposed. Electromechanical systems are nonlinear and can be multivariable. One solves challenging problems performing the systematic design, analysis, and optimization in order to design high-performance systems. The automated synthesis can be attained implementing sound design flows and taxonomies. The design of electromechanical systems is a process that starts from the specification and requirements progressively proceeding to a functional design and optimization that are gradually refined through a sequence of steps. Specifications typically include the performance requirements derived from systems functionality, operating envelope, sizing features, cost, and so on. Both device-to-system and system-to-device hardware integration, as well as hardware-software integration, should be combined studying hierarchy, integrity, regularity, modularity, compliance, matching, and completeness. The electromechanical systems synthesis should guarantee an eventual consensus between functional, behavioral, and structural (organizational) domains. The descriptive and integrative features must be ensured through quantitative and qualitative analysis. The unified analysis of advanced hardware (actuators, sensors, power electronics, ICs, microprocessors, and other components) and software are carried out.

One strives to achieve the synergistic combination of precision engineering, electronic control, sound analysis paradigms, enabling optimization concepts, and advanced hardware within the functional, structural, and behavioral designs. For electromechanical systems (robots, electric drives, servomechanisms, pointing systems, etc.), accurate actuation, sensing, and control are challenging problems. Actuators and sensors must be designed (or chosen) and integrated with the corresponding power electronics. The principles of matching and compliance are general design principles that require that the systems should be synthesized soundly integrating all components. The matching conditions have to be determined and guaranteed, and actuators–sensors–ICs–power electronics compliance must be satisfied. Electromechanical systems must be controlled, and controllers are utilized. For closed-loop systems, various analog and digital control laws can be synthesized, examined, tested, verified, and implemented with the objective to derive an optimal controller, which should ensure optimal performance. The control tasks aim to derive and implement controllers, find feedback gains, optimize system performance, enhance system capabilities, and so on. To implement digital controllers, microprocessors and DSPs with ICs (input-output devices, A/D and D/A converters, optocouplers, transistor drivers, etc.) are used. Analog controllers are also commonly used. The schematics of the controlled closed-loop electromechanical system is documented in Figure 1.6 for the aircraft emphasizing a stand-alone actuator. In aircraft and other flight and marine vehicles, various control surfaces must be properly displaced. The system organization is different if analog or digital controllers are used. For example, analog controllers can be realized using operational amplifiers, whereas digital controllers are implemented using microcontrollers, microprocessors, and DSPs.

As given in Figure 1.6, each actuator (electromechanical motion device) is regulated by using the state variables $x(t)$, as well as the difference between the desired reference input $r(t)$ and the output $y(t)$. The tracking error $e(t) = r(t) - y(t)$

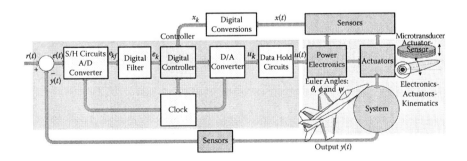

FIGURE 1.6
Block diagram of electromechanical systems (flight actuators to displace control surfaces) with a digital controller.

is used. For robots, aircraft, submarines, and other systems, the Euler angles θ, ϕ, and ψ are commonly considered as the outputs. That is, the output vector is

$$y(t) = \begin{cases} \theta(t) \\ \phi(t). \\ \psi(t) \end{cases}$$

The reference inputs for an aircraft, depicted in Figure 1.6, are the desired Euler angles r_θ, r_ϕ and r_ψ, yielding the reference vector

$$r(t) = \begin{cases} r_\theta(t) \\ r_\phi(t). \\ r_\psi(t) \end{cases}$$

To control the aircraft, one deflects various control surfaces utilizing the control surface servos. In particular, by applying the voltage, one changes the angular or linear displacement of actuators thereby accomplishing control of aircraft. For each rotational or translational actuator, the desired (reference) deflection $r(t)$ is compared with the actual displacement $y(t)$ resulting in $e(t)$. One distinguishes the variables $r(t)$, $y(t)$, and $e(t)$ used for an actuator, with the vectors $\mathbf{r}(t)$, $\mathbf{y}(t)$, and $\mathbf{e}(t)$ used for an entire system. Advanced fighters are controlled by displacing hundreds of rotational and translational actuators with outputs y_i as well as varying the engine thrust.

Figure 1.7 depicts the underwater vehicle hull with four actuators (to displace fins) and power electronics. As illustrated, one controls the underwater vehicle path by using four actuators that displace control surfaces.

FIGURE 1.7
Underwater vehicle hull with actuators to displace fins. The actuators are controlled using power electronics with embedded controllers.

Microprocessors and DSPs are widely used to control electromechanical systems. Specifically, DSPs are used to derive control signals based on the control algorithms, perform data acquisition, implement filters, attain decision making, and so on. For single-input/single-output systems, assuming that the continuous reference and output are measured by sensors, the continuous-time error signal $e(t) = r(t) - y(t)$ is converted into the digital form to perform digital filtering and control. As illustrated in Figure 1.6, the sample-and-hold circuit (S/H circuit) receives the continuous-time (analog) signal and holds this signal at the constant value for the specified period of time that is related to the sampling period. Analog-to-digital converter (A/D converter) converts this piecewise or continuous-time signal to the digital (binary) form. The conversion of continuous-time signals to discrete-time signals is called sampling or discretization. The input signal to the filter is the sampled version of the continuous-time error $e(t)$. The input to a digital controller (microcontroller or DSP) is the digital filter output signal. Analog filters are also widely used. At each sampling, the discretized value of the error signal e_k in binary form is used by a digital controller to generate the control signal, which must be converted to analog form to be fed to the driving circuitry of power converter. The digital-to-analog conversion (decoding) is performed by the digital-to-analog converter (D/A converter) and the data hold circuit. Coding and decoding are synchronized by using the clock. This brief description illustrates that there are various signal conversions, for example, multiplexing, demultiplexing, sample and hold, analog-to-digital (quantizing and encoding) conversion, digital-to-analog (decoding) conversion, and so on. Electromechanical systems are designed using advanced ICs, microprocessors, DSPs, power electronics, actuators, and sensors. Advanced hardware is usually applied in electromechanical systems.

The single-input/single-output electromechanical system is reported in Figure 1.8. In particular, we consider an electromechanical system which consists of pointing system–geared motor–PWM amplifier–ICs–DSP. Using the reference r and actual θ angular displacements (measured by the encoder), the DSP (with embedded controller) develops PWM signals to drive high-frequency IGBTs or MOSFETs. The number of the PWM outputs needed depends on the converter output stage topology. A three-phase permanent-magnet synchronous motor is illustrated in Figure 1.8. Usually, six PWM outputs drive six transistors to vary the phase voltages u_{as}, u_{bs}, and u_{cs}. The magnitude of the output voltage of the PWM amplifier is controlled by changing the duty cycle, and the Hall-effect sensors measure the rotor angular displacement θ_r to generate the balanced phase voltages u_{as}, u_{bs}, and u_{cs}. A fully integrated electromechanical system hardware is shown in Figure 1.8.

Figure 1.9 illustrates the functional block-diagram of the electromechanical system (energy source is not illustrated) that comprise different components.

We formulate a definition to define an electromechanical system as: The electromechanical system, which comprises of electromechanical motion

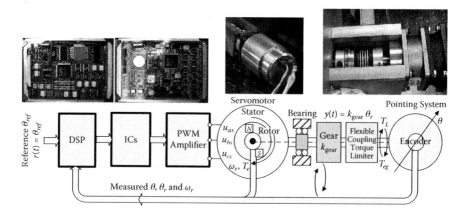

FIGURE 1.8
Schematic diagram and hardware of a digital electromechanical system.

device–sensors–power electronics–controlling/processing and driving/sensing circuitry (ICs and DSPs):

1. Performs energy conversion, actuation, and sensing;
2. Converts physical stimuli and events to electrical and mechanical quantities and vice versa;
3. Comprises control, diagnostics, data acquisition, and other features.

The functional and structural designs of high-performance electromechanical systems frequently imply the component (modules, devices, etc.) developments. Electric machines (electromechanical motion devices), used

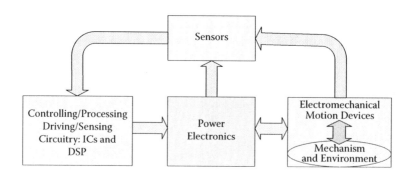

FIGURE 1.9
Functional block-diagram of electromechanical system.

as actuators, generators, and sensors, are one of the major components. The following problems are usually emphasized:

- Design and optimization of electric machines (actuators, generators, and sensors) according to their applications and overall systems requirements;
- Design of high-performance actuators, generators, and sensors for specific applications;
- Integration of actuators and generators with sensors, power electronics, and ICs (one emphasizes integrity, regularity, modularity, compliance, matching, and completeness);
- Control of actuators, generators, and sensors.

We primarily focus on high-performance electric machines. The device physics, functionality, and operation will be examined. Optimization, synthesis, modeling, and simulation are complementary activities performed to ensure sound design and analysis. Simulation starts with the model developments, whereas synthesis starts with the specifications imposed on the behavior and analysis of the system performance. The designer studies, analyzes, and evaluates the system behavior using state, output, performance, control, events, disturbance, and other variables. Modeling, simulation, analysis, and prototyping are critical and urgently important aspects for developing advanced electromechanical systems. As a flexible high-performance simulation environment, MATLAB® has become a standard software tool. Competition has prompted cost and product cycle reductions. To accelerate design and analysis, facilitate significant gains in productivity and creativity, integrate control and signal processing using advanced microprocessors/DSPs, accelerate prototyping features, generate real-time C code, visualize the results, perform data acquisition, as well as ensure data-intensive analysis, MATLAB and Simulink® are used. Within MATLAB, the following toolboxes can be applied: Real-Time Workshop, Control System, Nonlinear Control Design, Optimization, Robust Control, Signal Processing, Symbolic Math, System Identification, Partial Differential Equations, Neural Networks, and other application-specific tools. We will demonstrate the MATLAB capabilities by solving practical examples enhancing the user competence. The MATLAB environment offers a rich set of capabilities to efficiently solve a variety of complex analysis, control, and optimization problems. The examples will provide the practice and educational experience within the highest degree of comprehensiveness and coverage.

Conventional and mini-/micro-scale electromechanical systems should be studied from the unified perspective. The operating features, basic phenomena, and dominant effects are based on electromagnetics and classical mechanics. Electromechanical systems integrate various components. No matter how well an individual component (actuator, motion device, sensor, power amplifier, or DSP) performs, the overall performance can be degraded if the designer fails to soundly integrate and optimize the system. Although actuators, generators, sensors, power electronics, and microcontrollers/ DSPs should be analyzed, designed, and optimized, the focus also should be centered on the hardware, software, and hardware-software integration and compliance. The designer sometimes fails to grasp and understand the global picture. Although the component-based *divide-and-solve* approach or *systems design* are applicable in the preliminary design phase, it is important that the design and analysis of an integrated electromechanical system be accomplished in the context of sound optimization with proper functionality, objectives, specifications, requirements, and limits. Enabling hardware (components, modules, and systems), advanced technologies, high-performance software, and software-hardware codesign tools must be applied. Excellent textbooks in electric machinery [1–8], power electronics [9–11], microelectronics and ICs [12], and sensors [13, 14] have been published. Educational examples in analysis and design of linear electromechanical systems are available in control books [15–21]. We focus on high-performance electromechanical systems covering a broad spectrum of the cornerstone problems. There is a need to further enhance the basic theory and practice, as well as provide the coherent coverage of the current state-of-the art developments reporting most notable results.

Homework Problems

Problem 1.1

Provide examples of electromechanical systems and electromechanical motion devices.

Problem 1.2

What is the difference between electromechanical systems and electromechanical motion devices?

Problem 1.3

Choose a specific electromechanical system or device (propulsion system, traction electric drive, control surface actuator, speaker, microphone, etc.) and explicitly identify problems that need to be addressed, studied, and solved to design and analyze the examined system or device. Formulate and report the specifications and requirements imposed on the system of your interest. Develop a high-level functional diagram with the major components. Formulate and report the structural and behavioral design tasks.

Problem 1.4

Explain why systems and devices must be examined in the functional, structural, and behavioral domains.

References

1. S.J. Chapman, *Electric Machinery Fundamentals*, McGraw-Hill, New York, 1999.
2. A.E. Fitzgerald, C. Kingsley, and S.D. Umans, *Electric Machinery*, McGraw-Hill, New York, 1990.
3. P.C. Krause and O. Wasynczuk, *Electromechanical Motion Devices*, McGraw-Hill, New York, 1989.
4. P.C. Krause, O. Wasynczuk, and S.D. Sudhoff, *Analysis of Electric Machinery*, IEEE Press, New York, 1995.
5. W. Leonhard, *Control of Electrical Drives*, Springer, Berlin, 1996.
6. S.E. Lyshevski, *Electromechanical Systems, Electric Machines, and Applied Mechatronics*, CRC Press, Boca Raton, FL, 1999.
7. D.W. Novotny and T.A. Lipo, *Vector Control and Dynamics of AC Drives*, Clarendon Press, Oxford, 1996.
8. G.R. Slemon, *Electric Machines and Drives*, Addison-Wesley Publishing Company, Reading, MA, 1992.
9. D.W. Hart, *Introduction to Power Electronics*, Prentice Hall, Upper Saddle River, NJ, 1997.
10. J.G. Kassakian, M.F. Schlecht, and G.C. Verghese, *Principles of Power Electronics*, Addison-Wesley Publishing Company, Reading, MA, 1991.
11. N.T. Mohan, M. Undeland, and W.P. Robbins, *Power Electronics: Converters, Applications, and Design*, John Wiley and Sons, New York, 1995.
12. A.S. Sedra and K.C. Smith, *Microelectronic Circuits*, Oxford University Press, New York, 1997.
13. J. Fraden, *Handbook of Modern Sensors: Physics, Design, and Applications*, AIP Press, Woodbury, NY, 1997.
14. G.T.A. Kovacs, *Micromachined Transducers Sourcebook*, McGraw-Hill, New York, 1998.
15. R.C. Dorf and R.H. Bishop, *Modern Control Systems*, Addison-Wesley Publishing Company, Reading, MA, 1995.
16. J.F. Franklin, J.D. Powell, and A. Emami-Naeini, *Feedback Control of Dynamic Systems*, Addison-Wesley Publishing Company, Reading, MA, 1994.
17. B.C. Kuo, *Automatic Control Systems*, Prentice Hall, Englewood Cliffs, NJ, 1995.
18. S.E. Lyshevski, *Control Systems Theory with Engineering Applications*, Birkhäuser, Boston, MA, 2001.
19. K. Ogata, *Discrete-Time Control Systems*, Prentice-Hall, Upper Saddle River, NJ, 1995.
20. K. Ogata, *Modern Control Engineering*, Prentice-Hall, Upper Saddle River, NJ, 1997.
21. C.L. Phillips and R.D. Harbor, *Feedback Control Systems*, Prentice Hall, Englewood Cliffs, NJ, 1996.

2

Analysis of Electromechanical Systems and Devices

2.1 Introduction to Analysis and Modeling

Electromechanical systems must be analyzed with the ultimate objective to examine, assess, evaluate, and optimize their performance approaching the *achievable* capabilities under the physical (electromagnetic, mechanical, thermal, etc.), technological, and other limits. Various design tasks should be performed, for example, modeling, simulation, optimization, and so on. The design and analysis of electromechanical systems and devices are challenging problems because complex electromagnetic, mechanical, thermodynamic, vibroacoustic, and other phenomena and effects must be studied and described. High-fidelity and lamped-parameter mathematical models, developed utilizing physical laws, can be applied. For moderate complexity electromechanical systems, experienced designers usually can accomplish a near-optimal design utilizing their experience, expertise, and practice. Using the cornerstone physical laws, some performance estimates (efficiency, force/torque and power densities, settling time, etc.) can be estimated avoiding high-fidelity modeling and heterogeneous simulations. However, many key performance measures and capabilities (efficiency, stability, robustness, sensitivity, dynamic and steady-state accuracy, acceleration rate, dynamic responses, sampling period, etc.) may not be accurately determined by even experienced designers without a coherent analysis which is centered on describing and modeling of various physical phenomena taking into account device physics and limits imposed. To perform sound quantitative and qualitative analysis and design, applying laws of physics, mathematical models (system and device descriptions in the form of differential and constitutive equations) must be developed with minimum level of simplifications and assumptions. These models must coherently describe the baseline phenomena (time-varying fields, energy conversion, torque or force development, voltage induction, etc.), effects, and processes accurately describing the system behavior and device evolution. The device-level physics and system organization must be examined.

The application of physical laws to derive equations of motion that describe system behavior are reported in this chapter applying electromagnetics and classical mechanics. It is impractical to apply *generic* approaches, *model-free* concepts, *linguistic* models, *descriptive* techniques, and other *systems design* (systems engineering) tools to perform coherent synthesis, design, and analysis. Therefore, they are of limited applicability, practicality, and soundness for the above-mentioned concepts. Utilizing well-known cornerstone physical laws, one can straightforwardly develop accurate and coherent descriptions of electromechanical systems and motion devices. As the model is obtained, analysis and control tasks are carried out. For example, one may derive the sound control concepts and examine the closed-loop system performance and capabilities. The electromechanical system organization is reported in Chapter 1. The evolutionary analysis and design taxonomy, applied at the device and system levels, integrate:

- Device physics analysis;
- Components matching, compliance, and completeness;
- Assessment of system organization and evaluation of system capabilities;
- Data-intensive high-fidelity electromagnetic and mechanical analysis and optimization to assess the estimated *achievable* performance and capabilities, while avoiding costly and time-consuming hardware testing to accomplish these preliminary evaluation tasks;
- Rapid design that has innumerable features to assist the user to coherently formulate the problem and assess the system performance;
- Software and hardware synergy;
- Hardware and software testing, characterization, and evaluation with possible redesign tasks.

All components and devices, which constitute electromechanical systems, are very important. For example, one cannot achieve the specified angular displacement accuracy if the sensor resolution (number of pulses per revolution in resolver or optical encoder) is not adequate (see Figures 1.7 and 1.8). However, the sensor accuracy is limited. As the electromechanical system is synthesized and organization is found, one focuses on various analysis and design activities. For all components (actuators, sensors, ICs, etc.), compliance and matching must be guaranteed. The overall system performance and capabilities are **largely defined by chosen actuators, sensors, power electronics, and ICs**. Various actuators have been devised and applied. For example, electromagnetic versus electrostatic, permanent-magnet versus variable reluctance- or inductance-centered, and so on. We examine a great number of electromagnetic-based actuators (motors), generators, and sensors. Maxwell's equations, classical mechanics, energy conversion principles, and other concepts are reported and applied. The major emphasis is placed on the electromagnetic devices because, compared to the electrostatic devices, they ensure the superior force/torque and power densities, efficiency, reliability, affordability, and so on.

The stored electric and magnetic volume energy densities ρ_{We} and ρ_{Wm} are

- $\rho_{We} = \frac{1}{2}\varepsilon E^2$ for electric (electrostatic) transducers,

- $\rho_{Wm} = \frac{1}{2}\dfrac{B^2}{\mu} = \frac{1}{2}\mu H^2$ for magnetic (electromagnetic) transducers,

where ε is the permittivity, $\varepsilon = \varepsilon_0 \varepsilon_r$; ε_0 and ε_r are the permittivity of free space and relative permittivity, $\varepsilon_0 = 8.85 \times 10^{-12}$ F/m; E is the electric field intensity; μ is the permeability, $\mu = \mu_0 \mu_r$; μ_0 and μ_r are the permeability of free space and relative permeability, $\mu_0 = 4\pi \times 10^{-7}$ T-m/A; B and H are the magnetic field density and intensity.

The electrostatic actuators are commonly utilized in micro-electromechanical systems (MEMS) due to the applicability of the well-established complementary metal-oxide-semiconductor (CMOS) technology to fabricate MEMS using surface and bulk micromachining. The maximum energy density of electrostatic actuators is limited by the maximum field (voltage) that can be applied before electrostatic breakdown occurs. In mini- and microstructures, the maximum electric field is constrained resulting in the maximum energy density ρ_{Wemax}. For example, in 100×100 μm to millimetre size structures with a few μm airgap, $E < 3 \times 10^6$ V/m. Thus, one estimates ρ_{Wemax} to be ~40 J/m³. In contrast, for electromagnetic actuators, the maximum energy density ρ_{Wmmax} is limited by saturation flux density B_{sat} (B_{sat} could be ~2.5 T) and material permeability (the relative permeability μ_r varies from 100 to 1,000,000). Thus, the resulting magnetic energy density ρ_{Wmmax} could be ~100,000 J/m³. We conclude that $\rho_{We} \ll \rho_{Wm}$, and the electromagnetic transducers can store energy at least 1000 times larger than electrostatic. The ratio $\rho_{Wm}/\rho_{We} > 100$ is guaranteed even for the most favorable microscale actuators dimensions if the soft micromagnets (Fe, Ni, or NiFe) with low B_{sat} are used. The application of hard magnets ensures $\rho_{Wm}/\rho_{We} > 1000$.

REMARK

One must perform a coherent mechanical design and technology assessment integrating actuators and mechanisms to actuate. For example, electrostatic actuators may ensure favorable kinematics, fabrication, actuator-ICs integration, and packaging solutions. This may result in significant reduction in the force or torque required. Therefore, in some MEMS, electrostatic actuators could be a favorable solution ensuring the desired performance. The translational and rotational electrostatic actuators are studied in Examples 2.15 and 2.16. For example, ~1,000,000 micromirrors (each 10×10 μm), controlled by a DSP, are repositioned in the digital light processing (DLP) module used in high-definition displays. Each electrostatic microactuator is controlled, and the settling time is ~0.0001 sec. To actuate the mechanism, the developed force or torque must be greater than the load force or torque. In addition, the

TABLE 2.1

Characteristics of Some Soft Magnetic Materials

Material	B_{sat} [T]	μ_r	$\rho_e \times 10^{-6}$ [ohm-m]	T_C [°C]
Silicon iron	2	5,000	0.25–0.55	800
Iron	2.1	5,500	0.1	760
Hiperco 27	2.3	2,800	0.58	925
Molybdenum 4–79 permalloy	0.8	400,000	0.55	454
Ferrites	0.2–0.5	150–10,000	$0.1 \times 10^6 - 5 \times 10^6$	140–480

actuator integration, packaging, housing, and other aspects must be examined. Hence, although in the electromechanical systems the electromagnetic actuators ensure the superiority, in MEMS, the electrostatic, thermal, and piezoelectric actuators may be a preferable solution.

Different material can be used to fabricate electromechanical devices. The device performance is affected by the material constant and characteristics. Various materials with different characteristics can be considered. The saturation flux density B_{sat}, maximum relative permeability μ_r, electrical resistivity, and Curie temperature (T_C) for some commonly used soft magnetic materials are reported in Table 2.1. Other useful material data is provided in Section 2.2.

Studying electromechanical systems, the emphasis is placed on:

- Devising and design of high-performance systems by applying advanced (or discovering innovative) components, modules and devices, for example, actuators, power electronics, sensors, controllers, driving/sensing circuitry, ICs, and so on;
- Analysis and optimization of rotational and translational motion devices (actuators, motors, sensors, transducers, etc.);
- Development of high-performance signal processing and controlling ICs;
- Design and implementation of optimal control algorithms;
- Development of advanced software and hardware to attain the highest degree of synergy, integration, efficiency, and performance.

To assess the system performance and examine system capabilities, one performs testing, characterization, and evaluation at the systems and device levels. For sound analysis and design of electromechanical motion devices, one needs to model (describe) electromagnetic-mechanical dynamics, perform optimization, design closed-loop systems, and carry out data-intensive analysis. We focus on modeling, simulation, and optimization of high-performance systems and devices. The integrated multidisciplinary features approach quickly. The complexity of electromechanical systems and devices has been increased drastically due to hardware and software advancements as well as stringent performance requirements. Answering the demands of the rising systems complexity and performance specifications, the fundamental theory must be coherently applied.

2.2 Energy Conversion and Force Production in Electromechanical Motion Devices

Energy conversion takes place in electromechanical motion devices that convert electrical energy to mechanical energy and vice versa [1-8]. The device physics is based on the specific phenomena (effects) exhibited and utilized affecting the energy conversion and force/torque production. The performance estimates for the stored energy for electromagnetic and electrostatic devices were reported in Section 2.1. One optimizes the energy conversion and force production performing various tasks with the ultimate objective to guarantee optimal overall performance. Design and analysis of high-performance electromechanical systems require coherent studies of electromagnetics, mechanics, and energy conversion. Analytical and numerical studies ensure quantitative and qualitative analysis. In order to analyze and optimize steady-state and dynamic system behavior, one integrates phenomena exhibited, energy conversion, force production, and control. Fundamental principles of energy conversion are studied to provide basic fundamentals.

The cornerstone principle is formulated as: for a lossless electromechanical motion device (in the conservative system no energy is lost through friction, heat, or other irreversible energy conversion), the sum of the instantaneous kinetic and potential energies of the system remains constant. The energy conversion is represented in Figure 2.1.

The general equation that describes the energy conversion is

$$\underset{\text{Electrical Energy Input}}{E_E} - \underset{\text{Ohmic Losses}}{L_E} - \underset{\text{Magnetic Losses}}{L_M} = \underset{\text{Mechanical Energy}}{E_M} + \underset{\text{Friction Losses}}{L_E} + \underset{\text{Stored Energy}}{L_S} \ .$$

For conservative (lossless) energy conversion, one can write

$$\underset{\text{Change in Electrical Energy Input}}{\Delta W_E} = \underset{\text{Change in Mechanical Energy}}{\Delta W_M} + \underset{\text{Change in Electromagnetic Energy}}{\Delta W_m} \ .$$

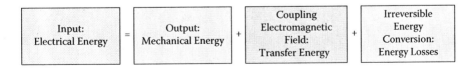

FIGURE 2.1
Energy conversion and transfer in electromechanical devices.

Using these equations one can perform the analysis of energy conversion. The electrical energy, mechanical energy, and energy losses must be explicitly defined and examined. As was emphasized in Section 2.1, electromagnetic motion devices ensure superior performance. Very high power and force densities are achieved by using permanent magnet actuators and electric machines. The total energy stored in the magnetic field is

$$W_m = \frac{1}{2} \int_v \vec{B} \cdot \vec{H} dv.$$

The material becomes magnetized in response to the external field $\vec{H}$, and the dimensionless magnetic susceptibility χ_m or relative permeability μ_r are used. We have

$$\vec{B} = \mu \vec{H} = \mu_0 (1 + \chi_m) \vec{H} = \mu_0 \mu_r \vec{H}.$$

Using the magnetic susceptibility χ_m, the magnetization is expressed as $\vec{M} = \chi_m \vec{H}$. Based upon the value of the magnetic susceptibility χ_m, the materials are classified as

- Nonmagnetic, $\chi_m = 0$, and, thus, $\mu_r = 1$;
- Diamagnetic, $\chi_m \approx -1 \times 10^{-5}$ ($\chi_m = -9.5 \times 10^{-6}$ for copper, $\chi_m = -3.2 \times 10^{-5}$ for gold, and $\chi_m = -2.6 \times 10^{-5}$ for silver);
- Paramagnetic, $\chi_m \approx 1 \times 10^{-4}$ ($\chi_m = 1.4 \times 10^{-3}$ for Fe_2O_3, and $\chi_m = -1.7 \times 10^{-3}$ for Cr_2O_3);
- Ferromagnetic, $|\chi_m| \gg 1$ (iron, nickel, cobalt, neodymium-iron-boron, samarium-cobalt, and other permanent magnets). Ferromagnetic materials exhibit high magnetizability, and materials are classified as hard (alnico, rare-earth elements, copper-nickel, and other alloys) and soft (iron, nickel, cobalt, and their alloys) materials.

The relative permeability μ_r of some bulk materials is reported in Table 2.2. The permeability **strongly** depends on the fabrication technologies and dimensionality (sizing). Therefore, one cannot assume that the superpermalloy will always guarantee that μ_r is ~1,000,000.

The magnetization behavior of the ferromagnetic materials is described by the magnetization curve, where H is the externally applied magnetic field, and B is the total magnetic flux density in the medium. Typical B-H curves for hard and soft ferromagnetic materials are shown in Figure 2.2.

Assume that initially $B_0 = 0$ and $H_0 = 0$. Let H increases form $H_0 = 0$ to H_{max}. Then, B increases from $B_0 = 0$ until the maximum value of B, denoted as B_{max}, is reached. If then H decreases to H_{min}, B decreases to B_{min} through

TABLE 2.2

Relative Permeability of Some Diamagnetic, Paramagnetic, Ferromagnetic, and Ferrimagnetic Materials

Material	Relative Permeability [μ_r]
Diamagnetic	
Silver	0.9999736
Copper	0.9999905
Paramagnetic	
Aluminum	1.000021
Tungsten	1.00008
Platinum	1.0003
Manganese	1.001
Ferromagnetic	
Purified iron (99.96% Fe)	280,000
Electric steel (99.6% Fe)	5,000
Permalloy ($Ni_{78.5\%}Fe_{21.5\%}$)	70,000
Superpermalloy ($Ni_{79\%}Fe_{15\%}Mo_{5\%}Mn_{0.5\%}$)	1,000,000
Ferrimagnetic	
Nickel-zinc ferrite	600–1,000
Manganese-zinc ferrite	700–1,500

the remanent value B_r (the so-called the residual magnetic flux density) along the different curve as illustrated in Figure 2.2. For variations of H, $H \in [H_{min}\ H_{max}]$, B changes within the *hysteresis loop*, and $B \in [B_{min}\ B_{max}]$. Figure 2.2 reports typical curves representing the dependence of magnetic induction B on magnetic field H for ferromagnetic materials. When H is first applied, B follows curve a as the favorably oriented magnetic domains grow.

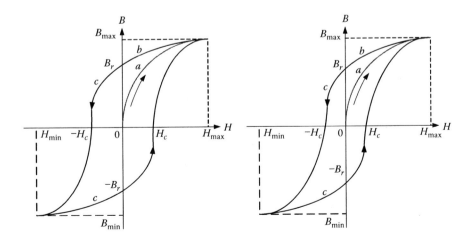

FIGURE 2.2

B-H curves for hard and soft ferromagnetic materials.

This curve reaches the saturation. When H is then reduced, B follows curve b, but retains a finite value (the remanence B_r) at $H=0$. In order to demagnetize the material, a negative field $-H_c$ must be applied. Here, H_c is called the coercive field or coercivity. As H is further decreased, and, then, increased to complete the cycle (curve c), a hysteresis loop is formed.

The area within this loop is a measure of the energy loss per cycle for a unit volume of the material. The B-H curve allows one to establish the energy analysis. In the per-unit volume, the applied field energy is $W_F = \oint_B H dB$, whereas the stored energy is expressed as $W_c = \oint_H B dH$. The equations for field and stored energies represent the areas enclosed by the corresponding curve.

In the volume v, we have the following expressions for the field and stored energies

$$W_F = v\oint_B H dB \text{ and } W_c = v\oint_H B dH.$$

In ferromagnetic materials, time-varying magnetic flux produces core losses that consist of hysteresis losses (as a result of the hysteresis loop of the B-H curve) and the eddy-current losses, which are proportional to the current frequency and lamination thickness. The area of the hysteresis loop is related to the hysteresis losses. Soft ferromagnetic materials have a narrow hysteresis loop and they are easily magnetized and demagnetized. Therefore, the lower hysteresis losses, compared with hard ferromagnetic materials, result. Different *soft* and *hard* magnets have been utilized. The following magnets have been commonly used in electromechanical motion devices: neodymium iron boron ($Nd_2Fe_{14}B$), samarium cobalt (usually Sm_1Co_5 and Sm_2Co_{17}), ceramic (ferrite), and alnico (AlNiCo). The term *soft* is used to describe those magnets that have high saturation magnetization and a low coercivity (narrow B-H curve). Another property of these magnets is their low magnetostriction. The *soft* micromagnets have been widely used in magnetic recording heads. The *hard* magnets have wide B-H curves (high coercivity), and, therefore, high-energy storage capacity. These magnets are widely used in electric machines in order to attain high force, torque, and power densities. As reported, the energy density is given as the area enclosed by the B-H curve, and the magnetic volume energy density is

$$w_m = \frac{1}{2}\vec{B}\cdot\vec{H} \quad \text{or} \quad w_m = \frac{1}{2}\mathbf{B}\cdot\mathbf{H} \ [J/m^3].$$

Most *hard* magnets are fabricated using the metallurgical processes. In particular, sintering (creating a solid but porous material from a powder), pressure bonding, injection molding, casting, and extruding.

When permanent magnets are used in electric machines and electromagnetic actuators, the demagnetization curve (second quadrant of the B-H curve) is studied. Permanent magnets store, exchange, and convert energy. In particular, permanent magnets produce stationary magnetic field without external energy sources. The operating point is determined by the permanent magnet geometry

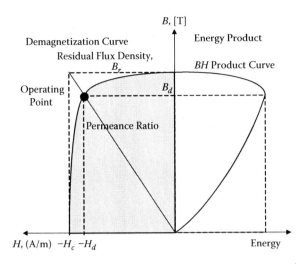

FIGURE 2.3
B-H demagnetization and energy product curves.

and properties, and the hysteresis minor loop occurs. The second-quadrant *B-H* characteristic is given in Figure 2.3 (permanent magnets operate on the demagnetization curve of the hysteresis loop). Figure 2.3 also illustrates the energy product curve for a permanent magnet. The demagnetization and energy product curves are in mutual correspondence.

The operating point is denoted by H_d and B_d. Given an air gap where the magnetic flux exists in machines or actuators, according to Ampere's law, we have

$$H_d l_m = H_{ag} l_{ag},$$

where l_m is the length of the magnet; l_{ag} is the length of the air gap parallel to the flux lines; H_{ag} is the magnetic field intensity in the air gap.

The cross-sectional area of a magnet required to produce a specific flux density in the air gap is

$$A_m = \frac{B_{ag} A_{ag}}{B_d},$$

where A_{ag} is the airgap area; B_{ag} is the flux density in the air gap.

The flux linkages caused by the magnet in the air gap are $\psi = N\Phi = NB_{ag}A_{ag}$, and the co-energy is given by

$$W = \int_\psi i \cdot d\psi = \int_i \psi \cdot di.$$

In a lossless system, an informative energy equation for the air gap is

$$Vol_{ag} B_{ag} H_{ag} = A_{ag} l_{ag} \frac{A_m l_m B_d H_d}{A_{ag} l_{ag}} = A_m l_m B_d H_d = \psi i.$$

The flux density at position **r** can be derived. In a one-dimensional case for cylindrical magnets (length l_m and radius r_m), which have *near-linear* demagnetization curves, the flux density at a distance x can be approximated as

$$B = \frac{B_r}{2}\left(\frac{l_m}{\sqrt{r_m^2 + (l_m + x)^2}} - \frac{x}{\sqrt{r_m^2 + x^2}}\right).$$

The properties of magnetic and ferromagnetic materials, as well as composites, commonly used are of interest. Table 2.3 reports the initial permeability μ_i, maximum relative permeability $\mu_{r\,max}$, coercivity (coercive force) H_c, saturation polarization J_s, hysteresis loss per cycle W_h, and Curie temperature T_C for high-permeability bulk metals and alloys. The designer must be aware that, in particular motion devices, the reported parameters can significantly vary.

Table 2.4 reports the remanence B_r, flux coercivity H_{Fc}, intrinsic coercivity H_{Ic}, maximum energy product BH_{max}, Curie temperature, and the maximum operating temperature T_{max} for *hard* permanent magnets. These characteristics are strongly affected by the dimensions, temperature, fabrication, and so on.

For electromechanical motion devices, the flux linkages are plotted versus current because i and ψ are commonly used as the state variables rather than the field intensity and density. In actuators and electric machines almost all energy is stored in the air gap. Using the fact that the air is a conservative medium, one concludes that the coupling filed is lossless. Figure 2.4 illustrates the nonlinear magnetizing characteristic (normal magnetization curve). The energy stored in the magnetic field is

$$W_F = \oint_\psi i\,d\psi,$$

whereas the coenergy is found as

$$W_c = \oint_i \psi\,di.$$

The total energy is

$$W_F + W_c = \oint_\psi i\,d\psi + \oint_i \psi\,di = \psi i.$$

The flux linkage is the function of the current i and position x (for translational motion) or angular displacement θ (for rotational motion). That is, $\psi = f(i,x)$ or $\psi = f(i,\theta)$. The current can be found as the nonlinear function of the flux linkages and position or angular displacement. Hence,

$$d\psi = \frac{\partial\psi(i,x)}{\partial i}di + \frac{\partial\psi(i,x)}{\partial x}dx \quad \text{or} \quad d\psi = \frac{\partial\psi(i,\theta)}{\partial i}di + \frac{\partial\psi(i,\theta)}{\partial\theta}d\theta,$$

TABLE 2.3

Magnetic Properties of High-Permeability Soft Metals and Alloys

Material	Composition (%)	μ_i	$\mu_{r\,max}$	H_c [A/m]	J_s [T]	W_h [J/m³]	T_C [K]
Iron	$Fe_{99\%}$	200	6000	70	2.16	500	1043
Iron	$Fe_{99.9\%}$	25000	350000	0.8	2.16	60	1043
Silicon-iron	$Fe_{96\%}Si_{4\%}$	500	7000	40	1.95	50–150	1008
Silicon-iron (110) [001]	$Fe_{97\%}Si_{3\%}$	9000	40000	12	2.01	35–140	1015
Silicon-iron {100} <100>	$Fe_{97\%}Si_{3\%}$		100000	6	2.01		1015
Steel	$Fe_{99.4\%}C_{0.1\%}Si_{0.1\%}Mn_{0.4\%}$	800	1100	200			
Hypernik	$Fe_{50\%}Ni_{50\%}$	4000	70000	4	1.60	22	753
Deltamax {100} <100>	$Fe_{50\%}Ni_{50\%}$	500	200000	16	1.55		773
Isoperm {100} <100>	$Fe_{50\%}Ni_{50\%}$	90	100	480	1.60		
78 Permalloy	$Ni_{78\%}Fe_{22\%}$	4000	100000	4	1.05	50	651
Supermalloy	$Ni_{79\%}Fe_{16\%}Mo_{5\%}$	100000	1000000	0.15	0.79	2	673
Mumetal	$Ni_{77\%}Fe_{16\%}Cu_{5\%}Cr_{2\%}$	20000	100000	4	0.75	20	673
Hyperco	$Fe_{64\%}Co_{35\%}Cr_{0.5\%}$	650	10000	80	2.42	300	1243
Permendur	$Fe_{50\%}Co_{50\%}$	500	6000	160	2.46	1200	1253
2V Permendur	$Fe_{49\%}Co_{49\%}V_{2\%}$	800	4000	160	2.45	600	1253
Supermendur	$Fe_{49\%}Co_{49\%}V_{2\%}$		60000	16	2.40	1150	1253
25 Perminvar	$Ni_{45\%}Fe_{30\%}Co_{25\%}$	400	2000	100	1.55		
7 Perminvar	$Ni_{70\%}Fe_{23\%}Co_{7\%}$	850	4000	50	1.25		
Perminvar (magnetically annealed)	$Ni_{43\%}Fe_{34\%}Co_{23\%}$		400000	2.4	1.50		
Alfenol (Alperm)	$Fe_{84\%}Al_{16\%}$	3000	55000	3.2	0.8		723
Alfer	$Fe_{87\%}Al_{13\%}$	700	3700	53	1.20		673
Aluminum-Iron	$Fe_{96.5\%}Al_{3.5\%}$	500	19000	24	1.90		
Sendust	$Fe_{85\%}Si_{10\%}Al_{5\%}$	36000	120000	1.6	0.89		753

(Data from *Handbook of Chemistry and Physics,* 83rd Edition, Ed. D. R. Lide, CRC Press, Boca Raton, FL, 2002; *Handbook of Engineering Tables,* Ed. R. C. Dorf, CRC Press, Boca Raton, FL, 2003; and S. E. Lyshevski, *Nano- and Micro-Electromechanical Systems: Fundamentals of Nano- and Micro-engineering,* CRC Press, Boca Raton, FL, 2004.)

and

$$di = \frac{\partial i(\psi, x)}{\partial \psi} d\psi + \frac{\partial i(\psi, x)}{\partial x} dx \quad \text{or} \quad di = \frac{\partial i(\psi, \theta)}{\partial \psi} d\psi + \frac{\partial i(\psi, \theta)}{\partial \theta} d\theta.$$

TABLE 2.4

Magnetic Properties of High-Permeability Hard Metals and Alloys

Composite and Composition	B_r [T]	H_{Fc} [A/m]	H_{lc} [A/m]	BH_{max} [kJ/m³]	T_C [°C]	T_{max} [°C]
Alnico1: 20Ni,12Al,5Co	0.72		35	25		
Alnico2: 17Ni,10Al,12.5Co,6Cu	0.72		40–50	13–14		
Alnico3: 24–30Ni,12–14Al,0–3Cu	0.5–0.6		40–54	10		
Alnico4: 21–28Ni, 11–13Al, 3–5Co, 2–4Cu	0.55–0.75		36–56	11–12		
Alnico5: 14Ni,8Al,24Co,3Cu	1.25	53	54	40	850	520
Alnico6: 16Ni,8Al,24Co,3Cu,2Ti	1.05		75	52		
Alnico8: 15Ni,7Al,35Co,4Cu,5Ti	0.83	1.6	160	45		
Alnico9: 5Ni,7Al,35Co,4Cu,5Ti	1.10	1.45	1.45	75	850	520
Alnico12: 3.5Ni,8Al,24.5Co,2Nb	1.20		64	76.8		
Ferroxdur: $BaFe_{12}O_{19}$	0.4	1.6	192	29	450	400
$SrFe_{12}O_{19}$	0.4	2.95	3.3	30	450	400
$LaCo_5$	0.91			164	567	
$CeCo_5$	0.77			117	380	
$PrCo_5$	1.20			286	620	
$NdCo_5$	1.22			295	637	
$SmCo_5$	1.00	7.9	696	196	700	250
$Sm(Co_{0.76}Fe_{0.10}Cu_{0.14})_{6.8}$	1.04	4.8	5	212	800	300
$Sm(Co_{0.65}Fe_{0.28}Cu_{0.05}Zr_{0.02})_{7.7}$	1.2	10	16	264	800	300
$Nd_2Fe_{14}B$ (sintered)	1.22	8.4	1120	280	300	100
Vicalloy II: Fe,52Co,14V	1.0	42		28	700	500
Fe,24Cr,15Co,3Mo (anisotropic)	1.54	67		76	630	500
Chromindur II: Fe,28Cr,10.5Co	0.98	32		16	630	500
Fe,23Cr,15Co,3V,2Ti	1.35	4		44	630	500
Fe, 36Co	1.04		18	8		
Co (rare-earth)	0.87		638	144		
Cunife: Cu,20Ni,20Fe	0.55	4		12	410	350
Cunico: Cu,21Ni,29Fe	0.34	0.5		8		
Pt,23Co	0.64	4		76	480	350
Mn,29.5Al,0.5C (anisotropic)	0.61	2.16	2.4	56	300	120

(Data from *Handbook of Chemistry and Physics*, 83rd Edition, Ed. D. R. Lide, CRC Press, Boca Raton, FL, 2002; *Handbook of Engineering Tables*, Ed. R. C. Dorf, CRC Press, Boca Raton, FL, 2003; and S. E. Lyshevski, *Nano- and Micro-Electromechanical Systems: Fundamentals of Nano- and Micro-engineering*, CRC Press, Boca Raton, FL, 2004.)

Therefore, we have

$$W_F = \oint_\psi i\, d\psi = \oint_i i \frac{\partial \psi(i,x)}{\partial i} di + \oint_x i \frac{\partial \psi(i,x)}{\partial x} dx,$$

$$W_F = \oint_\psi i\, d\psi = \oint_i i \frac{\partial \psi(i,\theta)}{\partial i} di + \oint_\theta i \frac{\partial \psi(i,\theta)}{\partial \theta} d\theta,$$

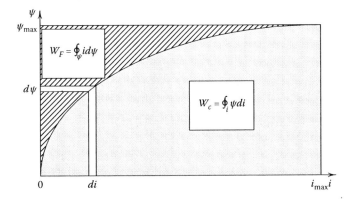

FIGURE 2.4
Magnetization curve and energies.

and

$$W_c = \oint_i \psi di = \oint_\psi \psi \frac{\partial i(\psi,x)}{\partial \psi} d\psi + \oint_x \psi \frac{\partial i(\psi,x)}{\partial x} dx,$$

$$W_c = \oint_i \psi di = \oint_\psi \psi \frac{\partial i(\psi,\theta)}{\partial \psi} d\psi + \oint_\theta \psi \frac{\partial i(\psi,\theta)}{\partial \theta} d\theta.$$

Assuming that the coupling field is lossless, the differential change in the mechanical energy (which is found using the differential displacement $\vec{dl}$ as $dW_{mec} = \vec{F}_m \cdot \vec{dl}$) is related to the differential change of the coenergy. For the displacement dx at constant current, one obtains $dW_{mec} = dW_c$. Hence, for a one-dimensional case, the electromagnetic force is

$$F_e(i,x) = \frac{\partial W_c(i,x)}{\partial x}.$$

For rotational motion, the electromagnetic torque is given as

$$T_e(i,\theta) = \frac{\partial W_c(i,\theta)}{\partial \theta}.$$

The reported equations for electromagnetic (electrostatic) force and torque are important because the ultimate objective is to control electromechanical motion devices and systems. To control the velocity and displacement, one should change the electromagnetic field, varying E, D, H or B. The torque and force are derived in the terms of the electromagnetic field quantities as reported in Section 2.4.

REMARK

We study the electromechanical motion devices that profoundly different as compared to biological motion devices. For example, *Escherichia coli* (*E. coli*) bacteria possess biomotors made from proteins. The motors rotate flagella, which varies its geometry to perform propulsion. The motor diameter is 45–50 nm, and these motors likely do not utilize magnets. One may expect that biosystems do not use magnets. In 1962, Professor Heinz A. Lowenstam discovered magnetite (Fe_3O_4) biomineralization in the teeth of chitons (mollusks of the class *Polyplacophora*) demonstrating that living organisms were able to precipitate the mineral magnetite. The next intriguing finding was the discovery of the magnetotactic bacteria by Richard Blakemore in 1975. Three-billion-year-evolved magnetotactic bacteria contain magnetosomes (magnetic mineral particles) enclosed in the protein-based membranes. In most cases the magnetosomes are arranged in a chain or chains fixed within the cell. In many magnetotactic bacteria, the magnetosome mineral particles are either 30 to 100 nm magnetite (Fe_3O_4), or, in marine and sulfidic environments, greigite (Fe_3S_4). These permanent magnets sense the magnetic field, and bacteria swim (migrate) along the magnetic field lines. The magnetosome chain are usually oriented so that a 111 crystallographic axis of each particle lies along the chain direction, whereas the chained greigite particles are usually oriented so that a 100 crystallographic axis of each particle is oriented along the chain direction. Whether the magnetic mineral particles are magnetite or greigite, the chain of magnetosome particles constitutes a permanent magnetic dipole fixed within the bacterium. Therefore, magnetotactic bacteria have two magnetic poles, depending on the orientation of the magnetic dipole within the cell. The poles can remagnetized by a magnetic pulse, which is greater than the coercive force of the chain of particles. The magnetosome particles are uniformly magnetized forming permanent magnetic domains. All particles are arranged along the chain axis such that the crystallographic magnetic easy axes are aligned. The size specificity and crystallographic orientation of the chain assembly is optimally designed for magnetotaxis in the geomagnetic field. Magnetosome particles occur in at least three different crystal forms. The simplest form, found in *M. magnetotacticum*, is cubo-octahedral, which preserves the cubic crystal symmetry of magnetite. A second type, found in coccoid and vibrioid strains, is an elongated hexagonal prism with the axis of elongation parallel to the 111 crystal direction. A third type, observed in some uncultured cells, is an elongated cubo-octahedral form producing cylindrical, bullet-shaped, tear-drop, and arrowhead particles (see Figure 2.5). The growth (synthesis) mechanisms for these forms are unknown, but particle shapes may be related to anisotropic ion flux through the magnetosome membrane or from constraints imposed by the surrounding membrane structure. Whereas the cubo-octahedral form is common in

FIGURE 2.5
Magnetotactic bacterium and image of a chain of 60–100 nm diameter cylindrical magneto-some mineral magnetic particles (rectangular, octahedral, prismatic, and other shapes of 30–100 nm magnetosome particles exist).

inorganic magnetites, the prevalence of elongated hexagonal forms in magnetosomes appears to be a unique feature of the biomineralization process.

2.3 Introduction to Electromagnetics

To study various electromechanical motion devices, one applies the electromagnetic field theory and classical mechanics. Electrostatic interaction was investigated by Charles Coulomb. For two charges q_1 and q_2, separated by a distance x in free space, the magnitude of the electric force is

$$F = \frac{|q_1 q_2|}{4\pi\varepsilon_0 x^2},$$

where using the permittivity of free space $\varepsilon_0 = 8.85 \times 10^{-12}$ F/m or C^2/N-m^2, we have $1/4\pi\varepsilon_0 = 9 \times 10^9$ N-m^2/C. The unit for the force is the newton [N], whereas the charges are given in coulombs [C].

The force is the vector. Therefore, in general, we have

$$\vec{F} = \frac{q_1 q_2}{4\pi\varepsilon_0 x^2} \vec{a}_x,$$

where $\vec{a}_x$ is the unit vector that is directed along the line joining these two charges.

The elegance and uniformity of electromagnetics arise from related fundamental laws that allow one to study the field quantities. We denote the vector of electric flux density as $\vec{D}$ [F/m] and the vector of electric field intensity as $\vec{E}$ [V/m or N/C]. Using the Gauss law, the total electric flux Φ [C] through a closed surface is found to be equal to the total force charge enclosed by the surface. That is,

$$\Phi = \oint_s \vec{D} \cdot d\vec{s} = Q_s, \quad \vec{D} = \varepsilon \vec{E},$$

where $d\vec{s}$ is the vector surface area, $d\vec{s} = ds\vec{a}_n$; $\vec{a}_n$ is the unit vector which is normal to the surface; Q_s is the total charge enclosed by the surface.

Ohm's law for circuits is $V = ir$. However, for a media, the Ohm law relates the volume charge density $\vec{J}$ and electric field intensity $\vec{E}$ using conductivity σ. In particular,

$$\vec{J} = \sigma \vec{E},$$

where σ is the conductivity [A/V-m], and for copper $\sigma = 5.8 \times 10^7$, whereas for aluminum $\sigma = 3.5 \times 10^7$.

The application of Ohm's law to the Ampere-Maxwell equation

$$\nabla \times \vec{B} = \mu \vec{J} + \mu \varepsilon \frac{\partial \vec{E}}{\partial t},$$

results in the wave equation

$$\nabla^2 \vec{B} = \mu \varepsilon \frac{\partial^2 \vec{B}}{\partial t^2} + \mu \sigma \frac{\partial \vec{B}}{\partial t}.$$

This equation can be solved. For example, the one-dimensional equation

$$\frac{\partial^2 B_x}{\partial x^2} = \mu \varepsilon \frac{\partial^2 B_x}{\partial t^2} + \mu \sigma \frac{\partial B_x}{\partial t},$$

has the solution

$$B_x(t, x) = B_{x0} e^{i(kx - \omega t)}.$$

Taking note of the derivatives

$$\frac{\partial^2 B_x}{\partial x^2} = -k^2 B_{x0}, \quad \frac{\partial^2 B_x}{\partial t} = -i\omega B_{x0} \quad \text{and} \quad \frac{\partial^2 B_x}{\partial t^2} = -\omega^2 B_{x0},$$

one finds $k^2 = \mu \varepsilon \omega^2 + \mu \sigma i \omega$.

We documented that the derived partial differential equations can be solved. There exist a variety of numerical and analytical methods to solve partial differential equations. For example, the three-dimensional wave equation (if $\sigma \approx 0$)

$$\nabla^2 B(t,x,y,z) = \mu\varepsilon \frac{\partial^2 B(t,x,y,z)}{\partial t^2}$$

with boundary conditions $B = 0$ when $x = 0$, $x = a$, $y = 0$, $y = b$, $z = 0$, $z = d$, has the following solution

$$B(t,x,y,z) = \sin\left(n\pi \frac{x}{a} \right) \sin\left(m\pi \frac{y}{b} \right) \sin\left(l\pi \frac{z}{d} \right) e^{i\omega t},$$

where integers n, m, and l are the so-called mode numbers, and the lowest frequency mode is the (1 1 1) mode.

Taking note of the derivatives

$$\nabla^2 B = -\left[\left(n\pi \frac{x}{a} \right)^2 + \left(m\pi \frac{y}{b} \right)^2 + \left(l\pi \frac{z}{d} \right)^2 \right] B$$

and

$$\frac{\partial^2 B}{\partial t^2} = -\omega^2 B,$$

one obtains the frequency of oscillation

$$\omega = \pi \sqrt{\frac{1}{\mu\varepsilon}} \sqrt{\left(\frac{n}{a} \right)^2 + \left(\frac{m}{b} \right)^2 + \left(\frac{l}{d} \right)^2}.$$

The equation of the electric field in the conducting media is

$$\nabla^2 \vec{E} = \mu\varepsilon \frac{\partial^2 \vec{E}}{\partial t^2} + \mu\sigma \frac{\partial \vec{E}}{\partial t},$$

and the solution for one-dimensional case is

$$E_x(t,x) = E_{x0} e^{i(kx-\omega t)}.$$

Taking note of

$$k^2 = \mu\varepsilon\omega^2 + \mu\sigma i\omega,$$

we have

$$\text{Re}(k) = \omega\sqrt{\frac{1}{2}\mu\varepsilon}\sqrt{\sqrt{1+\left(\frac{\sigma}{\omega\varepsilon}\right)^2}+1}.$$

and

$$\text{Im}(k) = \omega\sqrt{\frac{1}{2}\mu\varepsilon}\sqrt{\sqrt{1+\left(\frac{\sigma}{\omega\varepsilon}\right)^2}-1}.$$

In vacuum $\sigma = 0$, yielding $\text{Re}(k) = \omega\sqrt{\mu_0\varepsilon_0}$ and $\text{Im}(k) = 0$. In a perfect conductor, the charges and currents are all at the surface. In a real conductor, the surface currents penetrate a finite distance, called the skin depth, and

$$d_{skin} = \frac{1}{\text{Im}(k)} = \left(\omega\sqrt{\frac{1}{2}\mu\varepsilon}\sqrt{\sqrt{1+\left(\frac{\sigma}{\omega\varepsilon}\right)^2}-1}\right)^{-1}.$$

As the frequency increases, the skin depth decreases. At high frequencies (microwaves), the attenuation becomes significant. Thus, the resistance increases with frequency. One can examine the real and imaginary parts for high conductivity analyzing the physics of high-conductivity media. The ratio $\sigma/\varepsilon\omega$ defines whether the conductivity is large or small.

Consider the current density equation

$$\nabla \cdot \vec{j} = -\frac{\partial\rho}{\partial t}.$$

Using Ohm's law, we have

$$\nabla \cdot (\sigma\vec{E}) = -\frac{\partial\rho}{\partial t}.$$

From the Gauss law

$$\nabla \cdot \vec{E} = \frac{\rho}{\varepsilon}$$

and assuming constant conductivity, one finds

$$\frac{\partial\rho}{\partial t} = -\frac{\sigma}{\varepsilon}\rho.$$

The solution is

$$\rho(t,x,y,z) = \rho_0(x,y,z)e^{-\frac{\sigma}{\varepsilon}t}.$$

Correspondingly, the attenuation rate is defined by σ/ε.

The current i is proportional to the potential difference. The resistivity ρ of the conductor is the ratio between the electric field $\vec{E}$ and the current density $\vec{J}$. Thus, $\rho = \vec{E}/\vec{J}$.

The resistance r of the conductor is related to the resistivity and conductivity by the following formulas

$$r = \frac{\rho l}{A} \quad \text{and} \quad r = \frac{l}{\sigma A},$$

where l is the length; A is the cross-sectional area.

The resistivity, as well as other parameters, vary. For example, the resistivity depends on temperature T, and

$$\rho(T) = \rho_0 = [1 + \alpha_{p1}(T - T_0) + \alpha_{p2}(T - T_0)^2 + \ldots],$$

where α_{p1} and α_{p2} are the coefficients. In the small temperature range (up to 160°C) for copper at $T_0 = 20$°C, we have $\rho(T) = 1.7 \times 10^{-3}[1 + 0.0039(T - 20)]$.

2.4 Fundamentals of Electromagnetics

Electromagnetic theory and classical mechanics form the basis to examine (and devise) the device physics, study the inherent phenomena exhibited that are utilized, as well as derive the equations of motion that describe dynamic. The electrostatic and magnetostatic equations in linear isotropic media are found using the vectors of the electric field intensity $\vec{E}$, electric flux density $\vec{D}$, magnetic field intensity $\vec{H}$, and magnetic flux density $\vec{B}$. In addition, one uses the constitutive equations $\vec{D} = \varepsilon\vec{E}$ and $\vec{B} = \mu\vec{H}$. The basic equations in the Cartesian coordinate system are reported in Table 2.5.

In the static (time-invariant) fields, electric and magnetic field vectors form separate and independent pairs. That is, $\vec{E}$ and $\vec{D}$ are not related to $\vec{H}$ and $\vec{B}$, and vice versa. However, the electric and magnetic fields are time-varying. The changes of magnetic field influence the electric field, and vice versa. The partial differential equations are found using Maxwell's equations.

TABLE 2.5

Fundamental Equations of Electrostatic and Magnetostatic Fields in Media

	Electrostatic Equations	**Magnetostatic Equations**
Governing equations	$\nabla \times \vec{E}(x,y,z,t) = 0$ $\nabla \cdot \vec{E}(x,y,z,t) = \dfrac{\rho_v(x,y,z,t)}{\varepsilon}$	$\nabla \times \vec{H}(x,y,z,t) = 0$ $\nabla \cdot \vec{H}(x,y,z,t) = 0$
Constitutive equations	$\vec{D} = \varepsilon \vec{E}$	$\vec{B} = \mu \vec{H}$

Four Maxwell's equations in the differential form for time-varying fields are

$$\nabla \times \vec{E}(x,y,z,t) = -\mu \frac{\partial \vec{H}(x,y,z,t)}{\partial t}, \text{ (Faraday's law)}$$

$$\nabla \times \vec{H}(x,y,z,t) = \sigma \vec{E}(x,y,z,t) + \vec{J}(x,y,z,t) = \sigma \vec{E}(x,y,z,t) + \varepsilon \frac{\partial \vec{E}(x,y,z,t)}{\partial t},$$

$$\nabla \cdot \vec{E}(x,y,z,t) = \frac{\rho_v(x,y,z,t)}{\varepsilon} \text{ (Gauss's law)},$$

$$\nabla \cdot \vec{H}(x,y,z,t) = 0,$$

where $\vec{E}$ is the electric field intensity, and using the permittivity ε, the electric flux density is $\vec{D} = \varepsilon \vec{E}$; $\vec{H}$ is the magnetic field intensity, and using the permeability μ, the magnetic flux density is $\vec{B} = \mu \vec{H}$; $\vec{J}$ is the current density, and using the conductivity σ, we have $\vec{J} = \sigma \vec{E}$; ρ_v is the volume charge density, and the total electric flux through a closed surface is $\Phi = \oint_s \vec{D} \cdot d\vec{s} = \oint_v \rho_v dv = Q$ (Gauss's law), whereas the magnetic flux crossing surface is $\Phi = \oint_s \vec{B} \cdot d\vec{s}$.

The second equation

$$\nabla \times \vec{H}(x,y,z,t) = \sigma \vec{E}(x,y,z,t) + \vec{J}(x,y,z,t) = \sigma \vec{E}(x,y,z,t) + \varepsilon \frac{\partial \vec{E}(x,y,z,t)}{\partial t}$$

was derived by Maxwell adding the term

$$\vec{J}(x,y,z,t) = \varepsilon \frac{\partial \vec{E}(x,y,z,t)}{\partial t}$$

in the Ampere law.

The constitutive (auxiliary) equations are given using the permittivity, permeability, and conductivity tensors ε, μ, and σ. In particular, one has

$$\vec{D} = \varepsilon \vec{E} \quad \text{or} \quad \vec{D} = \varepsilon \vec{E} + \vec{P},$$

$$\vec{B} = \mu\vec{H} \quad \text{or} \quad \vec{B} = \mu(\vec{H} + \vec{M}),$$

$$\vec{J} = \sigma\vec{E} \quad \text{or} \quad \vec{J} = \rho_v\vec{v}.$$

The Maxwell's equations can be solved using the boundary conditions on the field vectors. In two-region media, we have

$$\vec{a}_N \times (\vec{E}_2 - \vec{E}_1) = 0, \quad \vec{a}_N \times (\vec{H}_2 - \vec{H}_1) = \vec{J}_s, \quad \vec{a}_N \cdot (\vec{D}_2 - \vec{D}_1) = \rho_s, \quad \vec{a}_N \cdot (\vec{B}_2 - \vec{B}_1) = 0,$$

where $\vec{J}_s$ is the surface current density vector; $\vec{a}_N$ is the surface normal unit vector at the boundary from region 2 into region 1; ρ_s is the surface charge density.

The constitutive relations that describe media can be integrated with Maxwell's equations, which relate the fields in order to find two partial differential equations. Using the electric and magnetic field intensities $\vec{E}$ and $\vec{H}$, the electromagnetic fields are described as

$$\nabla \times (\nabla \times \vec{E}) = \nabla(\nabla \cdot \vec{E}) - \nabla^2\vec{E} = -\mu\frac{\partial \vec{J}}{\partial t} - \mu\frac{\partial^2 \vec{D}}{\partial t^2} = -\mu\sigma\frac{\partial \vec{E}}{\partial t} - \mu\varepsilon\frac{\partial^2 \vec{E}}{\partial t^2},$$

$$\nabla \times (\nabla \times \vec{H}) = \nabla(\nabla \cdot \vec{H}) - \nabla^2\vec{H} = -\mu\sigma\frac{\partial \vec{H}}{\partial t} - \mu\varepsilon\frac{\partial^2 \vec{H}}{\partial t^2}.$$

The following pair of homogeneous and inhomogeneous wave equations

$$\nabla^2\vec{E} - \mu\sigma\frac{\partial \vec{E}}{\partial t} - \mu\varepsilon\frac{\partial^2 \vec{E}}{\partial t^2} = \nabla\left(\frac{\rho_v}{\varepsilon}\right),$$

$$\nabla^2\vec{H} - \mu\sigma\frac{\partial \vec{H}}{\partial t} - \mu\varepsilon\frac{\partial^2 \vec{H}}{\partial t^2} = 0$$

is equivalent to four Maxwell's equations and constitutive relations. For some cases, these two equations can be solved independently. It is not always possible to use the boundary conditions using only $\vec{E}$ and $\vec{H}$. Hence, the problem cannot always be simplified to two electromagnetic field vectors. The electric scalar and magnetic vector potentials are used. Denoting the magnetic vector potential as $\vec{A}$ and the electric scalar potential as V, we have

$$\nabla \times \vec{A} = \vec{B} = \mu\vec{H} \text{ and } \vec{E} = -\frac{\partial \vec{A}}{\partial t} - \nabla V.$$

The electromagnetic field is derivative from the potentials. Using the Lorentz equation

$$\nabla \cdot \vec{A} = -\frac{\partial V}{\partial t}$$

the inhomogeneous vector potential wave equation to be solved is

$$-\nabla^2 \vec{A} + \mu\sigma \frac{\partial \vec{A}}{\partial t} + \mu\varepsilon \frac{\partial^2 \vec{A}}{\partial t^2} = -\mu\sigma\nabla V.$$

Using equation $\vec{B} = \nabla \times \vec{A}$, one has the following nonhomogeneous vector wave equation

$$\nabla^2 \times \vec{A} - \mu\varepsilon \frac{\partial^2 \vec{A}}{\partial t^2} = -\mu\vec{J}.$$

The solution of this equation gives the waves traveling with the velocity $1/\sqrt{\mu\varepsilon}$.

Examining actuators and other electromechanical motion devices, one also concentrates on the deriving the expressions for force, torque, *electromotive* and *magnetomotive* forces, and so on. Using the volume charge density ρ_v, the Lorenz force, which relates the electromagnetic and mechanical variables, is

$$\vec{F} = \rho_v(\vec{E} + \vec{v} \times \vec{B}) = \rho_v \vec{E} + \vec{J} \times \vec{B}.$$

In particular, the Lorentz force law is

$$\vec{F} = \frac{d\vec{p}}{dt} = q(\vec{E} + \vec{v} \times \vec{B}) = q\left[-\nabla V - \frac{\partial \vec{A}}{\partial t} + \vec{v} \times (\nabla \times \vec{A}) \right],$$

where for the canonical momentum $\vec{p}$ we have

$$\frac{d\vec{p}}{dt} = -\nabla\Pi.$$

The Maxwell equations illustrate how charges produce the electromagnetic fields, and the Lorentz force law demonstrates how the electromagnetic fields affect charges. The energy per unit time per unit area, transported by the electromagnetic fields is called the Poynting vector, and

$$\vec{S} = \frac{1}{\mu}(\vec{E} \times \vec{B}).$$

The electromagnetic force can be found by applying the Maxwell stress tensor. This concept employs a volume integral to obtain the stored energy, and stress at all points of a bounding surface can be determined. The sum of local stresses gives the net force. In particular, the electromagnetic stress is

$$\vec{F} = \int_v \rho_v(\vec{E} + \vec{v} \times \vec{B})dv = \int_v (\rho_v\vec{E} + \vec{J} \times \vec{B})dv = \frac{1}{\mu}\oint_s \vec{\vec{T}} \cdot d\vec{s} - \varepsilon\mu \frac{d}{dt} \int_v \vec{S}dv.$$

The force per unit volume is

$$\vec{F}_u = \rho_v \vec{E} + \vec{J} \times \vec{B} = \nabla \cdot \ddot{T} - \varepsilon\mu \frac{\partial \vec{S}}{\partial t}.$$

The electromagnetic stress energy tensor $\ddot{T}$ (the second Maxwell stress tensor) is

$$(\vec{a} \cdot \ddot{T})_j = \sum_{i=x,y,z} a_i T_{ij}.$$

We have

$$(\nabla \cdot \ddot{T})_j = \varepsilon \left[(\nabla \cdot \vec{E})E_j + (\vec{E} \cdot \nabla)E_j - \frac{1}{2}\nabla_j E^2 \right] + \frac{1}{\mu} \left[(\nabla \cdot \vec{B})B_j + (\vec{B} \cdot \nabla)B_j - \frac{1}{2}\nabla_j B^2 \right]$$

or

$$T_{ij} = \varepsilon \left(E_i E_j - \frac{1}{2}\delta_{ij} E^2 \right) + \frac{1}{\mu} \left(B_i B_j - \frac{1}{2}\delta_{ij} B^2 \right),$$

where i and j are the indexes that refer to the coordinates x, y, and z, (the stress tensor $\ddot{T}$ has nine components T_{xx}, T_{xy}, T_{xz}, ..., T_{zy} and T_{zz}); δ_{ij} is the Kronecker delta-function, which is defined to be 1 if the indexes are the same and 0 otherwise,

$$\delta_{ij} = \begin{cases} 1 & \text{if } i = j \\ 0 & \text{if } i \neq j \end{cases}, \quad \delta_{xx} = \delta_{yy} = \delta_{zz} = 1 \quad \text{and} \quad \delta_{xy} = \delta_{xz} = \delta_{yz} = 0.$$

The electromagnetic torque developed by the actuator is found using the electromagnetic field, and the electromagnetic stress tensor is given as

$$T_s = T_s^E + T_s^M = \begin{bmatrix} E_1 D_1 - \frac{1}{2}E_j D_j & E_1 D_2 & E_1 D_3 \\ E_2 D_1 & E_2 D_2 - \frac{1}{2}E_j D_j & E_2 D_3 \\ E_3 D_1 & E_3 D_2 & E_3 D_3 - \frac{1}{2}E_j D_j \end{bmatrix}$$

$$+ \begin{bmatrix} B_1 H_1 - \frac{1}{2}B_j H_j & B_1 H_2 & B_1 H_3 \\ B_2 H_1 & B_2 H_2 - \frac{1}{2}B_j H_j & B_2 H_3 \\ B_3 H_1 & B_3 H_2 & B_3 H_3 - \frac{1}{2}B_j H_j \end{bmatrix}$$

For the Cartesian, cylindrical, and spherical coordinate systems, we have

$$E_x = E_1, E_y = E_2, E_z = E_3, D_x = D_1, D_y = D_2, D_z = D_3,$$

$$H_x = H_1, H_y = H_2, H_z = H_3, B_x = B_1, B_y = B_2, B_z = B_3;$$

$$E_r = E_1, E_\theta = E_2, E_z = E_3, D_r = D_1, D_\theta = D_2, D_z = D_3,$$
$$H_r = H_1, H_\theta = H_2, H_z = H_3, B_r = B_1, B_\theta = B_2, B_z = B_3;$$
$$E_\rho = E_1, E_\theta = E_2, E_\phi = E_3, D_\rho = D_1, D_\theta = D_2, D_\phi = D_3,$$
$$H_\rho = H_1, H_\theta = H_2, H_\phi = H_3, B_\rho = B_1, B_\theta = B_2, B_\phi = B_3.$$

The results derived can be viewed using the energy analysis, and one has

$$\sum \vec{F}(\vec{r}) = -\nabla \Pi(\vec{r}), \quad \Pi(\vec{r}) = \frac{\varepsilon_0 \varepsilon_r}{2} \int_v \vec{E} \cdot \vec{E} dv + \frac{1}{2\mu_0 \mu_r} \int_v \vec{H} \cdot \vec{H} dv.$$

One may relax the complexity of the tensor calculus in the analysis of forces. The expressions for energies stored in electrostatic and magnetic fields in terms of field quantities should be derived. The total potential energy stored in the electrostatic field is obtained using the potential difference V as

$$W_e = \frac{1}{2} \int_v \rho_v V dv,$$

where the volume charge density is given as $\rho_v = \vec{\nabla} \cdot \vec{D}$; $\vec{\nabla}$ is the curl operator.

In the Gauss form, using $\rho_v = \vec{\nabla} \cdot \vec{D}$ and $\vec{E} = -\vec{\nabla} V$, one obtains the following expression for the energy stored in the electrostatic field

$$W_e = \frac{1}{2} \int_v \vec{D} \cdot \vec{E} dv,$$

and the electrostatic volume energy density is $\frac{1}{2} \vec{D} \cdot \vec{E}$. For a linear isotropic medium, one finds

$$W_e = \frac{1}{2} \int_v \varepsilon |\vec{E}|^2 \, dv = \frac{1}{2} \int_v \frac{1}{\varepsilon} |\vec{D}|^2 \, dv.$$

The electric field $\vec{E}(x,y,z)$ is found using the scalar electrostatic potential function $V(x, y, z)$ as

$$\vec{E}(x,y,z) = -\vec{\nabla} V(x,y,z).$$

In the cylindrical and spherical coordinate systems, we have

$$\vec{E}(r,\phi,z) = -\vec{\nabla} V(r,\phi,z) \quad \text{and} \quad \vec{E}(r,\theta,\phi) = -\vec{\nabla} V(r,\theta,\phi).$$

Using

$$W_e = \frac{1}{2} \int_v \rho_v V dv,$$

the potential energy that is stored in the electric field between two surfaces (for example, in capacitors) is found to be

$$W_e = \frac{1}{2} QV = \frac{1}{2} CV^2.$$

Using the principle of virtual work, for the lossless conservative system, the differential change of the electrostatic energy dW_e is equal to the differential change of mechanical energy dW_{mec} That is

$$dW_e = dW_{mec}.$$

For translational motion, one has $dW_{mec} = \vec{F}_e \cdot d\vec{l}$, where $d\vec{l}$ is the differential displacement.

From $dW_e = \vec{\nabla} W_e \cdot d\vec{l}$ one concludes that the force is the gradient of the stored electrostatic energy, and $\vec{F}_e = \vec{\nabla} W_e$.

In the Cartesian coordinates, we have

$$F_{ex} = \frac{\partial W_e}{\partial x}, \quad F_{ey} = \frac{\partial W_e}{\partial y} \quad \text{and} \quad F_{ez} = \frac{\partial W_e}{\partial z}.$$

To find the stored energy in the magnetostatic field in terms of field quantities, the following formula is used

$$W_m = \frac{1}{2} \int_v \vec{B} \cdot \vec{H} dv, \quad \text{or} \quad W_m = \frac{1}{2} \int_v \mu |\vec{H}|^2 \, dv = \frac{1}{2} \int_v \frac{|\vec{B}|^2}{\mu} \, dv.$$

To show how the energy concept studied is applied to electromechanical devices, we find the energy stored in inductors. To approach this problem, we substitute $\vec{B} = \vec{\nabla} \times \vec{A}$. Using the following vector identity

$$\vec{H} \cdot \vec{\nabla} \times \vec{A} = \vec{\nabla} \cdot (\vec{A} \times \vec{H}) + \vec{A} \cdot \vec{\nabla} \times \vec{H},$$

one obtains

$$W_m = \frac{1}{2} \int_v \vec{B} \cdot \vec{H} dv = \frac{1}{2} \int_v \vec{\nabla} \cdot (\vec{A} \times \vec{H}) dv + \frac{1}{2} \int_v \vec{A} \cdot \vec{\nabla} \times \vec{H} dv$$

$$= \frac{1}{2} \int_s (\vec{A} \times \vec{H}) \cdot d\vec{s} + \frac{1}{2} \int_v \vec{A} \cdot \vec{J} dv = \frac{1}{2} \int_v \vec{A} \cdot \vec{J} dv.$$

Using the general expression for the vector magnetic potential $\vec{A}(\vec{r})$ [Wb/m], as given by

$$\vec{A}(\vec{r}) = \frac{\mu_0}{4\pi} \int_{v_A} \frac{\vec{J}(\vec{r}_A)}{x} dv_j, \qquad \vec{\nabla} \cdot \vec{A} = 0,$$

we have

$$W_m = \frac{\mu}{8\pi} \int_v \int_{v_j} \frac{\vec{J}(\vec{r}_A) \cdot \vec{J}(\vec{r})}{x} dv_j dv,$$

where v_j is the volume of the medium where $\vec{J}$ exists.

The general formula for the self-inductance $i = j$ and the mutual inductance $i \neq j$ of loops i and j is

$$L_{ij} = \frac{N_i \Phi_{ij}}{i_j} = \frac{\psi_{ij}}{i_j},$$

where ψ_{ij} is the flux linkage through ith coil due to the current in jth coil; i_j is the current in jth coil.

The *Neumann* formula

$$L_{ij} = L_{ji} = \frac{\mu}{4\pi} \oint_{l_i} \oint_{l_j} \frac{d\vec{l}_j \cdot d\vec{l}_i}{x_{ij}}, \quad i \neq j$$

is applied to find the mutual inductance. Using

$$W_m = \frac{\mu}{8\pi} \int_v \int_{v_j} \frac{\vec{J}(\vec{r}_A) \cdot \vec{J}(\vec{r})}{x} dv_j dv,$$

one obtains

$$W_m = \frac{\mu}{8\pi} \int_{l_i} \int_{l_j} \frac{i_j d\vec{l}_j \cdot i_i d\vec{l}_i}{x_{ij}}.$$

Hence, the energy stored in the magnetic field is found to be

$$W_m = \frac{1}{2} i_i L_{ij} i_j.$$

Using the current vector $\mathbf{i} = [i_1, i_2, \ldots, i_{n-1}, i_n]$ and the inductance matrix $L \in \mathbb{R}^{n \times n}$, we have

$$W_m = \frac{1}{2} \mathbf{i}^T L \mathbf{i},$$

where T denotes the transpose symbol.

As an example, the magnetic energy, stored in the inductor with a single winding is

$$W_m = \frac{1}{2} Li^2.$$

The differential change in the stored magnetic energy should be found. Using

$$\frac{dW_m}{dt} = \frac{1}{2} \left(L_{ij} i_j \frac{di_i}{dt} + L_{ij} i_i \frac{di_j}{dt} + i_i i_j \frac{dL_{ij}}{dt} \right),$$

we have

$$dW_m = \frac{1}{2} (L_{ij} i_j di_i + L_{ij} i_i di_j + i_i i_j dL_{ij}).$$

For translational motion, the differential change in the mechanical energy is expressed by $dW_{mec} = \vec{F}_m \cdot d\vec{l}$. Assuming that the system is conservative (for lossless systems $dW_{mec} = dW_m$), in the rectangular coordinate system we obtain the following equation

$$dW_m = \frac{\partial W_m}{\partial x} dx + \frac{\partial W_m}{\partial y} dy + \frac{\partial W_m}{\partial z} dz = \vec{\nabla} W_m \cdot d\vec{l}.$$

Hence, the force is the gradient of the stored magnetic energy, and

$$\vec{F}_m = \vec{\nabla} W_m.$$

In the XYZ coordinate system for the translational motion, we have

$$F_{mx} = \frac{\partial W_m}{\partial x}, \quad F_{my} = \frac{\partial W_m}{\partial y} \quad \text{and} \quad F_{mz} = \frac{\partial W_m}{\partial z}.$$

For the rotational motion, the torque should be used. Using the differential change in the mechanical energy as a function of the angular displacement θ, the following formula results if the rigid body (rotor) is constrained to rotate around the z-axis

$$dW_{mec} = T_e d\theta,$$

where T_e is the z-component of the electromagnetic torque.

Assuming that the system is lossless, one obtains the following expression for the electromagnetic torque

$$T_e = \frac{\partial W_m}{\partial \theta}.$$

The *electromotive* and *magnetomotive* forces (*emf* and *mmf*) are found as

$$emf = \oint_l \vec{E} \cdot d\vec{l} = \underbrace{\oint_l (\vec{v} \times \vec{B}) \cdot d\vec{l}}_{\text{motional induction (generation)}} - \underbrace{\oint_s \frac{\partial \vec{B}}{\partial t} d\vec{s}}_{\text{transformer induction}}$$

and

$$mmf = \oint_l \vec{H} \cdot d\vec{l} = \oint_s \vec{J} \cdot d\vec{s} + \oint_s \frac{\partial \vec{D}}{\partial t} d\vec{s}.$$

The motional *emf* is a function of the velocity and the magnetic flux density, whereas the *emf* induced in a stationary closed circuit is equal to the negative rate of increase of the magnetic flux (transformer induction). The induced *mmf* is the sum of the induced current and the rate of change of the flux penetrating the surface bounded by the contour. To show that, we apply Stoke's theorem to find the integral form of Ampere's law (second Maxwell's equation), as

$$\oint_l \vec{H}(t) \cdot d\vec{l} = \oint_s \vec{J}(t) \cdot d\vec{s} + \oint_s \frac{d\vec{D}(t)}{dt} d\vec{s},$$

where $\vec{J}(t)$ is the time-varying current density vector.

This section covered the basics of electromagnetics, which are applicable to electromechanical motion devices. The relative complexity of the results reported is evident. The electromagnetics provides sound approaches to examine the device physics, accomplish coherent structural design of electromechanical motion devices, perform sound analysis, study performance, and so on. For a great number of analysis and design problems, the reported results can be simplified. The engineering application of the electromagnetics and practical examples are covered in Section 2.6.

2.5   Classical Mechanics and Its Application

Distinct electromechanical systems and motion devices can be studied applying classical mechanics. Newtonian mechanics, Lagrange's concept, and Hamilton's method. These concepts provide meaningful approaches to derive the governing equations thereby allowing one to accomplish the quantitative and qualitative analysis. The application of Lagrange's and Hamilton's concepts results in a coherent analysis integrating electromagnetics, mechanics, and circuits.

2.5.1   Newtonian Mechanics

2.5.1.1   Newtonian Mechanics, Energy Analysis, Generalized Coordinates, and Lagrange Equations: Translational Motion

We study the system behavior with the corresponding analysis of forces that cause motion. The equations of motion for mechanical systems can be found using Newton's second law of motion. In particular, using the position (displacement) vector $\vec{r}$ the Newton equation in the vector form is given as

$$\sum \vec{F}(t,\vec{r}) = m\vec{a}, \qquad (2.1)$$

where $\Sigma \vec{F}(t,\vec{r})$ is the vector sum of all forces applied to the body ($\Sigma \vec{F}$ is the *net* force); $\vec{a}$ is the vector of acceleration of the body with respect to an inertial reference frame; m is the mass of the body.

In (2.1), $m\vec{a}$ represents the magnitude and direction of the applied net force acting on the object. Hence, $m\vec{a}$ is not a force. A body is at equilibrium (the object is at rest or is moving with constant speed) if $\Sigma \vec{F} = 0$; in other words, there is no acceleration if $\Sigma \vec{F} = 0$. Using (2.1), in the Cartesian system, we have the mechanical equations of motion in the xyz coordinates

$$\sum \vec{F}(t,\vec{r}) = m\vec{a} = m\frac{d\vec{r}^2}{dt^2} = m \begin{bmatrix} \dfrac{d\vec{x}^2}{dt^2} \\[2mm] \dfrac{d\vec{y}^2}{dt^2} \\[2mm] \dfrac{d\vec{z}^2}{dt^2} \end{bmatrix}, \quad \begin{bmatrix} \vec{a}_x \\[1mm] \vec{a}_y \\[1mm] \vec{a}_z \end{bmatrix} = \begin{bmatrix} \dfrac{d\vec{x}^2}{dt^2} \\[2mm] \dfrac{d\vec{y}^2}{dt^2} \\[2mm] \dfrac{d\vec{z}^2}{dt^2} \end{bmatrix}.$$

Hence, one obtains the second-order ordinary differential equations that are the rigid-body mechanical equations of motion. The forces to control the motion should be developed by actuators. The force can be a function of current (electromagnetic actuators), voltage (electrostatic actuators), or pressure (hydraulic actuators). Correspondingly, as electromechanical actuators are under our consideration, equation (2.1) must be complemented by the electromagnetic dynamics to achieve the completeness and coherence.

In the Cartesian coordinate system, Newton's second law is expressed as

$$\sum \vec{F}_x = m\vec{a}_x, \quad \sum \vec{F}_y = m\vec{a}_y, \quad \sum \vec{F}_z = m\vec{a}_z.$$

Newton's second law in terms of the linear momentum, which is found as $\vec{p} = m\vec{v}$, is given by

$$\sum \vec{F} = \frac{d\vec{p}}{dt} = \frac{d(m\vec{v})}{dt},$$

where $\vec{v}$ is the vector of the object velocity.

Thus, the force is equal to the rate of change of the momentum. The object or particle moves uniformly if $\frac{d\vec{p}}{dt} = 0$, which implies $\vec{p} = \text{const}$.

Taking note of the expression for the energy $\Pi(\vec{r})$, for the conservative mechanical system, we have

$$\sum \vec{F}(\vec{r}) = -\nabla \Pi(\vec{r}).$$

Therefore, the work done per unit time is

$$\frac{dW}{dt} = \sum \vec{F}(\vec{r}) \frac{d\vec{r}}{dt} = -\nabla \Pi(\vec{r}) \frac{d\vec{r}}{dt} = -\frac{d\Pi(\vec{r})}{dt}.$$

From Newton's second law one obtains

$$m\vec{a} - \sum \vec{F}(\vec{r}) = 0 \quad \text{or} \quad m\frac{d^2\vec{r}}{dt^2} - \sum \vec{F}(\vec{r}) = 0.$$

Hence, for a conservative system the following equation results

$$m\frac{d^2\vec{r}}{dt^2} + \nabla \Pi(\vec{r}) = 0.$$

Having documented the Newtonian mechanics and the use of the kinetic and potential energies, let us introduce one of the most general concept in the modeling of dynamic systems that is based on the Lagrangian mechanics. Instead of applying the displacement, velocity, or acceleration, we will utilize the generalized coordinates. By using the generalized coordinates $(q_1, ..., q_n)$ and generalized velocities

$$\left(\frac{dq_1}{dt}, ..., \frac{dq_n}{dt} \right),$$

one finds the total kinetic

$$\Gamma\left(q_1, ..., q_n, \frac{dq_1}{dt}, ..., \frac{dq_n}{dt} \right)$$

and potential $\Pi(q_1, ..., q_n)$ energies. Using the expressions for the total kinetic and potential energies, Newton's second law of motion can be given in the following form

$$\frac{d}{dt}\left(\frac{\partial \Gamma}{\partial \dot{q}_1} \right) + \frac{\partial \Pi}{\partial q_i} = 0.$$

Example 2.1

Consider a positioning table actuated by a motor. Let us find how much work is required to accelerate a 20 g payload ($m = 20$ g) from $v_0 = 0$ m/sec to $v_f = 1$ m/sec.

The work needed is calculated as

$$W = \frac{1}{2}\left(mv_f^2 - mv_0^2\right) = \frac{1}{2}\,20 \times 10^{-3} \times 1^2 = 0.01 \text{ J.}$$

□

Example 2.2

Consider a body of mass m in the XY coordinate system. The force $\vec{F}_a$ is applied in the x direction. Let us find the equations of motion neglecting *Coulomb* and static frictions. We assume that the viscous friction force is

$$F_{fr} = B_v\frac{dx}{dt},$$

where B_v is the viscous friction coefficient.

The free-body diagram is illustrated in Figure 2.6.

The sum of the forces, acting in the y direction, is expressed as

$$\sum \vec{F}_Y = \vec{F}_N - \vec{F}_g,$$

where $\vec{F}_g = mg$ is the gravitational force acting on the mass m; $\vec{F}_N$ is the normal force that is equal and opposite to the gravitational force.

From (2.1), the equation of motion in the y direction is

$$\vec{F}_N - \vec{F}_g = ma_y = m\frac{d^2y}{dt^2},$$

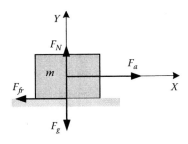

FIGURE 2.6
Free-body diagram.

where a_y is the acceleration in the y direction,

$$a_y = \frac{d^2y}{dt^2}.$$

Making use of $\vec{F}_N = \vec{F}_g$, we have the resulting equation

$$\frac{d^2y}{dt^2} = 0.$$

The sum of the forces acting in the x direction is found using the applied force $\vec{F}_a$ and the friction force $\vec{F}_{fr}$. We have

$$\sum \vec{F}_X = \vec{F}_a - \vec{F}_{fr}.$$

The applied force can be time-invariant $\vec{F}_a = \text{const}$ or nonlinear time-varying $\vec{F}_a(t) = f(t, x, y, z)$. Let

$$\vec{F}_a(t, x) = x \sin(6t - 4)e^{-0.5t} + \frac{dx}{dt}t^2 + x^3 \cos\left(\frac{dx}{dt}t - x^2t^4\right).$$

Using (2.1), the equation motion in the x direction is found to be

$$\vec{F}_a - \vec{F}_{fr} = ma_x = m\frac{d^2x}{dt^2},$$

where a_x is the acceleration in the x direction,

$$a_x = \frac{d^2x}{dt^2},$$

and the velocity in the x direction is

$$v = \frac{dx}{dt}.$$

Assume that the *Coulomb* and static frictions can be neglected. The friction force, as a function of the viscous friction coefficient B_v and velocity v, is given as

$$F_{fr} = B_v \frac{dx}{dt} = B_v v$$

Hence, one obtains the second-order nonlinear differential equation to describe the rigid-body dynamics in the x direction

$$\frac{d^2x}{dt^2} = \frac{1}{m}\left(F_a - B_v \frac{dx}{dt} \right)$$

$$= \frac{1}{m}\left[x\sin(6t-4)e^{-0.5t} + \frac{dx}{dt}t^2 + x^3\cos\left(\frac{dx}{dt}t - x^2t^4\right) - B_v\frac{dx}{dt} \right].$$

From the derived equation of motion, a set of two first-order linear differential equations results. In particular,

$$\frac{dx}{dt} = v,$$

$$\frac{dv}{dt} = \frac{1}{m}[x\sin(6t-4)e^{-0.5t} + vt^2 + x^3\cos(vt - x^2t^4) - B_v v], \quad t \geq 0. \qquad \square$$

2.5.1.2 Newtonian Mechanics: Rotational Motion

For rotational devices, instead of linear displacement and acceleration, the angular displacement and acceleration are used. The net torque is considered. The rotational Newton's second law for a rigid body is

$$\sum \vec{T}(t,\vec{\theta}) = J\vec{\alpha}, \tag{2.2}$$

where $\sum \vec{T}$ is the *net* torque; J is the moment of inertia (*rotational inertia*); $\vec{\alpha}$ is the angular acceleration vector,

$$\vec{\alpha} = \frac{d}{dt}\frac{d\vec{\theta}}{dt} = \frac{d^2\vec{\theta}}{dt^2} = \frac{d\vec{\omega}}{dt};$$

$\vec{\theta}$ is the angular displacement; $\vec{\omega}$ is the angular velocity.

The angular momentum of the system $\vec{L}_M$ is $\vec{L}_M = \vec{R} \times \vec{p} = \vec{R} \times m\vec{v}$ and

$$\sum \vec{T} = \frac{d\vec{L}_M}{dt} = \vec{R} \times \vec{F},$$

where $\vec{R}$ is the position vector with respect to the origin.

For the rigid body, rotating around the axis of symmetry, we have

$$\vec{L}_M = J\vec{\omega}.$$

For one-dimensional rotational systems, Newton's second law of motion is also expressed as

$$M = J\alpha.$$

where M is the sum of all moments about the center of mass of a body, (N-m); J is the moment of inertia about its center of mass, (kg-m^2); α is the angular acceleration of the body, (rad/sec^2).

Example 2.3

A motor has the equivalent moment of inertia $J = 0.5$ kg-m^2. When the motor accelerates, the angular velocity of the rotor is $\omega_r = 10t^3$, $t \geq 0$. One can find the angular momentum and the developed electromagnetic torque as functions of time. Assume that the load and friction torques are zero. The angular momentum is $L_M = J\omega_r = 5t^3$. The developed electromagnetic torque is

$$T_e = \frac{dL_M}{dt} = 15t^2 \ [\text{N-m}].$$ □

From Newtonian mechanics, one concludes that the applied net force or torque play a key role in qualitatively and quantitatively describing the motion as well as controlling the dynamics. The analysis of motion can be performed using the energy or momentum quantities, which are conserved. The principle of conservation of energy states: energy can be only converted from one form to another.

Kinetic energy is associated with motion, while potential energy is associated with position. The sum of the kinetic (Γ), potential (Π), and dissipated (D) energies is called the total energy of the system (Σ_T), which is conserved, and the total amount of energy remains constant; for example, $\Sigma_T = \Gamma + \Pi + D = \text{const}$.

Example 2.4

Consider the translational motion of a body, which is attached to an ideal spring that exerts the force which obeys ideal Hooke's law. Neglecting friction, one obtains the following expression for the total energy

$$\Sigma_T = \Gamma + \Pi = \frac{1}{2}(mv^2 + k_s x^2) = \text{const}.$$

The translational kinetic energy is

$$\Gamma = \frac{1}{2}mv^2,$$

whereas the elastic potential energy of the spring is

$$\Pi = \frac{1}{2}k_s x^2,$$

where k_s is the force constant of the spring.

For rotational motion and torsional spring, we have

$$\Sigma_T = \Gamma + \Pi = \frac{1}{2}(J\omega^2 + k_s\theta^2) = \text{const},$$

where the rotational kinetic energy and the elastic potential energy are

$$\Gamma = \frac{1}{2}J\omega^2 \quad \text{and} \quad \Pi = \frac{1}{2}k_s\theta^2. \qquad \square$$

The kinetic energy of a rigid body having translational and rotational components of motion is

$$\Gamma = \frac{1}{2}(mv^2 + J\omega^2).$$

That is, motion of the rigid body is represented as a combination of translational motion of the center of mass and rotational motion about the axis through the center of mass. The moment of inertia J depends on how the mass is distributed with respect to the axis, and J is different for different axes of rotation. If the body is uniform in density, J can be calculated for regularly shaped bodies using their dimensions. For example, a rigid cylinder of mass m (which is uniformly distributed), radius R, and length l, has the following horizontal and vertical moments of inertia

$$J_{horizontal} = \frac{1}{2}mR^2 \quad \text{and} \quad J_{vertical} = \frac{1}{4}mR^2 + \frac{1}{12}ml^2.$$

The *radius of gyration* can be found for irregularly shaped objects, and the moment of inertia can be easily obtained.

In electromechanical motion devices, the force and torque are of a great interest. Assuming that the body is rigid and the moment of inertia is constant, one has

$$\vec{T}d\vec{\theta} = J\vec{\alpha}d\vec{\theta} = J\frac{d\vec{\omega}}{dt}d\vec{\theta} = J\frac{d\vec{\theta}}{dt}d\vec{\omega} = J\vec{\omega}d\vec{\omega}.$$

The total work, as given by

$$W = \int_{\theta_0}^{\theta_f} \vec{T}d\vec{\theta} = \int_{\omega_0}^{\omega_f} J\vec{\omega}d\vec{\omega} = \frac{1}{2}\left(J\omega_f^2 - J\omega_0^2\right),$$

represents the change of the kinetic energy. Furthermore,

$$\frac{dW}{dt} = \vec{T}\frac{d\vec{\theta}}{dt} = \vec{T} \times \vec{\omega},$$

and the power is $P = \vec{T} \times \vec{\omega}$.

This equation is an analog of $P = \vec{F} \times \vec{v}$, which is applied for translational motion.

Example 2.5

Assume that the rated power and angular velocity of a motor are 1 W and 1000 rad/sec. The rated electromagnetic torque is found to be

$$T_e = \frac{P}{\omega_r} = \frac{1}{1000} = 1 \times 10^{-3}\ [\text{N - m}]. \qquad \square$$

Example 2.6

Given a point mass m suspended by a massless unstretchable string of length l (see Figure 2. 7), one can derive the equations of motion for a simple pendulum.

The restoring force, which is $-mg\sin\theta$, is the tangential component of the net force. Therefore, the sum of the moments about the pivot point O is

$$\sum M = -mgl\sin\theta + T_a,$$

where T_a is the applied torque; l is the length of the pendulum measured from the point of rotation.

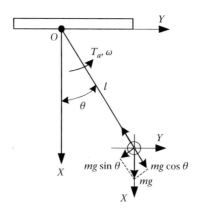

FIGURE 2.7
A simple pendulum.

Using (2.2), one obtains the equation of motion

$$J\alpha = J\frac{d^2\theta}{dt^2} = -mgl\sin\theta + T_a,$$

where J is the moment of inertial of the mass about the point O. Hence, the second-order differential equation is

$$\frac{d^2\theta}{dt^2} = \frac{1}{J}(-mgl\sin\theta + T_a).$$

Using the differential equation for the angular displacement

$$\frac{d\theta}{dt} = \omega,$$

one obtains a set of two first-order differential equations

$$\frac{d\omega}{dt} = \frac{1}{J}(-mgl\sin\theta + T_a), \quad \frac{d\theta}{dt} = \omega.$$

The moment of inertia is $J = ml^2$. Hence, we have the following differential equations to describe the dynamics of a simple pendulum

$$\frac{d\omega}{dt} = -\frac{g}{l}\sin\theta + \frac{1}{ml^2}T_a,$$

$$\frac{d\theta}{dt} = \omega.$$ □

Example 2.7 Friction in Motion Devices

A thorough consideration of friction is essential. Friction is a very complex nonlinear phenomenon that is difficult to describe. The classical *Coulomb* friction is a retarding frictional force (for translational motion) or torque (for rotational motion) that changes its sign with the reversal of the direction of motion, and the amplitude of the frictional force or torque is constant. For translational and rotational motions, the *Coulomb* friction force and torque are

$$F_{Couloumb} = k_{Fc}\,\text{sgn}(v) = k_{Fc}\,\text{sgn}\left(\frac{dx}{dt}\right) \quad \text{and} \quad T_{Coulomb} = k_{Tc}\,\text{sgn}(\omega) = k_{Tc}\,\text{sgn}\left(\frac{d\theta}{dt}\right),$$

where k_{Fc} and k_{Tc} are the *Coulomb* friction coefficients.

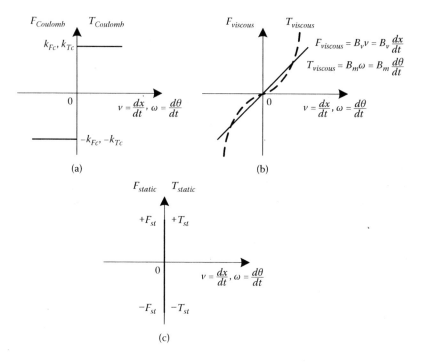

FIGURE 2.8
(a) Coulomb friction; (b) viscous friction; (c) static friction.

Figure 2.8a illustrates the *Coulomb* friction.

Viscous friction is a retarding force or torque that is a linear (or nonlinear) function of linear or angular velocity. The viscous friction force and torque versus linear and angular velocities are shown in Figure 2.8b. The following expressions are commonly used to describe the viscous friction

$$F_{viscous} = B_v v = B_v \frac{dx}{dt} \quad \text{or} \quad F_{viscous} = \sum_{n=1}^{\infty} B_{vn} v^{2n-1} \qquad \text{for translational motion}$$

$$T_{viscous} = B_m \omega = B_m \frac{d\theta}{dt} \quad \text{or} \quad T_{viscous} = \sum_{n=1}^{\infty} B_{mn} \omega^{2n-1} \qquad \text{for rotational motion,}$$

where B_v and B_m are the viscous friction coefficients.

The static friction exists only when the body is stationary and vanishes as motion begins. The static friction is a force F_{static} or torque T_{static}. One can apply the following expressions

$$F_{static} = \pm F_{st}\big|_{v=\frac{dx}{dt}=0} \quad \text{and} \quad T_{static} = \pm T_{st}\big|_{\omega=\frac{d\theta}{dt}=0}.$$

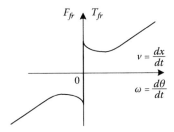

FIGURE 2.9
Friction force and torque as functions of linear and angular velocities.

We conclude that the static friction is a retarding force or torque that tends to prevent the initial translational or rotational motion at beginning (see Figure 2.8c).

The friction force and torque are nonlinear functions that are modeled using frictional memory, presliding conditions, and so on. The empirical formulas, commonly used to express F_{fr} and T_{fr} are

$$F_{fr} = (k_{fr1} - k_{fr2}e^{-k|v|} + k_{fr3}|v|)\mathrm{sgn}(v) \text{ and } T_{fr} = (k_{fr1} - k_{fr2}e^{-k|\omega|} + k_{fr3}|\omega|)\mathrm{sgn}(\omega).$$

The typifying F_{fr} and T_{fr} are shown in Figure 2.9. □

2.5.2 Lagrange Equations of Motion

Electromechanical systems integrate mechanical, electromagnetic, circuit, and electronic components. Therefore, one studies mechanical, electromagnetic, and circuitry transients. The designer may apply Newtonian's dynamics deriving the rigid-body dynamics (translational and *torsional-mechanical* dynamics for linear and rotational actuators), and, then, using the co-energy concept derive the expression for the electromagnetic (or electrostatic) force or torque, which are functions of current, voltage, or electromagnetic field quantities. Thus, the energy conversion, force/torque production, and electromagnetic dynamics, obtained using laws of electromagnetics, must be integrated. The examples on this approach will be documented in Section 2.6 and Chapters 4, 5, and 6.

In contrast, the Lagrange and Hamilton concepts are based on the energy analysis of an entire system. Using the Lagrange equations, one can integrate the rigid-body (mechanical) dynamics and circuitry-electromagnetic equations of motion. Hence, the Lagrange and Hamilton concepts are more general approaches to be applied. Using the system variables, one finds the total kinetic, dissipation, and potential energies, which are denoted as Γ, D and Π. Taking note of the total

- Kinetic

$$\Gamma\left(t, q_1, \ldots, q_n, \frac{dq_1}{dt}, \ldots, \frac{dq_n}{dt}\right),$$

- Dissipation

$$D\left(t, q_1, \ldots, q_n, \frac{dq_1}{dt}, \ldots, \frac{dq_n}{dt}\right),$$

- Potential

$$\Pi(t, q_1, \ldots, q_n)$$

energies, the Lagrange equations of motion are

$$\frac{d}{dt}\left(\frac{\partial \Gamma}{\partial \dot{q}_i}\right) - \frac{\partial \Gamma}{\partial q_i} + \frac{\partial D}{\partial \dot{q}_i} + \frac{\partial \Pi}{\partial q_i} = Q_i. \qquad (2.3)$$

Here, q_i and Q_i are the generalized coordinates and the generalized forces (applied forces and disturbances). These generalized coordinates q_i are used to derive expressions for energies

$$\Gamma\left(t, q_1, \ldots, q_n, \frac{dq_1}{dt}, \ldots, \frac{dq_n}{dt}\right), \quad D\left(t, q_1, \ldots, q_n, \frac{dq_1}{dt}, \ldots, \frac{dq_n}{dt}\right) \text{ and } \Pi(t, q_1, \ldots, q_n).$$

The generalized coordinates are assigned to explicitly and coherently express the energies of the system. As the generalized coordinates, one uses

- Linear or angular displacement (for translational and rotational devices),
- Charges (the charge is commonly denoted using the symbol q).

For conservative (lossless) systems $D = 0$, and we have the following Lagrange's equations of motion

$$\frac{d}{dt}\left(\frac{\partial \Gamma}{\partial \dot{q}_i}\right) - \frac{\partial \Gamma}{\partial q_i} + \frac{\partial \Pi}{\partial q_i} = Q_i.$$

To illustrate the application of the Lagrange equation of motion we consider familiar examples.

Example 2.8   Simple Pendulum

Our goal is to derive the equations of motion for a simple pendulum as depicted in Figure 2.7. We will apply the Lagrange equations of motion. The equations of motion for the simple pendulum were derived in Example 2.6 using the Newtonian mechanics.

For the studied conservative (lossless) system we have $D = 0$. The Lagrange equations of motion are

$$\frac{d}{dt}\left(\frac{\partial \Gamma}{\partial \dot{q}_i}\right) - \frac{\partial \Gamma}{\partial q_i} + \frac{\partial \Pi}{\partial q_i} = Q_i.$$

The kinetic energy of the pendulum bob is

$$\Gamma = \frac{1}{2}m(l\dot{\theta})^2.$$

The potential energy is found as $\Pi = mgl(1 - \cos\theta)$.

The angular displacement is the generalized coordinate. We have only one generalized coordinate, $q_i = q_1 = \theta$.

The generalized force is the applied torque, hence, $Q_i = T_a$.

We found the expressions for kinetic and potential energies as

$$\Gamma = \frac{1}{2}m(l\dot{q})^2 \quad \text{and} \quad \Pi = mgl(1 - \cos q).$$

One obtains the following expressions for the derivatives

$$\frac{\partial \Gamma}{\partial \dot{q}_i} = \frac{\partial \Gamma}{\partial \dot{\theta}} = ml^2\dot{\theta}, \quad \frac{\partial \Gamma}{\partial q_i} = \frac{\partial \Gamma}{\partial \theta} = 0 \quad \text{and} \quad \frac{\partial \Pi}{\partial q_i} = \frac{\partial \Pi}{\partial \theta} = mgl\sin\theta.$$

The first term of the Lagrange equation is

$$\frac{d}{dt}\left(\frac{\partial \Gamma}{\partial \dot{\theta}}\right) = ml^2\frac{d^2\theta}{dt^2} + 2ml\frac{dl}{dt}\frac{d\theta}{dt}.$$

Assuming that the string is unstretchable, we have

$$\frac{dl}{dt} = 0.$$

If this assumption is not valid, one should use the appropriate expression for the length as a function of the generalized coordinate q. For

$$\frac{dl}{dt} = 0,$$

we obtain

$$ml^2 \frac{d^2\theta}{dt^2} + mgl \sin\theta = T_a.$$

Thus, one obtains

$$\frac{d^2\theta}{dt^2} = \frac{1}{ml^2}(-mgl\sin\theta + T_a).$$

Recall that the equation of motion, derived by using Newtonian mechanics, is

$$\frac{d^2\theta}{dt^2} = \frac{1}{J}(-mgl\sin\theta + T_a), \quad \text{where } J = ml^2.$$

One concludes that the results are the same, and the equations are

$$\frac{d\omega}{dt} = -\frac{g}{l}\sin\theta + \frac{1}{ml^2}T_a, \quad \frac{d\theta}{dt} = \omega.$$

The Lagrange equations of motion provide more general results. We will illustrate that electromechanical devices can be modeled using the Lagrange equations of motion. Newton's laws can be used only to model the rigid-body dynamics unless electromechanical "analogies" are applied. Furthermore, it was illustrated that the coordinate-dependent system parameters can be accounted for ensuring coherency and accuracy. For example, l can be a function of θ.   □

Example 2.9   Double Pendulum

Consider a double pendulum of two degrees of freedom with no external forces applied to the system (see Figure 2.10). Using the Lagrange equations of motion, we should derive the differential equations.

The angular displacements θ_1 and θ_2 are the independent generalized coordinates q_1 and q_2. In the xy plane, let (x_1, y_1) and (x_2, y_2) be the rectangular coordinates of m_1 and m_2. We obtain

$$x_1 = l_1\cos\theta_1, \quad x_2 = l_1\cos\theta_1 + l_2\cos\theta_2,$$

$$y_1 = l_1\sin\theta_1, \quad y_2 = l_1\sin\theta_1 + l_2\sin\theta_2.$$

The total kinetic energy Γ is found to be a nonlinear function of the displacements, and

$$\Gamma = \frac{1}{2}m_1\left(\dot{x}_1^2 + \dot{y}_1^2\right) + \frac{1}{2}m_2\left(\dot{x}_2^2 + \dot{y}_2^2\right)$$

$$= \frac{1}{2}(m_1 + m_2)l_1^2\dot{\theta}_1^2 + m_2l_1l_2\dot{\theta}_1\dot{\theta}_2\cos(\theta_2 - \theta_1) + \frac{1}{2}m_2l_2^2\dot{\theta}_2^2.$$

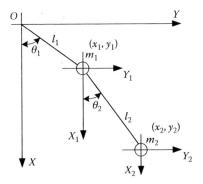

FIGURE 2.10
Double pendulum.

One obtains

$$\frac{\partial \Gamma}{\partial \theta_1} = m_2 l_1 l_2 \sin(\theta_2 - \theta_1)\dot{\theta}_1\dot{\theta}_2, \quad \frac{\partial \Gamma}{\partial \dot{\theta}_1} = (m_1 + m_2)l_1^2\dot{\theta}_1 + m_2 l_1 l_2 \cos(\theta_2 - \theta_1)\dot{\theta}_2,$$

$$\frac{\partial \Gamma}{\partial \theta_2} = -m_2 l_1 l_2 \sin(\theta_1 - \theta_2)\dot{\theta}_1\dot{\theta}_2, \quad \frac{\partial \Gamma}{\partial \dot{\theta}_2} = m_2 l_1 l_2 \cos(\theta_2 - \theta_1)\dot{\theta}_1 + m_2 l_1^2\dot{\theta}_2.$$

The total potential energy is

$$\Pi = m_1 g x_1 + m_2 g x_2 = (m_1 + m_2)g l_1 \cos\theta_1 + m_2 g l_2 \cos\theta_2,$$

yielding

$$\frac{\partial \Pi}{\partial \theta_1} = -(m_1 + m_2)g l_1 \sin\theta_1 \quad \text{and} \quad \frac{\partial \Pi}{\partial \theta_2} = -m_2 g l_2 \sin\theta_2.$$

The Lagrange equations of motion are

$$\frac{d}{dt}\left(\frac{\partial \Gamma}{\partial \dot{\theta}_1}\right) - \frac{\partial \Gamma}{\partial \theta_1} + \frac{\partial \Pi}{\partial \theta_1} = 0, \quad \frac{d}{dt}\left(\frac{\partial \Gamma}{\partial \dot{\theta}_2}\right) - \frac{\partial \Gamma}{\partial \theta_2} + \frac{\partial \Pi}{\partial \theta_2} = 0.$$

Hence, the differential equations that describe the motion are

$$l_1 \Big[(m_1 + m_2)l_1\ddot{\theta}_1 + m_2l_2 \cos(\theta_2 - \theta_1)\ddot{\theta}_2 - m_2l_2 \sin(\theta_2 - \theta_1)\dot{\theta}_2^2$$
$$- m_2l_2 \sin(\theta_2 - \theta_1)\dot{\theta}_1\dot{\theta}_2 - (m_1 + m_2)g \sin\theta_1 \Big] = 0,$$

$$m_2l_2 \Big[l_2\ddot{\theta}_2 + l_1 \cos(\theta_2 - \theta_1)\ddot{\theta}_1 + l_1 \sin(\theta_2 - \theta_1)\dot{\theta}_1^2 + l_1 \sin(\theta_2 - \theta_1)\dot{\theta}_1\dot{\theta}_2 - g \sin\theta_2 \Big] = 0.$$

If the torques T_1 and T_2 are applied to the first and second joints (two-degree-of-freedom robot), the following equations of motions result

$$l_1 \Big[(m_1 + m_2)l_1\ddot{\theta}_1 + m_2l_2 \cos(\theta_2 - \theta_1)\ddot{\theta}_2 - m_2l_2 \sin(\theta_2 - \theta_1)\dot{\theta}_2^2$$
$$- m_2l_2 \sin(\theta_2 - \theta_1)\dot{\theta}_1\dot{\theta}_2 - (m_1 + m_2)g \sin\theta_1 \Big] = T_1,$$

$$m_2l_2 \Big[l_2\ddot{\theta}_2 + l_1 \cos(\theta_2 - \theta_1)\ddot{\theta}_1 + l_1 \sin(\theta_2 - \theta_1)\dot{\theta}_1^2 + l_1 \sin(\theta_2 - \theta_1)\dot{\theta}_1\dot{\theta}_2 - g \sin\theta_2 \Big] = T_2.$$

These torques T_1 and T_2 can be time-varying functions to be controlled by actuators. Furthermore, the load torques, as were illustrated, can be added to the equations derived. $\square$

We have illustrated the use of the Lagrange equations of motion for mechanical systems. The following examples will demonstrate the use of Lagrangian mechanics for electric circuits.

Example 2.10   Circuit Network

Consider a two-mesh electric circuit as shown in Figure 2.11. Our goal is to derive the equations which describe the circuitry dynamics.

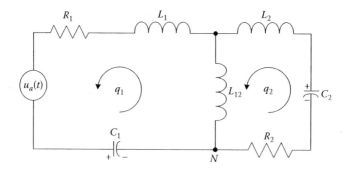

FIGURE 2.11
Electric circuit.

We use the electric charges as the generalized coordinates. That is, q_1 and q_2 are the independent generalized coordinates (variables) as shown in Figure 2.11. Here, q_1 is the electric charge in the first loop, and q_2 represents the electric charge in the second loop.

The generalized force (applied voltage u_a), applied to the system, is denoted as Q_1. That is, $u_a(t) = Q_1$.

The generalized coordinates are related to the currents. In particular, the currents i_1 and i_2 are found in terms of charges as $i_1 = \dot{q}_1$ and $i_2 = \dot{q}_2$. That is, we have

$$q_1 = \frac{i_1}{s} \quad \text{and} \quad q_2 = \frac{i_2}{s}.$$

The total magnetic energy (kinetic energy) is expressed by

$$\Gamma = \frac{1}{2}L_1\dot{q}_1^2 + \frac{1}{2}L_{12}(\dot{q}_1 - \dot{q}_2)^2 + \frac{1}{2}L_2\dot{q}_2^2.$$

By using this equation for Γ, we have

$$\frac{\partial \Gamma}{\partial q_1} = 0, \quad \frac{\partial \Gamma}{\partial \dot{q}_1} = (L_1 + L_{12})\dot{q}_1 - L_{12}\dot{q}_2,$$

$$\frac{\partial \Gamma}{\partial q_2} = 0, \quad \frac{\partial \Gamma}{\partial \dot{q}_2} = -L_{12}\dot{q}_1 + (L_2 + L_{12})\dot{q}_2.$$

Using the equation for the total electric energy (potential energy)

$$\Pi = \frac{1}{2}\frac{q_1^2}{C_1} + \frac{1}{2}\frac{q_2^2}{C_2},$$

one finds

$$\frac{\partial \Pi}{\partial q_1} = \frac{q_1}{C_1} \quad \text{and} \quad \frac{\partial \Pi}{\partial q_2} = \frac{q_2}{C_2}.$$

The total heat energy dissipated is

$$D = \frac{1}{2}R_1\dot{q}_1^2 + \frac{1}{2}R_2\dot{q}_2^2.$$

Hence,

$$\frac{\partial D}{\partial \dot{q}_1} = R_1\dot{q}_1 \quad \text{and} \quad \frac{\partial D}{\partial \dot{q}_2} = R_2\dot{q}_2.$$

The Lagrange equations of motion are expressed using the independent coordinates. In particular,

$$\frac{d}{dt}\left(\frac{\partial \Gamma}{\partial \dot{q}_1}\right) - \frac{\partial \Gamma}{\partial q_1} + \frac{\partial D}{\partial \dot{q}_1} + \frac{\partial \Pi}{\partial q_1} = Q_1, \quad \frac{d}{dt}\left(\frac{\partial \Gamma}{\partial \dot{q}_2}\right) - \frac{\partial \Gamma}{\partial q_2} + \frac{\partial D}{\partial \dot{q}_2} + \frac{\partial \Pi}{\partial q_2} = 0.$$

Hence, the differential equations for the circuit are found to be

$$(L_1 + L_{12})\ddot{q}_1 - L_{12}\ddot{q}_2 + R_1\dot{q}_1 + \frac{q_1}{C_1} = u_a,$$

$$-L_{12}\ddot{q}_1 + (L_2 + L_{12})\ddot{q}_2 + R_2\dot{q}_2 + \frac{q_2}{C_2} = 0.$$

To assess the transient dynamics, one can solve the resulting equations. Analytic and numerical methods are used. To perform numerical simulations, the MATLAB environment will be introduced and used in this chapter. Two second-order nonlinear differential equations, which describe the circuit dynamics, can be simulated. Using the differential equations derived

$$\ddot{q}_1 = \frac{1}{(L_1 + L_{12})}\left(-\frac{q_1}{C_1} - R_1\dot{q}_1 + L_{12}\ddot{q}_2 + u_a\right),$$

$$\ddot{q}_2 = \frac{1}{(L_2 + L_{12})}\left(L_{12}\ddot{q}_1 - \frac{q_2}{C_2} - R_2\dot{q}_2\right),$$

the resulting Simulink diagram is shown in Figure 2.12a.

To perform simulations, the circuitry parameters are assigned: $L_1 = 0.01$ H, $L_2 = 0.005$ H, $L_{12} = 0.0025$ H, $C_1 = 0.02$ F, $C_2 = 0.1$ F, $R_1 = 10$ ohm, $R_2 = 5$ ohm, and $u_a = 100 \sin(200t)$ V. These parameters are uploaded in the Command Window. The simulation results, which provide the evolution of the variables in the time domain $q_1(t)$, $q_2(t)$, $i_1(t)$, and $i_2(t)$, are documented in Figure 2.12b. The currents i_1 and i_2 are expressed in terms of charges as $i_1 = \dot{q}_1$ and $i_2 = \dot{q}_2$.   □

Example 2.11   Electric Circuit

Using the Lagrange equations of motion, we can derive the corresponding differential equations for the circuit depicted in Figure 2.13. The equations of motion derived using the Lagrange concept should be equivalent to the model developed using Kirchhoff's law.

We use q_1 and q_2 as the independent generalized coordinates (charges in the first and second loops). Here, $i_a = \dot{q}_1$ and $i_L = \dot{q}_2$. The generalized force, applied to the system, is Q_1, and $u_a(t) = Q_1$.

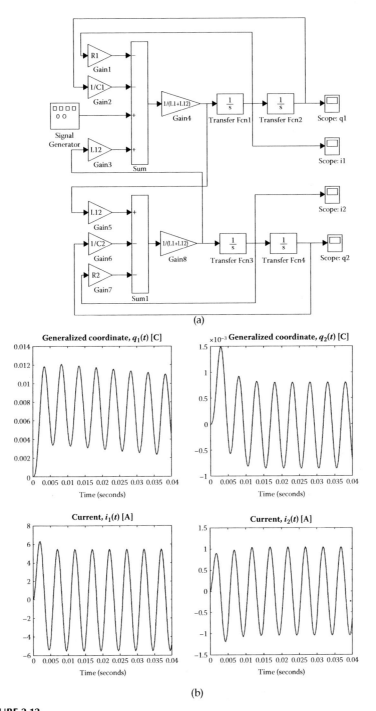

FIGURE 2.12
(a) Simulink diagram. (b) Circuit dynamics: evolution of the generalized coordinates and currents.

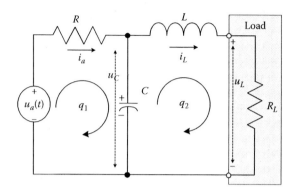

FIGURE 2.13
Electric circuit.

The total kinetic energy is given as

$$\Gamma = \frac{1}{2} L \dot{q}_2^2.$$

Therefore, we have

$$\frac{\partial \Gamma}{\partial q_1} = 0, \quad \frac{\partial \Gamma}{\partial \dot{q}_1} = 0 \quad \text{and} \quad \frac{d}{dt}\left(\frac{\partial \Gamma}{\partial \dot{q}_1}\right) = 0,$$

$$\frac{\partial \Gamma}{\partial q_2} = 0, \quad \frac{\partial \Gamma}{\partial \dot{q}_2} = L \dot{q}_2 \quad \text{and} \quad \frac{d}{dt}\left(\frac{\partial \Gamma}{\partial \dot{q}_2}\right) = L \ddot{q}_2.$$

The total potential energy is expressed as

$$\Pi = \frac{1}{2} \frac{(q_1 - q_2)^2}{C},$$

yielding

$$\frac{\partial \Pi}{\partial q_1} = \frac{q_1 - q_2}{C} \quad \text{and} \quad \frac{\partial \Pi}{\partial q_2} = \frac{-q_1 + q_2}{C}.$$

The total dissipated energy is

$$D = \frac{1}{2} R \dot{q}_1^2 + \frac{1}{2} R_L \dot{q}_2^2.$$

Therefore,

$$\frac{\partial D}{\partial \dot{q}_1} = R\dot{q}_1 \quad \text{and} \quad \frac{\partial D}{\partial \dot{q}_2} = R_L \dot{q}_2.$$

The Lagrange equations of motion

$$\frac{d}{dt}\left(\frac{\partial \Gamma}{\partial \dot{q}_1}\right) - \frac{\partial \Gamma}{\partial q_1} + \frac{\partial D}{\partial \dot{q}_1} + \frac{\partial \Pi}{\partial q_1} = Q_1, \quad \text{and} \quad \frac{d}{dt}\left(\frac{\partial \Gamma}{\partial \dot{q}_2}\right) - \frac{\partial \Gamma}{\partial q_2} + \frac{\partial D}{\partial \dot{q}_2} + \frac{\partial \Pi}{\partial q_2} = 0$$

are applied. One obtains two differential equations

$$R\dot{q}_1 + \frac{q_1 - q_2}{C} = u_a,$$

$$L\ddot{q}_2 + R_L \dot{q}_2 + \frac{-q_1 + q_2}{C} = 0.$$

We found a set of first- and second-order differential equations

$$\dot{q}_1 = \frac{1}{R}\left(\frac{-q_1 + q_2}{C} + u_a\right),$$

$$\ddot{q}_2 = \frac{1}{L}\left(-R_L \dot{q}_2 + \frac{q_1 - q_2}{C}\right).$$

By using Kirchhoff's law, the following two differential equations result

$$\frac{du_C}{dt} = \frac{1}{C}\left(-\frac{u_C}{R} - i_L + \frac{u_a(t)}{R}\right),$$

$$\frac{di_L}{dt} = \frac{1}{L}(u_C - R_L i_L).$$

From $i_a = \dot{q}_1$ and $i_L = \dot{q}_2$, using

$$C\frac{du_C}{dt} = i_a - i_L,$$

we obtain

$$u_C = \frac{q_1 - q_2}{C}.$$

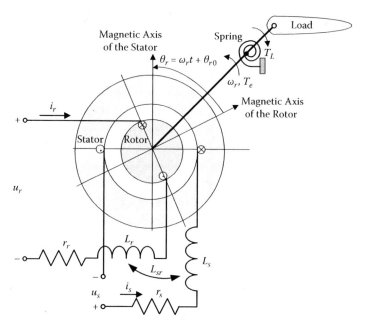

FIGURE 2.14
Actuator with stator and rotor windings.

The equivalence of the differential equations derived using the Lagrange equations of motion and Kirchhoff's law is proven.   □

Applying the Lagrange equations of motion, one straightforwardly models and examines various circuitry topologies reported in References 5 and 9–12. Chapter 3 provides examples and details.

Example 2.12   Electromechanical Actuator

Consider an electromechanical motion device that actuates the load (robotic arm, pointer, etc.). The actuator has two independently excited stator and rotor windings, as reported in Figure 2.14. One needs to derive the differential equations to describe the system dynamics.

The following notations are used for the actuator variables and parameters: i_s and i_r are the currents in the stator and rotor windings; u_s and u_r are the applied voltages to the stator and rotor windings; ω_r and θ_r are the rotor angular velocity and displacement; T_e and T_L are the electromagnetic and load torques; r_s and r_r are the resistances of the stator and rotor windings; L_s and L_r are the self-inductances of the stator and rotor windings; L_{sr} is the mutual inductance of the stator and rotor windings; $\Re_m$ is the reluctance of the magnetizing path; N_s and N_r are the number of turns in the stator and rotor windings; J is the equivalent moment of inertia of the rotor and attached load; B_m is the viscous friction coefficient; k_s is the spring constant.

The magnetic fluxes that cross the air gap produce an electromagnetic force. The developed electromagnetic torque T_e is countered by the torsional spring which causes a *counterclockwise* rotation. The load torque T_L is considered.

By using the Lagrange concept, the independent generalized coordinates are q_1, q_2, and q_3, where q_1 and q_2 are the electric charges in the stator and rotor windings, whereas q_3 is the rotor angular displacement.

We denote the generalized forces, applied to a system, as Q_1, Q_2 and Q_3, where Q_1 and Q_2 are the applied voltages to the stator and rotor windings, whereas Q_3 is the load torque.

The first derivative of the generalized coordinates $\dot{q}_1$ and $\dot{q}_2$ represent the stator and rotor currents i_s and i_r, whereas $\dot{q}_3$ is the angular velocity of the rotor ω_r. We have

$$q_1 = \frac{i_s}{s}, \quad q_2 = \frac{i_r}{s}, \quad q_3 = \theta_r, \quad \dot{q}_1 = i_s, \quad \dot{q}_2 = i_r, \quad \dot{q}_3 = \omega_r,$$

$$Q_1 = u_s, \quad Q_2 = u_r \quad \text{and} \quad Q_3 = -T_L.$$

The Lagrange equations are expressed in terms of each independent coordinate, yielding

$$\frac{d}{dt}\left(\frac{\partial \Gamma}{\partial \dot{q}_1}\right) - \frac{\partial \Gamma}{\partial q_1} + \frac{\partial D}{\partial \dot{q}_1} + \frac{\partial \Pi}{\partial q_1} = Q_1,$$

$$\frac{d}{dt}\left(\frac{\partial \Gamma}{\partial \dot{q}_2}\right) - \frac{\partial \Gamma}{\partial q_2} + \frac{\partial D}{\partial \dot{q}_2} + \frac{\partial \Pi}{\partial q_2} = Q_2,$$

$$\frac{d}{dt}\left(\frac{\partial \Gamma}{\partial \dot{q}_3}\right) - \frac{\partial \Gamma}{\partial q_3} + \frac{\partial D}{\partial \dot{q}_3} + \frac{\partial \Pi}{\partial q_3} = Q_3.$$

The total kinetic energy of electrical and mechanical systems is found as a sum of the total magnetic (electrical) Γ_E and mechanical Γ_M energies. The total kinetic energy of the stator and rotor circuitry is

$$\Gamma_E = \frac{1}{2}L_s\dot{q}_1^2 + L_{sr}\dot{q}_1\dot{q}_2 + \frac{1}{2}L_r\dot{q}_2^2.$$

The total kinetic energy of the mechanical system is

$$\Gamma_M = \frac{1}{2}J\dot{q}_3^2.$$

Therefore,

$$\Gamma = \Gamma_E + \Gamma_M = \frac{1}{2}L_s\dot{q}_1^2 + L_{sr}\dot{q}_1\dot{q}_2 + \frac{1}{2}L_r\dot{q}_2^2 + \frac{1}{2}J\dot{q}_3^2.$$

The mutual inductance depends on the displacement of the rotor winding with respect to the stator winding. It is obvious that $L_{sr}(\theta_r)$ is a periodic function of the angular rotor displacement, and $L_{sr\,min} \leq L_{sr}(\theta_r) \leq L_{sr\,max}$. The amplitude of the mutual inductance between the stator and rotor windings is denoted as L_M, and $L_M = L_{sr\,max}$. If the windings are orthogonal, we have $L_{sr} = 0$. We conclude that the mutual inductance L_{sr} can be approximated as

$$L_{sr}(\theta_r) = L_M \cos \theta_r = L_M \cos q_3.$$

One obtains an explicit expression for the total kinetic energy as

$$\Gamma = \frac{1}{2} L_s \dot{q}_1^2 + L_M \dot{q}_1 \dot{q}_2 \cos q_3 + \frac{1}{2} L_r \dot{q}_2^2 + \frac{1}{2} J \dot{q}_3^2.$$

The following partial derivatives result

$$\frac{\partial \Gamma}{\partial q_1} = 0, \quad \frac{\partial \Gamma}{\partial \dot{q}_1} = L_s \dot{q}_1 + L_M \dot{q}_2 \cos q_3,$$

$$\frac{\partial \Gamma}{\partial q_2} = 0, \quad \frac{\partial \Gamma}{\partial \dot{q}_2} = L_M \dot{q}_1 \cos q_3 + L_r \dot{q}_2,$$

$$\frac{\partial \Gamma}{\partial q_3} = -L_M \dot{q}_1 \dot{q}_2 \sin q_3, \quad \frac{\partial \Gamma}{\partial \dot{q}_3} = J \dot{q}_3.$$

The potential energy of the spring with constant k_s is

$$\Pi = \frac{1}{2} k_s q_3^2.$$

Therefore,

$$\frac{\partial \Pi}{\partial q_1} = 0, \quad \frac{\partial \Pi}{\partial q_2} = 0 \quad \text{and} \quad \frac{\partial \Pi}{\partial q_3} = k_s q_3.$$

The total heat energy dissipated is

$$D = D_E + D_M,$$

where D_E is the heat energy dissipated in the stator and rotor windings,

$$D_E = \frac{1}{2} r_s \dot{q}_1^2 + \frac{1}{2} r_r \dot{q}_2^2;$$

D_M is the heat energy dissipated by mechanical system,

$$D_M = \frac{1}{2} B_m \dot{q}_3^2.$$

The derived

$$D = \frac{1}{2} r_s \dot{q}_1^2 + \frac{1}{2} r_r \dot{q}_2^2 + \frac{1}{2} B_m \dot{q}_3^2$$

yields

$$\frac{\partial D}{\partial \dot{q}_1} = r_s \dot{q}_1, \quad \frac{\partial D}{\partial \dot{q}_2} = r_r \dot{q}_2 \quad \text{and} \quad \frac{\partial D}{\partial \dot{q}_3} = B_m \dot{q}_3.$$

Using the following relationships between the generalized coordinates and state variables

$$q_1 = \frac{i_s}{s}, q_2 = \frac{i_r}{s}, q_3 = \theta_r, \dot{q}_1 = i_s, \dot{q}_2 = i_r, \dot{q}_3 = \omega_r, Q_1 = u_s, Q_2 = u_r \quad \text{and} \quad Q_3 = -T_L,$$

we have three differential equations for the considered actuator as

$$L_s \frac{di_s}{dt} + L_M \cos\theta_r \frac{di_r}{dt} - L_M i_r \sin\theta_r \frac{d\theta_r}{dt} + r_s i_s = u_s,$$

$$L_r \frac{di_r}{dt} + L_M \cos\theta_r \frac{di_s}{dt} - L_M i_s \sin\theta_r \frac{d\theta_r}{dt} + r_r i_r = u_r,$$

$$J \frac{d^2\theta_r}{dt^2} + L_M i_s i_r \sin\theta_r + B_m \frac{d\theta_r}{dt} + k_s \theta_r = -T_L.$$

The last equation may be rewritten recalling that

$$\frac{d\theta_r}{dt} = \omega_r.$$

Using the stator and rotor currents, angular velocity, and displacement as the state variables, the nonlinear differential equations in Cauchy's form are

$$\frac{di_s}{dt} = \frac{-r_s L_r i_s - \frac{1}{2} L_M^2 i_s \omega_r \sin 2\theta_r + r_r L_M i_r \cos\theta_r + L_r L_M i_r \omega_r \sin\theta_r + L_r u_s - L_M \cos\theta_r u_r}{L_s L_r - L_M^2 \cos^2\theta_r},$$

$$\frac{di_r}{dt} = \frac{r_s L_M i_s \cos\theta_r + L_s L_M i_s \omega_r \sin\theta_r - r_r L_s i_r - \frac{1}{2} L_M^2 i_r \omega_r \sin 2\theta_r - L_M \cos\theta_r u_s + L_s u_r}{L_s L_r - L_M^2 \cos^2\theta_r},$$

$$\frac{d\omega_r}{dt} = \frac{1}{J}(-L_M i_s i_r \sin\theta_r - B_m \omega_r - k_s \theta_r - T_L),$$

$$\frac{d\theta_r}{dt} = \omega_r.$$

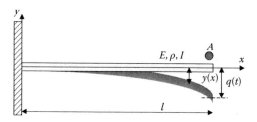

FIGURE 2.15
Beam in the *xy* plane.

The developed nonlinear differential equations cannot be linearized. One must analyze the actuator and examine the system performance using a complete set of equations of motion.   □

Example 2.13   Beam Equations of Motion

Consider an elastic beam of length *l* with constant cross-sectional area *A* and uniform weight per unit volume (density) ρ. We denote the static vertical displacement at the free end of the beam, which is illustrated in Figure 2.15, as *q*.

One needs to find the kinetic and potential energies. We will make some assumption to illustrate the application of the Lagrange equations of motion. Let the third-order deflection polynomial be

$$y(x) = \frac{1}{2}\left(3\frac{x^2}{l^2} - \frac{x^3}{l^3}\right)q.$$

Using $q(t)$, we have

$$y(t,x) = \frac{1}{2}\left(3\frac{x^2}{l^2} - \frac{x^3}{l^3}\right)q(t).$$

The kinetic energy is

$$\Gamma(q) = \frac{1}{2}\int_0^l \dot{y}^2 dm = \frac{1}{2}A\rho\int_0^l \frac{1}{4}\left(3\frac{x^2}{l^2} - \frac{x^3}{l^3}\right)^2 \dot{q}^2 dx = \frac{33}{280}A\rho l\dot{q}^2,$$

whereas the potential energy of elastic deformation is

$$\Pi(q) = \frac{1}{2}\int_0^l EI\left(\frac{\partial^2 y}{\partial x^2}\right)^2 dx = \frac{1}{2}EI\int_0^l \frac{3}{2l^3}\left(1 - \frac{x}{l}\right)^2 d\left(\frac{x}{l}\right) = \frac{3}{2}\frac{EI}{l^3}q^2.$$

Here, *E* is Young's modulus of elasticity; *I* is the moment of inertia of the cross section about its neural axis.

From the Lagrange equation

$$\frac{d}{dt}\left(\frac{\partial \Gamma}{\partial \dot{q}}\right) - \frac{\partial \Gamma}{\partial q} + \frac{\partial \Pi}{\partial q} = Q,$$

the beam equation of motion yields as

$$\frac{d^2 q}{dt^2} = -12.7\frac{EI}{A\rho l^4}q + F_q(t,x).$$

The reported simplified concept results in the second-order differential equation, which describes the beam dynamics. In general, the potential energy of elastic beam is

$$\Pi = \frac{1}{2}\int_{\mathbf{r}}\sigma_{ij}\varepsilon_{ij}d\mathbf{r} + \int_{\mathbf{r}}T(\mathbf{r})w(\mathbf{r})d\mathbf{r} + \int_{\mathbf{r}}F(\mathbf{r})w(\mathbf{r})d\mathbf{r},$$

where $T(\mathbf{r})$ and $F(\mathbf{r})$ are the beam surface traction and force. The term

$$\frac{1}{2}\sigma_{ij}\varepsilon_{ij}$$

gives the strain energy stored. For the laterally distributed load $T(x)$, the equation for the beam bending is

$$a_b\frac{d^4 w}{dx^4} = T(x), \quad a_b = EI.$$

This equation is related to the commonly applied equation for the beam motion which vibrates with initial displacement and velocity

$$y(0,x) = f_{0d}(x) \quad \text{and} \quad \left.\frac{\partial y}{\partial t}\right|_{0,x} = f_{0v}(x).$$

In particular, the displacement at any point can be described by the partial differential equation

$$a^2\frac{\partial^4 y(t,x)}{\partial x^4} = -\frac{\partial^2 y(t,x)}{\partial t^2}, \quad a = \sqrt{\frac{EIg}{A\rho}}.$$

Let the boundary conditions for the deflection $y(t,x)$ be

$$y(t,0) = 0, \left.\frac{\partial y}{\partial x}\right|_{t,0} = 0, \left.\frac{\partial^2 y}{\partial x^2}\right|_{t,l} = 0, \quad \text{and} \quad \left.\frac{\partial^3 y}{\partial x^3}\right|_{t,l} = 0.$$

That is, the displacement, slope, bending moment

$$EI\frac{\partial^2 y}{\partial x^2},$$

and shear

$$\frac{\partial}{\partial x}\left(EI\frac{\partial^2 y}{\partial x^2}\right)$$

at the corresponding beam ends (fixed and free) are zeros.

Using the boundary conditions, the solution of the partial differential equation is given as

$$y(t,x)=\sum_{n=1}^{\infty}X_n(x)(A_n\cos\lambda_n t+B_n\sin\lambda_n t),$$

$$X_n(x)=(\sinh z_n+\sin z_n)\left(\cos\frac{z_n x}{l}-\cosh\frac{z_n x}{l}\right)$$

$$-(\cosh z_n+\cos z_n)\left(\sin\frac{z_n x}{l}-\sinh\frac{z_n x}{l}\right),$$

$$A_n=\frac{\int_0^l f_{0d}(x)X_n(x)dx}{\int_0^l X_n^2(x)dx},\quad B_n=\frac{\int_0^l f_{0v}(x)X_n(x)dx}{\lambda_n\int_0^l X_n^2(x)dx},\quad \lambda_n=\frac{a}{l^2}\sum_{n=1}^{\infty}z_n^2.$$

To satisfy the initial displacement and initial velocity conditions, one has

$$y(0,x)=f_{0d}(x)\sum_{n=1}^{\infty}A_n X_n(x)\text{ and }\frac{\partial y}{\partial t}\bigg|_{0,x}=f_{0v}(x)=\sum_{n=1}^{\infty}\lambda_n B_n X_n(x).$$

Thus, the analytic solution is derived using the beam equation of motions. Another commonly used equation is

$$\xi EI(x)\frac{\partial^5 y(t,x)}{\partial x^4\partial t}+EI(x)\frac{\partial^4 y(t,x)}{\partial x^4}+m_0(x)\frac{\partial^2 y(t,x)}{\partial t^2}+m(x)\frac{d^2\varphi}{dt^2}=F(t,x),$$

where $F(t,x)$ is the distributed force through the beam.   □

2.5.3   Hamilton Equations of Motion

The Hamilton concept allows one to describe the system dynamics. The differential equations are found using the generalized momenta p_i,

$$p_i=\frac{\partial L}{\partial\dot{q}_i}.$$

The generalized coordinates were used in the Lagrange equations of motion. The Lagrangian function

$$L\left(t, q_1, \ldots, q_n, \frac{dq_1}{dt}, \ldots, \frac{dq_n}{dt}\right)$$

for the conservative systems is the difference between the total kinetic and potential energies. We have

$$L\left(t, q_1, \ldots, q_n, \frac{dq_1}{dt}, \ldots, \frac{dq_n}{dt}\right) = \Gamma\left(t, q_1, \ldots, q_n, \frac{dq_1}{dt}, \ldots, \frac{dq_n}{dt}\right) - \Pi(t, q_1, \ldots, q_n).$$

One concludes that

$$L\left(t, q_1, \ldots, q_n, \frac{dq_1}{dt}, \ldots, \frac{dq_n}{dt}\right)$$

is the function of $2n$ independent variables, and

$$dL = \sum_{i=1}^{n}\left(\frac{\partial L}{\partial q_i} dq_i + \frac{\partial L}{\partial \dot{q}_i} d\dot{q}_i\right) = \sum_{i=1}^{n}(p_i dq_i + p_i d\dot{q}_i) + \frac{\partial L}{\partial t} dt.$$

We define the Hamiltonian function as

$$H(t, q_1, \ldots, q_n, p_1, \ldots, p_n) = -L\left(t, q_1, \ldots, q_n, \frac{dq_1}{dt}, \ldots, \frac{dq_n}{dt}\right) + \sum_{i=1}^{n} p_i \dot{q}_i,$$

$$dH = \sum_{i=1}^{n}(-\dot{p}_i dq_i + \dot{q}_i dp_i) - \frac{\partial L}{\partial t} dt.$$

where

$$\sum_{i=1}^{n} p_i \dot{q}_i = \sum_{i=1}^{n} \frac{\partial L}{\partial \dot{q}_i} \dot{q}_i = 2\Gamma.$$

Thus, we have

$$H\left(t, q_1, \ldots, q_n, \frac{dq_1}{dt}, \ldots, \frac{dq_n}{dt}\right) = \Gamma\left(t, q_1, \ldots, q_n, \frac{dq_1}{dt}, \ldots, \frac{dq_n}{dt}\right) + \Pi(t, q_1, \ldots, q_n)$$

or

$$H(t, q_1, \ldots, q_n, p_1, \ldots, p_n) = \Gamma(t, q_1, \ldots, q_n, p_1, \ldots, p_n) + \Pi(t, q_1, \ldots, q_n).$$

One concludes that the Hamiltonian, which represents the total energy, is expressed as a function of the generalized coordinates and generalized momenta. The equations of motion are governed by the equations

$$\dot{p}_i = -\frac{\partial H}{\partial q_i},$$

$$\dot{q}_i = \frac{\partial H}{\partial p_i}. \tag{2.4}$$

These equations are called the Hamiltonian equations of motion. Using the Hamiltonian mechanics, one obtains the system of $2n$ first-order differential equations to describe the system dynamics. In contrast, using the Lagrange equations of motion, the system of n second-order differential equations results. However, the derived differential equations are equivalent.

Example 2.14

Consider the harmonic oscillator, which is formed by the sliding mass m attached to the spring assuming that there is no viscous friction. The total energy is given as the sum of the kinetic and potential energies. That is,

$$\Sigma_T = \Gamma + \Pi = \frac{1}{2}(mv^2 + k_s x^2).$$

One can find the equations of motion using the Lagrange and Hamilton concepts.

Recall that $q = x$. The Lagrangian function is

$$L\left(x, \frac{dx}{dt}\right) = \Gamma - \Pi = \frac{1}{2}(mv^2 - k_s x^2) = \frac{1}{2}(m\dot{x}^2 - k_s x^2).$$

From the Lagrange equations of motion

$$\frac{d}{dt}\frac{\partial L}{\partial \dot{x}} - \frac{\partial L}{\partial x} = 0,$$

where $q = x$, we have the following second-order differential equation

$$m\frac{d^2 x}{dt^2} + k_s x = 0.$$

From Newton's second law, the second-order differential equation of motion is given as

$$m\frac{d^2x}{dt^2} + k_s x = 0.$$

The Hamiltonian function is expressed as

$$H(x,p) = \Gamma + \Pi = \frac{1}{2}\left(mv^2 + k_s x^2\right) = \frac{1}{2}\left(\frac{1}{m}p^2 + k_s x^2\right).$$

From the Hamiltonian equations of motion

$$\dot{p}_i = -\frac{\partial H}{\partial q_i} \quad \text{and} \quad \dot{q}_i = \frac{\partial H}{\partial p_i},$$

as given by (2.4), one obtains

$$\dot{p} = -\frac{\partial H}{\partial x} = -k_s x, \quad \dot{x} = \dot{q} = \frac{\partial H}{\partial p} = \frac{p}{m}.$$

The equivalence of the resulting equations of motion is obvious. □

2.6 Application of Electromagnetics and Classical Mechanics to Electromechanical Systems

The cornerstone laws of electromagnetics and mechanics are applied to examine device physics as well as describe physical phenomena, effects, and processes in systems. In Sections 2.2 to 2.5, differential equations that describe the system (device) behavior were derived. Differential (or difference) equations with the corresponding constitutive equations, which describe the system dynamics, are called mathematical models. The term *modeling* means the deviations of these equations (models). The quantitative and quantitative analyses are carried out by performing modeling, simulations, and result assessment. High-fidelity modeling, as three-dimensional Maxwell's equations and tensor calculus are applied, results in data-intensive analysis. However, the complexity is evident. The complexity may be relaxed by applying sound engineering methods without loss of generality, thereby ensuring accuracy and coherency. As the soundness of functional and structural designs is guaranteed, the designer focuses on the qualitative and quantitative analysis. Performing modeling tasks, one can integrate and study different components, for example, power electronics and actuators, ICs and motion devices, and so on. The functional, structural, and behavioral designs can be related to the analysis and modeling problems within the following generic taxonomy. The first step is to accomplish various design and analysis tasks. In particular,

- Examine and analyze electromechanical system using a multi-level hierarchy concept: develop multivariable input-output pairs (devices and components), for example, motion and motionless devices (actuators [1–8], sensors [13, 14], transducers)–electronics–ICs–controller-input/output devices.
- Assess the soundness of system organization and functionality.
- Collect and evaluate the data and information.
- Develop input-output variable pairs, identify the independent and dependent control, disturbance, output, reference (command), state, and performance variables (examples are available in References 15–21).
- Make accurate assumptions and simplify the problem to ensure tractability. Any descriptive models are the idealization of physical systems, phenomena, effects, processes, and so on. Mathematical models are never absolutely accurate, but models must allow the designer to support coherent analysis and design.

The second step is to derive equations that relate the variables and events. In particular,

- Define and specify the basic laws (Kirchhoff, Maxwell, Newton, Lagrange, and other) to be used to obtain the equations of motion. Electromagnetic, electronic, and mechanical components should be examined to describe electromagnetics, mechanics, and circuits deriving mathematical models of a system using defined variables.
- Derive and examine mathematical models in the form of differential (or difference) and constitutive equations.

The third step is simulation and validation:

- Identify numerical and analytic methods to be used.
- Analytically and/or numerically solve equations that describe system behavior (e.g., differential or difference equations with constitutive relationships).
- Define the criteria to assess the correspondence and accuracy. Using the information variables (measured or observed) and events, synthesize the fitting and mismatch functionals.

- Verify the results through the comprehensive comparison of the solution (modeled input-state-output-event mapping sets) versus the experimental data (experimental input-state-output-event mapping sets). The steady-state and dynamic behavior can be described by the behavioral modeling (*m*) and experimental (*e*) sixruples $(\mathbf{r}_m, \mathbf{x}_m, \mathbf{y}_m, \mathbf{u}_m, \mathbf{d}_m, \mathbf{e}_m)$ and $(\mathbf{r}_e, \mathbf{x}_e, \mathbf{y}_e, \mathbf{u}_e, \mathbf{d}_e, \mathbf{e}_e)$, where

$$\mathbf{r} \in \mathbb{R}^b, \ \mathbf{x} \in \mathbb{R}^n, \ \mathbf{y} \in \mathbb{R}^k, \ \mathbf{u} \in \mathbb{R}^m, \ \mathbf{d} \in \mathbb{R}^l, \ \text{and} \ \mathbf{e} \in \mathbb{R}^s$$

are the reference (input), state, output, control, disturbance, and events vectors. One examines the system in

$$R_{m,e} \times X_{m,e} \times Y_{m,e} \times U_{m,e} \times D_{m,e} \times E_{m,e}.$$

The measured sets are expressed as

$$M_m = \{(\mathbf{r}_m, \mathbf{x}_m, \mathbf{y}_m, \mathbf{u}_m, \mathbf{d}_m, \mathbf{e}_m) \in R_m \times X_m \times Y_m \times U_m \times D_m \times E_m, \ \forall t \in T\}$$
$$\text{and } M_e = \{(\mathbf{r}_e, \mathbf{x}_e, \mathbf{y}_e, \mathbf{u}_e, \mathbf{d}_e, \mathbf{e}_e) \in R_e \times X_e \times Y_e \times U_e \times D_e \times E_e, \ \forall t \in T\}.$$

The strong and weak behavior matching are guaranteed if $M_m = M_e$ and $M_m \subseteq M_e$, respectively.
- Calculate the fitting and mismatch functionals.
- Examine the analytical and/or numerical data against new experimental data and evidence.
- Evaluate the fitness and accuracy with the ultimate objective to refine the approaches and models derived in order to ensure coherency and matching.

Until the specified accuracy and coherency are guaranteed, the designer returns to the specific task (step) within the design cycle again.

Our goal is to integrate equations of motion derived and report meaningful fundamentals with practical examples. It was illustrated that electromechanical devices can be described (modeled) using Maxwell's equations and mechanical equations of motion. Forces and torques were found using the Maxwell stress tensor and energies. The nonlinear partial and ordinary differential equations were found. In modeling, simulation and control, the lumped-parameter models can be used. Consider the motor's rotor (bar magnet, current loop, and solenoid) in a uniform magnetic field as illustrated in Figure 2.16.

The torque tends to align the magnetic moment $\vec{m}$ with $\vec{B}$, and it is known that $\vec{T} = \vec{m} \times \vec{B}$.

For a magnetic bar with the length *l*, the pole strength is *Q*. The magnetic moment is given as $m = Ql$, whereas the force is $F = QB$. The electromagnetic torque is found to be

$$T = 2F \frac{1}{2} l \sin \alpha = QlB \sin \alpha = mB \sin \alpha.$$

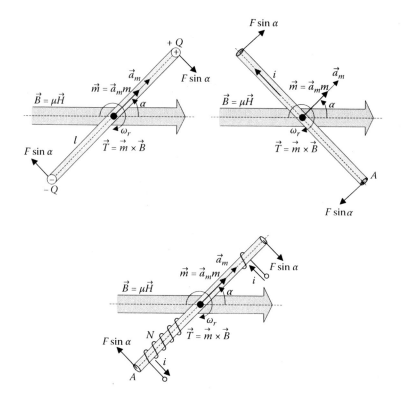

FIGURE 2.16
Clockwise rotation of magnetic bar, current loop, and solenoid.

Using the vector notations, one obtains

$$\vec{T} = \vec{m} \times \vec{B} = \vec{a}_m m \times \vec{B} = Q l \vec{a}_m \times \vec{B},$$

where $\vec{a}_m$ is the unit vector in the magnetic moment direction.

For a current loop with the cross-sectional area A, the torque is

$$\vec{T} = \vec{m} \times \vec{B} = \vec{a}_m m \times \vec{B} = iA\vec{a}_m \times \vec{B}.$$

For a solenoid with N turns, one obtains

$$\vec{T} = \vec{m} \times \vec{B} = \vec{a}_m m \times \vec{B} = iAN\vec{a}_m \times \vec{B}.$$

The expression for the electromagnetic torque developed should be integrated with the *torsional-mechanical* dynamics given by the Newton second law for the rotational motion

$$J\frac{d\omega_r}{dt} = \sum \vec{T}_\Sigma.$$

One recalls that $\Sigma \vec{T}_\Sigma$ is the *net* torque, ω_r is the angular velocity, and J is the equivalent moment of inertia. The transient evolution of the angular displacement θ_r is described as

$$\frac{d\theta_r}{dt} = \omega_r.$$

Combining the equations for the electromagnetic torque (found in terms of the magnetic field variables $\vec{m}$ and $\vec{B}$, or current i and $\vec{B}$) and the *torsional-mechanical* dynamics (the state variables are the angular velocity ω_r and displacement θ_r), the mathematical model results.

For translational motion, Newton's second law states that the *net* force acting on the object is related to its acceleration as $\Sigma \vec{F} = m\vec{a}$. In the XYZ coordinate system, one obtains

$$\sum F_x = ma_x, \quad \sum F_y = ma_y \quad \text{and} \quad \sum F_z = ma_z.$$

The force is the gradient of the stored magnetic energy W_m. That is, $\vec{F}_m = \vec{\nabla} W_m$. Hence, in the xyz directions, we have

$$F_{mx} = \frac{\partial W_m}{\partial x}, \quad F_{my} = \frac{\partial W_m}{\partial y} \quad \text{and} \quad F_{mz} = \frac{\partial W_m}{\partial z}.$$

The total magnetic flux through the surface is given by $\Phi = \int \vec{B} \cdot d\vec{s}$. The Ampere circuital law is

$$\oint_l \vec{B} \cdot d\vec{l} = \mu_0 \int_s \vec{J} \cdot d\vec{s}.$$

For the filamentary current, Ampere's law connects the magnetic flux with the algebraic sum of the enclosed (linked) currents (*net current*) i_n, and $\oint_l \vec{B} \cdot d\vec{l} = \mu_0 i_n$.

The time-varying magnetic field produces the electromotive force (*emf*), denoted as $\mathscr{E}$, which induces the current in the closed circuit. Faraday's law relates the *emf* (induced voltage due to conductor motion in the magnetic field) to the rate of change of the magnetic flux Φ penetrating the loop. Lenz's law should be used to find the direction of *emf* and the current induced. In particular, the *emf* is in such a direction as to produce a current whose flux, if added to the original flux, would reduce the magnitude of the *emf*. According to Faraday's law, the induced *emf* in a closed-loop circuit is defined in terms of the rate of change of the magnetic flux Φ. One has the following equation for the induced *emf* (induced voltage)

$$\mathscr{E} = \oint_l \vec{E}(t) \cdot d\vec{l} = -\frac{d}{dt} \int_s \vec{B}(t) \cdot d\vec{s} = -N\frac{d\Phi}{dt} = -\frac{d\psi}{dt},$$

where N is the number of turns; ψ is the flux linkages.

The equation

$$\mathcal{E} = -\frac{d\psi}{dt} = -N\frac{d\Phi}{dt}$$

represents the Faraday law of induction. The current flows in an opposite direction to the flux linkages. The unit for the *emf* is volts. The *emf* (*energy-per-unit-charge quantity*) represents a magnitude of the potential difference V in a circuit carrying a current. We have

$$V = -ir + \mathcal{E} = -ir - \frac{d\psi}{dt}.$$

The Kirchhoff voltage law states that around a closed path in an electric circuit, the algebraic sum of the *emf* is equal to the algebraic sum of the voltage drop across the resistance. This formulation will be used to examine various electromagnetic actuators. Another formulation is the algebraic sum of the voltages around any closed path in a circuit is zero. The Kirchhoff current law states that the algebraic sum of the currents at any node in a circuit is zero.

The magnetomotive force (*mmf*) is the line integral of the time-varying magnetic field intensity $\vec{H}(t)$. That is, $mmf = \oint_l \vec{H}(t)\cdot d\vec{l}$. The unit for the *mmf* is amperes or ampere-turns. The duality of the *emf* and *mmf* can be observed using the following two equations given in terms of the electric and magnetic field intensity vectors

$$\mathcal{E} = \oint_l \vec{E}(t)\cdot d\vec{l} \quad \text{and} \quad mmf = \oint_l \vec{H}(t)\cdot d\vec{l}.$$

The inductance is the ratio of the total flux linkages to the current, which they link,

$$L = \frac{N\Phi}{i}.$$

The reluctance is the ratio of the *mmf* to the total flux, $\mathbb{R} = mmf/\phi$. Hence, *emf* and *mmf* are used to find inductance and reluctance. The equation $L = \psi/i$ yields

$$\mathcal{E} = -\frac{d\psi}{dt} = -\frac{d(Li)}{dt} = -L\frac{di}{dt} - i\frac{dL}{dt}.$$

If $L = $ const, one obtains

$$\mathcal{E} = -L\frac{di}{dt}.$$

That is, the self-inductance is the magnitude of the self-induced *emf* per unit rate of change of current.

The basic principles of electromagnetic theory are reviewed below with illustrative examples. We examine various electromagnetic and electro-static actuators. The force- and torque-energy relations are emphasized. The energy stored in the capacitor is

$$\frac{1}{2}CV^2$$

whereas the energy stored in the inductor is

$$\frac{1}{2}Li^2.$$

The energy in the capacitor is stored in the electric field between plates, whereas the energy in the inductor is stored in the magnetic field within the coils. In fact, using

$$W_e = \frac{1}{2}\int_v \rho_v V dv,$$

the potential energy, which is stored in the electric field between two surfaces (for example, in capacitors), is found to be

$$W_e = \frac{1}{2}QV = \frac{1}{2}CV^2.$$

Example 2.15

Consider the capacitor (the plates have area A and they are separated by x), which is charged to a voltage V. The permittivity of the dielectric is ε. Find the stored electrostatic energy and the force F_{ex} in the x direction.

Neglecting the fringing effect at the edges, one concludes that the electric field is uniform, and $E = V/x$. Therefore,

$$W_e = \frac{1}{2}\int_v \varepsilon |\vec{E}|^2 \, dv = \frac{1}{2}\int_v \varepsilon \left(\frac{V}{x}\right)^2 dv = \frac{1}{2}\varepsilon\frac{V^2}{x^2}Ax = \frac{1}{2}\varepsilon\frac{A}{x}V^2 = \frac{1}{2}C(x)V^2,$$

where

$$C(x) = \varepsilon\frac{A}{x}.$$

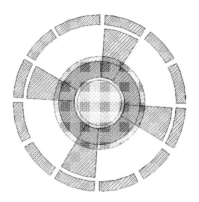

FIGURE 2.17
Electrostatic motor.

The force is

$$F_{ex} = \frac{\partial W_e}{\partial x} = \frac{\partial\left(\frac{1}{2}C(x)V^2\right)}{\partial x} = \frac{1}{2}V^2\frac{\partial C(x)}{\partial x}.$$

From $C(x)$, one finds

$$F_{ex} = \frac{1}{2}V^2\frac{\partial C(x)}{\partial x} = -\frac{1}{2}\varepsilon A\frac{1}{x^2}V^2. \qquad\qquad \square$$

Example 2.16

Rotational electrostatic motors have been widely examined as microelectromechanical motion devices. The cross-sectional view of the electrostatic motor is shown in Figure 2.17.

As the voltage V is applied to the parallel conducting rotor and stator plates, the charge is $Q = CV$, where C is the capacitance,

$$C = \varepsilon\frac{A}{g} = \varepsilon\frac{WL}{g};$$

A is the overlapping area of the plates, $A = WL$; W and L are the width and length of the plates respectively; ε is the permittivity of the media between the plates, $\varepsilon = \varepsilon_0\varepsilon_r$, $\varepsilon_0 = 8.85 \times 10^{-12}$ C^2/N-m^2 = 8.85 $\times$ 10^{-12} F/m, $\varepsilon_r = 1$; g is the airgap between the plates.

The energy associated with electrical potential is

$$W_e = \frac{1}{2}CV^2.$$

The electrostatic force at each overlapping plate segment

$$F_{el} = \frac{\partial W_e}{\partial g} = -\frac{1}{2}\frac{\varepsilon WL}{g^2}V^2$$

is balanced by the opposite segment (we assume an ideal fabrication for which W, L, and g are the same for all segments).

The tangential force due to misalignment is

$$F_t = \frac{\partial W_e}{\partial x} = \frac{1}{2}\frac{\varepsilon}{g}\frac{\partial(WL)}{\partial x}V^2,$$

where x is the direction in which misalignment potentially occurs. If the misalignment occurs in the width direction,

$$F_{t,w} = \frac{\partial W_e}{\partial x} = \frac{1}{2}\frac{\varepsilon L}{g}\frac{\partial W(x)}{\partial x}V^2.$$

The capacitance of a cylindrical capacitor is needed to be found to derive the electrostatic torque that drives the rotor. The voltage between the cylinders can be obtained by integrating the electric field. The electric field at a distance r from a conducting cylinder has only a radial component denoted as E_r. We have

$$E_r = \frac{\rho}{2\pi\varepsilon r},$$

where ρ is the linear charge density, and $Q=\rho L$. Hence, the potential difference is

$$\Delta V = V_a - V_b = \int_a^b \vec{E}\cdot d\vec{l} = \int_a^b E_r \cdot dr = \frac{\rho}{2\pi\varepsilon}\int_{r_1}^{r_2}\frac{1}{r}dr = \frac{\rho}{2\pi\varepsilon}\ln\frac{r_2}{r_1}.$$

Thus,

$$C = \frac{Q}{\Delta V} = \frac{2\pi\varepsilon L}{\ln(r_2/r_1)}.$$

The capacitance per unit length is

$$\frac{C}{L} = \frac{\rho}{\Delta V} = \frac{2\pi\varepsilon}{\ln(r_2/r_1)}.$$

Using the stator-rotor plates overlap, for the rotational electrostatic motor, the capacitance can be expressed as a function of the angular displacement as

$$C(\theta_r) = N\frac{2\pi\varepsilon}{\ln(r_2/r_1)}\theta_r,$$

where N is the number of overlapping stator-rotor plates; r_1 and r_2 are the radii of the rotor and stator where the plates are positioned.

From

$$C(\theta_r) = N \frac{2\pi\varepsilon}{\ln(r_2 / r_1)} \theta_r,$$

the electrostatic torque developed is

$$T_e = \frac{1}{2} \frac{\partial C(\theta_r)}{\partial \theta_r} V^2 = N \frac{\pi\varepsilon}{\ln(r_2 / r_1)} V^2.$$

Other expressions for capacitances $C(\theta_r)$ can be found resulting in alternative expressions for T_e.

The *torsional-mechanical* equations of motion are

$$\frac{d\omega_r}{dt} = \frac{1}{J}(T_e - B_m\omega - T_L) = \frac{1}{J}\left(N \frac{\pi\varepsilon}{\ln(r_2 / r_1)} V^2 - B_m\omega_r - T_L \right),$$

$$\frac{d\theta_r}{dt} = \omega_r.$$

These differential equations describe the dynamics of the studied electrostatic motor. Let us carry out and report some performance estimates and design steps.

To rotate the motor, inequality $T_e > T_L + T_{\text{friction}}$ must be guaranteed. That is, the motor must develop the electrostatic torque

$$T_e = N \frac{\pi\varepsilon}{\ln(r_2 / r_1)} V^2$$

higher than the maximum (or rated) load torque. Estimating the load torque $T_{L\max}$ and assigning the desired acceleration capabilities

$$\frac{\Delta\omega_r}{\Delta t} = \frac{1}{J}(T_e - B_m\omega - T_L),$$

taking note of the estimated J, one obtains the desired T_e. One evaluates the following motor parameters: N, r_1 r_2, and J. Taking note of those parameters, the motor sizing estimates are derived, and the fabrication technologies (processes and materials) are assessed. The applied voltage V is bounded as discussed in Section 2.1. The designer refines the design as needed. The fabrication technologies and processes significantly affect the motor dimensions and parameters. For example, one may attempt to minimize the air gap to attain the minimum (possible) value of $(r_2 - r_1)$

minimizing the expression $\ln(r_2/r_1)$ thereby maximizing T_e. The moment of inertia J can be minimized reducing the rotor mass using cavities and plastic materials. However, there are physical limits on the maximum V and E, which were emphasized in Section 2.1. The technological limits result in the specific r_2/r_1 ratio. Other critical issue is the need to form a mechanical contact (brushes) with the rotating rotor to apply the voltage to the rotor plates. These features reduce the overall performance of electrostatic-centered actuators. Most important, the torque density of electrostatic actuators is much lower than electromagnetic. The above-mentioned factors limit the application of the electrostatic rotational machines in conventional applications. □

Having documented translational and rotational electrostatic devices in Examples 2.15 and 2.16, the translational electromagnetic devices can be examined. These devices are related to the well-known relays and solenoids, which are the electromagnetic devices with the varying reluctance. The varying reluctance results in the electromagnetic torque production. In contract, high-performance electromagnetic motion devices utilize a coupling (magnetic interaction) between windings that are carrying currents and the stationary magnetic field developed by permanent magnets or electromagnets. In separately exited DC and induction machines, there is a magnetic coupling between windings as a result of their mutual inductances. The electromagnetics-centered device physics, energy conversion, torque production, *emf* induction, and other topics are covered in Chapters 4, 5, and 6. To derive the electromagnetic force or torque, one applies $\vec{T} = \vec{m} \times \vec{B}$, or uses the expressions for the coenergy $W_c[i,L(x)]$ (translational motion) or $W_c[i,L(\theta)]$ (rotational motion). The developed electromagnetic force and torque are given as

$$F_e(i,x) = \frac{\partial W_c[i,L(x)]}{\partial x} \quad \text{and} \quad T_e(i,x) = \frac{\partial W_c[i,L(\theta)]}{\partial \theta}, \quad \text{or} \quad T_e = m \times B.$$

Example 2.17   Solenoid

Solenoids usually integrate movable (plunger) and stationary (fixed) members made from high-permeability ferromagnetic materials. The windings wound within a helical pattern. These solenoids, as electromechanical devices, convert electrical energy to mechanical energy. Solenoids and relays operate because of the varying reluctance, and the force is generated because of the changes of reluctance. Performance of solenoids is strongly affected by the magnetic system, materials, geometry, relative permeability, winding resistance, friction, and so on. The plunger moves in the center of the stationary member as shown in Figure 2.18. When the voltage is applied to the winding, current flows in the winding, and the electromagnetic force is developed causing the plunger to move. When the applied voltage becomes zero, the plunger resumes its original position due to the returning spring

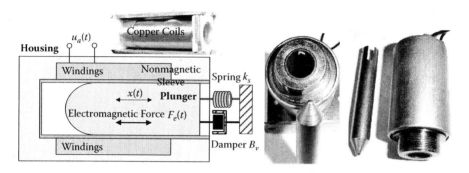

FIGURE 2.18
Schematic and image of a solenoid.

(assuming that the static and *Coulomb* frictions are negligibly small). The undesirable phenomena such as residual magnetism and friction must be minimized. Different materials for the central guide (nonmagnetic sleeve) and plunger coating (plating) should be chosen to attain minimum friction and minimize wear. Glass-filled nylon and brass (for the guide), silver, copper, aluminum, tungsten, platinum plating, or other low friction coatings (for the plunger) are possible candidates. The friction coefficients of lubricated (solid film and oil) and unlubricated for different possible candidates are: tungsten on tungsten 0.04–0.1 and 0.3; copper on copper 0.04–0.08 and 1.2–1.5; aluminum on aluminum 0.04–0.12 and 1; platinum on platinum 0.04–0.25 and 1.2; and titanium on titanium 0.04–0.1 and 0.6. The design, analysis, and optimization of solenoids require application of basic physics (electromagnetics, mechanics, thermodynamics, etc.) considering different fabrication options. In many cases, it is essential to make tradeoffs among a variety of electromagnetic, mechanical, thermal, acoustical, and other physical properties. The image of the solenoid is reported in Figure 2.18.   □

Example 2.18

We can find the inductances of a solenoid with air-core ($\mu_r = 1$) and filled-core ($\mu_r = 10,000$). The solenoid has 100 turns ($N = 100$), the length is 5 cm ($l = 0.05$ m), and the uniform circular cross-sectional area is 1×10^{-4} m² ($A = 1 \times 10^{-4}$ m²).

The magnetic field inside a solenoid is

$$B = \frac{\mu N i}{l}, \quad \mu = \mu_0 \mu_r.$$

From

$$\mathscr{E} = -N \frac{d\Phi}{dt} - L \frac{di}{dt},$$

by applying

$$\Phi = BA = \frac{\mu NiA}{l},$$

one obtains the following expression for the inductance

$$L = \frac{\mu N^2 A}{l}.$$

For the solenoid with air-core one obtains $L = 2.5 \times 10^{-5}$ H.
The MATLAB statement to calculate the numerical value of L is

```
mu0=4*pi*1e-7; mur=1; N=100; A=1e-4; l=5e-2; L=mu0*mur*N*N*A/l
```

The result displayed in the Command Window is $L = 2.5133e-005$
If the solenoid is filled with a ferromagnetic material, we have $L=0.25$ H. $\square$

Example 2.19
We can derive a formula for the self-inductance of a toroidal solenoid, which has a rectangular cross-section ($2a \times b$) and mean radius r. The magnetic flux through a cross-section is

$$\Phi = \int_{r-a}^{r+a} Bb\,dr = \int_{r-a}^{r+a} \frac{\mu Ni}{2\pi r}b\,dr = \frac{\mu Nib}{2\pi} \int_{r-a}^{r+a} \frac{1}{r}dr = \frac{\mu Nib}{2\pi} \ln\left(\frac{r+a}{r-a}\right)$$

yielding

$$L = \frac{N\Phi}{i} = \frac{\mu N^2 b}{2\pi} \ln\left(\frac{r+a}{r-a}\right). \qquad\qquad \square$$

Example 2.20
It is possible to calculate the stored magnetic energy of the toroidal solenoid. Assume the self-inductance is $L = 0.2$ H and the current is $i = 1 \times 10^{-3}$ A.
The stored field energy is

$$W_m = \frac{1}{2}Li^2.$$

Hence, $W_m = 1 \times 10^{-6}$ J. $\qquad\qquad \square$

Example 2.21
Derive the expression for the electromagnetic force developed by the relay, which is depicted in Figure 2.19. The current $i_a(t)$ in N coils produces the constant flux Φ.

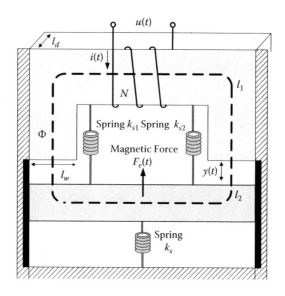

FIGURE 2.19
Relay with springs.

We assume that the flux is constant. The displacement (the virtual displacement is denoted as dy) changes only the magnetic energy stored in the air gaps. From

$$W_m = \frac{1}{2} \int_v \mu |\vec{H}|^2 \, dv = \frac{1}{2} \int_v \frac{|\vec{B}|^2}{\mu} \, dv,$$

we have

$$dW_m = dW_{m \, \text{air gap}} = 2 \frac{B^2}{2\mu_0} A dy = \frac{\Phi^2}{\mu_0 A} dy,$$

where A is the cross-sectional area, $A = l_w l_d$.

If $\Phi = \text{const}$ (the current is constant), one concludes that the increase of the air gap (dy) leads to increase of the stored magnetic energy. Using

$$F_e = \frac{\partial W_m}{\partial y},$$

one finds the expression for the electromagnetic force as

$$\vec{F}_e = -\vec{a}_y \frac{\Phi^2}{\mu_0 A}.$$

The result indicates that the force tends to reduce the airgap length (minimize the reluctance). The movable member (gravitational force is mg) is attached to the springs, which develop three forces in addition to the electromagnetic force.

The airgap reluctance (two air gaps are in series) is

$$\mathfrak{R}_g = \frac{2y}{\mu_0 A} = \frac{2y}{\mu_0 l_w l_d}.$$

The *fringing* effect may be integrated. The airgap reluctance can be approximated as

$$\mathfrak{R}_g = \frac{2y}{\mu_0(k_{g1}l_w l_d + k_{g2}y^2)},$$

where k_{g1} and k_{g2} are the nonlinear functions of the ferromagnetic material l_d/l_w ratio, *B-H* curve, load, and so on.

The reluctances of the ferromagnetic materials of stationary and movable members $\mathfrak{R}_1$ and $\mathfrak{R}_2$ are

$$\mathfrak{R}_1 = \frac{l_1}{\mu_0\mu_1 A} = \frac{l_1}{\mu_0\mu_1 l_w l_d} \quad \text{and} \quad \mathfrak{R}_2 = \frac{l_2}{\mu_0\mu_2 A} = \frac{l_2}{\mu_0\mu_2 l_w l_d}.$$

The magnetizing inductance is expressed as

$$L(y) = \frac{N^2}{\mathfrak{R}_g(y) + \mathfrak{R}_1 + \mathfrak{R}_2}.$$

The electromagnetic force is found as

$$F_e = \frac{1}{2}i^2\frac{dL(y)}{dy} = \frac{1}{2}i^2\frac{d[N^2/(\mathfrak{R}_g(y) + \mathfrak{R}_1 + \mathfrak{R}_2)]}{dy}.$$

Using two expressions for the airgap reluctance, one performs the differentiation deriving the expression for the electromagnetic force. The related results and differentiation is reported in Example 2.22. ☐

Example 2.22

Consider the actuator (solenoid), which has N turns (see Figure 2.20). The distance between the stationary and movable members is $x(t)$. The mean lengths of the stationary and movable members are l_1 and l_2, and the cross-sectional area is A. One can find the force exerted on the movable member as a function of the current $i_a(t)$ in the winding. The permeabilities of stationary and movable members are μ_1 and μ_2.

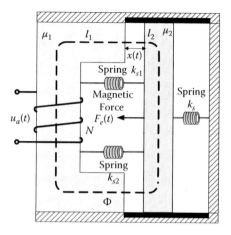

FIGURE 2.20
Schematic of an actuator.

The electromagnetic force is

$$F_e = \frac{\partial W_m}{\partial x}, \quad \text{where} \quad W_m = \frac{1}{2}Li_a^2(t).$$

The magnetizing inductance is

$$L = \frac{N\Phi}{i_a(t)} = \frac{\psi}{i_a(t)}.$$

The magnetic flux is found as

$$\Phi = \frac{Ni_a(t)}{\Re_1 + \Re_x + \Re_x + \Re_2}.$$

The reluctances of the ferromagnetic stationary and movable members $\Re_1$ and $\Re_2$, as well as the reluctance of the air gap $\Re_x$, are

$$\Re_1 = \frac{l_1}{\mu_0\mu_1 A}, \quad \Re_2 = \frac{l_2}{\mu_0\mu_2 A} \quad \text{and} \quad \Re_x = \frac{x(t)}{\mu_0 A}.$$

The equivalent magnetic circuit with the reluctances of the various paths is illustrated in Figure 2.21.

Using the expression for reluctances of the movable and stationary members and air gap, one obtains the following formula for the flux linkages

$$\psi = N\Phi = \frac{N^2 i_a(t)}{\frac{l_1}{\mu_0\mu_1 A} + \frac{2x(t)}{\mu_0 A} + \frac{l_2}{\mu_0\mu_2 A}}.$$

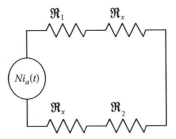

FIGURE 2.21
Equivalent magnetic circuit.

The magnetizing inductance is a nonlinear function of the displacement, and

$$L(x) = \frac{N^2}{\frac{l_1}{\mu_0\mu_1 A} + \frac{2x(t)}{\mu_0 A} + \frac{l_2}{\mu_0\mu_2 A}} = \frac{N^2\mu_0\mu_1\mu_2 A}{\mu_2 l_1 + 2\mu_1\mu_2 x(t) + \mu_1 l_2}.$$

Using

$$F_e = \frac{\partial W_m}{\partial x} = \frac{1}{2}\frac{\partial(L(x(t))i_a^2(t))}{\partial x},$$

the force in the x direction is found to be

$$F_e = -\frac{N^2\mu_0\mu_1^2\mu_2^2 A i_a^2}{(\mu_2 l_1 + 2\mu_1\mu_2 x + \mu_1 l_2)^2}.$$

The differential equations are derived to simulate and analyze the actuator. Using Newton's second law of motion, one obtains the following nonlinear differential equations to examine the performance and analyze the steady-state and dynamic behavior

$$\frac{dx}{dt} = v,$$

$$\frac{dv}{dt} = \frac{1}{m}\left(-\frac{N^2\mu_0\mu_1^2\mu_2^2 A i_a^2}{(\mu_2 l_1 + 2\mu_1\mu_2 x + \mu_1 l_2)^2} - k_s x + k_{s1} x + k_{s2} x\right).$$

This set of nonlinear differential equations describes the mechanical dynamics of a plunger. The voltage $u_a(t)$ is applied changing $i_a(t)$. The energy conversion, *emf* induction, and other electromagnetic phenomena and effects must be integrated. Kirchhoff's voltage law gives

$$u_a = ri_a + \frac{d\psi}{dt},$$

where the flux linkage ψ is expressed as $\psi = L(x)i_a$. Neglecting self and leakage inductances, from

$$u_a = ri_a + L(x)\frac{di_a}{dt} + i_a \frac{dL(x)}{dx}\frac{dx}{dt},$$

one finds

$$\frac{di_a}{dt} = \frac{1}{L(x)}\left[-ri_a + \frac{2N^2\mu_0\mu_1^2\mu_2^2 A}{(\mu_2 l_1 + 2\mu_1\mu_2 x + \mu_1 l_2)^2}i_a v + u_a\right].$$

We have a set of three differential equations

$$\frac{di_a}{dt} = \frac{\mu_2 l_1 + 2\mu_1\mu_2 x + \mu_1 l_2}{N^2\mu_0\mu_1\mu_2 A}\left[-ri_a + \frac{2N^2\mu_0\mu_1^2\mu_2^2 A}{(\mu_2 l_1 + 2\mu_1\mu_2 x + \mu_1 l_2)^2}i_a v + u_a\right],$$

$$\frac{dv}{dt} = \frac{1}{m}\left[-\frac{N^2\mu_0\mu_1^2\mu_2^2 Ai_a^2}{(\mu_2 l_1 + \mu_1\mu_2 2x(t) + \mu_1 l_2)^2} - k_s x + k_{s1} x + k_{s2} x\right],$$

$$\frac{dx}{dt} = v. \qquad\qquad\qquad\qquad\qquad\qquad\qquad\qquad\qquad \Box$$

Example 2.23

Two coils have a mutual inductance 0.00005 H ($L_{12} = 0.00005$ H). The current in the first coil is $i_1 = \sqrt{\sin 4t}$. One can find the induced *emf* in the second coil as

$$\mathcal{E}_2 = L_{12}\frac{di_1}{dt}.$$

By using the power rule for the time-varying current in the first coil

$$i_1 = \sqrt{\sin 4t},$$

we have

$$\frac{di_1}{dt} = \frac{2\cos 4t}{\sqrt{\sin 4t}}.$$

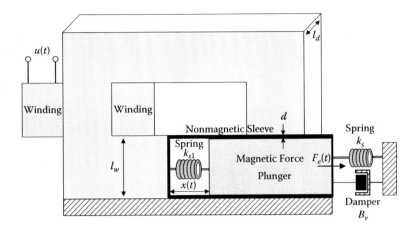

FIGURE 2.22
Solenoid schematics.

Hence,

$$\mathscr{E}_2 = \frac{0.0001\cos 4t}{\sqrt{\sin 4t}}.$$ □

Example 2.24

Figure 2.22 illustrates a solenoid with a stationary member and a movable plunger. Our goal is to derive the differential equations to perform analysis.

Applying Newton's second law of motion, one finds the differential equation to model the translational motion. In particular,

$$m\frac{d^2x}{dt^2} = F_e(t) - B_v\frac{dx}{dt} - (k_sx - k_{s1}x) - F_L(t).$$

The restoring/stretching forces exerted by the springs is $F_s = k_sx - k_{s1}x$.

Assuming that the magnetic system is linear. The coenergy is expressed as

$$W_c(i,x) = \frac{1}{2}L(x)i^2.$$

From

$$F_e(i,x) = \frac{\partial W_c(i,x)}{\partial x},$$

the electromagnetic force is

$$F_e(i,x) = \frac{1}{2} i^2 \frac{dL(x)}{dx}.$$

The magnetizing inductance is found as

$$L(x) = \frac{N^2}{\Re_f + \Re_x} = \frac{N^2 \mu_0 \mu_r A_f A_x}{A_x l_f + A_f \mu_r (x + 2d)},$$

where $\Re_f$ and $\Re_x$ are the reluctances of the ferromagnetic material and air gap; A_f and A_x are the associated cross section areas; l_f is the lengths of the magnetic material (though l_f varies as x changes, one may assume that $l_f = $ const because l_f variations do not significantly affect the results due to a high μ_r); $(x + 2d)$ is the effective air gap; d is the nonmagnetic sleeve thickness.

Hence,

$$\frac{dL}{dx} = -\frac{N^2 \mu_0 \mu_r^2 A_f^2 A_x}{[A_x l_f + A_f \mu_r (x + 2d)]^2}.$$

Using Kirchhoff's law

$$u = ri + \frac{d\psi}{dt}, \quad \psi = L(x)i,$$

one obtains

$$u = ri + L(x)\frac{di}{dt} + i\frac{dL(x)}{dx}\frac{dx}{dt}.$$

Thus, we have

$$\frac{di}{dt} = \frac{1}{L(x)}\left[-ri + \frac{N^2 \mu_0 \mu_r^2 A_f^2 A_x}{[A_x l_f + A_f \mu_r (x + 2d)]^2} iv + u\right].$$

Combining this equation with the second-order differential equation for the mechanical system, three first-order nonlinear differential equations are found as

$$\frac{di}{dt} = -\frac{r[A_x l_f + A_f \mu_r (x + 2d)]}{N^2 \mu_0 \mu_r A_f A_x} i + \frac{\mu_r A_f}{A_x l_f + A_f \mu_r (x + 2d)} iv + \frac{A_x l_f + A_f \mu_r (x + 2d)}{N^2 \mu_0 \mu_r A_f A_x} u,$$

$$\frac{dv}{dt} = \frac{N^2 \mu_0 \mu_r^2 A_f^2 A_x}{2m[A_x l_f + A_f \mu_r (x + 2d)]^2} i^2 - \frac{1}{m}(k_s x - k_{s1} x) - \frac{B_v}{m} v,$$

$$\frac{dx}{dt} = v.$$

Depending on the applications, requirements (accuracy, efficiency, power consumption, etc.), materials used, and other factors, the designer may refine the model developed enhancing its accuracy and applicability. For example, using the friction model as reported in Example 2.7, we have

$$\frac{dv}{dt} = \underbrace{\frac{N^2\mu_0\mu_f^2 A_f^2 A_x}{2m[A_x l_f + A_f \mu_f(x+2d)]^2}}_{\text{electromagnetic force developed}} i^2 - \frac{1}{m}(k_s x - k_{s1} x)$$

$$\underbrace{-\frac{1}{m}\left(k_{fr1} - k_{fr2}e^{-k\left|\frac{dx}{dt}\right|} + k_{fr3}\,|v|\right)\mathrm{sgn}(v)}_{\text{friction force}}.$$

The *fringing* effect may be integrated as were reported in Example 2.21. □

Example 2.25

The reported concepts can be applied in the design of electrical machines. For preliminary design, it is sufficiently accurate to apply Faraday's or Lenz's laws which provide the *emf* in term of the time-varying magnetic field changes. In particular,

$$emf = -\frac{d\psi}{dt} = -\frac{\partial\psi}{\partial t} - \frac{\partial\psi}{\partial\theta_r}\frac{\partial\theta_r}{dt} = -\frac{\partial\psi}{\partial t} - \frac{\partial\psi}{\partial\theta_r}\omega_r,$$

where

$$\frac{d\psi}{dt}$$

is the transformer term.

The total flux linkage in many electric machines can be expressed as

$$\psi = \frac{1}{4}\pi N_S \Phi_p,$$

where N_S is the number of turns; Φ_p is the flux per pole.

For radial topology machines, we have

$$\Phi_p = \frac{\mu i N_S}{P^2 g_e} R_{in\,st} L,$$

where i is the phase current in the winding; $R_{in\,st}$ is the inner stator radius; L is the inductance; P is the number of poles; g_e is the equivalent gap, which includes the air gap and radial thickness of the permanent magnet.

Denoting the number of turns per phase as N_s, one has

$$mmf = \frac{i N_s}{P}\cos P\theta_r.$$

The simplified estimate for the electromagnetic torque for radial topology motors with permanent magnets is

$$T = \frac{1}{2} P B_{ag} i_s N_s L_r D_r ,$$

where B_{ag} is the air gap flux density,

$$B_{ag} = \frac{\mu i N_S}{2 P g_e} \cos P\theta_r ;$$

i_s is the total current; L_r is the active length (rotor axial length); D_r is the outside rotor diameter.

One also may apply the following estimate $T = k_{ax} B_{ag} i_s N_S D_a^2$, where k_{ax} is the nonlinear coefficient which is found in terms of active conductors and permanent magnet length; D_a is the equivalent diameter, which is a function of windings and permanent-magnet topology.   □

2.7   Simulation of Systems in the MATLAB Environment

MATLAB® (MATrix LABoratory) is a high-performance interacting software environment for high-efficiency engineering and scientific numerical calculations. This environment can be applied to perform heterogeneous simulations and data-intensive analysis of complex electromechanical systems. MATLAB enables·users to solve a wide spectrum of analytical and numerical problems (control, optimization, identification, data acquisition, etc.) using matrix-based methods attaining excellent interactive capabilities. In addition, it allows compiling features with high-level programming languages, accessing and implementation of state-of-the-art numerical algorithms, powerful graphical and interface features, and so on. As a result of high flexibility and versatility, the MATLAB environment has been significantly enhanced and developed in recent years. A family of application-specific toolboxes, with a specialized collection of m-files for solving problems, guarantees comprehensiveness and effectiveness. For example, Simulink® is a companion graphical mouse-driven interactive environment enhancing MATLAB. A great number of books and MathWorks user manuals in MATLAB, Simulink, and different MATLAB toolboxes are available [22,23]. In addition, the MathWorks Inc. educational Web site can be used as references (e.g., http://education.mathworks.com and http://www.mathworks.com). This section introduces the MATLAB environment helping one to use this environment efficiently. The MATLAB environment (version 7.3) is used in this book, and the website http://www.mathworks.com/access/helpdesk/help/helpdesk.shtml can assist users to master MATLAB. MATLAB documentation and user manuals (thousands of pages each) are available in the Portable Document Format using the Help Desk. This book focuses on MATLAB applications to electromechanical systems, specifically how to solve practical problems using step-by-step instructions.

To start MATLAB, double-click the MATLAB R2006b icon. The MATLAB Command Window with Launch Pad and Command History appear on the screen as shown in Figure 2.23. Typing ver or demo, the available toolboxes are listed as reported in Figure 2.23.

The line

```
>>
```

is the MATLAB prompt.

Typing

```
>> a=1+2+3*4
```

and pressing the Enter (Return) key, we have the value for a. The following result is displayed

```
a = 15
```

To calculate and plot a function, for example $y = \sin(2x)$, if x varies from 0 to 3π, let the increment is 0.025π. The MATLAB statement, which is typed in the Command Window, is

```
>> x=0:0.025*pi:3*pi; y=sin(2*x); plot(x,y);
title('y=sin(2x)','FontSize',14)
```

By pressing the Enter key, the calculations are made, and the figure appears as shown in Figure 2.24. To capture this plot, one clicks the Edit icon and selects the Copy Figure option. The plot captured is illustrated as the second figure in Figure 2.24. If one refines the plotting statement to be plot(x,y,'o'), the resulting captured plot is shown in Figure 2.24.

To master the plotting, in the Command Window, type

```
>> help plot
```

and as the Enter key is pressed, detailed instructions are provided. At the very end, clicking on doc plot, makes the Microsoft Word file accessible.

The differential equations, which describe the dynamics of electromechanical systems, usually cannot be solved analytically. However, for simple equations, analytic solutions can be derived using MATLAB. Considering the translational and rotational rigid-body one-dimensional mechanical systems, it was shown in Section 2.5 that the application of Newton's second law of motion results in the second-order differential equations

$$m\frac{d^2x}{dt^2} + B_v\frac{dx}{dt} + k_sx = F_a(t) \quad \text{and} \quad J\frac{d^2\theta}{dt^2} + B_m\frac{d\theta}{dt} + k_s\theta = T_a(t),$$

where $F_a(t)$ and $T_a(t)$ are the time-varying applied force and torque.

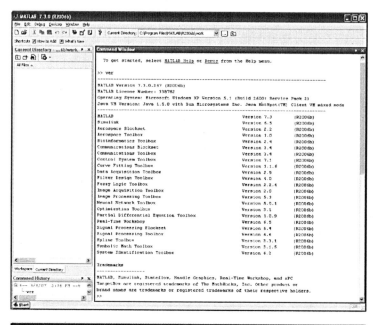

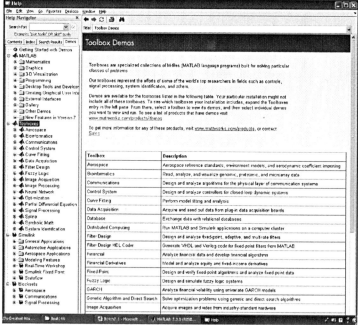

FIGURE 2.23
MATLAB Command Window and MATLAB toolboxes.

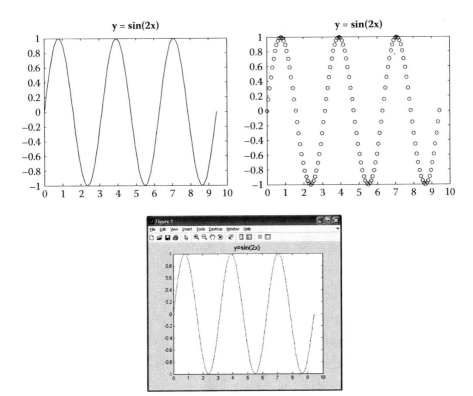

FIGURE 2.24
Plot for $y = \sin(2x)$.

For parallel and series RLC circuits, illustrated in Figure 2.25, one obtains the following equations

$$C\frac{d^2u}{dt^2} + \frac{1}{R}\frac{du}{dt} + \frac{1}{L}u = \frac{di_a}{dt} \quad \text{or} \quad \frac{d^2u}{dt^2} + \frac{1}{RC}\frac{du}{dt} + \frac{1}{LC}u = \frac{1}{C}\frac{di_a}{dt},$$

and

$$L\frac{d^2i}{dt^2} + R\frac{di}{dt} + \frac{1}{C}i = \frac{du_a}{dt} \quad \text{or} \quad \frac{d^2i}{dt^2} + \frac{R}{L}\frac{di}{dt} + \frac{1}{LC}i = \frac{1}{L}\frac{du_a}{dt}.$$

The analytic solution of linear differential equations with constant coefficients can be easily derived. The general solution of the second-order linear differential equation is found by using the characteristic roots (eigenvalues) of the characteristic equation. The damping coefficient ξ and the resonant

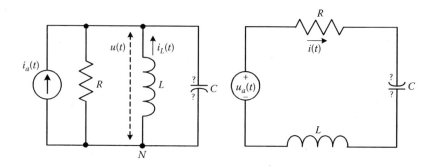

FIGURE 2.25
Parallel and series *RLC* circuits.

frequency ω_0 for the *RLC* circuits (parallel and series) and translational motion are

$$\xi = \frac{1}{2RC}, \ \omega_0 = \frac{1}{\sqrt{LC}}; \ \xi = \frac{R}{2L}, \ \omega_0 = \frac{1}{\sqrt{LC}}; \ \text{and} \ \xi = \frac{B_v}{2\sqrt{k_s m}}, \ \omega_0 = \sqrt{\frac{k_s}{m}}.$$

For a linear second-order differential equation

$$\frac{d^2x}{dt^2} + 2\xi\frac{dx}{dt} + \omega_0 x = f(t),$$

to find three possible solutions one studies the characteristic equation $s^2 + 2\xi s + \omega_0^2 = (s - s_1)(s - s_2) = 0$. This characteristic equation is found by using the Laplace operator $s = d/dt$. Furthermore, $s^2 = d^2/dt^2$. The characteristic roots (eigenvalues) are $s_{1,2} = -\xi \pm \sqrt{\xi^2 - \omega_0^2}$.

Case 1. If $\xi^2 > \omega_0^2$, the real distinct characteristic roots s_1 and s_2 result. The general solution is $x(t) = ae^{s_1 t} + be^{s_2 t} + c_f$, where coefficients a and b are obtained using the initial conditions; c_f is the solution as a result of the *forcing* function f, and for the *RLC* circuits f is $i_a(t)$ or $u_a(t)$.

Case 2. For $\xi^2 = \omega_0^2$, the characteristic roots are real and identical, and $s_1 = s_2 = -\xi$. The solution of the second-order differential equation is given as $x(t) = (a + b)e^{-\xi t} + c_f$.

Case 3. If $\xi^2 < \omega_0^2$, the complex distinct characteristic roots are $s_{1,2} = -\xi \pm j\sqrt{\omega_0^2 - \xi^2}$. The general solution is

$$x(t) = e^{-\xi t}\left[a\cos\left(\sqrt{\omega_0^2 - \xi^2}t\right) + b\sin\left(\sqrt{\omega_0^2 - \xi^2}t\right)\right] + c_f$$

$$= e^{-\xi t}\sqrt{a^2 + b^2}\cos\left[\left(\sqrt{\omega_0^2 - \xi^2}t\right) + \tan^{-1}\left(\frac{-b}{a}\right)\right] + c_f.$$

Example 2.26

Consider the series RLC circuit illustrated in Figure 2.25. We will derive and plot the transient response due to the unit step input with initial conditions. The parameters are $R = 0.5$ ohm, $L = 1$ H, $C = 2$ F, $a = 1$ and $b = -1$.

The series RLC circuit is described by the differential equation

$$\frac{d^2i}{dt^2} + \frac{R}{L}\frac{di}{dt} + \frac{1}{LC}i = \frac{1}{L}\frac{du_a}{dt},$$

which yields the following characteristic equation

$$s^2 + \frac{R}{L}s + \frac{1}{LC} = 0.$$

The characteristic roots are

$$s_1 = -\frac{R}{2L} - \sqrt{\left(\frac{R}{2L}\right)^2 - \frac{1}{LC}} \quad \text{and} \quad s_2 = -\frac{R}{2L} + \sqrt{\left(\frac{R}{2L}\right)^2 - \frac{1}{LC}}.$$

If $(\frac{R}{2L})^2 > \frac{1}{LC}$, the characteristic roots are real and distinct. For $(\frac{R}{2L})^2 = \frac{1}{LC}$, the characteristic roots are real and identical. Whereas, if $(\frac{R}{2L})^2 < \frac{1}{LC}$, the characteristic roots are complex.

For the assigned values for R, L, and C, one concludes that the characteristic roots are complex, and the dynamics is underdamped because the solution is

$$x(t) = e^{-\xi t}\left[a\cos\left(\sqrt{\omega_0^2 - \xi^2}\,t\right) + b\sin\left(\sqrt{\omega_0^2 - \xi^2}\,t\right)\right] + c_f$$

with

$$\xi = \frac{R}{2L} = 0.25 \quad \text{and} \quad \omega_0 = \frac{1}{\sqrt{LC}} = 0.71.$$

In the Command Window, we type the following statements:

```
R=0.5; L=1; C=2; a=1; b=-1; cf=1; e=R/2*L; w0=1/sqrt(L*C);
t=0:.01:30; x=exp(-e*t).*(a*cos(sqrt(w0^2-e^2)*t)+b*sin(sqrt(w0^2-e^2)*t))+cf;
plot(t,x); xlabel('Time (seconds)','FontSize',14);
title('Solution of Differential Equation x(t)','FontSize',14);
```

The resulting dynamics is documented in Figure 2.26.

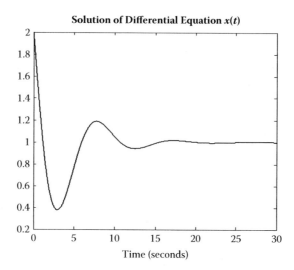

FIGURE 2.26
Dynamics caused by the unit step and initial conditions.

Example 2.27   Application of MATLAB to Analytically Solve Differential Equations

Let us analytically solve the third-order differential equation

$$\frac{d^3x}{dt^3} + 2\frac{dx}{dt} + 3x = 10f.$$

We will use the Symbolic Math Toolbox.

Using the dsolve command (analytic differential equation solver), we type in the Command Window

```
x=dsolve('D3x+2*Dx+3*x=10*f')
```

The resulting solution is

```
x =

10/3*f+C1*exp(-t)+C2*exp(1/2*t)*sin(1/2*11^(1/2)*t)
+C3*exp(1/2*t)*cos(1/2*11^(1/2)*t)
```

Using the pretty command, we find

```
>> pretty(x)
                                       1/2
10/3 f + C1 exp(-t) + C2 exp(1/2 t) sin(1/2 11    t)
                        1/2
+ C3 exp(1/2 t) cos(1/2 11    t)
```

Hence, we obtain

$$x(t) = \frac{10}{3}f + c_1 e^{-t} + c_2 e^{0.5t}\sin\left(\frac{1}{2}\sqrt{11}t\right) + c_3 e^{0.5t}\cos\left(\frac{1}{2}\sqrt{11}t\right).$$

Using the initial conditions, the unknown constants are found. As an example, we assign the following initial conditions

$$\left(\frac{d^2x}{dt^2}\right)_0 = 5, \quad \left(\frac{dx}{dt}\right)_0 = 15, \quad \text{and} \quad x_0 = -20.$$

Using the statement

```
x=dsolve('D3x+2*Dx+3*x=10*f','D2x(0)=5','Dx(0)=15','x(0)=-20');
pretty(x)
```

the resulting solution, with the derived c_1, c_2, and c_3, is

```
10/3 f + (-2 f - 14) exp(-t)
               1/2                        1/2
    - 8/33 11     (f - 3) exp(1/2 t) sin(1/2 11     t)
                                            1/2
    + (- 4/3 f - 6) exp(1/2 t) cos(1/2 11     t)
```

Thus,

$$x(t) = \frac{10}{3}f + (-2f - 14)e^{-t} - \frac{8}{33}\sqrt{11}(f-3)e^{0.5t}\sin\left(\frac{1}{2}\sqrt{11}t\right)$$

$$+ \left(-\frac{4}{3}f - 6\right)e^{0.5t}\cos\left(\frac{1}{2}\sqrt{11}t\right).$$

If the forcing function is time-varying, the analytic solution of

$$\frac{d^3x}{dt^3} + 2\frac{dx}{dt} + 3x = 10f(t)$$

is found by using the statement

```
x=dsolve('D3x+2*Dx+3*x=10*f(t)'); pretty(x)
```

If $f(t)$ is defined, and letting, $f(t)=5\cos(10t)$, by using

```
x=dsolve('D3x+2*Dx+3*x=10*5*cos(5*t)','D2x(0)=5','Dx(0)=15','x(0)=-20');
pretty(x)
```

we found $x(t)$ as

```
2875                75                 187
- ---- sin(5 t) + ---- cos(5 t)  -  --- exp(-t)
6617               6617               13

  5702                          1/2     1/2
+ ---- exp(1/2 t) sin(1/2 11   t) 11
  5599
  2864                         1/2
- ---- exp(1/2 t) cos(1/2 11   t)
  509
```

One concludes that

$$x(t) = -\frac{2875}{6617}\sin 5t + \frac{75}{6617}\cos 5t - \frac{187}{13}e^{-t} + \frac{5702}{5599}e^{0.5t}\sin(\tfrac{1}{2}\sqrt{11}t)\sqrt{11}$$

$$-\frac{2864}{509}e^{0.5t}\cos(\tfrac{1}{2}\sqrt{11}t)$$ □

Example 2.28

Consider the series RLC circuit as given in Figure 2.25. The goal is to find the analytical solution applying MATLAB. The aim is also to plot the circuitry dynamics assigning circuitry parameters and initial conditions.

Using the voltage across the capacitor and the current through the inductor as the state variables, and the supplied voltage $u_a(t)$ as the *forcing* function, we have the following set of first-order differential equations

$$C\frac{du_C}{dt} = i, \qquad L\frac{di}{dt} = -u_C - Ri + u_a(t).$$

Hence,

$$\frac{du_C}{dt} = \frac{1}{C}i, \quad \frac{di}{dt} = \frac{1}{L}(-u_C - Ri + u_a(t)).$$

The analytical solution is found using the Symbolic Math Toolbox. In particular, for time-varying $u_a(t)$, solving the set of differential equations by means of the statement

```
[V,I]=dsolve('DV=I/C','DI=(-V-R*I+Va(t))/L')
```

we obtain the resulting solution for the state variables $u_C(t)$ and $i(t)$. In particular,

```
V =
-1/2*(C^2*Int(Va(t)*exp(1/2*t*(R*C-(C*(R^2*C-
4*L))^(1/2))/C/L),t)*R*exp(1/2*t*(R*C+(C*(R^2*C-
4*L))^(1/2))/C/L-t*R/L)*f-C*Int(Va(t)*exp(1/2*t*(R*C-(C*(R^2*C-
4*L))^(1/2))/C/L),t)*exp(1/2*t*(R*C+(C*(R^2*C-4*L))^(1/2))/C/L-
t*R/L)*(C*(R^2*C-4*L))^(1/2)*f-
C^2*Int(Va(t)*exp(1/2*t*(R*C+(C*(R^2*C-
4*L))^(1/2))/C/L),t)*R*exp(1/2*t*(R*C-(C*(R^2*C-
4*L))^(1/2))/C/L-t*R/L)*f-C*Int(Va(t)*exp(1/2*t*(R*C+(C*(R^2*C-
```

```
4*L))^(1/2))/C/L),t)*exp(1/2*t*(R*C-(C*(R^2*C-4*L))^(1/2))/C/L-
t*R/L)*f*(C*(R^2*C-4*L))^(1/2)+exp(-1/2*t*(R*C-(C*(R^2*C-
4*L))^(1/2))/C/L)*C2*R*C*f^2*(C*(R^2*C-4*L))^(1/2)-
exp(-1/2*t*(R*C-(C*(R^2*C-4*L))^(1/2))/C/L)*C2*C*(R^2*C-
4*L)*f^2+exp(-1/2*t*(R*C+(C*(R^2*C-
4*L))^(1/2))/C/L)*C1*R*C*f^2*(C*(R^2*C-4*L))^(1/2)+
exp(-1/2*t*(R*C+(C*(R^2*C-4*L))^(1/2))/C/L)*C1*C*(R^2*C-
4*L)*f^2)/L/f^2/(C*(R^2*C-4*L))^(1/2)

I =
-(-exp(-1/2*t*(R*C-(C*(R^2*C-4*L))^(1/2))/C/L)*C2*(C*(R^2*C-
4*L))^(1/2)*f^2-exp(-1/2*t*(R*C+(C*(R^2*C-
4*L))^(1/2))/C/L)*C1*(C*(R^2*C-4*L))^(1/2)*f^2-
C*Int(Va(t)*exp(1/2*t*(R*C-(C*(R^2*C-
4*L))^(1/2))/C/L),t)*exp(1/2*t*(R*C+(C*(R^2*C-4*L))^(1/2))/C/L-
t*R/L)*f+C*Int(Va(t)*exp(1/   2*t*(R*C+(C*(R^2*C-
4*L))^(1/2))/C/L),t)*exp(1/2*t*(R*C-(C*(R^2*C-4*L))^(1/2))/C/L-
t*R/L)*f)/(C*(R^2*C-4*L))^(1/2)/f^2
```

For $u_a(t)$=const, from

```
[V,I]=dsolve('DV=I/C','DI=(-V-R*I+Va)/L')
```

one has

```
V =
(-1/2*(R*C-(R^2*C^2-4*C*L)^(1/2))/C/L*exp(-1/2*(R*C-(R^2*C^2-
4*C*L)^(1/2))/C/L*t)*C2-1/2*(R*C+(R^2*C^2-
4*C*L)^(1/2))/C/L*exp(-1/2*(R*C+(R^2*C^2-
4*C*L)^(1/2))/C/L*t)*C1)*C

I =
exp(-1/2*t*(R*C-(C*(R^2*C-4*L))^(1/2))/C/L)*C2+
exp(-1/2*t*(R*C+(C*(R^2*C-4*L))^(1/2))/C/L)*C1+Va
```

Assigning $R = 1$ ohm, $L = 0.1$ H, and $C = 0.01$ F, for $u_a(t) = 10$ V, the statement is

```
[V,I]=dsolve('DV=I/C','DI=(-V-R*I+Va)/L')
R=1, L=0.1, C=0.01, Va=10, [V,I]=dsolve('DV=I/0.01','DI=(-V-1*I+10)/0.1')
```

The solutions for the state variables $u_C(t)$ and $i(t)$ are displayed. In particular,

```
R = 1
L = 0.1000
C = 0.0100
Va= 10
V =
1/20*exp(-5*t)*(-sin(5*39^(1/2)*t)*C2+cos(5*39^(1/2)*t)*39^(1/2)*C2-
cos(5*39^(1/2)*t)*C1-sin(5*39^(1/2)*t)*39^(1/2)*C1)

I =
10+exp(-5*t)*(sin(5*39^(1/2)*t)*C2+cos(5*39^(1/2)*t)*C1)
```

The constants c_1 and c_2 must be found by using the initial conditions. Assigning the initial conditions to be $[20, -10]^T$, we modify the statement to be

```
[V,I]=dsolve('DV=I/C','DI=(-V-R*I+Va)/L'); R=1; L=0.1; C=0.01; Va=10;
[V,I]=dsolve('DV=I/0.01','DI=(-V-1*I+10)/0.1', 'V(0)=20, I(0)=-10')
```

The evolution of $u_C(t)$ and $i(t)$ are

```
V =
1/20*exp(-5*t)*(-200/39*sin(5*39^(1/2)*t)*39^(1/2)-200*cos(5*39^(1/2)*t))
I =
10+exp(-5*t)*(-190/39*sin(5*39^(1/2)*t)*39^(1/2)+10*cos(5*39^(1/2)*t))
```

The derived expressions can be simplified using the `simplify` command. In particular, `simplify(V)` and `simplify(I)` are used. From

```
V_simplify=simplify(V), I_simplify=simplify(I)
```

we have the expressions for $u_C(t)$ and $i(t)$ as

```
V_simplify =
-10/39*exp(-5*t)*(sin(5*39^(1/2)*t)*39^(1/2)+39*cos(5*39^(1/2)*t))
I_simplify =
10-190/39*exp(-5*t)*sin(5*39^(1/2)*t)*39^(1/2)+10*exp(-5*t)*cos(5*39^(1/2)*t)
```

Thus, one concludes that

$$u_C(t) = -\frac{10}{39}e^{-5t}\sin(5\sqrt{39}t)\sqrt{39} + 39\cos(5\sqrt{39}t),$$

$$i(t) = 10 - \frac{190}{39}e^{-5t}\sin(5\sqrt{39}t)\sqrt{39} + 10e^{-5t}\cos(5\sqrt{39}t).$$

Figure 2.27 documents the plots for $u_C(t)$ and $i(t)$, which are found using the following statement

```
t=0:0.001:1;
V_simplify=-10/39*exp(-5*t).*(sin(5*39^(1/2)*t)*39^(1/2)+39*cos(5*39^(1/2)*t));
I_simplify=10-190/39*exp(-5*t).*sin(5*39^(1/2)*t)*39^(1/2)+⋯
10*exp(-5*t).*cos(5*39^(1/2)*t);
plot(t,V_simplify,t,I_simplify); xlabel('Time (seconds)', 'FontSize',14;
title('Voltage and Current Dynamics in RLC Circuit, u_C(t) and i(t),' 'FontSize',14;);□
```

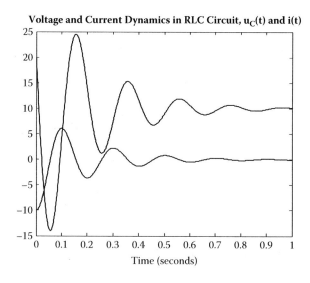

FIGURE 2.27
Dynamics for $u_C(t)$ and $i(t)$.

For nonlinear differential equations, which describe the dynamics of electromechanical systems, frequently analytic solutions cannot be derived. Therefore, numerical solutions must be found. The following example illustrates the application of MATLAB to numerically solve the ordinary differential equations.

Example 2.29

Using the MATLAB ode45 solver (built-in ode45 command), numerically solve a system of highly nonlinear differential equations

$$\frac{dx_1(t)}{dt} = -20x_1 + |x_2 x_3| + 10x_1 x_2 x_3, \, x_1(t_0) = x_{10},$$

$$\frac{dx_2(t)}{dt} = -5x_1 x_2 - 10\cos x_1 - \sqrt{|x_3|}, \, x_2(t_0) = x_{20},$$

$$\frac{dx_3(t)}{dt} = -5x_1 x_2 + 50x_2 \cos x_1 - 25x_3, \, x_3(t_0) = x_{30}.$$

The initial conditions are

$$x_0 = \begin{bmatrix} x_{10} \\ x_{20} \\ x_{30} \end{bmatrix} = \begin{bmatrix} 2 \\ 1 \\ -2 \end{bmatrix}.$$

Two m-files (ch2 _ 1.m and ch2 _ 2.m) are developed in order to numerically simulate this set of nonlinear differential equations. The evolution of the state variables $x_1(t)$, $x_2(t)$, and $x_3(t)$ must be plotted as the differential equations are solved. To illustrate the transient responses for $x_1(t)$, $x_2(t)$, and $x_3(t)$, the plot command is used. Comments, which are not executed, appear after the % symbol. These comments explain sequential steps. The MATLAB file (ch2 _ 1.m) with the ode45 solver, two-dimensional plotting statements using the plot command, as well as three-dimensional plotting statements using the plot3 command, is reported below

```
echo on; clear all
t0=0; tfinal=1; tspan=[t0 tfinal]; % initial and final time
y0=[2 1 -2]'; % initial conditions for state variables
[t,y]=ode45('ch2 _ 2',tspan,y0); %ode45 MATLAB solver using ode45 solver
% Plot of the transient dynamics of the state variables solving differential equations
% These differential equations are assigned in file ch2 _ 2.m
plot(t,y(:,1),'--',t,y(:,2),'-',t,y(:,3),':'); % plot the transient dynamics
xlabel('Time (seconds)','FontSize',14);
ylabel('State Variables','FontSize',14);
title('Solution of Differential Equations: x _ 1(t), x _ 2(t) and x _ 3(t)', 'FontSize',14);
pause
% 3-D plot using x1, x2 and x3
plot3(y(:,1),y(:,2),y(:,3));
```

```
xlabel('x_1','FontSize',14); ylabel('x_2','FontSize',14);
zlabel('x_3','FontSize',14);
title('Three-Dimensional State Evolutions: x_1(t), x_2(t) and x_3(t)', 'FontSize',14);
text(0,-2.5,2,'0 Origin','FontSize',14);
```

The second MATLAB file (ch2_2.m), with the specified set of differential equations to be numerically solved, is

```
% Simulation of the third-order differential equations
function yprime = difer(t,y);
% Differential equations parameters
a11=-20; a12=1; a13=10; a21=-5; a22=-10; a31=-5; a32=50; a33=-25;
% Three differential equations: System of three first-order differential equations
yprime=[a11*y(1,:)+a12*abs(y(2,:)*y(3,:))+a13*y(1,:)*y(2,:)*y(3,:);... % first
   differential equation
a21*y(1,:)*y(2,:)+a22*cos(y(1,:))+sqrt(abs(y(3,:)));... % second differential equation
a31*y(1,:)*y(2,:)+a32*cos(y(1,:))*y(2,:)+a33*y(3,:)]; % third differential equation
```

To calculate the transient dynamics and plot the transient dynamics, in the Command window one types

```
ch2_1
```

and presses the Enter key. The resulting transient behavior (two-dimensional plot) is documented in Figure 2.28 The three-dimensional evolution of the state variables is illustrated in Figure 2.28.

In the Command Window, type

```
who, x=[t,y]
```

so the variables used and arrays will be displayed. In particular, we have

```
Your variables are:
t t0 tfinal tspan y y0
```

The dynamics of $x_1(t)$, $x_2(t)$, and $x_3(t)$, are plotted in Figure 2.28. The resulting data for x, which is displayed in the Command Window, is reported below. In particular, we have four columns for time t, as well as for three state variables $x_1(t)$, $x_2(t)$, and $x_3(t)$. That is $x=[t, x_1, x_2, x_3]$.

```
x =
0          2.0000    1.0000    -2.0000
0.0013     1.9025    0.9941    -1.9724
0.0026     1.8108    0.9876    -1.9392
0.0039     1.7247    0.9807    -1.9009
0.0052     1.6438    0.9734    -1.8581
0.0106     1.3562    0.9377    -1.6405

......................................................................
0.9721     0.8522    -1.2536   -1.4325
0.9902     0.8536    -1.2542   -1.4335
0.9927     0.8538    -1.2542   -1.4336
0.9951     0.8539    -1.2543   -1.4336
0.9976     0.8541    -1.2543   -1.4336
1.0000     0.8543    -1.2543   -1.4336
```

One can perform plotting, data mining, filtering, and other advanced numerics.   □

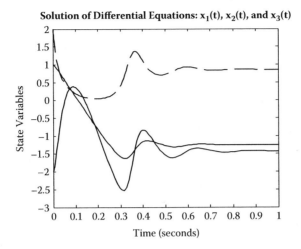

Solution of Differential Equations: $x_1(t)$, $x_2(t)$, and $x_3(t)$

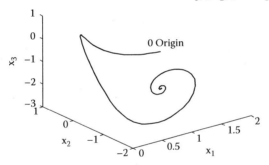

Three-Dimensional State Evolution: $x_1(t)$, $x_2(t)$, and $x_3(t)$

FIGURE 2.28
Dynamics and evolution of the state variables.

Simulink, as a part of the MATLAB environment, is an interactive computing package for simulation of differential equations and dynamic systems. Simulink is a graphical mouse-driven program that allows one to numerically simulate and analyze systems by developing and manipulating blocks and diagrams. It is applied to linear, nonlinear, continuous-time, discrete-time, multivariable, multirate, and hybrid systems. *Blocksets* are built-in blocks in Simulink that provide a full comprehensive block library for different system components, and C-code from block diagrams is generated using the *Real-time Workshop Toolbox*. Using a mouse-driven block-diagram interface, the Simulink diagrams (models) can be built. These block diagrams (mdl models) represent systems which are described by differential (difference) and constitutive equations. Hybrid and discrete-even systems can be simulated and examined. The distinct advantage is that Simulink provides a graphical user interface (GUI) for building models (block dia-grams) using "select-drag-connect-click" mouse-based operations.

A comprehensive library of sinks, sources, linear and nonlinear compo-nents (blocks), connectors, as well as customized blocks (S-functions) provide great flexibility, immense interactability, superior efficiency, and excellent prototyping features. For example, complex systems can be built using high- and low-level blocks. It was illustrated that systems can be numerically sim-ulated by solving differential equations using various MATLAB ode solvers. Different methods and algorithms are embedded and utilized in Simulink. However, one interacts using the Simulink menus rather than entering the commands and functions in the Command Window. The easy-to-use Simulink menus are very convenient for interactive design, simulations, analysis, and visualization.

To start Simulink, in the Command Window type

```
simulink
```

and press the Enter key. Alternatively, click on the Simulink icon ![icon]. The Simulink Library Browser window, shown in Figure 2.29, appears. To run various Simulink demonstration programs, type

```
demo simulink
```

The interactive Simulink demo window is documented in Figure 2.29.

To simulate and examine continuous- and discrete-time dynamic systems, block diagrams are used. Simulink notably extends the MATLAB environ-ment offering a large variety of ready-to-use building blocks to build dia-grams, which represent the models derived. One can learn and explore Simulink using the Simulink and MATLAB demos. Different MATLAB and Simulink releases are used. Although there are some differences, there is overall coherence between all releases. The Simulink manuals are supplied with the software, and available in the portable document format (pdf) [23]. These user-friendly manuals can be accessed, and this section does not aim to rewrite the excellent user manuals. With the ultimate goal of providing supplementary coverage and educate the reader on how to solve practical problems, we introduce Simulink with step-by-step instructions and prac-tical examples. Figure 2.30 reports the Simulink demo features with vari-ous simple, medium complexity, and advanced examples which are ready to be assessed and used. For example, the friction model was covered in Example 2.7. The Index and Search icon can be utilized. As illustrated in Figure 2.30, MATLAB offers the model of friction. Appealing examples, from aerospace to automotive applications, from electronics to mechanical systems, and others are available. However, the designer must coherently assess the fitness, applicability, and accuracy of MATLAB (Simulink and other toolboxes), as well as any other environment, for specific problems under consideration and objectives targeted. One may find that the presum-ably ready-to-use files, blocks, diagrams, and other tools may not be sound or require significant refinements.

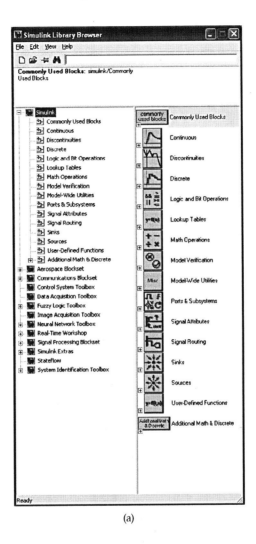

(a)

FIGURE 2.29
Simulink Library Browser and Simulink demo window.

Example 2.30   Van der Pol Differential Equations Simulations using Simulink

The van der Pol oscillator is described by the second-order nonlinear differential equation

$$\frac{d^2x}{dt^2} - k(1-x^2)\frac{dx}{dt} + x = d(t),$$

where $d(t)$ is the forcing function.

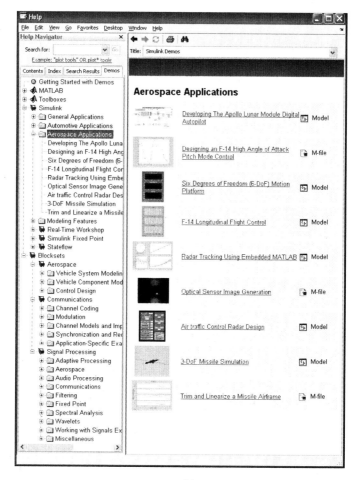

(b)

FIGURE 2.29
(Continued).

Let $k = 2$ and $d(t) = d_0 \text{rect}(\omega_0 t)$. The initial conditions are

$$x_0 = \begin{bmatrix} x_{10} \\ x_{20} \end{bmatrix} = \begin{bmatrix} 1 \\ -1 \end{bmatrix}.$$

The second-order van der Pol differential equation is rewritten as a system of two first-order differential equations

$$\frac{dx_1(t)}{dt} = x_2, \ x_1(t_0) = x_{10},$$

$$\frac{dx_2(t)}{dt} = -x_1 + kx_2 - kx_1^2 x_2 + d(t), \ x_2(t_0) = x_{20}.$$

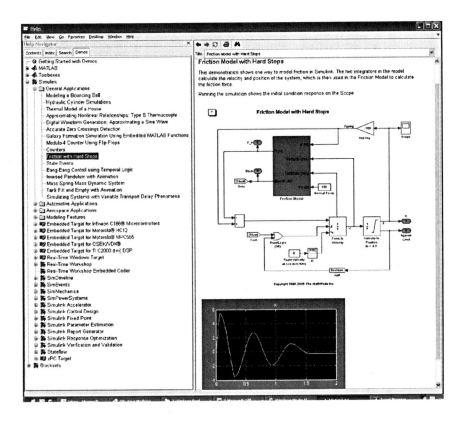

FIGURE 2.30
Simulink demo features.

In the literature, the differential equations for the van der Pol oscillator are also given as

$$\frac{dx_1(t)}{dt} = x_2, \quad \frac{dx_2(t)}{dt} = \mu\left[\left(1 - x_1^2\right)x_2 - x_1\right].$$

As shown in Figure 2.31, the Simulink diagram can be built using the following blocks: Function, Gain, Integrator, Mux, Signal Generator, Sum, and Scope. Simulation of the transient dynamics was performed assigning $d_0=0$ and $d_0 \neq 0$, $\omega_0 \neq 0$. The coefficients and initial conditions must be uploaded. The coefficient k can be assigned by double-clicking the Gain block and entering the value needed, and one can enter 2. Alternatively, one can type k, and in the Command Window we may type k = 2. By double-clicking the Signal Generator block, we select the square function and assign the corresponding magnitude d_0 and frequency ω_0. The initial conditions are set by double-clicking the Integrator blocks and typing x10 and x20. The values for x10 and x20 are entered in the Command Window by typing x10 = 1, x20 = -1. Hence, in the Command Window we upload

```
k=2; d0=0; w0=5; x10=1; x20=-1;
```

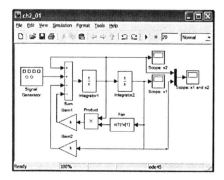

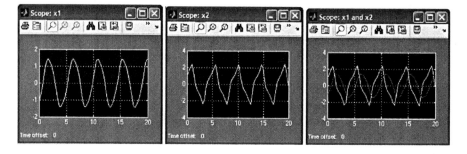

FIGURE 2.31
Simulink block diagram (ch2 _ 01.mdl), simulation configuration, and transient dynamics displayed in the scopes.

Specifying the simulation time to be 20 seconds (see Figure 2.31 where the simulation parameters window is illustrated), the Simulink model is run by clicking the ▶ icon. The simulation results are illustrated in Figure 2.31, which provides the behavior of two variables displayed by three Scopes.

The plotting statements can be effectively used. In the Scopes, in the General/Data History icon (the second icon to the left), one can assign the variable names. The variables x1, x2 and x12 are used in the first, second, and third scopes. By using

```
plot(x1(:,1),x1(:,2)); xlabel('Time (seconds)','FontSize',14);
ylabel('State Variable x _ 1','FontSize',14);
title('Solution of van der Pol Equation: x _ 1(t)','FontSize',14);
```

and

```
plot(x2(:,1),x2(:,2)); xlabel('Time (seconds)','FontSize',14);
ylabel('State Variable x _ 2','FontSize',14);
title('Solution of van der Pol Equation: x _ 2(t)','FontSize',14);
```

as well as

```
plot(x12(:,1),x12(:,2),x12(:,1),x12(:,3));
xlabel('Time (seconds)','FontSize',14);
ylabel('State Variables x _ 1 and x _ 2','FontSize',14);
title('Solution of van der Pol Equation: x _ 1(t) and x _ 2(t)', 'FontSize',14);
```

the resulting plots are illustrated in Figure 2.32.

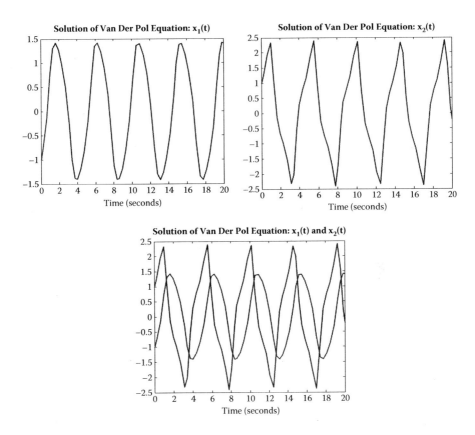

FIGURE 2.32
Dynamics of the state variables $x_1(t)$ and $x_2(t)$.

Many illustrative and valuable examples are given in the MATLAB and Simulink demos. The van der Pol equations simulations are covered. In the MATLAB demo, the following differential equations,

$$\frac{dx_1(t)}{dt} = x_2, \quad \frac{dx_2(t)}{dt} = -x_1 + x_2 - x_1^2 x_2$$

is simulated. □

All demo Simulink models can be modified for the specific problems to be solved. To start, stop, or pause the simulation, the Start, Stop, and Pause buttons are available in the Simulation menu (Start, Stop, and Pause buttons can be clicked using the toolbar commands as well). One can open Aerospace, Real Time Workshop, SimMechanics, and other toolboxes as documented in the Simulink demo features in Figure 2.30. The application-specific built-in blocks are available in the Simulink Library Browser. For example, the Commonly Used Blocks, Continuous, and Sources, which were used to build the Simulink diagram reported in Figure 2.31, are illustrated in

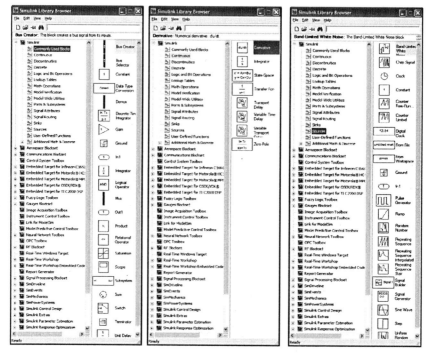

FIGURE 2.33
Commonly used Blocks, Continuous, and Sources Simulink libraries.

Figure 2.33. The designer can utilize other libraries depending on the problem under the consideration. One selects and drags those blocks to the Simulink diagram, and, then connects the blocks in order to perform simulations.

Example 2.31

Simulate the system described by the following two nonlinear differential equations

$$\frac{dx_1(t)}{dt} = -k_1 x_1 - k_2 x_2 + k_3 x_2^3 + k_4 \sin(k_5 x_1 + \pi) + k_6 x_1 x_2 + u(t), \, x_1(t_0) = x_{10},$$

$$\frac{dx_2(t)}{dt} = k_7 x_1, \, x_2(t_0) = x_{20}.$$

The input $u(t)$ is a train of periodic steps, and $u(t) = u_0 \text{rect}(\omega_0 t)$. Assume the amplitude is $u_0 = 0$ or $u_0 = 4$, and the frequency is $\omega_0 = 1$ rad/sec. The coefficients and initial conditions are $k_1 = 2$, $k_2 = 3$, $k_3 = -4$, $k_4 = -5$, $k_5 = 6$, $k_6 = -7$, $k_7 = 8$, $x_{10} = 2$, and $x_{20} = -2$.

We will use the Signal Generator, Sum, Gain, Integrator, Function, Mux, and Scope blocks. These blocks are dragged from the Simulink block libraries to the untitled mdl model, positioned (placed), and connected using the lines as shown in Figure 2.34. That is, by connecting the blocks

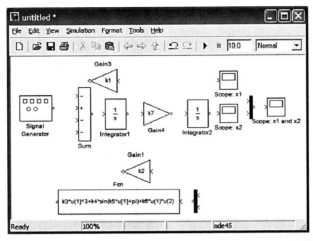

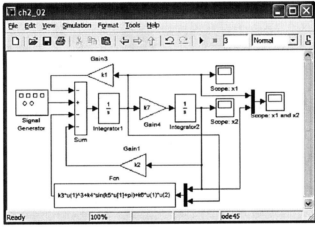

FIGURE 2.34
Simulink block diagram to simulate a set of two differential equations (`ch2 _ 02.mdl`).

and typing the coefficients and nonlinear term $k_3 x_2^3 + k_4 \sin(k_5 x_1 + \pi) + k_6 x_1 x_2$ in the Function block, the Simulink block diagram to be used results as shown in Figure 2.34.

The differential equations parameters and initial conditions are uploaded by typing in the Command window

```
k1=2; k2=3; k3=-4; k4=-5; k5=6; k6=-7; k7=8; u0=0; w0=1; x10=2; x20=-2;
```

The Signal Generator, Integrator, and Function blocks are used to generate the input $u(t)$, set up the specified initial conditions in the integrators, as well as implement the nonlinear functions. These blocks are illustrated in Figure 2.35.

For $u_0=0$, specifying the simulation time to be 2 sec, the transient behavior of the states $x_1(t)$ and $x_2(t)$, as provided by three scopes, are depicted in Figure 2.36.

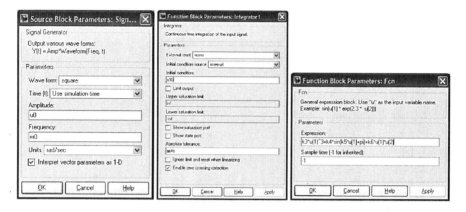

FIGURE 2.35
Signal Generator, Integrator, and Function blocks.

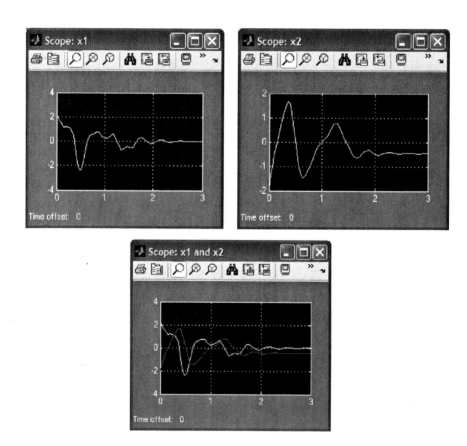

FIGURE 2.36
Simulation results displayed in the Scopes.

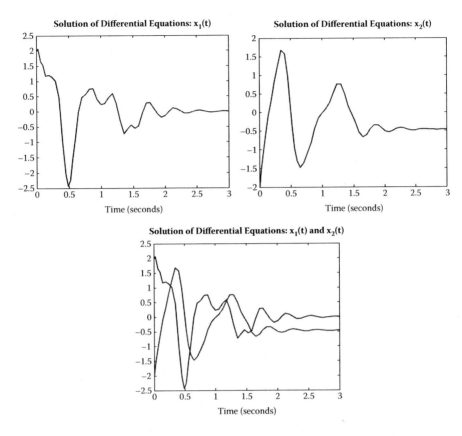

FIGURE 2.37
Dynamics of the state variables $x_1(t)$ and $x_2(t)$, $u_0 = 0$.

The plotting statements can be used. In the scopes, in the General/Data History icon (the second icon from the left), we assign the variable names as x1, x2, and x12. The statements

```
plot(x1(:,1),x1(:,2)); xlabel('Time (seconds)','FontSize',14);
ylabel('State Variable x _ 1','FontSize',14);
title('Solution of Differential Equations: x _ 1(t)','FontSize',14);
```

and

```
plot(x2(:,1),x2(:,2)); xlabel('Time (seconds)','FontSize',14);
ylabel('State Variable x _ 2','FontSize',14);
title('Solution of Differential Equations: x _ 2(t)','FontSize',14);
```

as well as

```
plot(x12(:,1),x12(:,2),x12(:,1),x12(:,3));
xlabel('Time (seconds)','FontSize',14);
ylabel('State Variables x _ 1 and x _ 2','FontSize',14);
title('Solution of Differential Equations: x _ 1(t) and x _ 2(t)','FontSize',14);
```

are used. The resulting plots are illustrated in Figure 2.37.

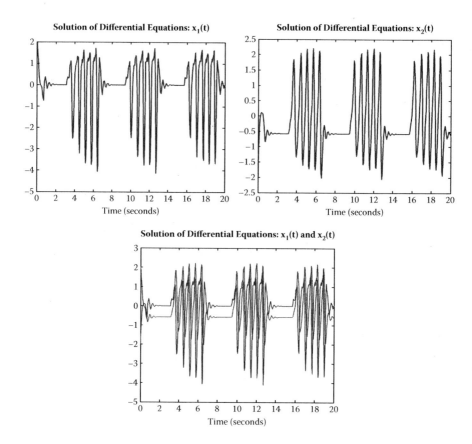

FIGURE 2.38
Dynamics of the state variables $x_1(t)$ and $x_2(t)$, $u_0 = 4$.

For $u_0 = 4$, the transient dynamics are depicted in Figure 2.38. The reader assesses the impact of input on the system behavior. The obtained dynamics can be examined from the prospects of stability. □

Homework Problems

Problem 2.1

A force $\mathbf{F} = 3\mathbf{i} + 2\mathbf{j} + 4\mathbf{k}$ acts through the point with position vector $\mathbf{r} = 2\mathbf{i} + \mathbf{j} + 3\mathbf{k}$. Derive a torque about a perpendicular axis. That is, find $\mathbf{T} = \mathbf{r} \times \mathbf{F}$.

Problem 2.2

A spherical electrostatic actuator, as documented in Figure 2.39, is designed using spherical conducting shells separated by the flexible material (for

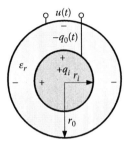

FIGURE 2.39
Electrostatic actuator.

example parylene, Teflon, and polyethylene have relative permittivity ~3). The inner shell has total charge $+q_i$ and diameter r_i. The charge of the outer shell $q_0(t)$ is seminegative and can be time-varying. The diameter of the outer shell is denoted as r_0.

2.2.1. For a spherical actuator, derive the expression for the capacitance $C(r)$. Calculate the numerical value for capacitance if $r_i = 1$ cm, $r_0 = 1.5$ cm, $q_i = 1$ C and $q_0(t) = [\sin(t) + 1]$ C.

2.2.2. Derive the expression for the electrostatic force using the coenergy $W_c[u, C(r)] = C(r)u^2/2$. Recall that the force is

$$F_e(u, r) = \frac{\partial W_c[u, C(r)]}{\partial r}.$$

Calculate the electrostatic force between the inner and outer shells assigning the various values for the applied voltage u which could be up to ~1000 V.

2.2.3. For a flexible materials (parylene, Teflon, or polyethylene), find the resulting displacements. Use the expression for the restoring force of the flexible media chosen. The approximation $F_s = k_s r$ can be applied, where $k_s = 1$ N/m.

2.2.4. Develop the differential equations that describe the spherical actuator dynamics. Examine the actuator performance and capabilities. The simulation in MATLAB should be performed.

Problem 2.3

The power generation device is illustrated in Figure 2.40. Assume that the magnetic field is uniform, and the top and/or low magnets or current loop vibrate such as the magnetic field relative the current loop vary as $B(t) = \sin(t) + \sin(2t) + \cos(5t) + \sin(t)\cos^2(t)$. That is, permanent magnets are stationary, whereas a

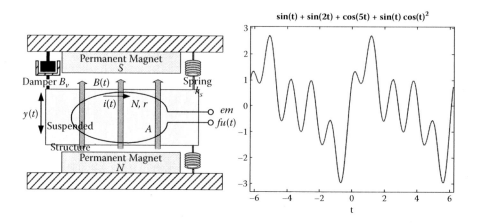

FIGURE 2.40
Power generation device and assumed variations of the magnetic field.

current loop is on the movable (suspended) structure. The variations of the magnetic field are shown in Figure 2.40. The area of the conducting current loop in the magnetic field is 1 mm². The resistance is 1 ohm.

2.3.1. Find the induced *emf* (generated voltage) and the current in the current loop. Assume that the number of turns N can be 1 or 100. Study how the *emf* changes if $N = 1$ and $N = 100$. The MATLAB Partial Differential Toolbox can be used to plot the resulting variables, perform differentiation, calculations, plotting, and so on. In particular, the MATLAB statement is

```
t=sym('t','positive');
B=sin(t)+sin(2*t)+cos(5*t)+sin(t).*cos(t)^2;
ezplot(B); dBdt=diff(B)
```

The resulting plot for the magnetic field and derivative for the magnetic field density are found. In the Command Window we have the result

```
dBdt = cos(t)+2*cos(2*t)-5*sin(5*t)+cos(t)^3-2*sin(t)^2*cos(t)
```

2.3.2. Assign the dimensions of the device to derive the parameters of the power generation system (mass of the suspended structure with the current loop m, spring constant k_s, etc.).

2.3.3. Derive differential equations for the device.

2.3.4. Simulate and examine the power generation system performance.

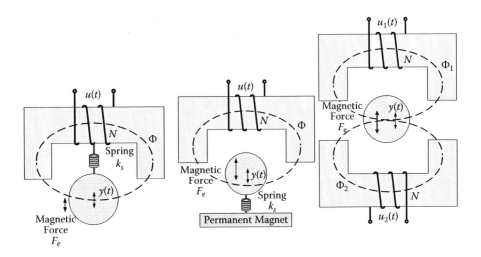

FIGURE 2.41
Magnetic levitation systems.

Problem 2.4

One-, two-, and three-dimensional magnetic levitation systems have been foreseen to be used in control of underwater and flight vehicles (so-called moving mass concept). Possible magnetic levitation systems with a ball are illustrated in Figure 2.41.

2.4.1. For the chosen device, derive a mathematical model by applying Newton's and Kirchhoff's laws. That is, derive the differential equations which describe the system dynamics. Find the expression for the electromagnetic force developed by using the coenergy. Note that the reluctance in the magnetic system varies. Develop the equivalent magnetic circuit.

2.4.2. Assign magnetic levitation system dimensions and derive the parameters. For example, assume the total length of the magnetic path is ~0.1 m. Assuming that the diameter of copper wire is 1 mm, one layer winding can include ~10 turns, but you may have multilayered winding. The geometry (shape) and diameter of the moving mass (ball) and its media density will result in the value for m, A, μ_r, and so on.

2.4.3. In MATLAB, develop the files and perform numerical simulations of the studied magnetic levitation system. Analyze the dynamics and assess the performance.

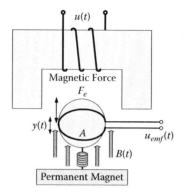

FIGURE 2.42
Magnetic levitation systems with the conducting current loop.

2.4.4. Examine what happens if in the suspended mass (coated with an insulator) will be surrounded by N turns conducting current loop as illustrated in Figure 2.42. Assume that the current loop has a resistance r and area A.

Problem 2.5

Let for a solenoid the inductance is $L(x) = 4\sin(3x)$. Derive the expression for the electromagnetic force and calculate the force for $i = 10$ A at $x = 0$.

Problem 2.6

Consider a solenoid as documented in Figure 2.18 and discussed in Example 2.17. This solenoid integrates movable (plunger) and stationary (fixed) members made from high-permeability ferromagnetic materials. The windings wound with a helical pattern. The performance of solenoids depends on the electromagnetic system, mechanical geometry, magnetic permeability, winging resistivity, inductance, friction, and so on.

2.6.1. Assign the dimensions of solenoid (say ~5 cm length and ~2 cm outer diameter) and make the estimates for all other dimensions. Chose the materials that can be used in the solenoid emphasizing high performance and capabilities.

2.6.2. Derive the magnetizing inductance, winding resistance, and other parameters of interest. Estimate friction coefficient, relative permeabilities of movable and stationary members, resistivity, spring constant, and so on.

2.6.3. Derive a mathematical model by applying Newton's and Kirchhoff's laws. Find the expression for the electromagnetic force developed using coenergy.

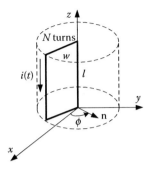

FIGURE 2.43
Rotor winding in the magnetic field.

2.6.4. In MATLAB, develop the files and perform numerical simulations of the studied solenoid. Analyze the results examining steady-state and dynamic performance.

Problem 2.7

Figure 2.43 illustrates a motion devise with windings on rotor. The electromagnetic torque is developed due to the interaction of the 20-turn rectangular coil (winding) in the yz plane and stationary magnetic field (established by magnets on stator windings).

2.7.1. Determine the magnetic moment and derive the electromagnetic torque acting on the coil. Assume the current in the coil ($l = 15$ cm and $w = 5$ cm) is $i = 10$ A, and the magnetic field density is

$$\mathbf{B} = 2 \times 10^{-2} (\mathbf{a}_x + 2\mathbf{a}_y) \text{ T.}$$

2.7.2. Derive at what angle ϕ $T_e = 0$. At what angle ϕ T_e is maximum? Determine the value of $T_{e\max}$. Derive the solution from: (a) physical (electromagnetic) viewpoint; (b) mathematical (mini-max problem) viewpoint.

References

1. S.J. Chapman, *Electric Machinery Fundamentals*, McGraw-Hill, New York, 1999.
2. A.E. Fitzgerald, C. Kingsley, and S.D. Umans, *Electric Machinery*, McGraw-Hill, New York, 1990.
3. P.C. Krause and O. Wasynczuk, *Electromechanical Motion Devices*, McGraw-Hill, New York, 1989.

4. P.C. Krause, O. Wasynczuk, and S.D. Sudhoff, *Analysis of Electric Machinery*, IEEE Press, New York, 1995.

5. W. Leonhard, *Control of Electrical Drives*, Springer, Berlin, 1996.

6. S.E. Lyshevski, *Electromechanical Systems, Electric Machines, and Applied Mechatronics*, CRC Press, Boca Raton, FL, 1999.

7. D.W. Novotny and T.A. Lipo, *Vector Control and Dynamics of AC Drives*, Clarendon Press, Oxford, 1996.

8. G.R. Slemon, *Electric Machines and Drives*, Addison-Wesley Publishing Company, Reading, MA, 1992.

9. D.W. Hart, *Introduction to Power Electronics*, Prentice Hall, Upper Saddle River, NJ, 1997.

10. J.G. Kassakian, M.F. Schlecht, and G.C. Verghese, *Principles of Power Electronics*, Addison-Wesley Publishing Company, Reading, MA, 1991.

11. N.T. Mohan, M. Undeland, and W.P. Robbins, *Power Electronics: Converters, Applications, and Design*, John Wiley and Sons, New York, 1995.

12. A.S. Sedra and K.C. Smith, *Microelectronic Circuits*, Oxford University Press, New York, 1997.

13. J. Fraden, *Handbook of Modern Sensors: Physics, Design, and Applications*, AIP Press, Woodbury, NY, 1997.

14. G.T.A. Kovacs, *Micromachined Transducers Sourcebook*, McGraw-Hill, New York, 1998.

15. R.C. Dorf and R.H. Bishop, *Modern Control Systems*, Addison-Wesley Publishing Company, Reading, MA, 1995.

16. J.F. Franklin, J.D. Powell, and A. Emami-Naeini, *Feedback Control of Dynamic Systems*, Addison-Wesley Publishing Company, Reading, MA, 1994.

17. B.C. Kuo, *Automatic Control Systems*, Prentice Hall, Englewood Cliffs, NJ, 1995.

18. S.E. Lyshevski, *Control Systems Theory with Engineering Applications*, Birkhäuser, Boston, MA, 2001.

19. K. Ogata, *Discrete-Time Control Systems*, Prentice-Hall, Upper Saddle River, NJ, 1995.

20. K. Ogata, *Modern Control Engineering*, Prentice-Hall, Upper Saddle River, NJ, 1997.

21. C.L. Phillips and R. D. Harbor, *Feedback Control Systems*, Prentice Hall, Englewood Cliffs, NJ, 1996.

22. S.E. Lyshevski, *Engineering and Scientific Computations Using MATLAB®*, Wiley, Hoboken, NJ, 2003.

23. *MATLAB R2006b*, CD-ROM, MathWorks, Inc., 2007.

3

Introduction to Power Electronics

3.1 Operational Amplifiers

In electromechanical systems, signal processing, signal conditioning, and other tasks are accomplished by ICs. The digital controllers and filters are implemented using microcontrollers and DSPs. Electromechanical systems are predominantly continuous, and analog controllers and filters can be implemented utilizing operational amplifiers and specialized ICs [1]. These controllers and filters are imbedded within power amplifiers. This section examines and documents the use of operational amplifiers to implement analog controller and filters with specified transfer functions. It was documented in Chapter 1 that controllers are used, and various operations on signals and variables may be required. For example, sensors convert time-varying physical quantities (displacement, velocity, acceleration, force, torque, pressure, temperature, and so on) in electric signals (voltage or current). The signal-level sensor output must be amplified and filtered. The single operational amplifier is a two-port network, which has *noninverting* and *inverting* input terminals (3 and 2) as well as one output terminal (6) as depicted in Figure 3.1. Two (or one) DC voltages are needed, and terminal 7 is connected to a positive voltage u_+, whereas a negative voltage (or ground) u_- is supplied to the terminal 4. The pin connections of the single, dual, and quad low-power operational amplifiers MC33171, MC33172, and MC33174 are reported in Figure 3.1, which also illustrates 8- and 14-pin plastic packages (cases 626 and 646). These operational amplifiers are also available in the surface mount packages (cases 751 and 948). Operational amplifiers, which consist of dozens of transistors, are fabricated using the complementary metal-oxide-semiconductor (CMOS) or bipolar-CMOS fabrication technologies [1]. Figure 3.1 depicts the representative schematic diagram.

The amplifier output is the difference between two input voltages, $u_1(t)$ and $u_2(t)$, applied to the *inverting* input terminal and the *noninverting* input terminal, multiplied by the differential open-loop coefficient k_{og}. That is, the resulting output voltage is

$$u_0(t) = k_{og}[u_2(t) - u_1(t)].$$

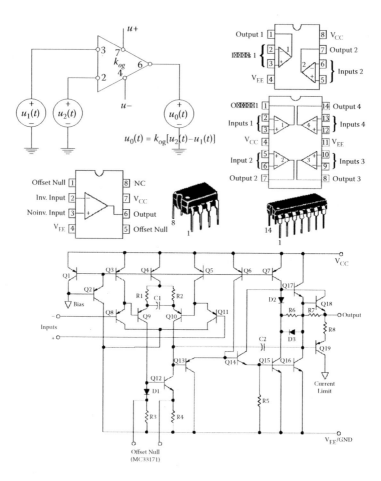

FIGURE 3.1
Operational amplifiers, pin connections, packages, and representative schematics (copyright of Motorola, used with permission) [2].

The differential open-loop coefficient is positive. The value of k_{og} is very large, and k_{og} varies as ~$[1 \times 10^5 \quad 1 \times 10^7]$. The general purpose operational amplifiers have input and output resistances ~$[1 \times 10^5 \quad 1 \times 10^{12}]$ and ~$[10 \quad 1000]$ ohm, respectively.

The *inverting* and *noninverting* input terminals are distinguished using "−" and "+" signs. Supplying the signal-level input voltage $u_1(t)$ to the *inverting* input terminal using external resistor R_1, and grounding the *noninverting* input terminal, one can find the differential closed-loop coefficient k_{cg} if a negative feedback is used. The output terminal is connected to the *inverting* input terminal, and the resistor R_2 is inserted as depicted in Figure 3.2.

To find the differential closed-loop coefficient k_{cg}, one has to obtain the ratio between the output and input voltages $u_0(t)$ and $u_1(t)$. The voltage between two input terminals is $u_0(t)/k_{og}$, and the voltage at the *inverting* input terminal

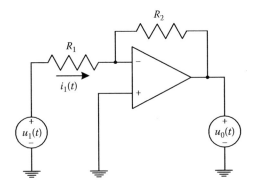

FIGURE 3.2
Inverting configuration of the operational amplifier.

is $u_0(t)/k_{og}$ because the *noninverting* input terminal is grounded. Therefore, we have

$$i_1(t) = \frac{u_1(t) + \dfrac{u_0(t)}{k_{og}}}{R_1},$$

and the output voltage is

$$u_0(t) = -\frac{u_0(t)}{k_{og}} - \frac{u_1(t) + \dfrac{u_0(t)}{k_{og}}}{R_1} R_2.$$

Hence, the differential closed-loop coefficient is found to be

$$k_{cg} = \frac{u_0(t)}{u_1(t)} = -\frac{\dfrac{R_2}{R_1}}{1 + \dfrac{1}{k_{og}} + \dfrac{R_2}{k_{og}R_1}}.$$

The configuration studied allows one to invert the signal-level input signal. Taking note that the differential open-loop coefficient k_{og} is very large (hundred thousands), we have

$$k_{cg} = \frac{u_0(t)}{u_1(t)} \approx -\frac{R_2}{R_1}.$$

The inverting summing amplifier (the so-called weighted summer) is shown in Figure 3.3.
We have an m input signals. In particular, the applied voltages are $u_{1,1}(t),\ldots,u_{1,m}(t)$. The currents $i_{1,1}(t),\ldots,i_{1,m}(t)$ are found to be $i_{1,1}(t) = u_{1,1}(t)/R_{1,1},\ldots,$

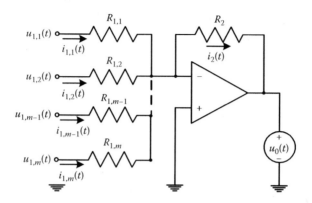

FIGURE 3.3
Summing amplifier.

$i_{1,m}(t) = u_{1,m}(t)/R_{1,m}$. The current in the feedback path is given by $i_2(t) = i_{1,1}(t) + \cdots + i_{1,m}(t)$. Hence, the amplifier output is

$$u_0(t) = -\left(\frac{R_2}{R_{1,1}} u_{1,1}(t) + \frac{R_2}{R_{1,2}} u_{1,2}(t) + \cdots + \frac{R_2}{R_{1,m-1}} u_{1,m-1}(t) + \frac{R_2}{R_{1,m}} u_{1,m}(t) \right).$$

To generalize the results derived studying the inverting configuration, we use the input impedance $Z_1(s)$ and the feedback path impedance $Z_2(s)$ as illustrated in Figure 3.4. The impedance is the ratio of the phasor voltage to the phasor current at the two terminals. In particular, the impedances of the resistor, capacitor, and inductor are

$$Z_R(s) = R, \quad Z_R(j\omega) = R, \quad Z_C(s) = \frac{1}{sC}, \quad Z_C(j\omega) = \frac{1}{j\omega C}, \quad \text{and}$$

$$Z_L(s) = sL, \quad Z_L(j\omega) = j\omega L.$$

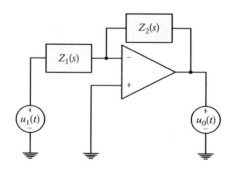

FIGURE 3.4
Inverting configuration of the operational amplifier with impedances $Z_1(s)$ and $Z_2(s)$.

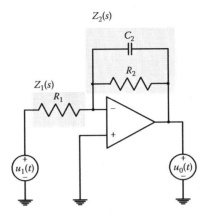

FIGURE 3.5
Inverting operational amplifier with $Z_1(s)$ and $Z_2(s)$.

The transfer function of the closed-loop amplifier configuration is

$$G(s) = \frac{U_0(s)}{U_1(s)} = -\frac{Z_2(s)}{Z_1(s)}.$$

As an example, consider the operational amplifier as shown in Figure 3.5. The transfer function is found by using the expressions for impedances

$$Z_1(s) = R_1 \quad \text{and} \quad Z_2(s) = \frac{R_2}{R_2 C_2 s + 1}.$$

We have

$$G(s) = \frac{U_0(s)}{U_1(s)} = -\frac{Z_2(s)}{Z_1(s)} = -\frac{R_2/R_1}{R_2 C_2 s + 1}.$$

The close-loop gain coefficient for the inverting operational amplifier studied is

$$k_{cg} = -\frac{R_2}{R_1},$$

whereas the time constant is $R_2 C_2$.

In the frequency domain, by substituting $s = j\omega$, we have

$$G(j\omega) = \frac{U_0(j\omega)}{U_1(j\omega)} = -\frac{Z_2(j\omega)}{Z_1(j\omega)} = -\frac{R_2/R_1}{R_2 C_2 j\omega + 1},$$

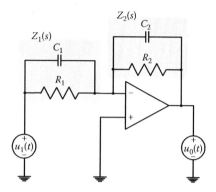

FIGURE 3.6
Inverting operational amplifier with $Z_1(s) = \frac{R_1}{(R_1C_1s+1)}$ and $Z_2(s) = \frac{R_2}{(R_2C_2s+1)}$.

and the Bode plots of the resulting low-pass filter can be easily found and plotted.

Filters with specific transfer functions are used to ensure the noise immunity performing the signal conditioning of signals, which are utilized to implement control algorithms or any other functions. In fact, sensor signals contain the noise which is due to different origins. The low, medium, and high frequency noise can be attenuated by filters. The filters must be properly designed and implemented. The possible filter configuration is demonstrated in Figure 3.6.

The transfer function is found using the input and feedback impedances. In particular,

$$G(s) = \frac{U_0(s)}{U_1(s)} = -\frac{Z_2(s)}{Z_1(s)} = -\frac{\frac{R_2}{R_2C_2s+1}}{\frac{R_1}{R_1C_1s+1}} = -\frac{\frac{R_2}{R_1}(R_1C_1s+1)}{R_2C_2s+1}.$$

Different transfer functions can be implemented by operational amplifiers utilizing passive elements in the input and feedback paths. Operational amplifiers are widely used to implement analog control laws. An inverting integrator is obtained by placing a capacitor C_2 in the feedback path (see Figure 3.7). The resulting transfer function is

$$G(s) = \frac{U_0(s)}{U_1(s)} = -\frac{Z_2(s)}{Z_1(s)} = -\frac{1}{R_1C_2s}.$$

Denoting the initial value of the capacitor voltage as $u_C(t_0)$, the amplifier output voltage is

$$u_0(t) = -u_C(t_0) - \frac{1}{R_1C_2}\int_{t_0}^{t_f} u_1(\tau)d\tau.$$

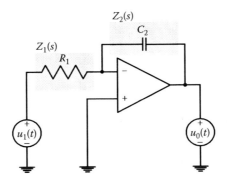

FIGURE 3.7
Inverting integrator.

The operational differentiator performs the differentiation of the input signal. The current through the input capacitor is

$$C_1 \frac{du_1(t)}{dt}$$

(see Figure 3.8). That is, the output voltage is proportional to the derivative of the input voltage with respect to time. Hence,

$$u_0(t) = -R_2 C_1 \frac{du_1(t)}{dt}.$$

The transfer function is

$$G(s) = \frac{U_0(s)}{U_1(s)} = -\frac{Z_2(s)}{Z_1(s)} = -R_2 C_1 s.$$

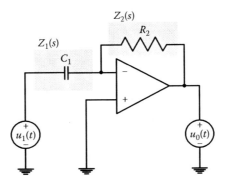

FIGURE 3.8
Inverting differentiator.

For various input and feedback path impedances $Z_1(s)$ and $Z_2(s)$, Table 3.1 provides the transfer functions of the inverting operational amplifier. The passive elements are utilized in the input and feedback paths.

The proportional-integral-derivative (PID) controller can be implemented using the configuration as depicted in Figure 3.9.

The transfer function realized by an inverting operational amplifier is

$$G(s) = \frac{U_0(s)}{U_1(s)} = -\frac{(R_1C_1s+1)(R_2C_2s+1)}{R_1C_2s} = -\frac{R_2C_1s^2 + \frac{R_1C_1+R_2C_2}{R_1C_2}s + \frac{1}{R_1C_2}}{s}.$$

The transfer function of the PID control law $G_{PID}(s)$ is given as

$$G_{PID}(s) = k_p + \frac{k_i}{s} + k_d s = \frac{k_d s^2 + k_p s + k_i}{s}.$$

For the derived transfer function $G(s)$, k_p, k_i, and k_d are defined by the values of resistors and capacitors of the input and feedback paths. In particular,

$$k_p = -\frac{R_1C_1 + R_2C_2}{R_1C_2}, \quad k_i = -\frac{1}{R_1C_2} \quad \text{and} \quad k_d = -R_2C_1.$$

To guarantee the stability, the proportional, integral, and derivative feedback gains (k_p, k_i and k_d) must be positive. As a result of the minus signs, these proportional, integral, and derivative feedback should be inverted. Hence, an additional inverting amplifier is needed. In practice, the configuration, which is illustrated in Figure 3.10, is commonly used to implement the PID controller.

One obtains the expression for the overall transfer function as

$$G(s) = \frac{U_0(s)}{U_1(s)} = \frac{R_{2p}}{R_{1p}} + \frac{1}{R_{1i}C_{2i}s} + R_{2d}C_{1d}s.$$

The proportional, integral, and derivative feedback coefficients are

$$k_p = \frac{R_{2p}}{R_{1p}}, \quad k_i = \frac{1}{R_{1i}C_{2i}} \quad \text{and} \quad k_d = R_{2d}C_{1d}.$$

The catalog data, schematics, characteristics, frequency response, and transient dynamics for low power, single supply, dual, and quad operational amplifiers manufactured by a great number of companies are available. The possible configurations (*noninverting*, *inverting*, notch, bandpass, and others) are easily examined.

TABLE 3.1

Transfer Functions of the Inverting Amplifier

Input Circuit with Impedance $Z_1(s)$	Feedback Circuit with Impedance $Z_2(s)$	Transfer Function
$Z_1(s)$ R_1 $Z_1(s) = R_1$	$Z_2(s)$ R_2 $Z_2(s) = R_2$	$G(s) = \dfrac{U_0(s)}{U_1(s)} = -\dfrac{R_2}{R_1}$
$Z_1(s)$ R_1 $Z_1(s) = R_1$	$Z_2(s)$ C_2 $Z_2(s) = \dfrac{1}{C_2 s}$	$G(s) = \dfrac{U_0(s)}{U_1(s)} = -\dfrac{1}{R_1 C_2 s}$
$Z_1(s)$ R_1 $Z_1(s) = R_1$	$Z_2(s)$ C_2 R_2 $Z_2(s) = \dfrac{R_2}{R_2 C_2 s + 1}$	$G(s) = \dfrac{U_0(s)}{U_1(s)} = -\dfrac{\dfrac{R_2}{R_1}}{R_2 C_2 s + 1}$
$Z_1(s)$ R_1 $Z_1(s) = R_1$	$Z_2(s)$ R_2 C_2 $Z_2(s) = \dfrac{R_2 C_2 s + 1}{C_2 s}$	$G(s) = \dfrac{U_0(s)}{U_1(s)} = -\dfrac{R_2 C_2 s + 1}{R_1 C_2 s}$
$Z_1(s)$ C_1 $Z_1(s) = \dfrac{1}{C_1 s}$	$Z_2(s)$ R_2 $Z_2(s) = R_2$	$G(s) = \dfrac{U_0(s)}{U_1(s)} = -R_2 C_1 s$
$Z_1(s)$ C_1 R_1 $Z_1(s) = \dfrac{R_1}{R_1 C_1 s + 1}$	$Z_2(s)$ R_2 $Z_2(s) = R_2$	$G(s) = \dfrac{U_0(s)}{U_1(s)} = -\dfrac{R_1 R_2 C_1 s + R_2}{R_1}$

(Continued)

TABLE 3.1

Transfer Functions of the Inverting Amplifier (Continued)

Input Circuit with Impedance $Z_1(s)$	Feedback Circuit with Impedance $Z_2(s)$	Transfer Function
$Z_1(s)$ C_1 R_1 $Z_1(s) = \dfrac{R_1}{R_1C_1s + 1}$	$Z_2(s)$ R_2 C_2 $Z_2(s) = \dfrac{R_2C_2\,s + 1}{C_2s}$	$G(s) = \dfrac{U_0(s)}{U_1(s)} = -\dfrac{(R_1C_1s + 1)(R_2C_2s + 1)}{R_1C_2s}$
$Z_1(s)$ C_1 R_1 $Z_1(s) = \dfrac{R_1}{R_1C_1s + 1}$	$Z_2(s)$ C_2 R_2 $Z_2(s) = \dfrac{R_2}{R_2C_2\,s + 1}$	$G(s) = \dfrac{U_0(s)}{U_1(s)} = -\dfrac{\dfrac{R_2}{R_1}(R_1C_1s + 1)}{R_2C_2s + 1}$

3.2 Power Amplifiers and Power Converters

3.2.1 Power Amplifier and Analog Controllers

Power amplifiers allow one to perform the amplification. Various classes of amplifiers are available. The most important part of power amplifiers is the output stage which deals with relatively large voltages and currents. For example, for 100 W (rated), 50 V permanent-magnet motors, the rated current is ~2 A, whereas the pear current could be ~20 A. The power dissipated in the output stage power transistors should be minimized to guarantee

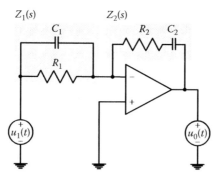

FIGURE 3.9
Implementation of the PID control law using an inverting operational amplifier.

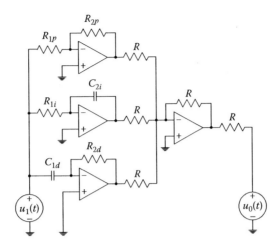

FIGURE 3.10
Implementation of an analog PID control law.

the efficiency. Depending upon the output stage topologies and operating concepts, output stages are classified as A, B, AB, C, and D classes. In electromechanical systems, D-class output stages are commonly used due to high efficiency, simplicity, reliability, low harmonic distortion, and so on. A possible largely simplified configuration with a standard push-pull D-class output stage to control the permanent-magnet DC motor is illustrated in Figure 3.11. The output stage topology and circuitry are much more complex as illustrated in Figures 3.12 and 3.13. The designer faces with many

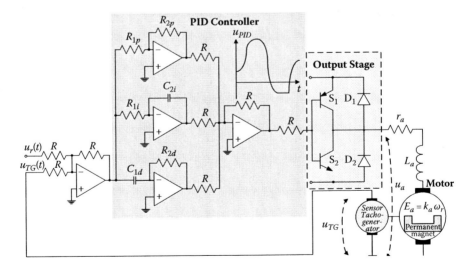

FIGURE 3.11
Application of the D-class power amplifier to control an electric drive with a permanent-magnet DC motor: Closed-loop configuration with a PID controller and sensor (tachogenerator) to measure the angular velocity ω_r.

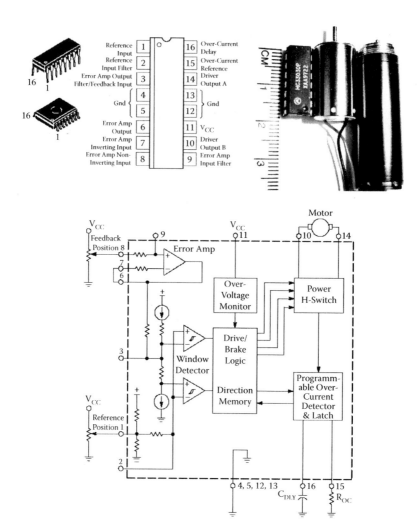

FIGURE 3.12
Pin connection and block diagram of the MC33030 DC servo motor controller/driver with permanent-magnet DC motor (copyright of Motorola, used with permission) [2].

challenges such as the back *emf*, current and voltage ripples, switching frequency matching the winding inductances, and so on. Figure 3.11 is reported from the illustrative purposes and is not applicable in practical applications as reported covering MC33030 DC servo motor controller/driver and one-quadrant converters. Using the PID controller, one switches the transistors, varying the average voltage u_a applied to the armature motor winding. The angular velocity ω_r is measured by a tachogenerator, and the measured angular velocity is compared with the desired velocity. The error between

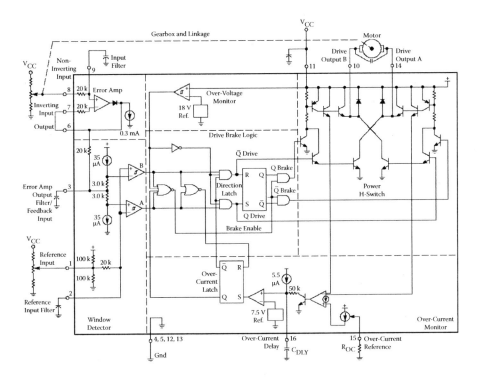

FIGURE 3.13
Schematics of the MC33030 DC servo motor controller/driver (copyright of Motorola, used with permission) [2].

two voltages (which correspond to angular velocities) is used as an input to the PID controller, which also can implement P, PI, and PD control laws. Usually, PI controllers are used because of the sensitivity of derivative feedback to the noise, which in practice may not be attenuated utilizing filters.

If the signal-level voltage u_{PID} of the output inverting operational amplifier of the PID controller is positive, transistor S_1 is *off* because the voltage between the base and emitter is zero. Transistor S_2 is *on*, and the negative voltage u_a is applied to the motor. If the voltage u_{PID} is negative, the S_1 is *on*, and S_2 is *off*, and the supplied voltage u_a is positive. The diodes D_1 and D_2 prevent damage to the transistors by the *back emf* from the motor. This simple D-class power stage can be modified to ensure the pulse-width-modulation (PWM) concept to vary the average value of u_a thereby controlling the angular velocity. The tachogenerator is used as a sensor to measure the angular velocity to be fed to the operational amplifier which compares the desired angular velocity $u_r(t)$ with the actual angular velocity of the motor measured as the voltage induced by the tachogenerator $u_{TG}(t)$.

The basic dc-dc power converters used in electromechanical systems are the switching converters. In particular, *buck, boost, buck-boost,* Cuk and other converters are commonly applied [2–5]. These converters should be considered and studied departing from the simplified schematic reported in Figure 3.11.

Small motors can be controlled using monolithic PWM amplifiers with corresponding ICs. For example, the MC33030 DC servo motor controller/ driver integrates on-chip operational amplifier and comparator, driving logics, PWM four-quadrant converter, and so on. The rated (peak) output voltage and current are 36 V and 1 A. Hence, one can use MC33030 for small ~10 W DC motors. It should be emphasized that for a short period of time permanent-magnet DC and synchronous electric machines can operate at higher voltage and current. For motors, T_{epeak}/T_{erated} and P_{peak}/P_{rated} could be ~10, e.g., the ratio i_{peak}/i_{rated} is ~10. However, for power electronics the ratio I_{peak}/I_{rated} is up to ~2. Therefore, power electronics should accommodate the peak motor current within the specific operating envelope. The monolithic MC33030 servo-motor driver contains 119 active transistors, and the catalog data provides a great deal of details with explicit description. The difference between the reference and actual angular velocity or displacement, linear velocity or position, is compared by the "error amplifier," and two comparators are used as shown in Figure 3.12 [2]. A *pnp* differential output power stage ensures driving and braking capabilities. The four-quadrant H-configured power stage guarantees high performance and efficiency. For a complete description of the MC33030 motor controller/driver, the reader is referenced to the Web site http://www.motorola.com.

The MC33030, as a PWM amplifier, can be utilized to drive ~10 W permanent-magnet DC motors in electric drives and servos application. For electric drives and servos the reference (command) and output are the angular velocity and displacement, which can be assigned and measured as voltages. Schematics of a drive/servo system with MC33030 is illustrated by a representative block diagram shown in Figure 3.13. One sets a voltage on the reference input (pin 1). The velocity or displacement sensor is installed to measure the output velocity or displacement. A tachogenerator or potentiometer can be used as sensors. The sensor voltage is supplied to pin 3. The reference voltage is compared with the sensor voltage. By using this difference, the system is controlled. For example, one can use the tracking error as was covered in Chapter 1, and $e(t)=r(t)-y(t)$. The "window detector" is composed of A and B comparators with hysteresis. The proportional controller $u(t)=k_p e(t)$ is implemented utilizing the "error amplifier." A *pnp* differential input stage ensures grounding. The four-quadrant power stage provides the armature voltage to rotate the motor. The permanent-magnet DC motor is connected to pins 10 and 14. The current limit is set on pin 15, and the voltage protection is ensured by the "over-voltage monitor," which is important to use because of the induced *back emf*. This schematics can be modified and enhanced by adding additional filters, control, motor, and other circuitry.

The dual power operational amplifiers can be utilized in various output stage topologies. Figure 3.14 documents the image of a 7-pin, 40 V, 1.5 A, heat-sink-mount dual power operational amplifier which can be used in half- and

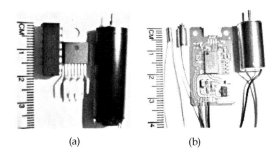

(a)   (b)

FIGURE 3.14
Power electronics and permanent-magnet DC electric machines: (a) MC30330 servo motor controller/driver and dual operational amplifier (~30 W peak) to control the armature voltage applied to permanent-magnet DC electric machines (3 W rated and ~30 W peak); (b) application-specific ICs-PWM amplifier and 2, 4 and 10 mm diameter permanent-magnet synchronous (2 and 4 mm) and DC (10 mm) electric motors.

full-bridge motor drivers. In general, the application-specific ICs (control, filtering, signal conditioning, and other tasks) and output stages need to be designed. For miniscale electric machines, though a state-of-the-art design is performed and enabling CMOS technology nodes are used to fabricate those integrated ICs-PWM amplifiers, the size of minimachines can be less than electronics. For example, 2 and 4 mm diameter electric machines, illustrated in Figure 3.14, are fabricated using CMOS and surface micromachining processes used in microelectronics.

3.2.2 Switching Converter: *Buck* Converter

Using a pulse-width-modulation (PWM) switching concept, the voltage at the load terminal can be effectively regulated ensuring high performance. A high-frequency *buck* (*step-down*) switching converter is shown in Figure 3.15.

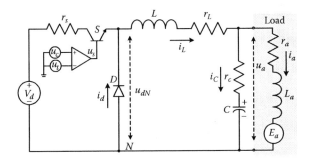

FIGURE 3.15
High-frequency *step-down* switching converter.

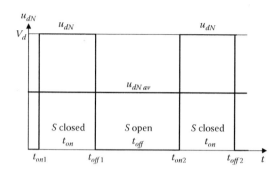

FIGURE 3.16
Voltage waveforms.

In the *step-down* switching converter, as documented in Figure 3.15, the switch S is opened and closed. The switching frequency is given as

$$f = \frac{1}{t_{on} + t_{off}},$$

where t_{on} and t_{off} are the switching *on* and *off* durations.

Assuming that the switch is lossless, the voltage u_{dN} is equal to the supplied voltage V_d when the switch is closed, and the output voltage is zero if the switch is open (see Figure 3.16). One concludes that the voltage u_{dN}, as well as the voltage applied to the load u_a, can be regulated by controlling the switching *on* and *off* durations (t_{on} and t_{off}). The average voltage, applied to the load, depends on t_{on} and t_{off}. In the steady-state, we have

$$u_{dN\ av} = \frac{t_{on}}{t_{on} + t_{off}} V_d = d_D V_d,$$

where d_D is the duty ratio (duty cycle), which is a function of the switching frequency and the time during which the switch is *on*,

$$d_D = \frac{t_{on}}{t_{on} + t_{off}}.$$

One has $d_D \in [0\ 1]$, and $d_D = 0$ if $t_{on} = 0$, while $d_D = 1$ if $t_{off} = 0$.

By changing the duty ratio d_D (switching activity), which is bounded by $d_D \in [0\ 1]$, the average voltage, supplied to the load u_a, is regulated. To establish the PWM switching, one can use a so-called control-triangle concept. The switching signal u_s, which drives the switch, is generated by comparing a signal-level control voltage u_c with a repetitive triangular signal u_t. A comparator is shown in Figures 3.15 and 3.17. The duration of the output pulses u_s represents the *weighted* value between the triangular voltage u_t with the assigned switching frequency and control signal u_c. The output voltage of the comparator u_s drives

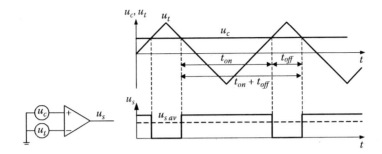

FIGURE 3.17
Comparator in the PWM concept and the voltage waveforms.

the switch S. Hence, the *on* and *off* durations result by comparing u_c and u_t. Figure 3.17 illustrates the voltage waveforms.

The Motorola dual operational amplifier and dual comparator MC3405 is depicted in Figure 3.18. The comparator schematics, operation, and waveforms, explained earlier, are reported.

For the high-frequency *step-down* switching converter (illustrated in Figure 3.15), a low-pass first-order filter with inductance L and capacitance C ensures the specified voltage ripple. We have

$$\frac{\Delta u_a}{u_a} = \frac{1-d_D}{8LCf^2}, \quad L_{min} = \frac{(1-d_D)r_a}{2f}.$$

Switches, inductors, and capacitors have resistances, which are denoted as r_s, r_L and r_c. Two circuits (one with switch closed and one with switch open) are illustrated in Figure 3.19.

Using Kirchhoff's laws, one finds the differential equations to describe the converter dynamics. If the switch is closed, the diode D is reverse biased. For the resulting circuit, as shown in Figure 3.19a, we have the following set of differential equations

$$\frac{du_C}{dt} = \frac{1}{C}(i_L - i_a),$$

$$\frac{di_L}{dt} = \frac{1}{L}(-u_C - (r_L + r_c)i_L + r_c i_a - r_s i_L + V_d),$$

$$\frac{di_a}{dt} = \frac{1}{L_a}(u_C + r_c i_L - (r_a + r_c)i_a - E_a).$$

One must distinguish the state variable u_C (voltage across the capacitor C) from comparator signal-level input u_c, which is the control input.

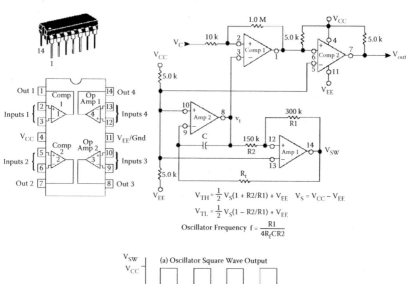

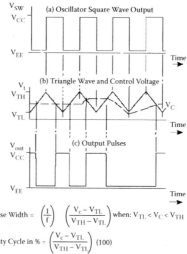

FIGURE 3.18

MC3405 comparator pin connections, schematics, and waveforms (copyright of Motorola, used with permission) [2].

If the switch is open, the diode D becomes forward biased, and $i_d = i_L$ (see Figure 3.19b). One obtains

$$\frac{du_C}{dt} = \frac{1}{C}(i_L - i_a),$$

$$\frac{di_L}{dt} = \frac{1}{L}(-u_C - (r_L + r_c)i_L + r_c i_a),$$

$$\frac{di_a}{dt} = \frac{1}{L_a}(u_C + r_c i_L - (r_a + r_c)i_a - E_a).$$

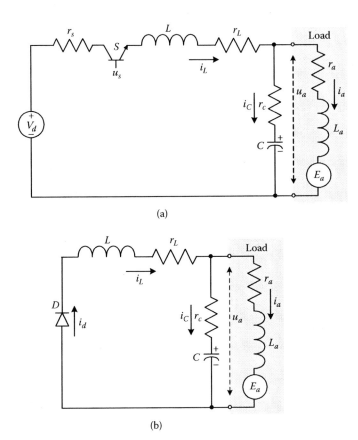

FIGURE 3.19
Circuits of the *buck* dc-dc converter: (a) switch is closed; (b) switch is open.

When the switch is closed, the duty ratio is $d_D = 1$. If the switch is open, the duty ratio is zero, e.g., $d_D = 0$. By using the *averaging* concept, from two sets of differential equations derived, one obtains the following nonlinear differential equations for the *buck* switching converter

$$\frac{du_C}{dt} = \frac{1}{C}(i_L - i_a),$$

$$\frac{di_L}{dt} = \frac{1}{L}(-u_C - (r_L + r_c)i_L + r_c i_a - r_s i_L d_D + V_d d_D),$$

$$\frac{di_a}{dt} = \frac{1}{L_a}(u_C + r_c i_L - (r_a + r_c)i_a - E_a).$$

The duty ratio is regulated by the bounded signal-level control voltage u_c. We obtain

$$d_D = \frac{u_c}{u_{t\,max}} \in [0 \quad 1], \quad u_c \in [0 \quad u_{c\,max}], \quad u_{c\,max} = u_{t\,max}.$$

Neglecting negligible small resistances of the switch, inductor, and capacitor, the analysis of the steady-state performance leads to the expression $\frac{u_a}{V_d} = d_D$.

The signal-level control voltage u_c is a control signal. The converter output is the voltage applied to the load terminal u_a. We have

$$u_a = u_C + r_c i_L - r_a i_a.$$

One concludes that the resulting *buck* converter is described by a set of nonlinear differential equations. In fact, having derived that the duty ratio is

$$d_D = \frac{u_c}{u_{t\,max}},$$

a nonlinear term

$$\frac{r_s}{L} i_L d_D = \frac{r_s}{L u_{t\,max}} i_L u_c$$

is the multiplication of the state variable i_L and control u_c. The hard control limit is imposed because

$$0 \le u_c \le u_{c\,max}, \quad u_c \in [0 \; u_{c\,max}].$$

Example 3.1
Simulate and examine the *step-down* converter steady-state and dynamic behavior applying MATLAB. The converter parameters are: $r_s = 0.025$ ohm, $r_L = 0.02$ ohm, $r_c = 0.15$ ohm, $r_a = 3$ ohm, $C = 0.003$ F, $L = 0.0007$ H, and $L_a = 0.005$ H. The duty ratio is 0.5. The supplied dc voltage is $V_d = 50$ V, and $E_a = 10$ V.

Using the differential equations derived, the following m-files are developed to perform the simulations using the ode45 differential equation solver (command). Two MATLAB files which solve the differential equations and perform plotting are reported here.

MATLAB file (ch3_01.m)

```
t0=0; tfinal=0.03; tspan=[t0 tfinal]; y0=[0 0 0]';
[t,y]=ode45('ch3 _ 02',tspan,y0);
subplot(2,2,1); plot(t,y);
xlabel('Time (seconds)','FontSize',10);
```

```
title('Transient Dynamics of State Variables','FontSize',10);
subplot(2,2,2); plot(t,y(:,1),'-');
xlabel('Time (seconds)','FontSize',10);
title('Voltage u _ C, [V]','FontSize',10);
subplot(2,2,3); plot(t,y(:,2),'-');
xlabel('Time (seconds)','FontSize',10);
title('Current i _ L, [A]','FontSize',10);
subplot(2,2,4); plot(t,y(:,3),'-');
xlabel('Time (seconds)','FontSize',10);
title('Current i _ a, [A]','FontSize',10);
```

MATLAB file (ch3_02.m)

```
% Dynamics of the buck converter
function yprime=difer(t,y);
% parameters
Vd=50; Ea=10; rs=0.025; rl=0.02; rc=0.15; ra=3;
C=0.003; L=0.0007; La=0.005; D=0.5;
% differential equations for buck converters
yprime=[(y(2,:)-y(3,:))/C;...
(-y(1,:)-(rl+rc)*y(2,:)+rc*y(3,:)-rs*y(2,:)*D+Vd*D)/L;...
(y(1,:)+rc*y(2,:)-(rc+ra)*y(3,:)-Ea)/La];
```

The transient dynamics for the state variables $u_C(t)$, $i_L(t)$ and $i_a(t)$ are illustrated in Figure 3.20, and the settling time is 0.025 sec. The steady-state value

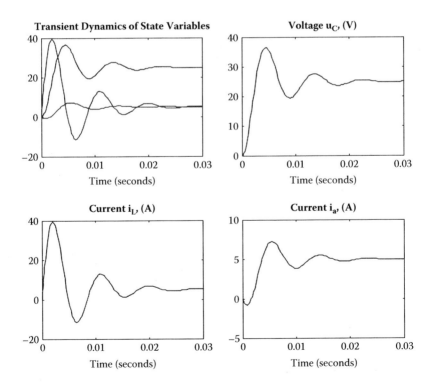

FIGURE 3.20
Transient dynamics of the *buck* converter, $V_d = 50$ V and $d_D = 0.5$.

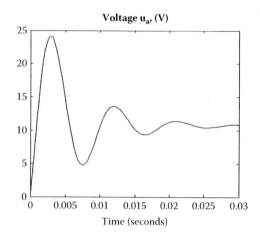

FIGURE 3.21
Voltage on the load terminal.

of the output voltage is 25 V because the applied voltage is 50 V and the duty
ratio is assigned $d_D = 0.5$.

Using $u_a = u_C + r_c i_L - r_a i_a$, the voltage at the load terminal u_a can be calculated
and plotted. In particular, in the Command Window type

```
rc=0.15; ra=3; plot(t,y(:,1)+rc*y(:,2)-ra*y(:,3),'-');
xlabel('Time (seconds)','FontSize',14);
title('Voltage u _ a, [V]','FontSize',14);
```

The plot $u_a(t)$ results as illustrated in Figure 3.21.   □

3.2.3 *Boost* Converter

A typical configuration of a one-quadrant *boost* (*step-up*) dc-dc switching converter is given in Figure 3.22.

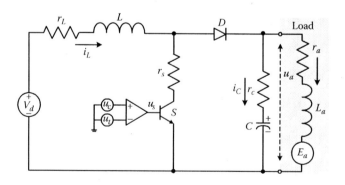

FIGURE 3.22
High-frequency *boost* converter.

When the switch S is closed, the diode D is reverse-biased. The following differential equations are found by applying Kirchhoff's laws

$$\frac{du_C}{dt} = -\frac{1}{C}i_a, \quad \frac{di_L}{dt} = \frac{1}{L}(-(r_L + r_s)i_L + V_d), \quad \frac{di_a}{dt} = \frac{1}{L_a}(u_C - (r_a + r_c)i_a - E_a).$$

If the switch is open, the diode is forward-biased because the direction of the current in the inductor i_L does not change instantly. Hence,

$$\frac{du_C}{dt} = \frac{1}{C}(i_L - i_a), \quad \frac{di_L}{dt} = \frac{1}{L}(-u_C - (r_L + r_c)i_L + r_c i_a + V_d),$$

$$\frac{di_a}{dt} = \frac{1}{L_a}(u_C + r_c i_L - (r_a + r_c)i_a - E_a).$$

Using the *averaging* concept, using d_D, one finds

$$\frac{du_C}{dt} = \frac{1}{C}(i_L - i_a - i_L d_D),$$

$$\frac{di_L}{dt} = \frac{1}{L}(-u_C - (r_L + r_c)i_L + r_c i_a + u_C d_D + (r_c - r_s)i_L d_D - r_c i_a d_D + V_d),$$

$$\frac{di_a}{dt} = \frac{1}{L_a}(u_C + r_c i_L - (r_a + r_c)i_a - r_c i_L d_D - E_a).$$

The steady-state analysis results in the following relationship

$$\frac{u_a}{V_d} = \frac{1}{1 - d_D}.$$

The expression for the voltage ripple is

$$\frac{\Delta u_a}{u_a} = \frac{d_D}{r_a C f^2}.$$

The minimum value of the inductance depends on the switching frequency and load resistance. One has

$$L_{min} = \frac{d_D(1 - d_D)^2 r_a}{2f}.$$

Example 3.2
Perform simulations of the *boost* converter if the parameters are: $r_s = 0.025$ ohm, $r_L = 0.02$ ohm, $r_c = 0.15$ ohm, $r_a = 3$ ohm, $C = 0.003$ F, $L = 0.0007$ H, and $L_a = 0.005$ H. Assume $d_D = 0.5$, $V_d = 50$ V, and $E_a = 10$ V.

Using the differential equations found, two m-files are developed. The first MATLAB file (ch3_03.m) is

```
t0=0; tfinal=0.04; tspan=[t0 tfinal]; y0=[0 0 0]';
[t,y]=ode45('ch3 _ 04',tspan,y0);
subplot(2,2,1); plot(t,y);
xlabel('Time (seconds)','FontSize',10);
title('Transient Dynamics of State Variables','FontSize',10);
subplot(2,2,2); plot(t,y(:,1),'-');
xlabel('Time (seconds)','FontSize',10);
title('Voltage u _ C, [V]','FontSize',10);
subplot(2,2,3); plot(t,y(:,2),'-');
xlabel('Time (seconds)','FontSize',10);
title('Current i _ L, [A]','FontSize',10);
subplot(2,2,4); plot(t,y(:,3),'-');
xlabel('Time (seconds)','FontSize',10);
title('Current i _ a, [A]','FontSize',10);
```

The second MATLAB file (ch3_04.m) is

```
% Dynamics of the boost converter
function yprime=difer(t,y);
% parameters
Vd=50; Ea=10; rs=0.025; rl=0.02; rc=0.15; ra=3;
C=0.003; L=0.0007; La=0.005; D=0.5;
% differential equations for boost converters
yprime=[(y(2,:)-y(3,:)-y(2,:)*D)/C;...
(-y(1,:)-(rl+rc)*y(2,:)+rc*y(3,:)+y(1,:)*D+(rc-rs)*y(2,:)*D-rc*y(3,:)*D+Vd)/L;...
(y(1,:)+rc*y(2,:)-(ra+rc)*y(3,:)-rc*y(2,:)*D-Ea)/La];
```

The transients for $u_C(t)$, $i_L(t)$, and $i_a(t)$ are plotted in Figure 3.23. The settling time is 0.038 sec. One concludes that the transient dynamics of the *boost* converters is fast. Therefore, the converter equations of motion are not usually integrated in the differential equations to analyze the transients of conventional electromechanical systems.

The analysis performed is very important to design and analyze the power electronics which affect the overall system performance such as efficiency, loading capabilities, and so on. We illustrate the correspondence and dependence between the motor and converter parameters. Therefore, the compliance, matching, and compatibility are very important issues. We demonstrated that the applied voltage to the motor windings may not be considered as the control input, and $u_a(t)$ is the converter output.

The three-dimensional plot, which illustrates the evolution of states $u_C(t)$, $i_L(t)$, and $i_a(t)$, can be plotted using the following statement

```
% 3-D plot using x1, x2 and x3 as the variables
plot3(y(:,1),y(:,2),y(:,3));
xlabel('Voltage \itu _ C','FontSize',14);
ylabel('Current \iti _ L','FontSize',14);
zlabel('Current \iti _ a','FontSize',14);
text(-50,5,8,'Initial Conditions, \itx _ 0','FontSize',14);
```

The resulting plot is depicted in Figure 3.24.   □

We applied Kirchhoff's laws to derive the mathematical model in the form of nonlinear differential equations for a one-quadrant *boost* dc-dc converter as illustrated in Figure 3.22. Alternatively, the Lagrange concept can be applied.

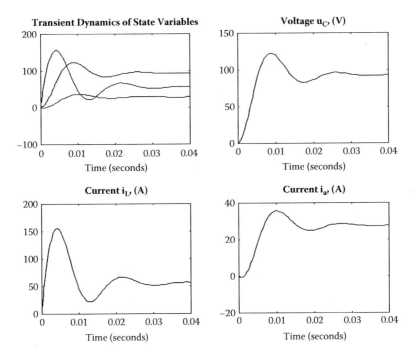

FIGURE 3.23
Transient dynamics of the *boost* converter.

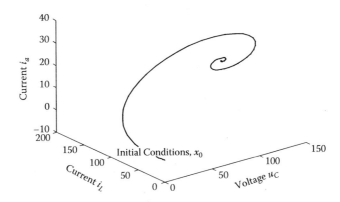

FIGURE 3.24
Evolution of the state variables.

The Lagrange equations of motion are

$$\frac{d}{dt}\left(\frac{\partial \Gamma}{\partial \dot{q}_1}\right) - \frac{\partial \Gamma}{\partial q_1} + \frac{\partial D}{\partial \dot{q}_1} + \frac{\partial \Pi}{\partial \dot{q}_1} = Q_1,$$

$$\frac{d}{dt}\left(\frac{\partial \Gamma}{\partial \dot{q}_2}\right) - \frac{\partial \Gamma}{\partial q_2} + \frac{\partial D}{\partial \dot{q}_2} + \frac{\partial \Pi}{\partial \dot{q}_2} = Q_2.$$

In these equations, the electric charges in the first and the second loops are denoted as q_1 and q_2. That is, we have $i_L = \dot{q}_1$ and $i_a = \dot{q}_2$. The generalized forces are denoted as Q_1 and Q_2, e.g., $Q_1 = V_d$ and $Q_2 = -E_a$.

When the switch is closed, the total kinetic Γ, potential Π, and dissipated D energies are

$$\Gamma = \frac{1}{2}\left(L\dot{q}_1^2 + L_a\dot{q}_2^2\right), \quad \Pi = \frac{1}{2}\frac{q_2^2}{C} \quad \text{and} \quad D = \frac{1}{2}\left((r_L + r_s)\dot{q}_1^2 + (r_c + r_a)\dot{q}_2^2\right).$$

Assuming that the resistances, inductances, and capacitance are not varying (time-constant), we have

$$\frac{\partial \Gamma}{\partial q_1} = 0, \quad \frac{\partial \Gamma}{\partial q_2} = 0, \quad \frac{\partial \Gamma}{\partial \dot{q}_1} = L\dot{q}_1, \quad \frac{\partial \Gamma}{\partial \dot{q}_2} = L_a\dot{q}_2,$$

$$\frac{d}{dt}\left(\frac{\partial \Gamma}{\partial \dot{q}_1}\right) = L\ddot{q}_1, \quad \frac{d}{dt}\left(\frac{\partial \Gamma}{\partial \dot{q}_2}\right) = L_a\ddot{q}_2,$$

$$\frac{\partial \Pi}{\partial q_1} = 0, \quad \frac{\partial \Pi}{\partial q_2} = \frac{q_2}{C},$$

and
$$\frac{\partial D}{\partial \dot{q}_1} = (r_L + r_s)\dot{q}_1, \quad \frac{\partial D}{\partial \dot{q}_2} = (r_c + r_a)\dot{q}_2.$$

The application of the Lagrange equations of motion yields

$$L\ddot{q}_1 + (r_L + r_s)\dot{q}_1 = Q_1,$$

$$L_a\ddot{q}_2 + (r_c + r_a)\dot{q}_2 + \frac{1}{C}q_2 = Q_2.$$

One obtains

$$\ddot{q}_1 = \frac{1}{L}\left(-(r_L + r_s)\dot{q}_1 + Q_1\right),$$

$$\ddot{q}_2 = \frac{1}{L_a}\left(-(r_c + r_a)\dot{q}_2 - \frac{1}{C}q_2 + Q_2\right).$$

If the switch is open, we have

$$\Gamma = \frac{1}{2}\left(L\dot{q}_1^2 + L_a\dot{q}_2^2\right), \quad \Pi = \frac{1}{2}\frac{(q_1 - q_2)^2}{C} \quad \text{and} \quad D = \frac{1}{2}\left(r_L\dot{q}_1^2 + r_c(\dot{q}_1 - \dot{q}_2)^2 + r_a\dot{q}_2^2\right).$$

Hence,

$$\frac{\partial \Gamma}{\partial q_1} = 0, \quad \frac{\partial \Gamma}{\partial q_2} = 0, \quad \frac{\partial \Gamma}{\partial \dot{q}_1} = L\dot{q}_1, \quad \frac{\partial \Gamma}{\partial \dot{q}_2} = L_a\dot{q}_2, \quad \frac{d}{dt}\left(\frac{\partial \Gamma}{\partial \dot{q}_1}\right) = L\ddot{q}_1,$$

$$\frac{d}{dt}\left(\frac{\partial \Gamma}{\partial \dot{q}_2}\right) = L_a\ddot{q}_1,$$

$$\frac{\partial \Pi}{\partial q_1} = \frac{q_1 - q_2}{C}, \quad \frac{\partial \Pi}{\partial q_2} = -\frac{q_1 - q_2}{C},$$

and

$$\frac{\partial D}{\partial \dot{q}_1} = (r_L + r_c)\dot{q}_1 - r_c\dot{q}_2, \quad \frac{\partial D}{\partial \dot{q}_1} = -r_c\dot{q}_1 + (r_c + r_a)\dot{q}_2.$$

The resulting equations are

$$L\ddot{q}_1 + (r_L + r_c)\dot{q}_1 - r_c\dot{q}_2 + \frac{q_1 - q_2}{C} = Q_1,$$

$$L_a\ddot{q}_2 - r_c\dot{q}_1 + (r_c + r_a)\dot{q}_2 - \frac{q_1 - q_2}{C} = Q_2.$$

Therefore,

$$\ddot{q}_1 = \frac{1}{L}\left(-(r_L + r_c)\dot{q}_1 + r_c\dot{q}_2 - \frac{q_1 - q_2}{C} + Q_1\right),$$

$$\ddot{q}_2 = \frac{1}{L_a}\left(r_c\dot{q}_1 - (r_c + r_a)\dot{q}_2 + \frac{q_1 - q_2}{C} + Q_2\right).$$

From the differential equations derived when the switch is closed and open, Cauchy's form of differential equations are found by using $i_L = \dot{q}_1$ and $i_a = \dot{q}_2$. That is,

$$\frac{dq_1}{dt} = i_L \quad \text{and} \quad \frac{dq_2}{dt} = i_a.$$

The voltage across the capacitor u_C can be expressed using the charges. That is, when the switch is closed

$$u_C = -\frac{q_2}{C},$$

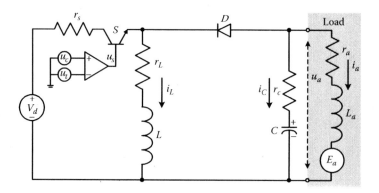

FIGURE 3.25
High-frequency *buck-boost* switching converter.

and if the switch is open, one obtains

$$u_C = \frac{q_1 - q_2}{C}.$$

The analysis of the differential equations derived using Kirchhoff's voltage law and the Lagrange equations of motion leads one to the conclusion that the mathematical model of the *boost* converter can be found using different state variables. In particular, u_C, i_L, i_a and q_1, i_L, q_2, i_a are used. However, the resulting differential equations, which describe the converter dynamics, are related, and the solutions are the same.

3.2.4 *Buck-Boost* Converters

The *buck-boost* switching converter is illustrated in Figure 3.25.

If the switch is closed, the diode is reverse-biased, and when the switch is open, the diode is forward-biased. One easily derives a set of differential equations using Kirchhoff's law or Lagrange equations of motion. The steady-state ratio between the supplied and terminal voltage is

$$\frac{u_a}{V_d} = \frac{-d_D}{1 - d_D}.$$

That is, depending on the duty ratio, the output voltage u_a is less or greater than V_d. The expressions for the voltage ripple and minimum inductance are

$$\frac{\Delta u_a}{u_a} = \frac{d_D}{r_a Cf}$$

and

$$L_{min} = \frac{(1 - d_D)^2 r_a}{2f}.$$

3.2.5 Cuk Converters

The Cuk converter is based on a capacitive energy transfer, whereas the *buck*, *buck-boost*, *boost*, and *flyback* converter topologies are based on the inductive energy transfer. If the switch is *on* or *off*, the currents in the input and output inductors L_1 and L are continuous. The output voltage, applied to the load can be either smaller or greater than the supplied voltage V_d. When the switch is turned *off*, the diode is forward-biased. The voltage V_d is supplied, and the capacitor C_1 is charged through the inductor L_1, (see Figure 3.26). To study how the converter operates, assume that the switch is turned *on*. The current through the inductor L_1 rises, and at the same time, the voltage of capacitor C_1 reverses bias the diode, and turns the diode *off*. The capacitor C_1 discharges the stored energy through the circuit formed by capacitors C_1, C, the load r_a--L_a--E_a, and the inductor L. Consider the situation if the switch is turned *off*. The voltage V_d is applied, and the capacitor C_1 charges. The energy, stored in the inductor L, transfers to the load. The diode and switch provide a synchronous switching action, and the capacitor C_1 is an element for transferring energy from the energy source to the load.

From Kirchhoff's laws, using the differential equations that describe the transient dynamics when the switch is opened and closed, one finds the following set of differential equations

$$\frac{du_{C1}}{dt} = \frac{1}{C_1}(i_{L1} - i_{L1}d_D + i_L d_D),$$

$$\frac{du_C}{dt} = \frac{1}{C}(i_L - i_a),$$

$$\frac{di_{L1}}{dt} = \frac{1}{L_1}(-u_{C1} - (r_{L1} + r_{c1})i_{L1} + u_{C1}d_D + (r_{c1} - r_s)i_{L1}d_D + r_s i_L d_D + V_d),$$

$$\frac{di_L}{dt} = \frac{1}{L}(-u_C - (r_L + r_c)i_L + r_c i_a - u_{C1}d_D + r_s i_{L1}d_D - (r_{c1} + r_s)i_L d_D),$$

$$\frac{di_a}{dt} = \frac{1}{L_a}(u_C + r_c i_L - (r_c + r_a)i_a - E_a).$$

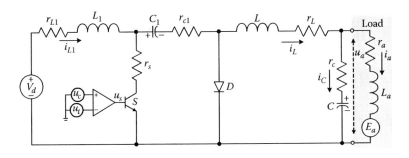

FIGURE 3.26
Cuk high-frequency switching converter.

From the differential equations derived, neglecting the resistances of the switch, inductors, and capacitors, one finds the following steady-state equations important in the converter design and converter-load matching

$$\frac{u_a}{V_d} = \frac{-d_D}{1-d_D}, \quad \frac{\Delta u_a}{u_a} = \frac{1-d_D}{8LCf^2}, \quad L_{1\min} = \frac{(1-d_D)^2 r_a}{2d_D f}, \quad \text{and} \quad L_{\min} = \frac{(1-d_D)r_a}{2f}$$

Example 3.3

Perform simulations of the Cuk converter if $V_d = 50$ V, $E_a = 10$ V, $r_{L1} = 0.035$ ohm, $r_L = 0.02$ ohm, $r_c = 0.15$ ohm, $r_s = 0.03$ ohm, $r_{c1} = 0.018$ ohm, $r_a = 3$ ohm, $C_1 = 2\times10^{-5}$ F, $C = 3.5\times10^{-6}$ F, $L_1 = 5\times10^{-6}$ H, $L = 7\times10^{-6}$ H, and $L_a = 0.005$ H. The duty ratio is $d_D = 0.5$.

Using the differential equations found, two m-files are developed. In particular, MATLAB files (ch3 _ 05.m and ch3 _ 06.m) are reported here. The first m-file is

```
t0=0; tfinal=0.002; tspan=[t0 tfinal]; y0=[0 0 0 0 0]';
[t,y]=ode45('ch3 _ 06',tspan,y0);
subplot(3,2,1); plot(t,y);
xlabel('Time (seconds)','FontSize',10);
title('Transient Dynamics of State Variables','FontSize',10);
subplot(3,2,2); plot(t,y(:,1),'-');
xlabel('Time (seconds)','FontSize',10);
title('Voltage u _ C _ 1, [V]','FontSize',10);
subplot(3,2,3); plot(t,y(:,2),'-');
xlabel('Time (seconds)','FontSize',10);
title('Voltage u _ C, [V]','FontSize',10);
subplot(3,2,4); plot(t,y(:,3),'-');
xlabel('Time (seconds)','FontSize',10);
title('Current i _ L _ 1, [A]','FontSize',10);
subplot(3,2,5); plot(t,y(:,4),'-');
xlabel('Time (seconds)','FontSize',10);
title('Current i _ L, [A]','FontSize',10);
subplot(3,2,6); plot(t,y(:,5),'-');
xlabel('Time (seconds)','FontSize',10);
title('Current i _ a, [A]','FontSize',10);
```

whereas the second m-file, where the parameters are provided and the differential equations are specified, is

```
% Dynamics of the Cuk converter
function yprime=difer(t,y);
Vd=50; Ea=10; rl1=0.035; rl=0.02; rc=0.15; rs=0.03; rc1=0.018; ra=3;
L1=5e-6; L=7e-6; La=0.005; C1=2e-5; C=3.5e-6; D=0.5;
yprime=[(y(3,:)-y(3,:)*D+y(4,:)*D)/C1; ...
(y(4,:)-y(5,:))/C; ...
(-y(1,:)-(rl1+rc1)*y(3,:)+y(1,:)*D+(rc1-rs)*y(3,:)*D+rs*y(4,:)*D+Vd)/L1; ...
(-y(2,:)-(rl+rc)*y(4,:)+rc*y(5,:)-y(1,:)*D+rs*y(3,:)*D-(rc1+rs)*y(4,:)*D)/L; ...
(y(2,:)+rc*y(4,:)-(rc+ra)*y(5,:)-Ea)/La];
```

The transient dynamics for voltages and currents $u_{C1}(t)$, $u_C(t)$, $i_{L1}(t)$, $i_L(t)$ and $i_a(t)$, which are considered as the converter state variables, are illustrated in Figure 3.27. The settling time is 0.0005 sec. The steady-state value of the converter variables are easily assessed and examined. □

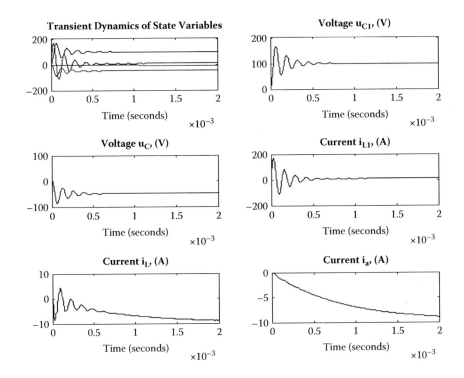

FIGURE 3.27
Transient dynamics of the Cuk converter.

3.2.6 *Flyback* and *Forward* Converters

To avoid the interference between the input and output, which is a serious disadvantage, *flyback* and *forward* converters are used to magnetically isolate the input and output using transformers in the switching scheme. The application of transformers increases the size and cost. However, *flyback* and *forward* converters are commonly used ensuring isolation between input and output. The energy is stored in the inductor when the switch is closed, and the stored energy is transformed to the load when the switch is open. The *flyback* and *forward* magnetically coupled dc-dc converters are illustrated in Figures 3.28 and 3.29.

When the switch is closed, the diode is reverse-biased. If the switch is open, the diode is forward-biased. The switch is closed for time $\frac{d_D}{f}$ and open for $\frac{1-d_D}{f}$. For the switch open

$$i_d = \frac{N_1}{N_2} i_L,$$

and using the duty ratio d_D, the differential equations can be found.

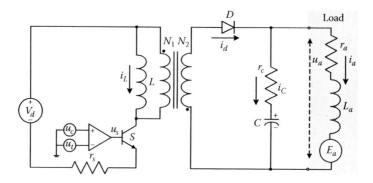

FIGURE 3.28
Flyback dc-dc converter.

The differential equations when the switch is closed are

$$\frac{du_C}{dt} = \frac{1}{C(r_c + r_a)}(-u_C + r_a i_L), \; \frac{di_L}{dt} = \frac{1}{L}\left(-\frac{r_a}{r_c + r_a}u_C + \left(\frac{r_a^2}{r_c + r_a} - r_L - r_a\right)i_L + \frac{N_2}{N_1}V_d\right).$$

If the switch is open, one finds

$$\frac{du_C}{dt} = \frac{1}{C(r_c + r_a)}(-u_C + r_a i_L), \; \frac{di_L}{dt} = \frac{1}{L}\left(-\frac{r_a}{r_c + r_a}u_C + \left(\frac{r_a^2}{r_c + r_a} - r_L - r_a\right)i_L\right).$$

Hence,

$$\frac{du_C}{dt} = \frac{1}{C(r_c + r_a)}(-u_C + r_a i_L),$$

$$\frac{di_L}{dt} = \frac{1}{L}\left(-\frac{r_a}{r_c + r_a}u_C + \left(\frac{r_a^2}{r_c + r_a} - r_L - r_a\right)i_L + \frac{N_2}{N_1}V_d d_D\right).$$

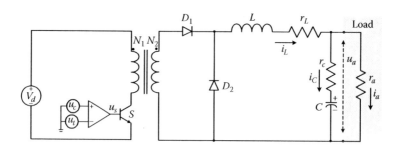

FIGURE 3.29
Forward dc-dc converter.

Example 3.4

Using a set of differential equations derived, simulate the *forward* converter if the applied voltage is $V_d = 50$ V and the duty ratio is $d_D = 0.5$. The parameters are: $r_L = 0.02$ ohm, $r_C = 0.01$ ohm, $r_a = 3$ ohm, $L = 0.000005$ H, $C = 0.003$ F, and $N_2/N_1 = 1$.

Two m-files are developed using the differential equations. The first MATLAB file (ch3_07.m) is

```
t0=0; tfinal=0.002; tspan=[t0 tfinal]; y0=[0 0]';
[t,y]=ode45('ch3 _ 08',tspan,y0);
subplot(2,2,1); plot(t,y);
xlabel('Time (seconds)','FontSize',10);
title('Transient Dynamics of State Variables','FontSize',10);
subplot(2,2,2); plot(t,y(:,1),'-');
xlabel('Time (seconds)','FontSize',10);
title('Voltage u _ C, [V]','FontSize',10);
subplot(2,2,3); plot(t,y(:,2),'-');
xlabel('Time (seconds)','FontSize',10);
title('Current i _ L, [A]','FontSize',10);
% 3-D plot using x1, x2 and x3 as the variables
subplot(2,2,4); plot3(y(:,1),y(:,2),t);
xlabel('Voltage \itu _ C','FontSize',10);
ylabel('Current \iti _ L','FontSize',10);
zlabel('Time \itt','FontSize',10);
```

The second MATLAB file (ch3_08.m) is

```
% Dynamics of the forward converter
function yprime=difer(t,y);
% parameters
Vd=50; rl=0.02; rc=0.01; ra=3; C=0.002; L=0.000005; D=0.5;
% differential equations for the forward converters
yprime=[(-y(1,:)+ra*y(2,:))/(C*(rc+ra));...
(-(ra/(rc+ra))*y(1,:)+(ra*ra/(rc+ra)-rl-ra)*y(2,:)+Vd*D)/L];
```

The transient dynamics for the state variables $u_C(t)$ and $i_L(t)$ are documented in Figure 3.30. The settling time is 0.0015 sec. The steady-state voltage, applied to the load terminal is 25 V. The three-dimensional plot for $u_C(t)$, $i_L(t)$, and time t is depicted in the last plot of Figure 3.30. □

3.2.7 Resonant and Switching Converters

The current trends in development of advanced switching converters have facilitated the unified activities in topology design, nonlinear analysis, optimization, and control. To attain high efficiency and power density, new topologies were developed. Nonlinear analysis and design must be performed to guarantee a spectrum of specifications and requirements imposed on the converter dynamics. It was documented that the output voltage of converters is regulated by changing the duty ratio, which is constrained by lower and upper limits. To approach the design tradeoffs and enhance converter performance (settling time, overshoot, stability, robustness, power losses, and other quantities), advanced concepts, topologies, and nonlinearities should be examined. The resonant converters have been widely used in

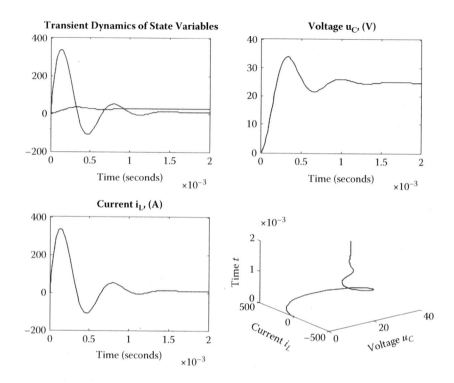

FIGURE 3.30
Transient dynamics of the *forward* converter and three-dimensional states evolution.

high-performance electromechanical systems. For example, the zero-voltage zero-current switching has become the enabling technology for the majority of converters to improve power density, efficiency, reliability, and other performance characteristics. Recent innovations include development of advanced converter topologies and control algorithms to maximize efficiency, minimize losses, increase power density, and so on. The problem of controlling high-frequency switching converters is a very important one in many applications. Nonlinear analysis and control are central issues to be solved to improve the steady-state and dynamic characteristics. A great variety of resonant converter topologies and filters have been developed. The resulting nonlinear dynamics cannot be linearized, and the hard bounds imposed cannot be neglected. Specific requirements are assigned, and the *absolute* limit of converters performance must be placed in the scope of practical design. The steady-state and dynamic characteristics of converters can be improved through coherent topology synthesis. There are a great number of converter topologies and filter configurations. Consider the resonant converter, which is shown in Figure 3.31.

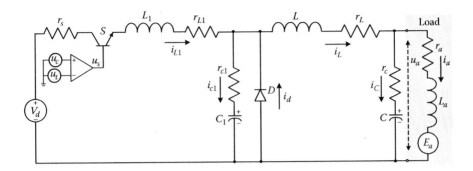

FIGURE 3.31
Resonant converter with zero-current switching.

The output voltage at the load terminal formed by a resistor r_a and inductor L_a is regulated by controlling the switching *on* and *off* durations t_{on} and t_{off}. The switch S is opened and closed, and the switching frequency is

$$\frac{1}{t_{on} + t_{off}}.$$

When the switch is open, the diode D is forward-biased to carry the output inductor current i_L, and the voltage across the capacitor C_1 is zero. When the switch is closed, the diode remains forward-biased while $i_{L1} < i_L$. As i_{L1} reaches i_L, the diode turns *off*. Hence, the switch turns *off* and *on* at zero-current (i_{L1} is zero). A control-triangle concept is used to establish the PWM switching. The switching signal u_s, which drives the switch, is generated by comparing a signal-level control voltage u_c with a repetitive triangular signal u_t. In resonant converters, the frequency usually is controlled to regulate the output voltage. A set of differential equations to describe the resonant converter dynamics is

$$\frac{du_{C1}}{dt} = \frac{1}{C_1}(i_{L1} - i_L)d_D,$$

$$\frac{du_C}{dt} = \frac{1}{C}(i_L - i_a),$$

$$\frac{di_{L1}}{dt} = \frac{1}{L_1}[-u_{C1} - (r_s + r_{L1} + r_{c1})i_{L1} + r_{c1}i_L + V_d]d_D,$$

$$\frac{di_L}{dt} = \frac{1}{L}[u_{C1} - u_C + r_{c1}i_{L1} - (r_{c1} + r_L + r_c)i_L + r_c i_a]d_D,$$

$$\frac{di_a}{dt} = \frac{1}{L_a}(u_C + r_c i_L - (r_c + r_a)i_a - E_a).$$

A nonlinear mathematical model results because of the multiplication of the state variables $u_{C1}(t)$, $u_C(t)$, $i_{L1}(t)$, $i_L(t)$, $i_a(t)$, and duty ratio d_D. From the differential equations obtained, we have the following nonlinear state-space model

$$
\begin{bmatrix} \dfrac{du_{C1}}{dt} \\[2mm] \dfrac{du_C}{dt} \\[2mm] \dfrac{di_{L1}}{dt} \\[2mm] \dfrac{di_L}{dt} \\[2mm] \dfrac{di_a}{dt} \end{bmatrix} =
\begin{bmatrix}
0 & 0 & 0 & 0 & 0 \\
0 & 0 & 0 & \dfrac{1}{C} & -\dfrac{1}{C} \\
0 & 0 & 0 & 0 & 0 \\
0 & 0 & 0 & 0 & 0 \\
0 & \dfrac{1}{L_a} & 0 & \dfrac{r_c}{L_a} & -\dfrac{r_c+r_a}{L_a}
\end{bmatrix}
\begin{bmatrix} u_{C1} \\ u_C \\ i_{L1} \\ i_L \\ i_a \end{bmatrix}
$$

$$
+ \begin{bmatrix}
\dfrac{1}{C_1}(i_{L1}-i_L) \\[3mm]
0 \\[3mm]
\dfrac{1}{L_1}(-u_{C1}-(r_s+r_{L1}+r_{c1})i_{L1}+r_{c1}i_L+V_d) \\[3mm]
\dfrac{1}{L}(u_{C1}-u_C+r_{c1}i_{L1}-(r_{c1}+r_L+r_c)i_L+r_c i_a) \\[3mm]
0
\end{bmatrix} d_D -
\begin{bmatrix} 0 \\ 0 \\ 0 \\ 0 \\ \dfrac{1}{L_a}E_a \end{bmatrix}.
$$

Example 3.5
Perform simulations and examine the resonant converter, which is illustrated in Figure 3.31. Let $V_d = 50$ V and $E_a = 10$ V. The parameters are: $r_s = 0.025$ ohm, $r_{L1} = 0.01$ ohm, $r_{c1} = 0.04$ ohm, $r_L = 0.02$ ohm, $r_c = 0.02$ ohm, $L_1 = 0.000005$ H, $L = 0.0007$ H, $C_1 = 0.000003$ F, and $C = 0.003$ F. The duty ratio is $d_D = 0.5$. The resistive-inductive load is formed by r_a and L_a, $r_a = 3$ ohm and $L_a = 0.005$ H.

Using the derived set of differential equations, two m-files are developed. The first MATLAB file (ch3_09.m) is

```
tspan=[0 0.04]; y0=[0 0 0 0 0]';
[t,y]=ode45('ch3 _ 10',tspan,y0);
subplot(3,2,1); plot(t,y);
xlabel('Time (seconds)','FontSize',10);
title('Transient Dynamics of State Variables','FontSize',10);
subplot(3,2,2); plot(t,y(:,1),'-');
xlabel('Time (seconds)','FontSize',10);
title('Voltage u _ C _ 1, [V]','FontSize',10);
subplot(3,2,3); plot(t,y(:,2),'-');
xlabel('Time (seconds)','FontSize',10);
title('Voltage u _ C, [V]','FontSize',10);
subplot(3,2,4); plot(t,y(:,3),'-');
```

```
xlabel('Time (seconds)','FontSize',10);
title('Current i _ L _ 1, [A]','FontSize',10);
subplot(3,2,5); plot(t,y(:,4),'-');
xlabel('Time (seconds)','FontSize',10);
title('Current i _ L, [A]','FontSize',10);
subplot(3,2,6); plot(t,y(:,5),'-');
xlabel('Time (seconds)','FontSize',10);
title('Current i _ a, [A]','FontSize',10);
```

The second MATLAB file (ch3_10.m) is

```
function yprime=difer(t,y);
Vd=50; Ea=10; rs=0.025; rl1=0.01; rc1=0.04; rl=0.02; rc=0.15; ra=3;
L1=0.000005; L=0.0007; La=0.004; C1=0.000003; C=0.003; D=0.5;
yprime=[D*(y(3,:)-y(4,:))/C1;...
(y(4,:)-y(5,:))/C;...
(Vd-(rs+rl1+rc1)*y(3,:)+rc1*y(4,:)-y(1,:))*D/L1;...
(y(1,:)-y(2,:)+rc1*y(3,:)-(rc1+rl+rc1)*y(4,:)-rc*y(5,:))*D/L;...
(y(2,:)+rc*y(4,:)-(rc+ra)*y(5,:)-Ea)/La];
```

Transient dynamics of the resonant converter with zero-current switching is studied. The transient responses for the state variables $u_{C1}(t)$, $u_C(t)$, $i_{L1}(t)$, $i_L(t)$, and $i_a(t)$ are plotted in Figure 3.32. □

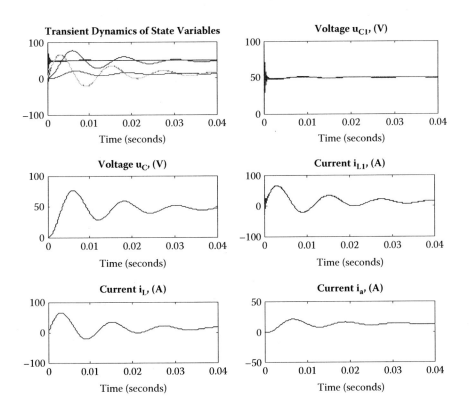

FIGURE 3.32
Transient dynamics of the resonant converter.

There is a great variety of high-performance PWM resonant and switching converters. We examined one-quadrant converters. In electromechanical systems, to ensure high performance, two- and four-quadrant converters are used. Those converters were also covered in this chapter (see Figures 3.12 and 3.13). As the converter topology, switching and energy storage mechanisms, filter circuitry, and other key components are developed, the corresponding data-intensive analysis is performed by deriving, solving, and examining nonlinear differential equations. The Kirchhoff's laws and Lagrange equations of motion are used to describe the converter and filter dynamics. Based on mathematical models derived, nonlinear analysis and design are performed as documented in this chapter.

Homework Problems

Problem 3.1

Using the operational amplifiers, develop the schematics to implement the proportional-integral controller. Report at least two possible schematics. Derive the transfer function $G(s)$ and express the feedback gains k_p and k_i using the circuitry parameters (resistances and capacitances).

Problem 3.2

Using the operational amplifiers, develop the schematics to implement the PID controller. Report at least two possible schematics. Derive the transfer functions $G(s)$ and express the feedback gains k_p, k_i, and k_d using the circuitry parameters. Note that there are many solutions which approximate an *ideal* $G_{PID}(s)$ to make the implementable circuit and relax some deficiencies of an *ideal* PID controller, including cases when the controller input is step-like signals or contain noise. Perform the trade-off studies for ideal and practical PID controllers. You may utilize the Laplace transform, Bode plots, frequency-domain, and other analyses.

Problem 3.3

Why should converters and power amplifiers be utilized in electromechanical systems?

Problem 3.4

Study the PWM switching concept applying the sinusoidal-like (not triangular u_t) signal to the comparator. For example, the switching signal u_s, which drives the switch S, can be generated by comparing a signal-level control voltage u_c with a repetitive sinusoidal-like signal u_{sin}. Propose the schematics to generate $u_{\sin}$ (transistors or operational amplifiers can be used). Report how to define and vary (if needed) the frequency of $u_{\sin}$. Report the voltage waveforms.

Problem 3.5

Report the four-quadrant H-configured power stage schematics. Explain how it operates.

References

1. A.S. Sedra and K.C. Smith, *Microelectronic Circuits*, Oxford University Press, New York, 1997.
2. S.E. Lyshevski, *Electromechanical Systems, Electric Machines, and Applied Mechatronics*, CRC Press, Boca Raton, FL, 1999.
3. D.W. Hart, *Introduction to Power Electronics*, Prentice Hall, Upper Saddle River, NJ, 1997.
4. J.G. Kassakian, M.F. Schlecht, and G.C. Verghese, *Principles of Power Electronics*, Addison-Wesley Publishing Company, Reading, MA, 1991.
5. N.T. Mohan, T.M. Undeland, and W.P. Robbins, *Power Electronics: Converters, Applications, and Design*, John Wiley and Sons, New York, 1995.

4

Direct-Current Electric Machines and Motion Devices

4.1 Permanent-Magnet Direct-Current Electric Machines

4.1.1 Radial Topology Permanent-Magnet Direct-Current Electric Machines

The basic electromagnetic principles and fundamental physical laws are used to devise, design, and examine various electromechanical motion devices [1–11]. Permanent-magnet direct-current (DC) electric machines guarantee high power and torques densities, efficiency, affordability, reliability, ruggedness, overloading capabilities, and other advantages. The power range of permanent-magnet DC electric machines (motors and generators) is from μW to ~100 kW, and the dimensions are from ~1 mm in diameter and ~5 mm long to ~1 m. The same permanent-magnet electric machine can be used as a motor or a generator. Due to the above-mentioned advantages and very high performance, permanent-magnet DC electric machines and motion devices are widely used in aerospace, automotive, marine, power, robotics, and other applications. Only permanent-magnet synchronous machines, which do not have brushes, surpass permanent-magnet DC machines. Therefore, depending on applications, in high-performance drives and servos, predominantly permanent-magnet DC and synchronous electric machines are utilized. One should be aware that to drive a computer/camera hard drive, household fan, passenger car, and 60 ton tank (track), ~1 W, ~10 W, ~10 kW, and ~100 kW machines are needed. Therefore, ~1 μW to ~100 kW power range covers the major consumer and industrial systems. In megawatt applications (ships, locomotives, high-power energy systems, etc.), induction and synchronous machines are used.

Permanent-magnet DC electric machines are rotating energy-transforming electromechanical motion devices that convert energies. Motors (actuators) convert electrical energy to mechanical energy, while generators convert mechanical energy to electrical energy. As was emphasized earlier, the same permanent-magnet electric machine can serve as the motor (if one applies the voltage) or as the generator (if the torque is applied to rotate the machine, the voltage is induced). Electric machines have stationary and rotating members, separated by an air gap. The armature winding is placed in the rotor

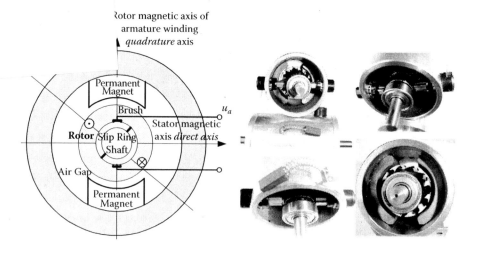

FIGURE 4.1
Permanent-magnet DC electric machine schematics and images.

slots and connected to a rotating commutator, which rectifies the voltage (see Figure 4.1). One supplies the armature voltage u_a to the rotor windings. The rotor windings and permanent magnets on stator are magnetically coupled. The brushes ride on the commutator which is connected to the armature windings. The armature winding consists of identical uniformly distributed coils. The excitation stationary magnetic field is produced by permanent magnets. The images of a permanent-magnet DC electric machine, with the above-mentioned components, are reported in Figure 4.1.

Because of the commutator (circular conducting segments as depicted in Figure 4.1), armature windings and permanent magnets produce stationary *mmfs,* which are displaced by 90 electrical degrees. The armature magnetic force is along the *quadrature* (rotor) magnetic axis, while the *direct* axis stands for a permanent magnet magnetic axis. The electromagnetic torque is produced as a result of the interaction of these *mmfs.* For motors, using Kirchhoff's law, one obtains the following steady-state equation for the armature voltage u_a and the *back emf* E_a

$$u_a - E_a = r_a i_a,$$

where r_a is the armature resistance, i_a is the currents in the armature winding.

The difference between the applied voltage and the *emf* is the voltage drop across the armature resistance r_a. The motor rotates at an angular velocity ω_r at which the *emf,* E_a, induced in the armature winding balances the armature voltage u_a supplied. If an electric machine operates as a motor, the induced *emf* is less than the voltage applied to the windings. If a machine operates as a generator, the generated (induced) *emf* is greater than the terminal voltage.

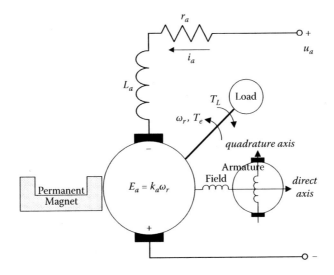

FIGURE 4.2
Schematic diagram of a permanent-magnet DC electric machine (current direction corresponds to the motor operation).

Furthermore, for generators, the armature current i_a is in the same direction as the induced *emf*, and the terminal voltage is $(E_a - r_a i_a)$.

The constant magnetic flux is produced by permanent-magnets, which are shown in Figure 4.1. For a permanent-magnet DC motor, a schematic diagram is depicted in Figure 4.2.

As documented in Chapter 2, the electromagnetic torque is found by using the coenergy $W_c = \oint_i \psi di$ as

$$T_e(i,\theta) = \frac{\partial W_c(i,\theta)}{\partial \theta}.$$

The magnetic flux crossing a surface is found to be $\Phi = \oint_s \vec{B} \cdot d\vec{s}$. The resulting expression for T_e agrees with the equation for a torque experienced by a current loop in the magnetic field, e.g., $\vec{T} = \vec{m} \times \vec{B} = \vec{a}_m m \times \vec{B} = iA\vec{a}_m \times \vec{B}$ (see Figure 2.15). In permanent-magnet electric machines, the stationary near-uniform magnetic field $\vec{B}$ is produced by permanent magnets, and A is constant.

One recalls that the expression for the *emf* and *mmf* are

$$emf = \oint_l \vec{E} \cdot d\vec{l} = \oint_l (\vec{v} \times \vec{B}) \cdot d\vec{l} - \oint_s \frac{\partial \vec{B}}{\partial t} d\vec{s} \text{ and } mmf = \oint_l \vec{H} \cdot d\vec{l} = \oint_l \vec{J} \times d\vec{s} + \oint_s \frac{\partial \vec{D}}{\partial t} d\vec{s}.$$

Denoting the *back emf* and *torque* constants as k_a, we have the following expressions for the *back emf* and the electromagnetic torque in the z-direction

$$E_a = k_a \omega_r \quad \text{and} \quad T_e = k_a i_a.$$

Using Kirchhoff's voltage law and Newton's second law of motion

$$u_a = r_a i_a + \frac{d\psi}{dt},$$

$$\frac{d\omega_r}{dt} = \frac{1}{J}(T_e - B_m\omega_r - T_L),$$

the differential equations for permanent-magnet DC motors are derived. Assume that the *susceptibility* of permanent magnets is constant (in general, there is Curie's constant, and the *susceptibility* varies as a function of temperature), one concludes that the flux, established by the permanent magnets, is constant. Thus, $k_a =$ const. The linear differential equations that describe the transient behavior of the armature current and angular velocity are

$$\frac{di_a}{dt} = -\frac{r_a}{L_a}i_a - \frac{k_a}{L_a}\omega_r + \frac{1}{L_a}u_a,$$

$$\frac{d\omega_r}{dt} = \frac{k_a}{J}i_a - \frac{B_m}{J}\omega_r - \frac{1}{J}T_L. \qquad (4.1)$$

In the state-space matrix form, we have

$$\frac{d\mathbf{x}}{dt} = A\mathbf{x} + B\mathbf{u}.$$

Taking note of $\mathbf{x} = [i_a\ \omega_r]^T$, $\mathbf{u} = u_a$, $A \in \mathbb{R}^{2\times2}$, and $B \in \mathbb{R}^{2\times1}$, from (4.1), we have

$$\begin{bmatrix} \dfrac{di_a}{dt} \\[2mm] \dfrac{d\omega_r}{dt} \end{bmatrix} = \begin{bmatrix} -\dfrac{r_a}{L_a} & -\dfrac{k_a}{L_a} \\[2mm] \dfrac{k_a}{J} & \dfrac{B_m}{J} \end{bmatrix} \begin{bmatrix} i_a \\[1mm] \omega_r \end{bmatrix} + \begin{bmatrix} \dfrac{1}{L_a} \\[1mm] 0 \end{bmatrix} u_a - \begin{bmatrix} 0 \\[1mm] \dfrac{1}{J} \end{bmatrix} T_L. \qquad (4.2)$$

An *s*-domain block diagram of permanent-magnet DC motors is illustrated in Figure 4.3.

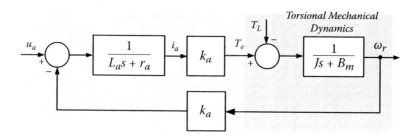

FIGURE 4.3
s-domain block diagram of permanent-magnet DC motors.

The angular velocity can be reversed if the polarity of the applied voltage is changed (the direction of the field flux cannot be changed).

From the differential equation

$$\frac{di_a}{dt} = -\frac{r_a}{L_a}i_a - \frac{k_a}{L_a}\omega_r + \frac{1}{L_a}u_a,$$

for the steady-state operation one has

$$0 = -r_a i_a - k_a \omega_r + u_a.$$

Hence,

$$\omega_r = \frac{u_a - r_a i_a}{k_a}.$$

The electromagnetic torque is $T_e = k_a i_a$, and in the steady-state operation $T_e = T_L$. Thus, the steady-state torque-speed characteristics are described by the following torque-speed equation

$$\omega_r = \frac{u_a - r_a i_a}{k_a} = \frac{u_a}{k_a} - \frac{r_a}{k_a^2}T_e. \tag{4.3}$$

Equation (4.3) illustrates that one changes the applied armature voltage u_a to vary the angular velocity. Furthermore, if the load is applied, the angular velocity reduces. The slope of the torque-speed characteristic is $-r_a/k_a^2$. The torque-speed characteristics are illustrated in Figure 4.4 for different u_a, and $|u_a| \le u_{a\,max}$, where $u_{a\,max}$ is the maximum (rated) voltage.

To reduce the angular velocity, one decreases u_a. The angular velocity at which a motor rotates is found as the intersection of the torque-speed characteristic

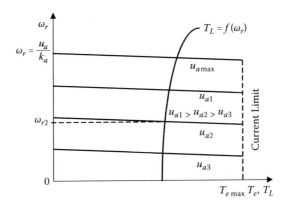

FIGURE 4.4
Torque-speed characteristics for permanent-magnet DC motors.

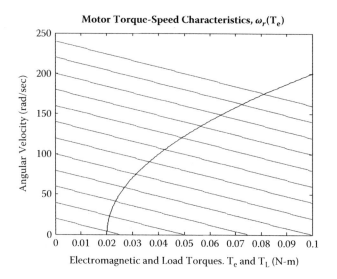

FIGURE 4.5
Torque-speed and load characteristics for a permanent-magnet DC motor.

and the load characteristic. For example, for u_{a2} applied, the angular velocity is ω_{r2}. In fact, from the Newton second law, neglecting friction, one has

$$\frac{d\omega_r}{dt} = \frac{1}{J}(T_e - T_L).$$

Thus, at $T_e = T_L$, a motor rotates at the constant angular velocity. At no load, from (4.3) one finds that the angular velocity is $\omega_r = u_a/k_a$.

Example 4.1

Calculate and plot the torque-speed characteristics for a 12 V (rated) permanent-magnet DC motor with the following parameters: $r_a = 2$ ohm and $k_a = 0.05$ V-sec/rad or N-m/A. The load is a nonlinear function of the angular velocity, and $T_L = f(\omega_r) = 0.02 + 0.000002\omega_r^2$ N-m.

The torque-speed characteristics are governed by equation (4.3). Using different values for the armature voltage u_a in (4.3), the steady-state characteristics are calculated and plotted as depicted in Figure 4.5. The load curve is also illustrated. The following MATLAB file to perform calculations and plotting is used.

```
% parameters of a permanent-magnet DC motor
ra=2; ka=0.05;
Te=0:0.001:0.1; % torque in N-m
for ua=1:1:12; % applied voltage
wr=ua/ka-(ra/ka^2)*Te; % angular velocity for different voltages
wrl=0:1:225; Tl=0.02+2e-6*wrl.^2; % load torque at different velocities
```

```
plot(Te,wr,'-',Tl,wrl,'-');
title('Motor Torque-Speed Characteristics, \omega _ r(T _ e)','FontSize',14);
xlabel('Electromagnetic and Load Torques. T _ e and T _ L [N-m]','FontSize',14);
ylabel('Angular Velocity [rad/sec]','FontSize',14);
hold on; axis([0, 0.1, 0, 250]);
end;                                                                    □
```

It was shown that to regulate the angular velocity, one changes u_a. The power electronics and PWM amplifiers were covered in Chapter 3. The four-quadrant H-configured power stages guarantee high performance and efficiency. To rotate motor clockwise and counterclockwise, the bipolar voltage u_a should be applied to the armature winding. Permanent-magnet DC electric machines are made in different size. For ~500 W permanent-magnet DC motors, see the image in Figure 4.1; the schematics of a four-quadrant 25A PWM servo amplifier (20–80 V, ±12.5 A continuous current, ±25 A peak current, 22 kHz, 129 × 76 × 25 mm dimensions) is documented in Figure 4.6. The motor armature winding is connected to P2-1 and P2-2. To control the angular velocity, one supplies the reference voltage to P1-4, and the voltage induced by the tachogenerator (proportional to the motor angular velocity) is supplied to P1-6. This amplifier can be used in the servosystem applications, and the angular (or linear) displacement should be supplied to P1-6. The proportional-integral analog controller is integrated in the amplifier. One can change the proportional and integral feedback gains adjusting the potentiometers (resistors). Various PWM amplifiers are available from Advanced Motion Controls and other companies. For example, the 12A8 amplifiers specifications are: 20–80 V, ±6 A continuous current, ±12 A peak current, 36 kHz, and 129 × 76 × 25 mm dimensions.

The size of permanent-magnet DC electric machine can be less that the size of operational amplifier. High-performance 2 and 4 mm in diameter permanent-magnet DC motors have been manufactured. One must estimate the load torque. To guarantee the rotation, the following condition $T_e > T_L$ must be met. Furthermore, the acceleration capability is found as $(T_e - T_l)/J$. Once the electromagnetic torque T_e and angular velocity ω_r required has been derived, the power is found as $P = T_e\omega_r$. The sizing and volumeric features can be estimated by taking note that the power density is ~1 W/cm³. The power density is significantly affected by the design, dimensionality, permanent-magnet used, magnet dimensions, angular velocity, and other features.

A small ~1–10 W permanent-magnet DC motors can be driven by dual operational amplifiers as reported in Section 3.2.1. The schematics is depicted in Figure 4.7. The motor angular velocity is controlled by changing u_a. However, monolithic PWM amplifiers with corresponding ICs are available. For example, the MC33030 DC servo motor controller/driver (36 V and 1 A) integrates on-chip operational amplifier and comparator, driving logics, PWM four-quadrant converter and other circuitry as reported in Section 3.2.1. The built-in proportional controller changes the armature voltage u_a using the

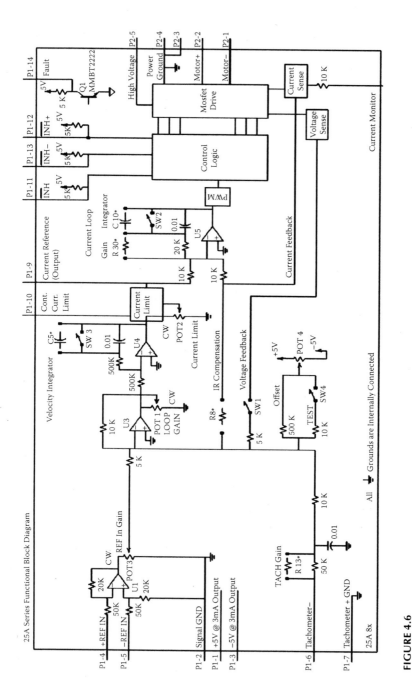

FIGURE 4.6
25A PWM servo amplifier (courtesy of Advanced Motion Controls, http://www.a-m-c.com) [6].

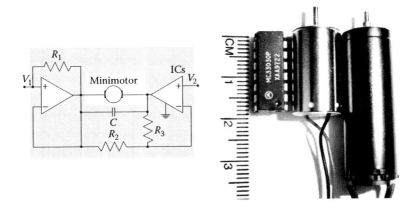

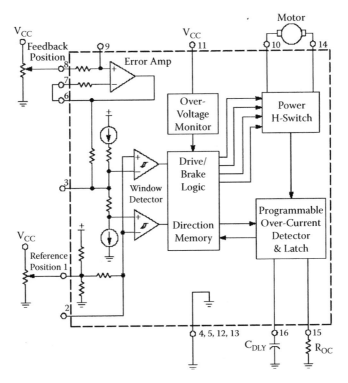

FIGURE 4.7
Application of dual operational amplifier to control permanent-magnet DC minimotors, and the MC33030 DC servo motor controller/driver with a permanent-magnet DC motor (copyright of Motorola, used with permission) [6].

difference between the reference and actual angular velocity (drive application) or displacement (servo application). We again document the MC33030 schematics in Figure 4.7 (MC33030 was reported in Figure 3.12).

If the permanent-magnet DC electric machine is used as a generator with a resistive load R_L, we have

$$\frac{di_a}{dt} = -\frac{r_a + R_L}{L_a} i_a + \frac{k_a}{L_a} \omega_r.$$

The induced *emf* is $E_a = k_a \omega_r$. In the steady-state operation, the induced terminal voltage is proportional to the angular velocity. In order to generate the voltage, one rotates a permanent-magnet DC electric machine by applying the prime mover torque T_{pm}. The applied torque could be the aerodynamic, hydrodynamic, thermal, or any other origin. The resulting differential equations of motion for a permanent-magnet DC generator are

$$\frac{di_a}{dt} = -\frac{r_a + R_L}{L_a} i_a + \frac{k_a}{L_a} \omega_r$$

$$\frac{d\omega_r}{dt} = -\frac{k_a}{J} i_a - \frac{B_m}{J} \omega_r + \frac{1}{J} T_{pm}. \qquad (4.4)$$

4.1.2 Simulation and Experimental Studies of Permanent-Magnet Direct-Current Machines

To accomplish a comprehensive analysis, analytical, numerical, and experimental studies should be carried out. One can perform simulation of permanent-magnet DC electric machines using different concepts. We start with the state-space linear differential equations.

Permanent-magnet DC electric machines are among a very limited class of electromechanical motion devices which can theoretically be described by linear differential equations. The majority of electric machines are described by nonlinear differential equations which can be solved using the differential equations solvers (in Chapters 2 and 3, we successfully applied ode45 command) and/or Simulink. Even though the equations of motion are described by the linear differential equations, the linear theory may not always be applied to permanent-magnet DC machines because there are constraints on the applied voltage $|u_a| \leq u_{a\,max}$.

The state-space model of permanent-magnet DC motors was found as given by (4.2), which is applied to perform simulations. The simulation parameters (final time, initial conditions and other) as well as electric machine parameters must be assigned. Let the initial conditions are zero and the final time is 0.5 sec. To carry out numerical studies, the coefficients of differential equations, which related to the parameters of electric machines, must be used. The following motor coefficients for a 12 V (rated) permanent-magnet DC motor were experimentally found (which also are given in the catalog): $r_a = 2$ ohm,

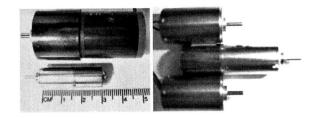

FIGURE 4.8
Geared and nongeared permanent-magnet DC motors with outputs $y = k_{gear}\omega_r$ and $y = \omega_r$.

$k_a = 0.05$ V-sec/rad (N-m/A), $L_a = 0.005$ H, $B_m = 0.0001$ N-m-sec/rad, and $J = 0.0001$ kg-m^2. In the Command Window we enter these parameters as

```
tfinal=0.5; x1initial=0; x2initial=0;
ra=2; ka=0.05; La=0.005; Bm=0.00001; J=0.0001;
```

The matrices of the state-space

$$\frac{d\mathbf{x}}{dt} = A\mathbf{x} + B\mathbf{u}, \quad \begin{bmatrix} \frac{di_a}{dt} \\ \frac{d\omega_r}{dt} \end{bmatrix} = \begin{bmatrix} -\frac{r_a}{L_a} & -\frac{k_a}{L_a} \\ \frac{k_a}{J} & \frac{B_m}{J} \end{bmatrix} \begin{bmatrix} i_a \\ \omega_r \end{bmatrix} + \begin{bmatrix} \frac{1}{L_a} \\ 0 \end{bmatrix} u_a - \begin{bmatrix} 0 \\ \frac{1}{J} \end{bmatrix} T_L$$

and output $\mathbf{y} = H\mathbf{x} + D\mathbf{u}$ equations $A \in \mathbb{R}^{2 \times 2}$, $B \in \mathbb{R}^{2 \times 1}$, $H \in \mathbb{R}^{1 \times 2}$ and $D \in \mathbb{R}^{1 \times 1}$ are uploaded as

```
A=[-ra/La -ka/La; ka/J -Bm/J]; B=[1/La; 0]; H=[0 1]; D=[0];
```

The state vector is $\mathbf{x} = [x_1 \ x_2]^T = [i_a \ \omega_r]^T$, while the output $y = \omega_r$ results in the output equation $\mathbf{y} = H\mathbf{x} + D\mathbf{u}$ with $H = [0 \ 1]$ and $D = [0]$. If the gears with k_{gear} are used, the output equation is $y = k_{gear}\omega_r$. The geared and nongeared permanent-magnet DC electric machines are documented in Figure 4.8. Using gears, one reduces (or increases) the output angular velocity ω_{rm} and increases (or reduces) the output torque T_{em}. Assuming that the efficiency of gear is 100 % (in practice, the maximum efficiency of a single gear stage is ~95%), one has $\omega_r T_e = \omega_{rm} T_{em}$.

Taking note of the motor parameters and output (angular velocity), we have the state-space model matrices

$$A = \begin{bmatrix} -400 & -10 \\ 500 & -0.1 \end{bmatrix}, \quad B \begin{bmatrix} 200 \\ 0 \end{bmatrix}, \quad H = [0 \ 1] \quad \text{and} \quad D = [0].$$

In particular,

```
A =
   -400.0000 -10.0000
    500.0000  -0.1000
B =
    200
      0
```

```
H =
    0  1
D =
    0
```

The following MATLAB file solves the simulation problem using the lsim command

```
t=0:.001: tfinal; x0=[x1initial x2initial];
Uaassigned=10; u=Uaassigned*ones(size(t));
[y,x]=lsim(A,B,H,D,u,t,x0);
plot(t,10*x(:,1),t,x(:,2),':');
xlabel('Time (seconds)','FontSize',14);
title('Angular Velocity \omega _ r [rad/sec] and Current i _ a [A]';
```

For the specified voltage applied (we let $u_a = 10$ V), the motor state variables are plotted in Figure 4.9. To visualize the dynamics for $i_a(t)$, the current is multiplied by factor of 10. The maximum armature current is 4.6 A. The simulation results illustrate that the motor reaches the final angular velocity ~200 rad/sec within ~0.5 sec. For $T_L = 0$, the steady-state value of the angular velocity should be $\omega_r = u_a/k_a = 200$ rad/sec; see equation (4.3). However, as evident from Figure 4.4, the steady-state ω_r is slightly less than 200 rad/sec. This is because of the friction torque, which is $B_m\omega_r \approx 0.002$ N-m. The armature current i_a will not decrease to zero because $T_{e\,\text{steady-state}} = B_m\omega_r = k_a i_a$.

The simulation of permanent-magnet DC electric machines also can be performed using Simulink. One needs to develop a Simulink block

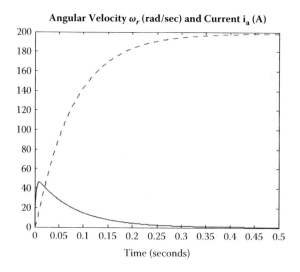

Angular Velocity ω_r (rad/sec) and Current i_a (A)

FIGURE 4.9
Dynamics of the state variables $i_a(t)$ (solid line) and $\omega_r(t)$ (dashed line).

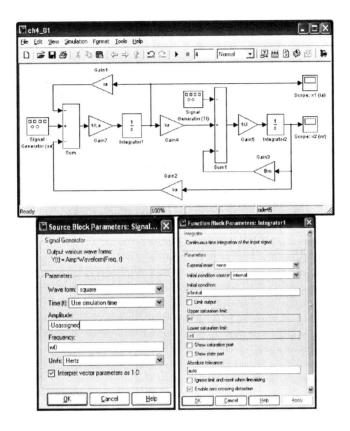

FIGURE 4.10
Simulink model to simulate permanent-magnet DC motors, Signal Generator and Integrator blocks.

diagram as covered in Section 2.7. An *s*-domain diagram for permanent-magnet DC motors was developed as given in Figure 4.3. Using this *s*-domain diagram, the corresponding Simulink model (ch4 _ 01.mdl) is built and represented in Figure 4.10. The initial conditions

$$x_0 = \begin{bmatrix} x_{10} \\ x_{20} \end{bmatrix} = \begin{bmatrix} 0 \\ 0 \end{bmatrix}$$

are applied, see the "Initial condition" line in the Integrator 1 (armature current) block, as shown in Figure 4.10. The Signal Generator block is used to set the applied voltage. As was emphasized in Section 2.7, the parameters can be assigned symbolically using symbols or relationships rather than using of numerical values. This ensures the greater flexibility. We enter the motor and simulation parameters as

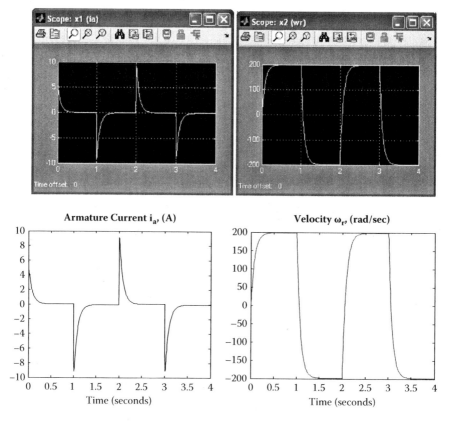

FIGURE 4.11
Permanent-magnet motor dynamics.

```
x1initial=0; x2initial=0; ra=2; ka=0.05; La=0.005; Bm=0.00001; J=0.0001;
Uaassigned=10; w0=0.5;
```

The applied armature voltage is u_a=10rect(0.5t) V. We set the load torque $T_L = 0$. The transient responses for two state variables, which are armature current $x_1(t) = i_a(t)$ and angular velocity $x_2(t) = \omega_r(t)$, are illustrated in Figure 4.11 using two scopes. To plot the motor dynamics, one may use the `plot` function. By using the stored data arrays x(:,1), x(:,2) and x1(:,1), x1(:,2) we type the following statements

```
plot(x(:,1),x(:,2)); xlabel('Time (seconds)','FontSize',14);
title('Armature Current i _ a, [A]','FontSize',14);
```

and

```
plot(x1(:,1),x1(:,2)); xlabel('Time (seconds)','FontSize',14);
title('Velocity \omega _ r, [rad/sec]','FontSize',14);
```

The resulting plots are documented in Figure 4.11.

As simulations are performed, the analysis can be accomplished. In particular, one analyzes the steady-state and dynamics responses of the state variables, settling time, overshoot, stability, and so on. For example, the efficiency and losses are of interest. The losses can be estimated as

$$P_l(t) = r_a i_a^2(t) + B_m \omega_r^2(t),$$

whereas the efficiency can be assessed using the input and output power as

$$\eta(t) = \frac{P_{output}(t)}{P_{input}(t)} = \frac{T_L(t)\omega_r(t)}{u_a(t)i_a(t)}, \ P_{input}(t) = P_l(t) + P_{output}(t) = r_a i_a^2(t) + B_m \omega_r^2(t) + T_L(t)\omega_r(t),$$

or

$$\eta(t) = \frac{P_{output}(t)}{P_{input}(t)} \times 100\% = \frac{T_L(t)\omega_r(t)}{u_a(t)i_a(t)} \times 100\%.$$

Meaningful steady-state and dynamic analysis of efficiency, power and losses should be accomplished. If electromechanical systems operate under varying inputs, loads and disturbances, this analysis should also be carried out. For motion devices that predominantly operate within the constant operating conditions, the steady-state analysis yields

$$\eta = \frac{P_{output}}{P_{input}} = \frac{T_L\Omega_r}{U_a I_a}, \quad P_{input} = P_l + P_{output} = r_a I_a^2 + B_m \Omega_r^2 + T_L\Omega_r,$$

or

$$\eta = \frac{P_{output}}{P_{input}} \times 100\% = \frac{T_L\Omega_r}{U_a I_a} \times 100\%.$$

Here, the steady-state angular velocity, load torque, voltage, and current are used. For many systems, the steady state-analysis can be accomplished from the dynamic analysis, but not vice versa. Therefore, the dynamic (behavioral) analysis is more general and should be prioritized. The losses increase significantly in the transient regimes. One may modify the Simulink model developed earlier to perform the analysis of power and losses. The resulting diagram (ch4 _ 011.mdl) is reported in Figure 4.12. The input power and losses are plotted using the following statements

```
plot(Pin(:,1),Pin(:,2));
xlabel('Time (seconds)','FontSize',14);
title('Input Power P_i_n [W]','FontSize',14);
```

and

```
plot(losses(:,1),losses(:,2));
xlabel('Time (seconds)','FontSize',14); title('Losses [W]','FontSize',14);
```

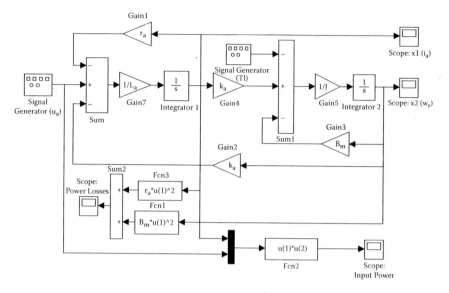

FIGURE 4.12
Simulink model to simulate permanent-magnet DC motors and perform the analysis of efficiency and losses.

The resulting plots are documented in Figure 4.13.

We demonstrated the application of MATLAB to simulate permanent-magnet DC motors. These simulations must be carried out in order to examine the system performance. However, the experimental studies are very important. The experimental results for JDH2250 permanent-magnet DC motors are

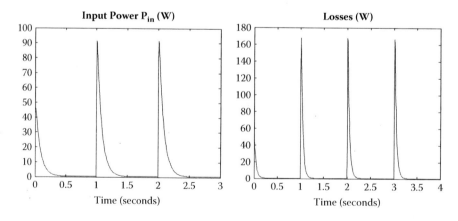

FIGURE 4.13
Input power and losses in a permanent-magnet DC motor.

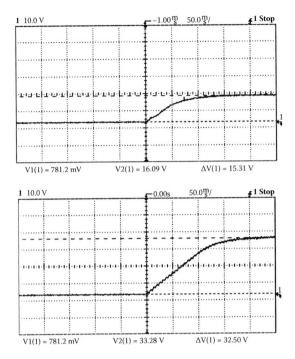

FIGURE 4.14
Motor acceleration to the angular velocity 150 and 300 rad/sec.

conducted. Figure 4.14 documents the acceleration of the unloaded motor if u_a is 7.5 and 15 V.

As the load torque T_L is applied, the angular velocity decreases. Figure 4.15 reports the acceleration and loading of the motor. The equation for toque-speed characteristics

$$\omega_r = \frac{u_a - r_a i_a}{k_a} = \frac{u_a}{k_a} - \frac{r_a}{k_a^2} T_e$$

illustrates the steady-state operation. The motor dynamics, as studied by solving the differential equations, provide general results. The experimental results are in a complete correspondence with the transient analysis performed using the equations of motion derived. The motor acceleration, transient dynamics waveforms, settling time, and other performance characteristics and capabilities were examined.

The disacceleration dynamics of the unloaded and loaded motor are depicted in Figure 4.16. The experimental results do not accurately match the simulated dynamics because of the complex friction phenomena emphasized in Example 2.7.

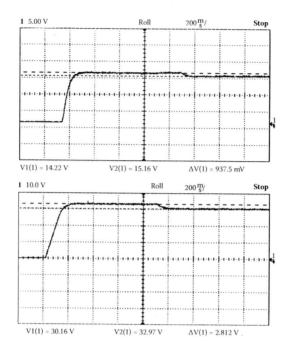

FIGURE 4.15
Motor acceleration and loading.

4.1.3 Permanent-Magnet Direct-Current Generator Driven by a Permanent-Magnet Direct-Current Motor

Our goal is to analyze and study two permanent-magnet DC electric machines integrated as the motor-generator system (see Figure 4.17). The prime mover (permanent-magnet DC motor) drives a generator.

For a permanent-magnet DC generator, if the resistive load is applied (R_L is inserted in series with the generator armature winding), the Kirchhoff voltage law yields

$$\frac{di_{ag}}{dt} = \frac{r_{ag} + R_L}{L_{ag}} i_{ag} + \frac{k_{ag}}{L_{ag}} \omega_{rpm},\tag{4.5}$$

where i_{ag} is the generator armature current; ω_{rpm} is the angular velocity of the prime mover and generator; r_{ag} and L_{ag} are the armature resistance and inductance of the generator winding; R_L is the load resistance; k_{ag} is the back *emf* (*torque*) constant of the generator.

Applying Newton's second law of motion, we obtain the *torsional-mechanical* dynamics of the generator-prime mover system as

$$\frac{d\omega_{rpm}}{dt} = \frac{1}{J_{pm} + J_g} [T_{epm} - (B_{mpm} + B_{mg})\omega_{rpm} - T_{eg}],$$

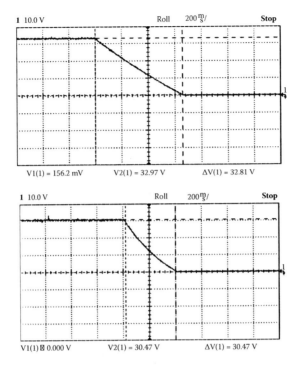

FIGURE 4.16
Unloaded and loaded motor disacceleration from 300 rad/sec to stall.

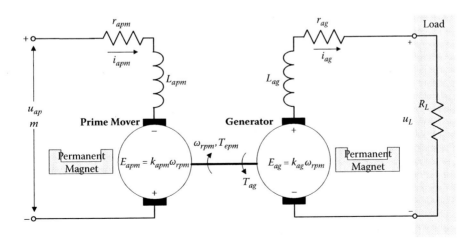

FIGURE 4.17
Permanent-magnet DC generator driven by a permanent-magnet DC motor.

where B_{mpm} and B_{mg} are the viscous friction coefficients; J_{pm} and J_g are the moments of inertia of the prime mover and generator.

The electromagnetic torque developed by the permanent-magnet DC motor is

$$T_{epm} = k_{apm} i_{apm},$$

where i_{apm} is the armature current in the prime mover winding; k_{apm} is the *torque* constant of the prime mover.

The load torque for the prime mover is the generator electromagnetic torque $T_{eg} = k_{ag} i_{ag}$. Thus, one obtains the *torsional-mechanical* dynamics as given by the following differential equation

$$\frac{d\omega_{rpm}}{dt} = \frac{k_{apm}}{J_{pm} + J_g} i_{apm} - \frac{B_{mpm} + B_{mg}}{J_{pm} + J_g} \omega_{rpm} - \frac{k_{ag}}{J_{pm} + J_g} i_{ag}. \qquad (4.6)$$

The dynamics of the prime mover armature current is described as

$$\frac{di_{rpm}}{dt} = \frac{r_{apm}}{L_{apm}} i_{apm} - \frac{k_{apm}}{L_{apm}} \omega_{rpm} + \frac{1}{L_{apm}} u_{apm}, \qquad (4.7)$$

where u_{apm} is the voltage applied to the armature winding of the prime mover; r_{apm} and L_{apm} are the armature resistance and inductance of the prime mover winding.

From (4.5), (4.6), and (4.7), we have the resulting set of equations

$$\frac{di_{apm}}{dt} = -\frac{r_{apm}}{L_{apm}} i_{apm} - \frac{k_{apm}}{L_{apm}} \omega_{rpm} + \frac{1}{L_{apm}} u_{apm},$$

$$\frac{di_{ag}}{dt} = -\frac{r_{ag} + R_L}{L_{ag}} i_{ag} + \frac{k_{ag}}{L_{ag}} \omega_{rpm},$$

$$\frac{d\omega_{rpm}}{dt} = \frac{k_{apm}}{J_{pm} + J_g} i_{apm} - \frac{B_{mpm} + B_{mg}}{J_{pm} + J_g} \omega_{rpm} - \frac{k_{ag}}{J_{pm} + J_g} i_{ag}. \qquad (4.8)$$

Example 4.2

The parameters of the electric machines are: $r_{apm} = 0.4$ ohm, $r_{ag} = 0.3$ ohm, $L_{apm} = 0.05$ H, $L_{ag} = 0.06$ H, $k_{apm} = 0.3$ V-sec/rad or N-m/A, $k_{ag} = 0.25$ V-sec/rad or N-m/A, $B_{mpm} = 0.0007$ N-m-sec/rad, $B_{mg} = 0.0008$ N-m-sec/rad, $J_{pm} = 0.04$ kg-m^2, and $J_g = 0.05$ kg-m^2. Our goal is to simulate and examine the steady-state and dynamic operation of a permanent-magnet DC generator driven by a 100 V permanent-magnet DC motor. We study the transient dynamics and the voltage generation for different resistive loads and angular velocities. The voltage u_{apm} is applied to change the angular velocity, and the R_L varies.

The rated angular velocity of the unloaded motor (if $R_L = \infty$) is

$$\omega_{rpm\,max} = u_{apm\,max}/k_{apm} = 100/0.3 = 333.3 \text{ rad/sec.}$$

We enter the parameters as

```
% Parameters of the permanent-magnet DC motor (prime mover)
rapm=0.4; Lapm=0.05; kapm=0.3; Jpm=0.04; Bmpm=0.0007;
% Parameters of the permanent-magnet DC generator
rag=0.3; Lag=0.06; kag=0.25; Jg=0.05; Bmg=0.0008;
% Load resistance
Rl=5;
```

A Simulink diagram is developed using a set of differential equations (4.8). The resulting diagram is illustrated in Figure 4.18 (ch4 _ 02.mdl).

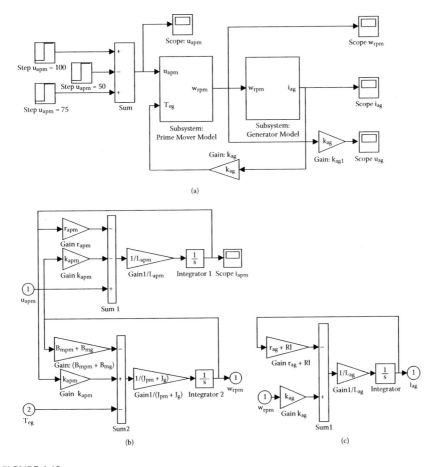

FIGURE 4.18
Simulink diagram to simulate a system with permanent-magnet DC motor and generator: (a) complete system model; (b) Simulink subsystem diagram of a permanent-magnet DC motor; (c) Simulink subsystem diagram of a permanent-magnet DC generator.

For distinct values of the inserted load resistor ($R_L = 5$, 25, and 100 ohm), by applying the following armature voltages to the motor (prime mover)

$$u_{apm} = \begin{cases} 100\,V, & \forall t \in [0\ 2) & \text{sec} \\ 50\,V, & \forall t \in [2\ 4) & \text{sec}, \\ 75\,V, & \forall t \in [4\ 6) & \text{sec} \end{cases}$$

the dynamics of the motor-generator system are simulated and studied. The transient behavior of the state variables, as well as the induced voltage waveforms, are documented in Figure 4.19.

As the motor starts from the stall, and the generator is loaded with $R_L = 5$ ohm, the settling time is 1.4 sec. For different steady-state angular velocities (315.3, 157, and 236.2 rad/sec) of the prime mover, which depend on the armature voltage applied (u_{apm} is 100, 50, and 75 V) and R_L, the steady-state values of the terminal voltage induced are 78.85, 39.3, and 59 V. One recalls that the

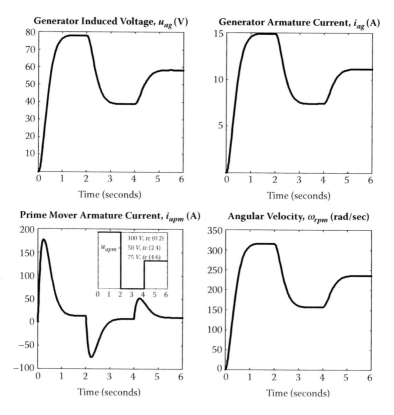

FIGURE 4.19
Motor-generator dynamics for $R_L = 5$, 25, and 100 ohm, respectively.

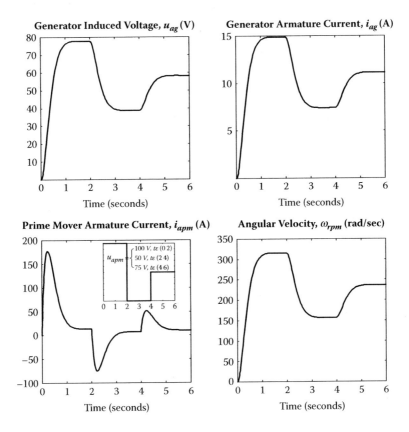

FIGURE 4.19
(Continued).

induced *emf* is $k_{ag}\omega_{rpm}$. The armature currents in the generator and prime mover are related to the load resistance.

Applying the specified armature voltage (u_{apm} is 50, 75, or 100 V), the values of the load resistance are assigned to be

$$R_L = \begin{cases} 100 \text{ ohm,} & \forall t \in [0\ 2) \quad \text{sec} \\ 50 \text{ ohm,} & \forall t \in [2\ 4) \quad \text{sec.} \\ 75 \text{ ohm,} & \forall t \in [4\ 6) \quad \text{sec} \end{cases}$$

A Simulink diagram is developed as documented in Figure 4.20 (ch4_03.mdl).

The dynamics of the permanent-magnet DC generator and motor are studied. The evolution of the induced *emf* $E_{ag}(t) = k_{ag}\omega_{rpm}(t)$, as well as the state variables $i_{apm}(t)$, $\omega_{rpm}(t)$ and $i_{ag}(t)$, are reported in Figure 4.21. One can analyze the acceleration capabilities, induced voltage variations due to loads, and efficiency, as well as other performance characteristics. □

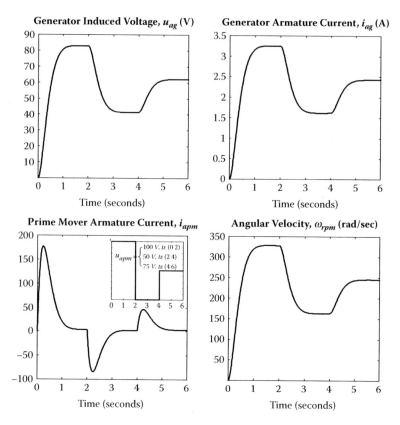

FIGURE 4.19
(Continued).

4.1.4 Electromechanical Systems with Power Electronics

The applied voltage to permanent-magnet DC motors is supplied by the high-frequency switching converters. One-quadrant converters were covered in Section 3.2. One can integrate and analyze electric machines controlled by PWM converters. For example, the schematics of a high-frequency *step-down* switching converter with a permanent-magnet DC motor is illustrated in Figure 4.22.

The voltage applied to the motor winding u_a is regulated by controlling the switching *on* and *off* durations t_{on} and t_{off}. One changes the duty ratio d_D, and

$$d_D = \frac{t_{on}}{t_{on} + t_{off}}.$$

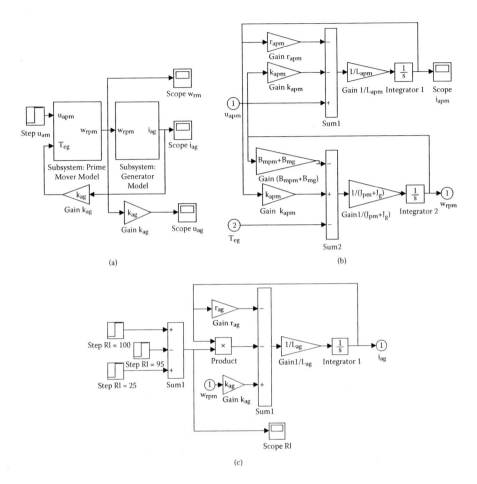

FIGURE 4.20
Simulink diagram to simulate the motor-generator system: (a) system model; (b) subsystem diagram to model a permanent-magnet DC motor; (c) subsystem diagram to model a permanent-magnet generator.

From (4.1), using the differential equations for the *buck* converter, as developed in Section 3.2.2 for the *RL* load

$$\frac{du_C}{dt} = \frac{1}{C}(i_L - i_a),$$

$$\frac{di_L}{dt} = \frac{1}{L}(-u_C - (r_L + r_c)i_L + r_c i_a - r_s i_L d_D + V_d d_D),$$

$$\frac{di_a}{dt} = \frac{1}{L_a}(u_C + r_c i_L - (r_a + r_c)i_a - E_a),$$

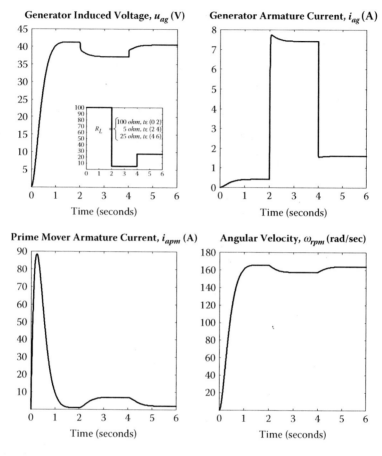

FIGURE 4.21
Dynamics of the system variables if

$$R_L = \begin{cases} 100 \text{ ohm}, & \forall t \in [0\ 2) \quad \text{sec} \\ 5 \text{ ohm}, & \forall t \in [2\ 4) \quad \text{sec} \\ 25 \text{ ohm}, & \forall t \in [4\ 6] \quad \text{sec} \end{cases},$$

whereas u_{apm} is 50, 75, and 100, respectively.

we have a set of five first-order differential equations. In particular,

$$\frac{du_C}{dt} = \frac{1}{C}(i_L - i_a),$$

$$\frac{di_L}{dt} = \frac{1}{L}(-u_C - (r_L + r_c)i_L + r_c i_a - r_s i_L d_D + V_d d_D),$$

$$\frac{di_a}{dt} = \frac{1}{L_a}(u_C + r_c i_L - (r_a + r_c)i_a - k_a \omega_r),$$

$$\frac{d\omega_r}{dt} = \frac{1}{J}(k_a i_a - B_m \omega_r - T_L).$$

(4.9)

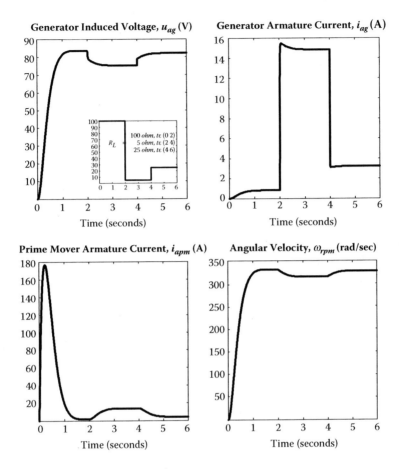

FIGURE 4.21
(Continued).

The duty ratio is regulated by changing the signal-level control voltage u_c, which cannot be greater than $u_{t\,max}$ and less than $u_{t\,min}$. That is, $u_{t\,min} \le u_c \le u_{t\,max}$. Hence, u_c is bounded. For $u_{t\,min} = 0$, we have

$$d_D = \frac{u_c}{u_{t\,max}} \in [0 \quad 1], \quad u_c \in [0 \quad u_{c\,max}], \quad u_{c\,max} = u_{t\,max}.$$

The signal-level control voltage u_c may be considered as a control input. From (4.9), we have

$$\frac{du_C}{dt} = \frac{1}{C}(i_L - i_a)$$

$$\frac{di_L}{dt} = \frac{1}{L}\left(-u_c - (r_L + r_c)i_L + r_c i_a - \frac{r_s}{u_{t\,max}}i_L u_c + \frac{V_d}{u_{t\,max}}u_c\right),$$

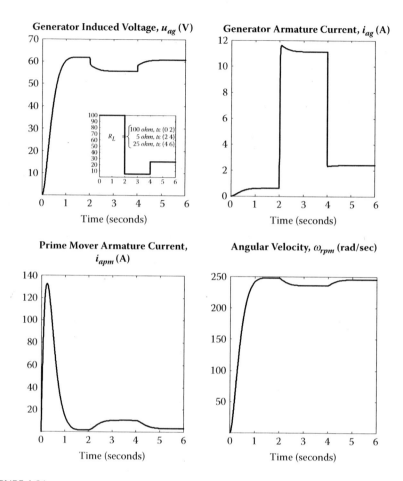

FIGURE 4.21
(Continued).

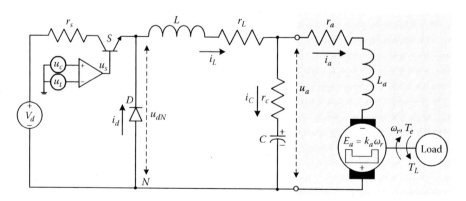

FIGURE 4.22
Permanent-magnet DC motor with a *step-down* switching converter.

$$\frac{di_a}{dt} = \frac{1}{L_a}(u_C + r_c i_L - (r_a + r_c)i_a - k_a \omega_r),$$

$$\frac{d\omega_r}{dt} = \frac{1}{J}(k_a i_a - B_m \omega_r - T_L).$$

The mathematical model of permanent-magnet DC motors with the *buck* converter is nonlinear due to a nonlinear term

$$\frac{r_s}{L u_{t\max}} i_L u_c.$$

Furthermore, the hard control bounds are imposed; in particular, $0 \le u_c \le u_{c\max}$, $u_c \in [0 \; u_{c\max}]$.

The armature voltage, applied to the motor windings, can be regulated by *boost (step-up)* dc-dc high-frequency switching converters. Using the equations of motion of the *boost* converter, as developed in Section 3.2.3, we have

$$\frac{du_C}{dt} = \frac{1}{C}\left(i_L - i_a - \frac{1}{u_{t\max}} i_L u_c\right),$$

$$\frac{di_L}{dt} = \frac{1}{L}\left(-u_c - (r_L + r_c)i_L + r_c i_a + \frac{1}{u_{t\max}}u_c u_c + \frac{(r_c - r_s)}{u_{t\max}}i_L u_c - \frac{r_c}{u_{t\max}}i_a u_c + V_d\right),$$

$$\frac{di_a}{dt} = \frac{1}{L_a}\left(u_C + r_c i_L - (r_a + r_c)i_a - \frac{r_c}{u_{t\max}}i_L u_c - k_a \omega_r\right),$$

$$\frac{d\omega_r}{dt} = \frac{1}{J}(k_a i_a - B_m \omega_r - T_L).$$

Section 3.2.5 examines the Cuk converter. Considering a permanent-magnet DC motor with the Cuk dc-dc converter, one obtains the following differential equations to be studied

$$\frac{du_{C1}}{dt} = \frac{1}{C_1}\left(i_{L1} - \frac{1}{u_{t\max}}i_{L1}u_c + \frac{1}{u_{t\max}}i_L u_c\right),$$

$$\frac{du_C}{dt} = \frac{1}{C}(i_L - i_a),$$

$$\frac{di_{L1}}{dt} = \frac{1}{L_1}\left(-u_{C1} - (r_{L1} + r_{c1})i_{L1} + \frac{1}{u_{t\max}}u_{C1}u_c + \frac{r_{c1} - r_s}{u_{t\max}}i_{L1}u_c + \frac{r_s}{u_{t\max}}i_L u_c + V_d\right),$$

$$\frac{di_L}{dt} = \frac{1}{L}\left(-u_C - (r_L + r_c)i_L + r_c i_a - \frac{1}{u_{t\max}}u_{C1}u_c + \frac{r_s}{u_{t\max}}i_{L1}u_c - \frac{r_{c1} + r_s}{u_{t\max}}i_L u_c\right),$$

$$\frac{di_a}{dt} = \frac{1}{L_a}(u_C + r_c i_L - (r_c + r_a)i_a - k_a \omega_r),$$

$$\frac{d\omega_r}{dt} = \frac{1}{J}(k_a i_a - B_m \omega_r - T_L).$$

We discussed the application of distinct high-frequency one-quadrant converters (*step-down*, *step-up*, Cuk, and others) to control electric motors. It was emphasized that two- and four-quadrant converters are commonly used. The corresponding analysis, design, and optimization tasks are straightforwardly performed by using the cornerstone foundations reported.

Example 4.3   Permanent-Magnet DC Motor with a *Step-Down* Converter

Consider a permanent-magnet DC motor with a *step-down* converter, as documented in Figure 4.22. A low-pass filter ensures ~5% voltage ripple. The converter parameters are: $r_s = 0.025$ ohm, $r_L = 0.02$ ohm, $r_c = 0.15$ ohm, $C = 0.003$ F, and $L = 0.0007$ H. Let $d_D = 0.5$ and $V_d = 50$ V. The motor coefficients are: $r_a = 2$ ohm, $k_a = 0.05$ V-sec/rad (N-m/A), $L_a = 0.005$ H, $B_m = 0.0001$ N-m-sec/rad, and $J = 0.0001$ kg-m^2.

In MATLAB we simulate the open-loop electromechanical system. Taking note of differential equations (4.9), two m-files are developed to perform the simulation assuming that the unloaded $(T_L = 0)$ motor accelerates from the stall. The MATLAB file (ch4 _ 1.m) is

```
t0=0; tfinal=0.4; tspan=[t0 tfinal]; y0=[0 0 0 0]';
[t,y]=ode45('ch4 _ 2',tspan,y0);
subplot(2,2,1); plot(t,y(:,1),'-');
xlabel('Time (seconds)','FontSize',12);
title('Voltage u _ C, [V]','FontSize',12);
subplot(2,2,2); plot(t,y(:,2),'-');
xlabel('Time (seconds)','FontSize',12);
title('Current i _ L, [A]','FontSize',12);
subplot(2,2,3); plot(t,y(:,3),'-');
xlabel('Time (seconds)','FontSize',12);
title('Current i _ a, [A]','FontSize',12);
subplot(2,2,4); plot(t,y(:,4),'-');
xlabel('Time (seconds)','FontSize',12);
title('Angular Velocity \omega _ r, [rad/sec]','FontSize',12);
```

whereas the second file (ch4 _ 2.m) is

```
% Dynamics of the PM DC motor with buck converter
function yprime=difer(t,y);
% parameters
Vd=50; D=0.5; rs=0.025; rl=0.02; rc=0.15; C=0.003; L=0.0007;
ra=2; ka=0.05; La=0.005; Bm=0.00001; J=0.0001; Tl=0;
% differential equations for PM DC Motor - Buck Converters
yprime=[(y(2,:)-y(3,:))/C;...
(-y(1,:)-(rl+rc)*y(2,:)+rc*y(3,:)-rs*y(2,:)*D+Vd*D)/L;...
(y(1,:)+rc*y(2,:)-(rc+ra)*y(3,:)-ka*y(4,:))/La;...
(ka*y(3,:)-Bm*y(4,:)-Tl)/J];
```

The transient dynamics for the state variables $u_C(t)$, $i_L(t)$, $i_a(t)$, and $\omega_r(t)$ are illustrated in Figure 4.23. The settling time is ~0.4 sec, and the motor reaches the steady-state angular velocity 496 rad/sec. This angular velocity can be obtained from equation (4.3). However, the analysis of transient dynamics is more general than the steady-state studies. For example, to evaluate efficiency, thermodynamics, acceleration capabilities and other important characteristics, the differential equations that describe the electromechanical system dynamics (and steady-state operation as well) must be solved.   □

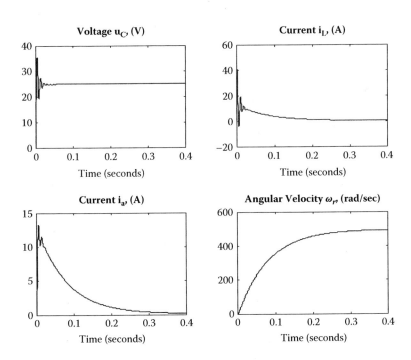

FIGURE 4.23
Transient dynamics of a permanent-magnet DC motor with a *buck* converter.

4.2 Axial Topology Permanent-Magnet Direct-Current Electric Machines

4.2.1 Fundamentals of Axial Topology Permanent-Magnet Machines

We have covered radial topology permanent-magnet DC electric machines, and now we will examine the axial topology electric machines commonly used as *disc* motors, hard drive actuators, *limited angle* motors, and so on. The planar segmented permanent magnets array is placed on the stator with the planar windings on the rotor. Brushes and commutator are used to supply the armature voltage to the windings on rotor. The planar windings on the rotor surface significantly simplify fabrication enabling one to fabricate affordable electromagnetic miniscale as well as conventional (high torque and power) machines. The concept and images of axial topology permanent-magnet DC electric machines are depicted in Figure 4.24.

Consider a current loop in the magnetic field that is produced by a permanent magnet. Assume that the magnetic flux is constant through the magnetic plane (current loop). The torque on a planar current loop of any size and shape in the uniform magnetic field is

$$\vec{T} = i\vec{s} \times \vec{B} = \vec{m} \times \vec{B},$$

where i is the current in the loop (winding); $\vec{m}$ is the magnetic dipole moment [A-m^2].

The torque is given as $\vec{T} = \vec{R} \times \vec{F}$, where for a filamentary closed loop we have $\vec{F} = -i\oint_l \vec{B} \times d\vec{l}$, which is simplified to $\vec{F} = -i\vec{B} \times \oint_l d\vec{l}$ for a uniform magnetic flux density distribution.

The torque on the current loop tends to turn the loop to align the magnetic field produced by the loop with permanent-magnet magnetic field causing the resulting electromagnetic torque. For example, a 10×20 cm current loop in a uniform magnetic field with the flux density $\vec{B} = -0.6\vec{a}_y + 0.8\vec{a}_z$ [T] is illustrated

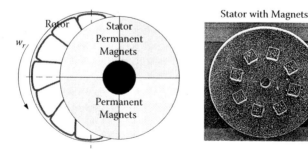

FIGURE 4.24
Axial topology permanent-magnet DC electric machine.

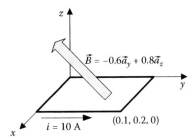

FIGURE 4.25
Rectangular planar current loop in an uniform magnetic field.

in Figure 4.25. The torque is

$$\vec{T} = i\vec{s} \times \vec{B} = \vec{m} \times \vec{B} = 10[(0.1)(0.2)\vec{a}_z] \times (-0.6\vec{a}_y + 0.8\vec{a}_z) = 0.12\vec{a}_x \text{ N-m}.$$

Hence, the loop tends to rotate around the axis parallel to the positive x axis. The electromagnetic force is

$$\vec{F} = \oint_l i d\vec{l} \times \vec{B} = -i \oint_l \vec{B} \times d\vec{l},$$

or, in the differential form, $d\vec{F} = id\vec{l} \times \vec{B}$. Furthermore, $\vec{T} = \vec{R} \times \vec{F}$. For a straight filament (conductor) in a uniform magnetic field the expression $\vec{F} = i\vec{l} \times \vec{B}$ is found from $\vec{F} = -i\vec{B} \times \oint_l d\vec{l}$. The interaction of the current (in the windings) and magnets result in the electromagnetic force and torque. Permanent magnets are magnetized in the specified direction to ensure the torque production resulting in the actuation (as a result of the force or torque developed). Each magnet produces the stationary magnetic field, and one considers the interactions of winding filaments with a number of magnets N_m ($N_m = 2m$, m is the integer) that are magnetized in the specific axis.

Consider a one-dimensional problem assuming that the resulting electromagnetic force, acting on the filament, is perpendicular to a filament producing the electromagnetic torque to rotate the rotor about a point O (center of the rotor). The expressions $\vec{F} = i\vec{l} \times \vec{B}$ and $\vec{T} = \vec{R} \times \vec{F}$ are simplified to a one-dimensional case assuming the ideal structural design.

We consider the rotational electromechanical motion devices. Under the sound simplifications and assumptions made, assuming an optimal structural design, one applies the *effective* flux density $B(\theta_r)$ which varies as a function of the angular displacement θ_r as a result of the angular displacement of rotor with windings relative to the stator with magnets (which produce the stationary field). Depending on the magnet magnetization, geometry, and shape, one applies distinct expressions for $B(\theta_r)$. For the permanent magnets illustrated in Figure 4.24, the flux density, as viewed from the windings, is a periodic function of θ_r. If there are no gaps between the magnets or strips of segmented magnets (which can be uniformly magnetized), or, if there is a spacing (or magnets geometry, shape or magnetization vary), one may use

$$B(\theta_r) = B_{max} \left| \sin\left(\frac{1}{2} N_m \theta_r\right) \right|$$

$$\text{or} \quad B(\theta_r) = B_{max} \sin^n\left(\frac{1}{2} N_m \theta_r\right), \quad n = 1,3,5,\ldots,$$

where B_{max} is the maximum effective flux density produced by the magnets as viewed from the winding (B_{max} depends on the magnets used, magnet-winding separation, temperature, etc.); N_m is the number of magnets (segments); n is the integer that is a function of the magnet magnetization, geometry, shape, width, thickness, separation between permanent-magnet segments, and so on.

Consider an illustrative example. For three $B(\theta_r)$ as given as

$$B(\theta_r) = B_{max} \sin\left(\frac{1}{2} N_m \theta_r\right), \quad B(\theta_r) = B_{max} \left| \sin\left(\frac{1}{2} N_m \theta_r\right) \right|, \quad \text{and}$$

$$B(\theta_r) = B_{max} \sin^5\left(\frac{1}{2} N_m \theta_r\right)$$

with $B_{max} = 0.9$ T and $N_m = 4$, we calculate and plot $B(\theta_r)$ using the following statements

```
th=0:.01:2*pi; Nm=4; Bmax=0.9; B=Bmax*sin(Nm*th/2); plot(th,B);
xlabel('Rotor Displacement, \theta _ r [rad]','FontSize',14);
ylabel('B(\theta _ r) [T]','FontSize',14);
title('Field as a Function on Displacement,B(\theta _ r)','FontSize',14);
```

and

```
th=0:.01:2*pi; Nm=4; Bmax=0.9; B=Bmax*sign(sin(Nm*th/2)); plot(th,B);
xlabel('Rotor Displacement, \theta _ r [rad]','FontSize',14);
ylabel('B(\theta _ r) [T]','FontSize',14);
title('Field as a Function on Displacement,B(\theta _ r)';
```

and

```
th=0:.01:2*pi; Nm=4; Bmax=0.9; B=Bmax*sin(Nm*th/2).^5; plot(th,B);
xlabel('Rotor Displacement, \theta _ r [rad]','FontSize',14);
ylabel('B(\theta _ r) [T]','FontSize',14);
title('Field as a Function on Displacement,B(\theta _ r)','FontSize',14);
```

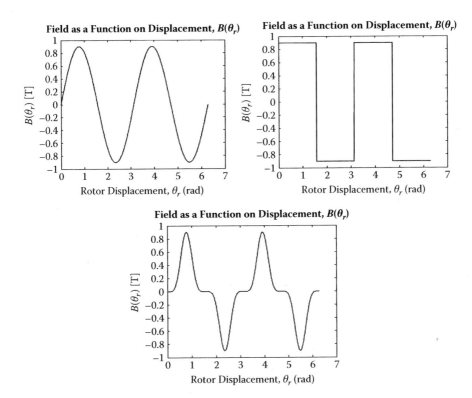

FIGURE 4.26

Plots of $B(\theta_r) = B_{\max} \sin(\frac{1}{2} N_m \theta_r)$, $B(\theta_r) = B_{\max} \left| \sin(\frac{1}{2} N_m \theta_r) \right|$ and $B(\theta_r) = B_{\max} \sin^5(\frac{1}{2} N_m \theta_r)$ $B_{\max} = 0.9$ T and $N_m = 4$.

The resulting plots for $B(\theta_r)$ are reported in Figure 4.26. One concludes that $B(\theta_r)$ can be described by

$$B(\theta_r) = B_{\max} \left| \sin\left(\frac{1}{2} N_m \theta_r \right) \right|, \qquad B(\theta_r) = \sum_{n=1}^{\infty} B_{\max n} \sin^{2n-1}\left(\frac{1}{2} N_m \theta_r \right),$$

as well as other periodic continuous or discontinuous functions. The analytic (or numerical) expression for $B(\theta_r)$ should be found in order to derive *emf* and T_e.

For a one-dimensional problem, the expressions $\vec{F} = i\vec{l} \times \vec{B}$ and $\vec{T} = \vec{R} \times \vec{F}$ result in

$$T_e = l_{eq} N i_a B(\theta_r),$$

where l_{eq} is the *effective* length, which includes the equivalent winding filament length and lever arm; N is the number of turns.

One also can use the expression for the coenergy $W_c(i_a,\theta_r) = A_{eq}(\theta_r)B(\theta_r)i_a$ to derive the electromagnetic torque. Here, A_{eq} is the *effective* area, which takes into account number of turns, magnetic field nonuniformity, etc.

Because of the use of brushes and commutator, the electromagnetic torque (produced by multiple filaments) is maximized by properly commutating coils by supplying u_a. For a great majority of axial and radial topology permanent-magnet DC motors, the expression for the electromagnetic torque is

$$T_e = k_a i_a,$$

where, for example, for

$$B(\theta_r) = B_{max}\left|\sin\left(\frac{1}{2}N_m\theta_r\right)\right|$$

one obtains

$$k_a = l_{eq}NB_{max}.$$

Using Kirchhoff's voltage law

$$u_a = r_a i_a + \frac{d\psi}{dt}$$

and Newton's second law of motion

$$\frac{d\omega_r}{dt} = \frac{1}{J}(T_e - B_m\omega_r - T_L)$$

the differential equations for the axial topology permanent-magnet DC motors are

$$\frac{di_a}{dt} = -\frac{r_a}{L_a}i_a - \frac{k_a}{L_a}\omega_r + \frac{1}{L_a}u_a, \qquad \frac{d\omega_r}{dt} = \frac{k_a}{J}i_a - \frac{B_m}{J}\omega_r - \frac{1}{J}T_L.$$

For the axial topology permanent-magnet DC generators, assuming the resistive load R_L, the induced *emf* is $E_a = k_a\omega_r$. The equations of motion for generators are

$$\frac{di_a}{dt} = -\frac{r_a + R_L}{L_a}i_a + \frac{k_a}{L_a}\omega_r$$

and

$$\frac{d\omega_r}{dt} = -\frac{k_a}{J}i_a - \frac{B_m}{J}\omega_r + \frac{1}{J}T_{pm}.$$

4.2.2 Axial Topology Hard Drive Actuator

Consider an axial topology permanent-magnet hard drive actuator with the segmented array formed as two permanent-magnet strips (see Figure 4.27).

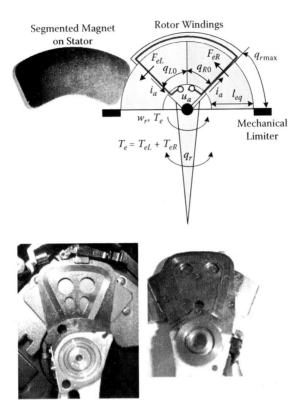

FIGURE 4.27
Axial topology hard-drive actuator.

To rotate this actuator clockwise or counterclockwise, the polarity of the applied voltage u_a is changed. This changes the direction of the electromagnetic force F_e developed by the left and right winding filaments. Because of different magnetization of left and right magnetic strips, the electromagnetic torque (force) is developed in the same direction. The mechanical limiters restrict the angular displacement to $-\theta_{rmax} \leq \theta_r \leq \theta_{rmax}$. As schematically illustrated in Figure 4.27, $-45 \leq \theta_r \leq 45$ deg. For the computer and camera hard drives illustrated in Figure 4.27, the displacement is $-10 \leq \theta_r \leq 10$ deg. The left filaments are on the rotor above the left magnetic strip. The interactions between the stationary magnetic field developed by the left magnet $B_L(\theta_r)$ and current i_a results in F_{eL} and T_{eL}. We assume that $B_R(\theta_r)$ produced by the right magnetic strip does not affect F_{eL} and T_{eL}. In addition, the left filaments are never above the right magnet. The same analysis is true for the right filaments. The images of two different actuators are reported in Figure 4.27, and one observes the similarity. We are considering the limited angle actuator, and the commutator is not required. If 360 degree rotation is required as in rotational motion devices, as the rotor rotates, the commutator is needed to change the polarity of dc voltage supplied to the filaments in order to develop the electromagnetic torque.

Kirchhoff's voltage law is

$$u_a = r_a i_a + \frac{d\psi}{dt}.$$

From $\psi = L_a i_a + A_{eq} B(\theta_r)$, we obtain

$$\frac{di_a}{dt} = \frac{1}{L_a}\left(-r_a i_a - A_{eq}\frac{dB(\theta_r)}{dt} + u_a\right).$$

There are two (left and right) filaments. Therefore, there are two *emf* terms. In particular,

$$\frac{di_a}{dt} = \frac{1}{L_a}\left(-r_a i_a - A_{eq}\frac{dB_L(\theta_r)}{d\theta_r}\omega_r - A_{eq}\frac{dB_R(\theta_r)}{d\theta_r}\omega_r + u_a\right).$$

Newton's second law of motion results in the following equations

$$\frac{d\omega_r}{dt} = \frac{1}{J}(T_e - B_m\omega_r - T_L), \quad T_e = T_{eL} + T_{eR},$$

$$\frac{d\theta_r}{dt} = \omega_r.$$

The analysis is straightforwardly performed as one takes note of $B(\theta_r)$. The distribution of $B(\theta_r)$ significantly affects the overall performance and capabilities. This is reflected by the model developed, which describes the transient and steady-state behavior. We consider two practical cases when two magnetic strips are magnetized to ensure

1. $B(\theta_r) = k\theta_r$, $k > 0$;
2. $B(\theta_r) = B_{max}\tanh(a\theta_r)$, $a > 0$.

For $B(\theta_r) = k\theta_r$, let $k = 1$, whereas for $B(\theta_r) = B_{max}\tanh(a\theta_r)$, we study $a = 10$ and $a = 100$ if $B_{max} = 0.7$ T.

Those $B(\theta_r)$ are plotted in Figure 4.28. To perform calculations and plotting, we have

```
th=-0.7:.01:0.7; k=1; Bmax=0.7; a1=10; a2=100;
B1=k*th; B21=Bmax*tanh(a1*th); B22=Bmax*tanh(a2*th);
plot(th,B1,':',th,B21,'-',th,B22,'--');
xlabel('Displacement, \theta_r [rad]','FontSize',14);
ylabel('B(\theta_r) [T]','FontSize',14);
title('Field as a Function of Displacement, B(\theta_r)','FontSize',14);
```

For $B(\theta_r) = k\theta_r$, we have

$$\frac{di_a}{dt} = \frac{1}{L_a}(-r_a i_a - 2A_{eq}k\omega_r + u_a).$$

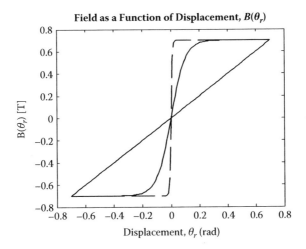

FIGURE 4.28
Plot of $B(\theta_r) = k\theta_r$, $k = 1$ and plots of $B(\theta_r) = B_{max}\tanh(a\theta_r)$, $B_{max} = 0.7$ T, $a = 10$, and $a = 100$.

The expression for the electromagnetic torque $T_e = T_{eL} + T_{eR}$ should be derived. We found $T_{ei} = l_{eq}Ni_aB_i(\theta_r)$. The angular displacements of two filament are $\theta_L(t)$ and $\theta_R(t)$. As documented in Figure 4.27, there are mechanical limits, and $-\theta_{rmax} \le \theta_r \le \theta_{rmax}$. For $-10 \le \theta_r \le 10$ deg, one has $-0.175 \le \theta_r \le 0.175$ rad; see the images of hard drives in Figure 4.27. Assuming $\theta_{r0} = 0$, $\theta_{L0} = 0.175$ rad and $\theta_{R0} = 0.175$ rad, for a symmetric kinematics studied, we have $\theta_L(t) = \theta_{L0} - \theta_r(t)$ and $\theta_R(t) = \theta_{R0} + \theta_r(t)$, $-0.175 \le \theta_r \le 0.175$ rad. It should be emphasized that the filament displacement angles $\theta_L(t)$ and $\theta_R(t)$ are constrained, but $\theta_L(t)$ and $\theta_R(t)$ are not the state variable.

As illustrated in Figure 4.27, the current i_a flows in different directions in the left and right filaments. The direction of the developed electromagnetic forces F_{eL} and F_{eR} is the same. Thus,

$$T_e = T_{eL} + T_{eR} = l_{eq}Nk(\theta_L + \theta_R)i_a, \ \theta_L(t) = \theta_{L0} - \theta_r(t), \ \theta_R(t) = \theta_{R0} + \theta_r(t), \ -\theta_{rmax} \le \theta_r \le \theta_{rmax}.$$

Using the nonideal Hook's law for the returning spring, we have

$$\frac{d\omega_r}{dt_r} = \frac{1}{J}[l_{eq}Nk(\theta_L + \theta_R)i_a - B_m\omega_r - k_s\theta_r - k_{s1}\theta_r^3 - T_{L\xi}], \quad \theta_L(t) = \theta_{L0} - \theta_r(t), \ \theta_R(t) = \theta_{R0} + \theta_r(t),$$

$$\frac{d\theta_r}{dt} = \omega_r, \quad -\theta_{rmax} \le \theta_r \le \theta_{rmax},$$

where $T_{L\xi}$ denotes the stochastic load torque.

The parameters can be estimated and experimentally measured. We have $k = 1$, $r_a = 35$ ohm, $L_a = 0.0041$ H, $l_{eq} = 0.015$ m, $N = 300$, $A_{eq} = 0.000045$, $B_m = 0.0005$ N-m-sec/rad, $k_s = 0.1$ N-m/rad, $k_{s1} = 0.05$ N-m/rad^3, and $J = 0.0000015$ kg-m^2.

The measured parameters can be easily justified. For example, for a copper winding with a radius of 0.05 mm and length $l_L + l_{top} + l_{bottom} + l_R$, we have

$$r_a = \frac{N(l_L + l_{top} + l_{bottom} + l_R)\sigma}{A} = \frac{300 \times 0.06 \times 1.72 \times 10^{-8}}{\pi (0.00005)^2} = 39 \text{ ohm.}$$

The circular loop self-inductance (loop radius is R_l and the wire diameter is d), is estimated as

$$L_a = N^2 R_1 \mu_0 \mu_r \left(\ln \frac{16 R_1}{d} - 2 \right)$$

$$= 300^2 \times 0.0075 \times 4\pi \times 10^{-7} \left(\ln \frac{8 \times 0.0075}{0.00005} - 2 \right) = 0.0043 \text{ H.}$$

The equation for the moment of inertia of thin disk $J = m R_{disk}^2$ overestimates the J value. We have the disk-centered geometry with the goal to minimize J in order to increase the acceleration capabilities. Therefore, the rotating pointer is made from plastic and has cavities, as illustrated in Figure 4.27. The estimated r_a, L_a and J are in the correspondence with the measured parameter values.

One uploads the parameters and constants as

```
k=1; N=300; ra=35; La=4.1e-3; leq=1.5e-2; Aeq=4.5e-5; Bm=5e-4;
ks=0.1; ks1=0.05; J=1.5e-6;
```

For the resulting equations of motion

$$\frac{di_a}{dt} = \frac{1}{L_a}(-r_a i_a - 2 A_{eq} k \omega_r + u_a),$$

$$\frac{d\omega_r}{dt} = \frac{1}{J}[l_{eq} N k (\theta_L + \theta_R) i_a - B_m \omega_r - k_s \theta_r - k_{s1} \theta_r^3 - T_{L\xi}], \quad \theta_L(t) = \theta_{L0} - \theta_r(t), \theta_R(t) = \theta_{R0} + \theta_r(t),$$

$$\frac{d\theta_r}{dt} = \omega_r, \quad -\theta_{r\max} \le \theta_r \le \theta_{r\max},$$

the corresponding Simulink diagram is depicted in Figure 4.29 (ch4 _ 04.mdl).

The transient dynamics for $i_a(t)$, $\omega_r(t)$, and $\theta_r(t)$ are reported in Figure 4.30. The plots are plotted using the saved scope data. The MATLAB statements to perform plotting are

```
plot(ia(:,1),ia(:,2)); xlabel('Time (seconds)','FontSize',14);
title('Current i_a, [A]','FontSize',14);
```

and

```
plot(wr(:,1),wr(:,2)); xlabel('Time (seconds)','FontSize',14);
title('Angular Velocity \omega_r, [rad/sec]','FontSize',14);
```

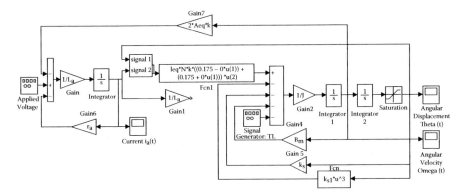

FIGURE 4.29
Simulink diagram to simulate the axial topology limited angle actuator when $B(\theta_r) = k\theta_r$.

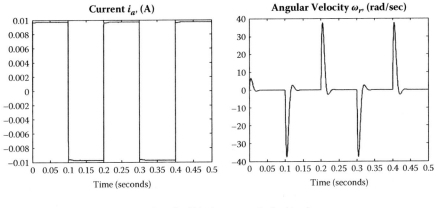

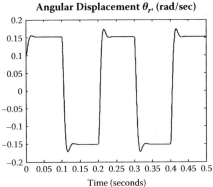

FIGURE 4.30
Transient dynamics of the state variables.

and

```
plot(th(:,1),th(:,2)); xlabel('Time (seconds)','FontSize',14);
title('Angular Displacement \theta _ r, [rad/sec]','FontSize',14);
```

We have examined the case when $B(\theta_r) = k\theta_r$. Consider the second case when the magnets are magnetized to ensure $B(\theta_r) = B_{max}\tanh(a\theta_r)$. From

$$\frac{di_a}{dt} = \frac{1}{L}\left(-r_a i_a - A_{eq}\frac{dB_L(\theta_r)}{d\theta_r}\omega_r - A_{eq}\frac{dB_R(\theta_r)}{d\theta_r}\omega_r + u_a\right),$$

we have

$$\frac{di_a}{dt} = \frac{1}{L_a}(-r_a i_a - A_{eq}aB_{max}\text{sech}^2(a\theta_L)\omega_r - A_{eq}aB_{max}\text{sech}^2(a\theta_R)\omega_r + u_a)$$

$$= \frac{1}{L_a}(-r_a i_a - A_{eq}aB_{max}\text{sech}^2 a(\theta_{L0} - \theta_r)\omega_r - A_{eq}aB_{max}\text{sech}^2 a(\theta_{R0} + \theta_r)\omega_r + u_a).$$

The electromagnetic torque is $T_e = T_{eL} + T_{eR}$.
Hence,

$$T_e = l_{eq}NB_{max}(\tanh a\theta_L + \tanh a\theta_R)i_a,$$

$$\theta_L(t) = \theta_{L0} - \theta_r(t),$$

$$\theta_R(t) = \theta_{R0} + \theta_r(t), \quad -\theta_{rmax} \le \theta_r \le \theta_{rmax}.$$

One obtains

$$\frac{d\omega_r}{dt} = \frac{1}{J}\left[l_{eq}NB_{max}(\tanh a\theta_L + \tanh a\theta_R)i_a - B_m\omega_r - k_s\theta_r - k_{s1}\theta_r^3 - T_{L\xi}\right]$$

$$= \frac{1}{J}\left\{l_{eq}NB_{max}[\tanh a(\theta_{L0} - \theta_r) + \tanh a(\theta_{R0} - \theta_r)]i_a - B_m\omega_r - k_s\theta_r - k_{s1}\theta_r^3 - T_{L\xi}\right\},$$

$$\frac{d\theta_r}{dt} = \omega_r, \quad -\theta_{rmax} \le \theta_r \le \theta_{rmax}.$$

The parameters and constants are uploaded as

```
Bmax=0.7; a=10; N=300; ra=35; La=4.1e-3; leq=1.5e-2;
Aeq=4.5e-5; Bm=5e-4; ks=0.1; ks1=0.05; J=1.5e-6;
```

The Simulink model (ch4 _ 05.mdl), which corresponds to the differential equations derived

$$\frac{di_a}{dt} = \frac{1}{L_a}(-r_a i_a - A_{eq}aB_{max}\text{sech}^2 a(\theta_{L0} - \theta_r)\omega_r - A_{eq}aB_{max}\text{sech}^2 a(\theta_{L0} + \theta_r)\omega_r + u_a),$$

$$\frac{d\omega_r}{dt} = \frac{1}{J}\{l_{eq}NB_{max}[\tanh a(\theta_{L0} - \theta_r) + \tanh a(\theta_{R0} - \theta_r)]i_a - B_m\omega_r - k_s\theta_r - k_{s1}\theta_r^3 - T_{L\xi}\},$$

$$\frac{d\theta_r}{dt} = \omega_r, \quad -\theta_{rmax} \le \theta_r \le \theta_{rmax}.$$

is reported in Figure 4.31.

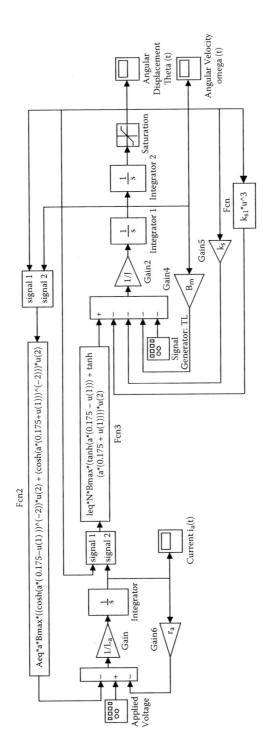

FIGURE 4.31
Simulink diagram to simulate the axial topology limited angle actuator when $B(\theta_r) = B_{max}\tanh(a\theta_r)$.

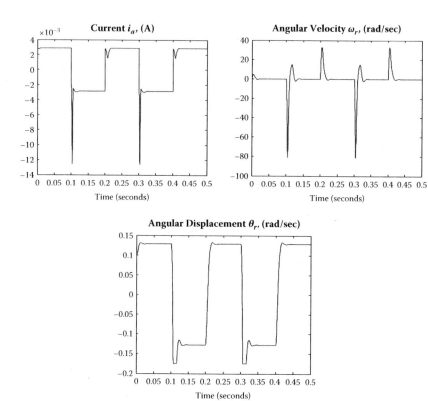

FIGURE 4.32
Transient dynamics of the state variables, $B(\theta_r) = B_{max}\tanh(a\theta_r)$, $a = 10$.

The dynamics of the state variables $i_a(t)$, $\omega_r(t)$ and $\theta_r(t)$ are represented in Figures 4.32 and 4.33 if in the expression $B(\theta_r) = B_{max}\tanh(a\theta_r)$ the constants are $a = 10$ and $a = 100$, respectively. The angular displacement is constrained by the mechanical limits $-0.175 \leq \theta_r \leq 0.175$ rad, and the Saturation block is inserted in the Simulink diagram to account for these bounds. One observes the effect of the return spring on the actuator dynamics when the pointer moves to the left and right. The performance of electromechanical motion devices is significantly affected by the magnet magnetization, parameters and constants, which depend on design, fabrication, materials, and so on. The closed-loop systems are used to optimize the system dynamics with the goal to obtain the optimal performance.

For $B(\theta_r) = B_{max}\tanh(a\theta_r)$, if $a = 100$, the resulting differential equations can be simplified by simplifying the expressions for the back *emf* and elec-tromagnetic torque. As shown in Figure 4.28, for even small displacement θ_r, $-\theta_{rmax} \leq \theta_r \leq \theta_{rmax}$, one has $B(\theta_r) = B_{max}\tanh(100\theta_r) \approx B_{max}\text{sgn}(\theta_r)$, yielding $B_L(\theta_r) \approx -B_{max}$ and $B_R(\theta_r) \approx B_{max}$. In fact, because of the mechanical limits and magnet magnetization, the variations of *emf* may appear only as the pointer moves at the very left and right. From $B_L(\theta_r) \approx$ const and $B_R(\theta_r) \approx$ const, one

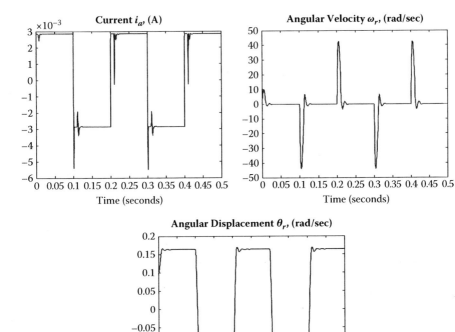

FIGURE 4.33
Transient dynamics of the state variables, $B(\theta_r) = B_{max}\tanh(a\theta_r)$, $a = 100$.

can assume $emf \approx 0$. Using

$$T_e = T_{eL} + T_{eR} = l_{eq}NB_{max}(\tanh a\theta_L + \tanh a\theta_R)i_a \approx 2l_{eq}NB_{max}i_a,$$

we have the equations of motion as

$$\frac{di_a}{dt} = \frac{1}{L_a}(-r_a i_a + u_a),$$

$$\frac{d\omega_r}{dt} = \frac{1}{J}[2l_{eq}NB_{max}i_a - B_m\omega_r - k_s\theta_r - k_{s1}\theta_r^3 - T_{L\xi}],$$

$$\frac{d\theta_r}{dt} = \omega_r, \quad -\theta_{r max} \leq \theta_r \leq \theta_{r max}.$$

The Simulink model (ch4 _ 06.mdl) is reported in Figure 4.34.

The dynamics of $i_a(t)$, $\omega_r(t)$ and $\theta_r(t)$ are illustrated in Figure 4.35. The comparison of the results reported in Figures 4.33 and 4.35 provides the evidence that though the overall dynamics is similar, for the data-intensive analysis, complete mathematical models must be used. For example, one uses $i_a(t)$ to

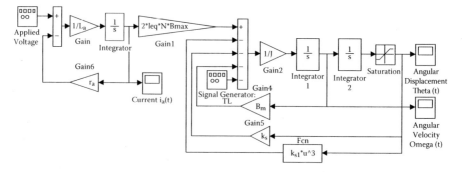

FIGURE 4.34
Simulink diagram to simulate the axial topology limited angle actuator described by a simplified set of differential equations.

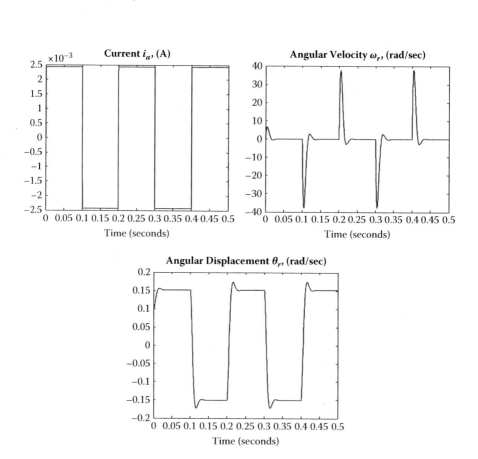

FIGURE 4.35
Transient dynamics of the state variables: simplified linear model.

evaluate the losses, thermodynamics, heating, and so on. An accurate model provides higher values of $i_a(t)$ to be used, ensuring accuracy and coherence.

4.3 Electromechanical Motion Devices: Synthesis and Classification

In Sections 4.1 and 4.2, we covered permanent-magnet DC electric machines and motion devices. The radial and axial topologies were studied. It was emphasized that ac electric machines (induction and synchronous) exist and will be covered in Chapters 5 and 6. Electric machines, reported in Chapters 4 to 6, are the electromagnetic devices. As illustrated in Chapter 2, electrostatic actuators exist and were examined. Furthermore, in addition to rotational devices, there exist translational (linear) transducers. Relays and solenoids, as electromagnetic variable-reluctance devices, were examined in Chapter 2 in details. This section documents the synthesis and classification aspects that lead not only to the understanding of structural design and device physics but also may contribute to devising novel electromechanical devices.

Consider a two-phase permanent-magnet synchronous slotless device, as shown in Figure 4.36. The electromagnetic system is *endless* (closed), and different geometries can be utilized as shown in Figures 4.36. In contrast, for translational (linear) synchronous devices, one yields the *open-ended* electromagnetic system.

The electromechanical motion devices may be studied not only as dc or ac (induction and synchronous), radial or axial, rotational or translational, but coherently classified by utilizing sound quantitative features. For example,

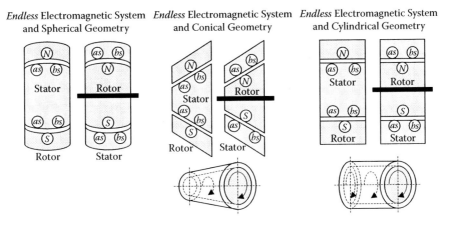

FIGURE 4.36
Permanent-magnet synchronous motion device with *endless* electromagnetic system and different geometry.

electromagnetic system, excitation, device and structure geometry (plate, spherical, torroidal, conical, cylindrical, etc.), winding configuration, and other descriptive and quantitative features may be integrated within the device physics prospective. This idea is useful not only to study the existing motion devices, but to synthesize sound high-performance devices that may have different electromagnetic systems (*endless, open-ended,* and *integrated*). The motion devices are categorized using a type classifier $Y = \{y: \in Y\}$. In addition, devices are distinguished using an electromagnetic system classifier (*endless E, open-ended O,* or *integrated I*) and a geometric classifier (plate *P,* spherical *S,* torroidal *T,* conical *N,* cylindrical *C,* or hybrid *A*). The classification concept, as visualized in Table 4.1, is represented by partitioning the descriptive features (electromagnetic system and geometry) by 3 horizontal and 6 vertical strips, yielding 18 sections. Each section is identified by the ordered pair of characters, for example, (*E, P*) or (*O, C*). In each ordered pair, the first entry is a letter chosen from the electromagnetic system set $M = \{E, O, I\}$.

TABLE 4.1

Classification of Motion Devices Using the Electromagnetic
System–Geometry Classifier

The second entry is a letter chosen from the geometric set $G = \{P, S, T, N, C, A\}$. For motion devises, the electromagnetic system–geometric set is

$$M \times G = \{(E, F), (E, S), (E, T), \ldots, (I, N), (I, C), (I, A)\}.$$

In general, $M \times G = \{(m, g) : m \in M \text{ and } g \in G\}$.

Other categorization can be applied. For example, single-, two-, three-, and multiphase devices are classified using a phase classifier $H = \{h : h \in H\}$. Therefore, we have

$$Y \times M \times G \times H = \{(y, m, g, h) : y \in Y, m \in M, g \in G, h \in H\}$$

Topology (radial or axial), permanent magnets shaping (strip, arc, disk, rectangular, triangular, etc.), permanent magnet characteristics (*BH* demagnetization curve, energy product, hysteresis minor loop, etc.), commutation, field distribution, cooling, power, torque, size, torque-speed characteristics, packaging, as well as other distinct features are easily classified.

Hence, the electromagnetic motion devices are classified as: {type, electromagnetic system, geometry, phase, topology, excitation, winding, connection, cooling, fabrication, materials, packaging, bearing, etc.}.

Using the *synthesis and classifier solver*, which is defined by Table 4.1 and the classification concept, the designer classifies the existing and synthesizes new motion devices. As an example, the *endless* electromagnetic system two-phase permanent-magnet synchronous machines with spherical, conical, and cylindrical geometries are illustrated in Figure 4.37. The spherical-conical and conical geometries ensure better performance and enhanced capabilities. These solutions guarantee higher power and torque densities, favorable thermodynamics, and other features valuable in automotive, aerospace, and marine power and propulsion systems.

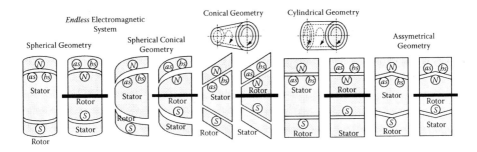

FIGURE 4.37
Two-phase permanent-magnet synchronous machines with *endless* electromagnetic system and distinct geometries.

Homework Problems

Problem 4.1

The torque-speed characteristic of a permanent-magnet DC motor is given in Figure 4.38 as one applies the rated voltage $u_a = 10$ V. Calculate the values for the *back emf* constant k_a, rated armature current i_a, and the armature resistance r_a. Assume that the viscous friction can be neglected.

Problem 4.2

Perform simulation (in MATLAB using Simulink) and analyze (study the steady-state and dynamic behavior, assess the losses for the unloaded and loaded motor, examine efficiency, etc.) of a 30 V, 10 A, 300 rad/sec permanent-magnet DC motor. The motor parameters are: $r_a = 1$ ohm, $k_a = 0.1$ V-sec/rad (N-m/A), $L_a = 0.005$ H, $B_m = 0.0001$ N-m-sec/rad and $J = 0.0001$ kg-m^2.

Problem 4.3

Develop the mathematical model for an axial topology limited angle actuator (computer hard drive) for which $-10^0 \leq \theta_r \leq 10^0$ or $-0.175 \leq \theta_r \leq 0.175$ rad. Let $B(\theta_r) = B_{max}\sin(a\theta_r)$, $a > 0$, $B_{max} = 1$ T.
 Perform simulation in Simulink and carry out analysis assigning $a = 1$, $a = 3$, or $a = 6$. Estimate and calculate the parameters of the actuator (or use the parameters reported in Section 4.2.2).
 Propose other sound expressions for $B(\theta_r)$ not reported in the chapter. For the proposed $B(\theta_r)$, derive the expression for T_e, develop a mathematical model, simulate the actuator, and analyze its performance.

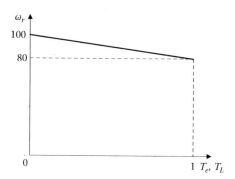

FIGURE 4.38
Torque-speed characteristic of a permanent-magnet DC motor.

References

1. S.J. Chapman, *Electric Machinery Fundamentals*, McGraw-Hill, Inc., New York, 1999.
2. A.E. Fitzgerald, C. Kingsley, and S.D. Umans, *Electric Machinery*, McGraw-Hill, Inc., New York, 1990.
3. P.C. Krause, O. Wasynczuk, and S.D. Sudhoff, *Analysis of Electric Machinery*, IEEE Press, New York, 1995.
4. P.C. Krause and O. Wasynczuk, *Electromechanical Motion Devices*, McGraw-Hill, New York, 1989.
5. W. Leonhard, *Control of Electrical Drives*, Springer, Berlin, 1996.
6. S.E. Lyshevski, *Electromechanical Systems, Electric Machines, and Applied Mechatronics*, CRC Press, Boca Raton, FL, 1999.
7. C.-M. Ong, *Dynamic Simulation of Electric Machines*, Prentice Hall, Upper Saddle River, NJ, 1998.
8. M.S. Sarma, *Electric Machines: Steady-State Theory and Dynamic Performance*, PWS Press, New York, 1996.
9. P.C. Sen, *Principles of Electric Machines and Power Electronics*, John Wiley and Sons, New York, 1989.
10. G.R. Slemon, *Electric Machines and Drives*, Addison-Wesley Publishing Company, Reading, MA, 1992.
11. D.C. White and H.H. Woodson, *Electromechanical Energy Conversion*, Wiley, New York, 1959.

5

Induction Machines

5.1 Fundamentals, Analysis, and Control of Induction Motors

5.1.1 Introduction

There are major torque production and energy conversion mechanisms that lead to the corresponding operation of electric machines as electromagnetic motion devices:

- *Induction Electromagnetics*—The phase voltages are induced in the rotor windings as a result of the time-varying stator magnetic field and motion of the rotor with respect to the stator. The electromagnetic torque results because of the interaction of time-varying electromagnetic fields.

- *Synchronous Electromagnetics*—The torque results because of the interaction of time-varying magnetic field established by the stator windings and stationary magnetic field established by the windings or magnets on the rotor.

- *Variable Reluctance Electromagnetics*—The torque is produced to minimize the reluctance of the electromagnetic system (solenoids, relays, etc.), and the torque is created by the magnetic system in an attempt to align the minimum-reluctance path of the rotor with the time-varying rotating airgap *mmf* (synchronous electric machine).

In high-performance drives and servos, permanent-magnet electric machines are used. For high-power (>200 kW) drives, three-phase induction machines are utilized. Fractional horse-power industrial and consumer drives use low-cost single- and three-phase induction machines [1–8]. Induction electric machines, compared with permanent-magnet motion devices, have much lower torque and power densities, may not be effectively used as generators, but may utilize simpler electronic and control solutions. Induction motors, as well as some other classes of electric machines, were invented and demonstrated by Nicola Tesla in 1888. In squirrel-cage induction electric motors, the phase voltages in the short-circuited rotor windings are induced as a result of

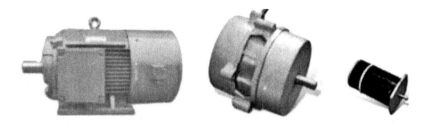

FIGURE 5.1
250 W induction and permanent-magnet synchronous (black) electric machines. (NEMA 56 frame size induction motors and NEMA 23 frame size permanent-magnet synchronous machine. NEMA 23 means 2.3 inches or 57 mm diameter.)

time-varying stator magnetic field as well as motion of the rotor with respect to the stator. The phase voltages are supplied to the stator windings. The electromagnetic torque results because of the interaction of the time-varying electromagnetic fields. The images of two 250 W induction motor are illustrated in Figure 5.1. For the illustrative purposes, a 250 W permanent-magnet synchronous machine (which is much smaller as compared to induction machines of the same rated power) is provided as the third image.

To analyze electric machines and evaluate their performance, one may perform the sequential steps starting from machine design and its optimization to modeling, simulations, testing, evaluation, characterization, and so on. It is unlikely that one may significantly improve the technology-centered designs of electric machines that have been successfully performed by different manufacturers. Stand-alone books concentrate on induction machine design, and those structural design tasks (three-dimensional electromagnetics, mechanical, thermal, vibroacoustic, etc.) are supported by sound CAD tools. In general, the machine design and optimization are of a great importance. Although three-dimensional tensor-centered electromagnetic analysis was covered in Chapter 2, the structural machine design tasks are beyond the scope of this book. For specific classes of electric machine, the design is covered in highly specialized books. The needs for structural machine design is obvious. However, only a very small fraction of engineers are involved because an optimal machine design is a narrow and highly specialized problem. Sound solutions are available and have been finalized within more than 100 years for existing broad lines and numerous series of DC, induction, and synchronous machines. In contrast, the power electronics, microelectronics, and sensors have been rapidly developed providing tremendous opportunities to advance electromechanical systems. Furthermore, the application and market for electric drives and servos have been notably expanded (avionics, bioengineering, electronics, information technologies, etc.) in addition to the conventional automotive, power, robotics and transportation areas. Therefore, we focus on the electromechanical system design issues considering electric machines as a key ready-to-use component rightly assuming the *optimal-performance* designed electric machines. In system design, modeling, simulation, and analysis are prioritized. For example, because *optimal-performance*

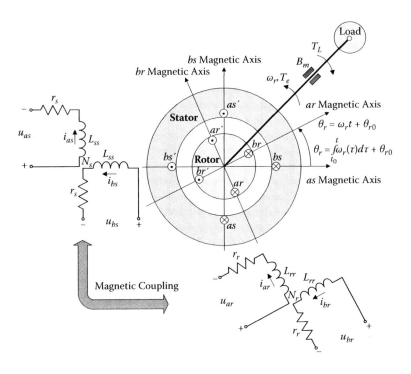

FIGURE 5.2
Two-phase symmetric induction motor.

electric machines exist, we concentrate on the systems design in order to guarantee the optimal *achievable* performance of electromechanical systems.

5.1.2 Two-Phase Induction Motors in Machine Variables

We study a two-phase induction machine as illustrated in Figure 5.2. To rotate induction machines and control the angular velocity, for squirrel-cage motors, one varies the frequency and magnitude of the phase voltages u_{as} and u_{bs} supplied to the stator windings. For the wound-rotor induction machines, in addition to controlling the magnitude and frequency of u_{as} and u_{bs}, one may vary the voltages u_{ar} and u_{br} supplied to the rotor windings. One should distinguish the voltages induced in the rotor windings (as the induced *emfs*) for the voltages supplied to the stator windings u_{ar} and u_{br}. The stator (*as* and *bs*) and rotor (*ar* and *br*) windings, as well as the stator-rotor magnetic coupling, are depicted in Figure 5.2. In squirrel-cage induction machines, the voltages are induced in the rotor windings due to the time-varying stator magnetic field and motion of the rotor with respect to the stator. The electromagnetic torque results due to the interaction of time-varying electromagnetic fields.

The analysis of induction machines starts from comprehension of device physics (energy conversion, torque production, etc.) progressing to modeling, simulation, performance evaluation, and capabilities assessment. The device physics of various electric machines was covered in previous chapters as well as in Section 5.1.1. To derive the governing equations for two-phase induction motors, we describe the stator and rotor circuitry-electromagnetic dynamics, model the *torsional-mechanical* behavior, and find the expression for electromagnetic torque. As the control and state variables, we use the voltages supplied to the *as, bs, ar,* and *br* windings, as well as the currents and flux linkages. Using Kirchhoff's voltage law, we have

$$u_{as} = r_s i_{as} + \frac{d\psi_{as}}{dt}, \qquad u_{bs} = r_s i_{bs} + \frac{d\psi_{bs}}{dt},$$

$$u_{ar} = r_r i_{ar} + \frac{d\psi_{ar}}{dt}, \qquad u_{br} = r_r i_{br} + \frac{d\psi_{br}}{dt}, \qquad (5.1)$$

where u_{as} and u_{bs} are the phase voltages supplied to the *as* and *bs* stator windings; u_{ar} and u_{br} are the phase voltages supplied to the *ar* and *br* rotor windings (for squirrel-cage motors, $u_{ar} = 0$ and $u_{br} = 0$); i_{as} and i_{bs} are the phase currents in the stator windings; i_{ar} and i_{br} are the phase currents in the rotor windings; ψ_{as} and ψ_{bs} are the stator flux linkages; ψ_{ar} and ψ_{br} are the rotor flux linkages; r_s and r_r are the resistances of the stator and rotor winding.

From (5.1), using the vector notations, one obtains the state-space form of Kirchhoff's voltage law as

$$\mathbf{u}_{abs} = \mathbf{r}_s \mathbf{i}_{abs} + \frac{d\psi_{abs}}{dt}, \qquad \mathbf{u}_{abr} = \mathbf{r}_r \mathbf{i}_{abr} + \frac{d\psi_{abr}}{dt}, \qquad (5.2)$$

where

$$\mathbf{u}_{abs} = \begin{bmatrix} u_{as} \\ u_{bs} \end{bmatrix}, \mathbf{u}_{abr} = \begin{bmatrix} u_{ar} \\ u_{br} \end{bmatrix}, \mathbf{i}_{abs} = \begin{bmatrix} i_{as} \\ i_{bs} \end{bmatrix}, \mathbf{i}_{abr} = \begin{bmatrix} i_{ar} \\ i_{br} \end{bmatrix}, \psi_{abs} = \begin{bmatrix} \psi_{as} \\ \psi_{bs} \end{bmatrix} \text{ and } \psi_{abr} = \begin{bmatrix} \psi_{ar} \\ \psi_{br} \end{bmatrix}$$

are the phase voltages, currents and flux linkages;

$$\mathbf{r}_s = \begin{bmatrix} r_s & 0 \\ 0 & r_s \end{bmatrix} \quad \text{and} \quad \mathbf{r}_r = \begin{bmatrix} r_r & 0 \\ 0 & r_r \end{bmatrix}$$

are the matrices of the stator and rotor resistances.

Assuming that the magnetic system is linear, the flux linkages are expressed using the self- and mutual inductances. In particular, we have

$$\psi_{as} = L_{asas} i_{as} + L_{asbs} i_{bs} + L_{asar} i_{ar} + L_{asbr} i_{br},$$

$$\psi_{bs} = L_{bsas} i_{as} + L_{bsbs} i_{bs} + L_{bsar} i_{ar} + L_{bsbr} i_{br},$$

$$\psi_{ar} = L_{aras} i_{as} + L_{arbs} i_{bs} + L_{arar} i_{ar} + L_{arbr} i_{br},$$

$$\psi_{br} = L_{bras} i_{as} + L_{brbs} i_{bs} + L_{brar} i_{ar} + L_{brbr} i_{br},$$

where $L_{asas}, L_{bsbs}, L_{arar},$ and L_{brbr} are the self-inductances of the stator and rotor windings; $L_{asbs}, L_{asar}, L_{asbr}, \dots, L_{bras}, L_{brbs}$ and L_{brar} are the mutual inductances between the corresponding stator and rotor windings, which denoted using the corresponding subscripts.

The stator and rotor self-inductances are denoted as L_{ss} and L_{rr}. One has

$$L_{ss} = L_{asas} = L_{bsbs} \quad \text{and} \quad L_{rr} = L_{arar} = L_{brbr}.$$

The stator (*as* and *bs*) and rotor (*ar* and *br*) windings are orthogonal. Therefore, there are no magnetic coupling between the *as* and *bs*, as well as between *ar* and *br* windings. Hence, the mutual inductances between the stator and rotor windings are

$$L_{asbs} = L_{bsas} = 0 \quad \text{and} \quad L_{arbr} = L_{brar} = 0.$$

The magnetic coupling between stator and rotor windings is studied. The mutual inductances are periodic functions of the electrical angular displacement of the rotor θ_r. Furthermore, the stator-rotor mutual inductances have minimum and maximum values. As illustrated in Figure 5.2, stator is the stationary member, and the rotor rotates with the electrical angular velocity ω_r. The stator-rotor winding coupling due to the mutual inductance is shown in Figure 5.2. Assuming that these variations obey the cosine law, we have the following expressions for the mutual inductances between stator and rotor windings

$$L_{asar} = L_{aras} = L_{sr}\cos\theta_r, \, L_{asbr} = L_{bras} = -L_{sr}\sin\theta_r, \text{ and}$$

$$L_{bsar} = L_{arbs} = L_{sr}\sin\theta_r, \, L_{bsbr} = L_{brbs} = L_{sr}\cos\theta_r.$$

For the magnetically coupled windings, we found the expressions for the flux linkages as

$$\psi_{as} = L_{ss}i_{as} + L_{sr}\cos\theta_r i_{ar} - L_{sr}\sin\theta_r i_{br},$$

$$\psi_{bs} = L_{ss}i_{bs} + L_{sr}\sin\theta_r i_{ar} + L_{sr}\cos\theta_r i_{br},$$

$$\psi_{ar} = L_{sr}\cos\theta_r i_{as} + L_{sr}\sin\theta_r i_{bs} + L_{rr}i_{ar},$$

$$\psi_{br} = -L_{sr}\sin\theta_r i_{as} + L_{sr}\cos\theta_r i_{bs} + L_{rr}i_{br}.$$

The following expression for the flux linkages results

$$\begin{bmatrix} \boldsymbol{\psi}_{abs} \\ \boldsymbol{\psi}_{abs} \end{bmatrix} = \begin{bmatrix} \mathbf{L}_s & \mathbf{L}_{sr}(\theta_r) \\ \mathbf{L}_{sr}^T(\theta_r) & \mathbf{L}_r \end{bmatrix} \begin{bmatrix} \mathbf{i}_{abs} \\ \mathbf{i}_{abr} \end{bmatrix},$$

where $\mathbf{L}_s$ is the matrix of the stator self-inductances,

$$\mathbf{L}_s = \begin{bmatrix} L_{ss} & 0 \\ 0 & L_{ss} \end{bmatrix}, \quad L_{ss} = L_{ls} + L_{ms}, \quad L_{ms} = \frac{N_s^2}{\Re_m};$$

$\mathbf{L}_r$ is the matrix of the rotor self-inductances,

$$\mathbf{L}_r = \begin{bmatrix} L_{rr} & 0 \\ 0 & L_{rr} \end{bmatrix}, \quad L_{rr} = L_{lr} + L_{mr}, \quad L_{mr} = \frac{N_r^2}{\mathcal{R}_m};$$

$\mathbf{L}_{sr}(\theta_r)$ is the stator-rotor mutual inductance mapping,

$$\mathbf{L}_{sr}(\theta_r) = \begin{bmatrix} L_{sr}\cos\theta_r & -L_{sr}\sin\theta_r \\ L_{sr}\sin\theta_r & L_{sr}\cos\theta_r \end{bmatrix}, \quad L_{sr} = \frac{N_s N_r}{\mathcal{R}_m};$$

L_{ms} and L_{mr} are the stator and rotor magnetizing inductances; L_{ls} and L_{lr} are the stator and rotor leakage inductances; N_s and N_r are the number of turns of the stator and rotor windings.

Using the number of turns in the stator and rotor windings, we have

$$\mathbf{i}'_{abr} = \frac{N_r}{N_s}\mathbf{i}_{abr}, \quad \mathbf{u}'_{abr} = \frac{N_s}{N_r}\mathbf{u}_{abr}, \quad \text{and} \quad \boldsymbol{\psi}'_{abr} = \frac{N_s}{N_r}\boldsymbol{\psi}_{abr}.$$

Applying the turn ratio, the flux linkages are given as

$$\begin{bmatrix} \boldsymbol{\psi}_{abs} \\ \boldsymbol{\psi}'_{abs} \end{bmatrix} = \begin{bmatrix} \mathbf{L}_s & \mathbf{L}'_{sr}(\theta_r) \\ \mathbf{L}'^T_{sr}(\theta_r) & \mathbf{L}'_s \end{bmatrix} \begin{bmatrix} \mathbf{i}_{abs} \\ \mathbf{i}'_{abr} \end{bmatrix},$$

$$\mathbf{L}'_r = \left(\frac{N_s}{N_r}\right)^2 \mathbf{L}_r = \begin{bmatrix} L'_{rr} & 0 \\ 0 & L'_{rr} \end{bmatrix}, \quad \mathbf{L}'_{sr}(\theta_r) = \left(\frac{N_s}{N_r}\right)\mathbf{L}_{sr}(\theta_r) = L_{ms} \begin{bmatrix} \cos\theta_r & -\sin\theta_r \\ \sin\theta_r & \cos\theta_r \end{bmatrix},$$

where

$$L'_{rr} = L'_{lr} + L'_{mr}, \quad L_{ms} = \frac{N_s}{N_r}L_{sr}, \quad L'_{mr} = \left(\frac{N_s}{N_r}\right)^2 L_{mr},$$

$$L'_{mr} = L_{ms} = \frac{N_s}{N_r}L_{sr} \quad \text{and} \quad L'_{rr} = L'_{lr} + L_{ms}.$$

From $\mathbf{L}_s$, $\mathbf{L}'_r$ and $\mathbf{L}'_{sr}(\theta_r)$, one obtains

$$\begin{bmatrix} \psi_{as} \\ \psi_{bs} \\ \psi'_{ar} \\ \psi'_{br} \end{bmatrix} = \begin{bmatrix} L_{ss} & 0 & L_{ms}\cos\theta_r & -L_{ms}\sin\theta_r \\ 0 & L_{ss} & L_{ms}\sin\theta_r & L_{ms}\cos\theta_r \\ L_{ms}\cos\theta_r & L_{ms}\sin\theta_r & L'_{rr} & 0 \\ -L_{ms}\sin\theta_r & L_{ms}\cos\theta_r & 0 & L'_{rr} \end{bmatrix} \begin{bmatrix} i_{as} \\ i_{bs} \\ i'_{ar} \\ i'_{br} \end{bmatrix}. \quad (5.3)$$

Therefore, the differential equations (5.1) and (5.2) are written as

$$\mathbf{u}_{abs} = \mathbf{r}_s \mathbf{i}_{abs} + \frac{d\boldsymbol{\psi}_{abs}}{dt}, \quad \mathbf{u}'_{abr} = \mathbf{r}'_r \mathbf{i}'_{abr} + \frac{d\boldsymbol{\psi}'_{abr}}{dt}, \quad (5.4)$$

where

$$\mathbf{r}_r' = \frac{N_s^2}{N_r^2} \mathbf{r}_r = \frac{N_s^2}{N_r^2} \begin{bmatrix} r_r & 0 \\ 0 & r_r \end{bmatrix}.$$

The self-inductances L_{ss} and L_{rr}' are time-invariant. Furthermore, L_{ms} is a constant. From (5.4), using the expressions for the flux linkages (5.3), one obtains a set of four nonlinear differential equations

$$L_{ss}\frac{di_{as}}{dt} + L_{ms}\frac{d(i_{ar}'\cos\theta_r)}{dt} - L_{ms}\frac{d(i_{br}'\sin\theta_r)}{dt} = -r_s i_{as} + u_{as},$$

$$L_{ss}\frac{di_{bs}}{dt} + L_{ms}\frac{d(i_{ar}'\sin\theta_r)}{dt} + L_{ms}\frac{d(i_{ar}'\cos\theta_r)}{dt} = -r_s i_{bs} + u_{bs},$$

$$L_{ms}\frac{d(i_{as}\cos\theta_r)}{dt} + L_{ms}\frac{d(i_{bs}\sin\theta_r)}{dt} + L_{rr}'\frac{di_{ar}'}{dt} = -r_r'i_{ar}' + u_{ar}',$$

$$-L_{ms}\frac{d(i_{as}\sin\theta_r)}{dt} + L_{ms}\frac{d(i_{bs}\cos\theta_r)}{dt} + L_{rr}'\frac{di_{br}'}{dt} = -r_{br}'i_{br}' + u_{br}'. \qquad (5.5)$$

The *emf* is given as

$$emf = \oint_l \vec{E}\cdot d\vec{l} = \underbrace{\oint_l (\vec{v}\times\vec{B})\cdot d\vec{l}}_{\substack{\text{motional induction}\\\text{(generation)}}} - \underbrace{\oint_s \frac{\partial\vec{B}}{\partial t}\,d\vec{s}}_{\substack{\text{transformer}\\\text{induction}}},$$

and the Faraday law of induction is

$$\mathcal{E} = \oint_l \vec{E}(t)\cdot d\vec{l} = -\frac{d}{dt}\int_s \vec{B}(t)\cdot d\vec{s} = -N\frac{d\Phi}{dt} = -\frac{d\psi}{dt}.$$

The unit for the *emf* is volts. From (5.5), the induced *emf* terms for the rotor windings are

$$emf_{ar} = -L_{ms}\frac{d(i_{as}\cos\theta_r)}{dt} - L_{ms}\frac{d(i_{bs}\sin\theta_r)}{dt} - L_{rr}'\frac{di_{ar}'}{dt},$$

$$emf_{br} = L_{ms}\frac{d(i_{as}\sin\theta_r)}{dt} - L_{ms}\frac{d(i_{bs}\cos\theta_r)}{dt} - L_{rr}'\frac{di_{br}'}{dt}.$$

Hence, for the rotor windings, the motional *emf* terms in the steady-state operation are

$$emf_{ar\omega} = L_{ms}(i_{as}\sin\theta_r - i_{bs}\cos\theta_r)\omega_r \quad \text{and} \quad emf_{br\omega} = L_{ms}(i_{as}\cos\theta_r + i_{bs}\sin\theta_r)\omega_r.$$

These *emfs* justify the statement that the voltages are induced in the rotor windings.

From (5.5), Cauchy's form of differential equations is found. In particular, we have the following nonlinear differential equations

$$
\frac{di_{as}}{dt} = -\frac{L'_{rr}r_s}{L_{ss}L'_{rr} - L^2_{ms}}i_{as} + \frac{L^2_{ms}}{L_{ss}L'_{rr} - L^2_{ms}}i_{bs}\omega_r + \frac{L_{ms}L'_{rr}}{L_{ss}L'_{rr} - L^2_{ms}}i'_{ar}\left(\omega_r\sin\theta_r + \frac{r'_r}{L'_{rr}}\cos\theta_r\right)
$$
$$
+ \frac{L_{ms}L'_{rr}}{L_{ss}L'_{rr} - L^2_{ms}}i'_{br}\left(\omega_r\cos\theta_r - \frac{r'_r}{L'_{rr}}\sin\theta_r\right) + \frac{L'_{rr}}{L_{ss}L'_{rr} - L^2_{ms}}u_{as}
$$
$$
- \frac{L_{ms}}{L_{ss}L'_{rr} - L^2_{ms}}\cos\theta_r u'_{ar} + \frac{L_{ms}}{L_{ss}L'_{rr} - L^2_{ms}}\sin\theta_r u'_{br},
$$

$$
\frac{di_{bs}}{dt} = -\frac{L'_{rr}r_s}{L_{ss}L'_{rr} - L^2_{ms}}i_{bs} - \frac{L^2_{ms}}{L_{ss}L'_{rr} - L^2_{ms}}i_{as}\omega_r - \frac{L_{ms}L'_{rr}}{L_{ss}L'_{rr} - L^2_{ms}}i'_{ar}\left(\omega_r\cos\theta_r - \frac{r'_r}{L'_{rr}}\sin\theta_r\right)
$$
$$
+ \frac{L_{ms}L'_{rr}}{L_{ss}L'_{rr} - L^2_{ms}}i'_{ar}\left(\omega_r\sin\theta_r + \frac{r'_r}{L'_{rr}}\cos\theta_r\right) + \frac{L'_{rr}}{L_{ss}L'_{rr} - L^2_{ms}}u_{bs}
$$
$$
- \frac{L_{ms}}{L_{ss}L'_{rr} - L^2_{ms}}\sin\theta_r u'_{ar} - \frac{L_{ms}}{L_{ss}L'_{rr} - L^2_{ms}}\cos\theta_r u'_{br},
$$

$$
\frac{di'_{ar}}{dt} = -\frac{L_{ss}r'_r}{L_{ss}L'_{rr} - L^2_{ms}}i'_{ar} + \frac{L_{ms}L_{ss}}{L_{ss}L'_{rr} - L^2_{ms}}i_{as}\left(\omega_r\sin\theta_r + \frac{r_s}{L_{ss}}\cos\theta_r\right)
$$
$$
- \frac{L_{ms}L_{ss}}{L_{ss}L'_{rr} - L^2_{ms}}i_{bs}\left(\omega_r\cos\theta_r - \frac{r_s}{L_{ss}}\sin\theta_r\right)
$$
$$
- \frac{L^2_{ms}}{L_{ss}L'_{rr} - L^2_{ms}}i'_{br}\omega_r - \frac{L_{ms}}{L_{ss}L'_{rr} - L^2_{ms}}\cos\theta_r u_{as}
$$
$$
- \frac{L_{ms}}{L_{ss}L'_{rr} - L^2_{ms}}\sin\theta_r u_{bs} + \frac{L_{ss}}{L_{ss}L'_{rr} - L^2_{ms}}u'_{ar},
$$

$$
\frac{di'_{br}}{dt} = -\frac{L_{ss}r'_r}{L_{ss}L'_{rr} - L^2_{ms}}i'_{br} + \frac{L_{ms}L_{ss}}{L_{ss}L'_{rr} - L^2_{ms}}i_{as}\left(\omega_r\cos\theta_r - \frac{r_s}{L_{ss}}\sin\theta_r\right)
$$
$$
+ \frac{L_{ms}L_{ss}}{L_{ss}L'_{rr} - L^2_{ms}}i_{bs}\left(\omega_r\sin\theta_r + \frac{r_s}{L_{ss}}\cos\theta_r\right)
$$
$$
+ \frac{L^2_{ms}}{L_{ss}L'_{rr} - L^2_{ms}}i'_{ar}\omega_r + \frac{L_{ms}}{L_{ss}L'_{rr} - L^2_{ms}}\sin\theta_r u_{as}
$$
$$
- \frac{L_{ms}}{L_{ss}L'_{rr} - L^2_{ms}}\cos\theta_r u_{bs} + \frac{L_{ss}}{L_{ss}L'_{rr} - L^2_{ms}}u'_{br}. \tag{5.6}
$$

The electrical angular velocity ω_r and displacement θ_r are used in (5.6) as the state variables. Therefore, the *torsional-mechanical* equation of motion must be derived describing the evolution of ω_r and θ_r. From Newton's second law, we have

$$\sum T = J \frac{d\omega_{rm}}{dt}, \quad \frac{d\theta_{rm}}{dt} = \omega_{rm},$$

where

$$\sum T = T_e - B_m \omega_{rm} - T_L.$$

Hence,

$$\frac{d\omega_{rm}}{dt} = \frac{1}{J}(T_e - B_m \omega_{rm} - T_L),$$

$$\frac{d\theta_{rm}}{dt} = \omega_{rm}.$$

The mechanical angular velocity of the rotor ω_{rm} is expressed by using the electrical angular velocity ω_r and the number of poles P. In particular,

$$\omega_{rm} = \frac{2}{p}\omega_r.$$

For the mechanical angular displacement θ_{rm}, we have

$$\theta_{rm} = \frac{2}{P}\theta_r.$$

It is convenient to derive the equations of motion using the electrical angular velocity ω_r and displacement θ_r in order to ease notations and ensure the generality of results. From Newton's second law of motion, one finds two differential equations

$$\frac{d\omega_r}{dt} = \frac{P}{2J}T_e - \frac{B_m}{J}\omega_r - \frac{P}{2J}T_L,$$

$$\frac{d\theta_r}{dt} = \omega_r. \tag{5.7}$$

The electromagnetic torque, developed by induction motors, is

$$T_e = \frac{P}{2}\frac{\partial W_c(\mathbf{i}_{abs}, \mathbf{i}'_{abr}, \theta_r)}{\partial \theta_r}.$$

Assuming that the magnetic system is linear, the coenergy is expressed as

$$W_c = W_f = \frac{1}{2}\mathbf{i}_{abs}^T(\mathbf{L}_s - L_{ls}\mathbf{I})\mathbf{i}_{abs} + \mathbf{i}_{abs}^T\mathbf{L}_{sr}'(\theta_r)\mathbf{i}_{abr}' + \frac{1}{2}\mathbf{i}_{abr}'^T(\mathbf{L}_r' - L_{lr}'\mathbf{I})\mathbf{i}_{abr}'.$$

The self-inductances L_{ss} and L_{rr}' as well as the leakage inductances L_{ls} and L_{lr}' are not functions of the angular displacement θ_r. From (5.3), the following expression for the stator-rotor mutual inductances is used assuming pure sinusoidal variations

$$\mathbf{L}_{sr}'(\theta_r) = L_{ms}\begin{bmatrix} \cos\theta_r & -\sin\theta_r \\ \sin\theta_r & \cos\theta_r \end{bmatrix}.$$

Hence, for P-pole two-phase induction motors, the electromagnetic torque T_e is

$$T_e = \frac{P}{2}\frac{\partial W_c(\mathbf{i}_{abs}, \mathbf{i}_{abr}', \theta_r)}{\partial\theta_r} = \frac{P}{2}\mathbf{i}_{abs}^T\frac{\partial\mathbf{L}_{sr}'(\theta_r)}{\partial\theta_r}\mathbf{i}_{abr}'$$

$$= \frac{P}{2}L_{ms}[i_{as} \quad i_{bs}]\begin{bmatrix} -\sin\theta_r & -\cos\theta_r \\ \cos\theta_r & -\sin\theta_r \end{bmatrix}\begin{bmatrix} i_{ar}' \\ i_{br}' \end{bmatrix}$$

$$= -\frac{P}{2}L_{ms}[(i_{as}i_{ar}' + i_{bs}i_{br}')\sin\theta_r + (i_{as}i_{br}' - i_{bs}i_{ar}')\cos\theta_r]. \tag{5.8}$$

Using (5.7) and (5.8), the *torsional-mechanical* equations are

$$\frac{d\omega_r}{dt} = -\frac{P^2}{4J}L_{ms}[(i_{as}i_{ar}' + i_{bs}i_{br}')\sin\theta_r + (i_{as}i_{br}' - i_{bs}i_{ar}')\cos\theta_r] - \frac{B_m}{J}\omega_r - \frac{P}{2J}T_L,$$

$$\frac{d\theta_r}{dt} = \omega_r. \tag{5.9}$$

Supplying the phase voltages u_{as} and u_{bs}, one rotates the motor in the desired direction (clockwise or counterclockwise). The electromagnetic torque T_e counteracts the load and friction torques. One recalls that the torques and forces are vectors. In actuators and motors, the friction torque acts against the electromagnetic torque, while the load torques usually opposes T_e. The signs for torques in (5.9) may require recalling the basic physics. Depending on the direction of rotation (clockwise and counterclockwise), the sign for T_e changes. The mutual inductances and expression

$$\mathbf{L}_{sr}'(\theta_r) = L_{ms}\begin{bmatrix} \cos\theta_r & -\sin\theta_r \\ \sin\theta_r & \cos\theta_r \end{bmatrix}$$

can be refined based on the direction of rotation and initial conditions (displacement of rotor with respect to stator).

From

$$\frac{d\omega_{rm}}{dt} = \frac{1}{J}(T_e - B_m\omega_{rm} - T_L),$$

one concludes that to ensure the rotation, $T_e > T_{\text{friction}} + T_L$ must be guaranteed.

The circuitry-electromagnetic and *torsional-mechanical* equations (5.6) and (5.9) are integrated obtaining the nonlinear differential equations, which describe the dynamics of two-phase induction motors

$$\frac{di_{as}}{dt} = -\frac{L'_{rr}r_s}{L_\Sigma}i_{as} + \frac{L^2_{ms}}{L_\Sigma}i_{bs}\omega_r + \frac{L_{ms}L'_{rr}}{L_\Sigma}i'_{ar}\left(\omega_r\sin\theta_r + \frac{r'_r}{L'_{rr}}\cos\theta_r\right)$$

$$+ \frac{L_{ms}L'_{rr}}{L_\Sigma}i'_{br}\left(\omega_r\cos\theta_r - \frac{r'_r}{L'_{rr}}\sin\theta_r\right) + \frac{L'_{rr}}{L_\Sigma}u_{as} - \frac{L_{ms}}{L_\Sigma}\cos\theta_r u'_{ar} + \frac{L_{ms}}{L_\Sigma}\sin\theta_r u'_{br},$$

$$\frac{di_{bs}}{dt} = -\frac{L'_{rr}r_s}{L_\Sigma}i_{bs} - \frac{L^2_{ms}}{L_\Sigma}i_{as}\omega_r - \frac{L_{ms}L'_{rr}}{L_\Sigma}i'_{ar}\left(\omega_r\cos\theta_r - \frac{r'_r}{L'_{rr}}\sin\theta_r\right)$$

$$+ \frac{L_{ms}L'_{rr}}{L_\Sigma}i'_{br}\left(\omega_r\sin\theta_r + \frac{r'_r}{L'_{rr}}\cos\theta_r\right) + \frac{L'_{rr}}{L_\Sigma}u_{bs} - \frac{L_{ms}}{L_\Sigma}\sin\theta_r u'_{ar} - \frac{L_{ms}}{L_\Sigma}\cos\theta_r u'_{br},$$

$$\frac{di'_{ar}}{dt} = -\frac{L_{ss}r'_r}{L_\Sigma}i'_{ar} + \frac{L_{ms}L_{ss}}{L_\Sigma}i_{as}\left(\omega_r\sin\theta_r + \frac{r_s}{L_{ss}}\cos\theta_r\right)$$

$$- \frac{L_{ms}L_{ss}}{L_\Sigma}i_{bs}\left(\omega_r\cos\theta_r - \frac{r_s}{L_{ss}}\sin\theta_r\right)$$

$$- \frac{L^2_{ms}}{L_\Sigma}i'_{br}\omega_r - \frac{L_{ms}}{L_\Sigma}\cos\theta_r u_{as} - \frac{L_{ms}}{L_\Sigma}\sin\theta_r u_{bs} + \frac{L_{ss}}{L_\Sigma}u'_{ar},$$

$$\frac{di'_{br}}{dt} = -\frac{L_{ss}r'_r}{L_\Sigma}i'_{br} + \frac{L_{ms}L_{ss}}{L_\Sigma}i_{as}\left(\omega_r\cos\theta_r - \frac{r_s}{L_{ss}}\sin\theta_r\right)$$

$$+ \frac{L_{ms}L_{ss}}{L_\Sigma}i_{bs}\left(\omega_r\sin\theta_r + \frac{r_s}{L_{ss}}\cos\theta_r\right)$$

$$+ \frac{L^2_{ms}}{L_\Sigma}i'_{ar}\omega_r + \frac{L_{ms}}{L_\Sigma}\sin\theta_r u_{as} - \frac{L_{ms}}{L_\Sigma}\cos\theta_r u_{bs} + \frac{L_{ss}}{L_\Sigma}u'_{br},$$

$$\frac{d\omega_r}{dt} = -\frac{P^2}{4J}L_{ms}[(i_{as}i'_{ar} + i_{bs}i'_{br})\sin\theta_r + (i_{as}i'_{br} - i_{bs}i'_{ar})\cos\theta_r] - \frac{B_m}{J}\omega_r - \frac{P}{2J}T_L,$$

$$\frac{d\theta_r}{dt} = \omega_r,$$

(5.10)

where $L_\Sigma = L_{ss}L'_{rr} - L^2_{ms}$.

From (5.10), in the state-space (matrix) form, a set of six highly coupled and nonlinear differential equations is

$$
\begin{bmatrix}
\dfrac{di_{as}}{dt} \\[2mm]
\dfrac{di_{bs}}{dt} \\[2mm]
\dfrac{di'_{ar}}{dt} \\[2mm]
\dfrac{di'_{br}}{dt} \\[2mm]
\dfrac{d\omega_r}{dt} \\[2mm]
\dfrac{d\theta_r}{dt}
\end{bmatrix}
=
\begin{bmatrix}
-\dfrac{L'_{rr}r_s}{L_\Sigma} & 0 & 0 & 0 & 0 & 0 \\[2mm]
0 & -\dfrac{L'_{rr}r_s}{L_\Sigma} & 0 & 0 & 0 & 0 \\[2mm]
0 & 0 & -\dfrac{L_{ss}r'_r}{L_\Sigma} & 0 & 0 & 0 \\[2mm]
0 & 0 & 0 & -\dfrac{L_{ss}r'}{L_\Sigma} & 0 & 0 \\[2mm]
0 & 0 & 0 & 0 & -\dfrac{B_m}{J} & 0 \\[2mm]
0 & 0 & 0 & 0 & 1 & 0
\end{bmatrix}
\begin{bmatrix}
i_{as} \\[2mm]
i_{bs} \\[2mm]
i'_{ar} \\[2mm]
i'_{br} \\[2mm]
\omega_r \\[2mm]
\theta_r
\end{bmatrix}
$$

$$
+
\begin{bmatrix}
\dfrac{L^2_{ms}}{L_\Sigma}i_{bs}\omega_r + \dfrac{L_{ms}L'_{rr}}{L_\Sigma}i'_{ar}\left(\omega_r\sin\theta_r + \dfrac{r'_r}{L'_{rr}}\cos\theta_r\right) + \dfrac{L_{ms}L'_{rr}}{L_\Sigma}i'_{br}\left(\omega_r\cos\theta_r - \dfrac{r'_r}{L'_{rr}}\sin\theta_r\right) \\[4mm]
-\dfrac{L^2_{ms}}{L_\Sigma}i_{as}\omega_r - \dfrac{L_{ms}L'_{rr}}{L_\Sigma}i'_{ar}\left(\omega_r\cos\theta_r - \dfrac{r'_r}{L'_{rr}}\sin\theta_r\right) + \dfrac{L_{ms}L'_{rr}}{L_\Sigma}i'_{br}\left(\omega_r\sin\theta_r + \dfrac{r'_r}{L'_{rr}}\cos\theta_r\right) \\[4mm]
\dfrac{L_{ms}L_{ss}}{L_\Sigma}i_{as}\left(\omega_r\sin\theta_r + \dfrac{r_s}{L_{ss}}\cos\theta_r\right) - \dfrac{L_{ms}L_{ss}}{L_\Sigma}i_{bs}\left(\omega_r\cos\theta_r - \dfrac{r_s}{L_{ss}}\sin\theta_r\right) - \dfrac{L^2_{ms}}{L_\Sigma}i'_{br}\omega_r \\[4mm]
\dfrac{L_{ms}L_{ss}}{L_\Sigma}i_{as}\left(\omega_r\cos\theta_r - \dfrac{r_s}{L_{ss}}\sin\theta_r\right) + \dfrac{L_{ms}L_{ss}}{L_\Sigma}i_{bs}\left(\omega_r\sin\theta_r + \dfrac{r_s}{L_{ss}}\cos\theta_r\right) + \dfrac{L^2_{ms}}{L_\Sigma}i'_{ar}\omega_r \\[4mm]
-\dfrac{P^2}{4J}L_{ms}[(i_{as}i'_{ar} + i_{bs}i'_{br})\sin\theta_r + (i_{as}i'_{br} - i_{bs}i'_{ar})\cos\theta_r] \\[4mm]
0
\end{bmatrix}
$$

$$
+\begin{bmatrix}
\dfrac{L'_{rr}}{L_\Sigma} & 0 & 0 & 0 \\[2mm]
0 & \dfrac{L'_{rr}}{L_\Sigma} & 0 & 0 \\[2mm]
0 & 0 & \dfrac{L_{ss}}{L_\Sigma} & 0 \\[2mm]
0 & 0 & 0 & \dfrac{L_{ss}}{L_\Sigma} \\[2mm]
0 & 0 & 0 & 0 \\[2mm]
0 & 0 & 0 & 0
\end{bmatrix}
\begin{bmatrix} u_{as} \\ u_{bs} \\ u'_{ar} \\ u'_{br} \end{bmatrix}
+\begin{bmatrix}
-\dfrac{L_{ms}}{L_\Sigma}\cos\theta_r u'_{ar} + \dfrac{L_{ms}}{L_\Sigma}\sin\theta_r u'_{br} \\[2mm]
-\dfrac{L_{ms}}{L_\Sigma}\sin\theta_r u'_{ar} - \dfrac{L_{ms}}{L_\Sigma}\cos\theta_r u'_{br} \\[2mm]
-\dfrac{L_{ms}}{L_\Sigma}\cos\theta_r u_{as} - \dfrac{L_{ms}}{L_\Sigma}\sin\theta_r u_{bs} \\[2mm]
\dfrac{L_{ms}}{L_\Sigma}\sin\theta_r u_{as} - \dfrac{L_{ms}}{L_\Sigma}\cos\theta_r u_{bs} \\[2mm]
0 \\[2mm]
0
\end{bmatrix}
-\begin{bmatrix} 0 \\ 0 \\ 0 \\ 0 \\ \dfrac{P}{2J} \\ 0 \end{bmatrix} T_L.
$$

$$\tag{5.11}$$

The *s*-domain block diagrams can be developed. For squirrel-cage induction motors, the rotor windings are short-circuited, and $u'_{ar} = u'_{br} = 0$. The block diagram, as reported in Figure 5.3, is derived using differential equations (5.10). However, the ability to derive the *s*-domain block diagram does not mean that transfer functions, Fourier transform or any other concepts of linear theory can be used to accomplish the analysis tasks. The differential equations (5.10) are nonlinear, and one cannot linearize or simplify them. The linear theory is virtually not applicable to induction machines as well as to a majority of other electromechanical motion devices.

5.1.3 Lagrange Equations of Motion for Induction Machines

The mathematical model can be derived using the Lagrange equations of motion. The generalized independent coordinates are four charges and the rotor angular displacement. Hence,

$$ q_1 = \frac{i_{as}}{s}, \qquad q_2 = \frac{i_{bs}}{s}, \qquad q_3 = \frac{i'_{ar}}{s}, \qquad q_4 = \frac{i'_{br}}{s} \qquad \text{and} \qquad q_5 = \theta_r. $$

The generalized forces are the voltages and load torque, for example,

$$ Q_1 = u_{as}, \qquad Q_2 = u_{bs}, \qquad Q_3 = u'_{ar}, \qquad Q_4 = u'_{br} \qquad \text{and} \qquad Q_5 = -T_L $$

The resulting five Lagrange equations

$$ \frac{d}{dt}\left(\frac{\partial \Gamma}{\partial \dot q_i}\right) - \frac{\partial \Gamma}{\partial q_i} + \frac{\partial D}{\partial \dot q_i} + \frac{\partial \Pi}{\partial q_i} = Q_i, $$

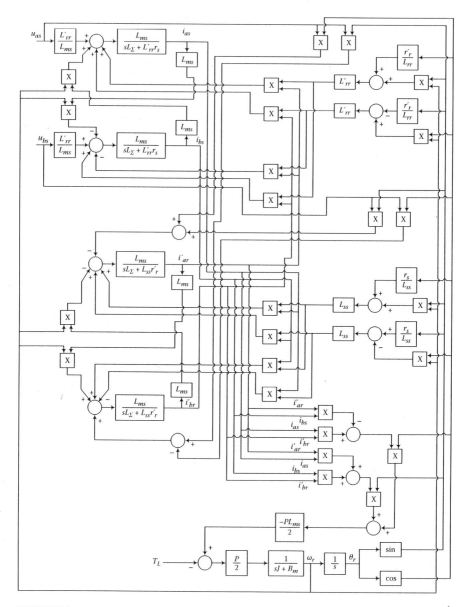

FIGURE 5.3
s-domain block diagram of squirrel-cage induction motors.

are

$$\frac{d}{dt}\left(\frac{\partial \Gamma}{\partial \dot{q}_1}\right) - \frac{\partial \Gamma}{\partial q_1} + \frac{\partial D}{\partial \dot{q}_1} + \frac{\partial \Pi}{\partial q_1} = Q_1, \qquad \frac{d}{dt}\left(\frac{\partial \Gamma}{\partial \dot{q}_2}\right) - \frac{\partial \Gamma}{\partial q_2} + \frac{\partial D}{\partial \dot{q}_2} + \frac{\partial \Pi}{\partial q_2} = Q_2,$$

$$\frac{d}{dt}\left(\frac{\partial\Gamma}{\partial\dot{q}_3}\right) - \frac{\partial\Gamma}{\partial q_3} + \frac{\partial D}{\partial\dot{q}_3} + \frac{\partial\Pi}{\partial q_3} = Q_3, \qquad \frac{d}{dt}\left(\frac{\partial\Gamma}{\partial\dot{q}_4}\right) - \frac{\partial\Gamma}{\partial q_4} + \frac{\partial D}{\partial\dot{q}_4} + \frac{\partial\Pi}{\partial q_4} = Q_4,$$

$$\frac{d}{dt}\left(\frac{\partial\Gamma}{\partial\dot{q}_5}\right) - \frac{\partial\Gamma}{\partial q_5} + \frac{\partial D}{\partial\dot{q}_5} + \frac{\partial\Pi}{\partial q_5} = Q_5. \tag{5.12}$$

Using the notations introduced, the total kinetic, potential and dissipated energies, to be used in equations (5.12), are

$$\Gamma = \frac{1}{2}L_{ss}\dot{q}_1^2 + L_{ms}\dot{q}_1\dot{q}_3\cos q_5 - L_{ms}\dot{q}_1\dot{q}_4\sin q_5 + \frac{1}{2}L_{ss}\dot{q}_2^2 + L_{ms}\dot{q}_2\dot{q}_3\sin q_5$$

$$+ L_{ms}\dot{q}_2\dot{q}_4\cos q_5 + \frac{1}{2}L'_{rr}\dot{q}_3^2 + \frac{1}{2}L'_{rr}\dot{q}_4^2 + \frac{1}{2}J\dot{q}_5^2,$$

$$\Pi = 0,$$

$$D = \frac{1}{2}\left(r_s\dot{q}_1^2 + r_s\dot{q}_2^2 + r'_r\dot{q}_3^2 + r'_r\dot{q}_4^2 + B_m\dot{q}_5^2\right).$$

The derivative terms of (5.12) are

$$\frac{\partial\Gamma}{\partial q_1} = 0, \quad \frac{\partial\Gamma}{\partial\dot{q}_1} = L_{ss}\dot{q}_1 + L_{ms}\dot{q}_3\cos q_5 - L_{ms}\dot{q}_4\sin q_5,$$

$$\frac{\partial\Gamma}{\partial q_2} = 0, \quad \frac{\partial\Gamma}{\partial\dot{q}_2} = L_{ss}\dot{q}_2 + L_{ms}\dot{q}_3\sin q_5 + L_{ms}\dot{q}_4\cos q_5,$$

$$\frac{\partial\Gamma}{\partial q_3} = 0, \quad \frac{\partial\Gamma}{\partial\dot{q}_3} = L'_{rr}\dot{q}_3 + L_{ms}\dot{q}_1\cos q_5 + L_{ms}\dot{q}_2\sin q_5,$$

$$\frac{\partial\Gamma}{\partial q_4} = 0, \quad \frac{\partial\Gamma}{\partial\dot{q}_4} = L'_{rr}\dot{q}_4 - L_{ms}\dot{q}_1\sin q_5 + L_{ms}\dot{q}_2\cos q_5,$$

$$\frac{\partial\Gamma}{\partial q_5} = -L_{ms}\dot{q}_1\dot{q}_3\sin q_5 - L_{ms}\dot{q}_1\dot{q}_4\cos q_5 + L_{ms}\dot{q}_2\dot{q}_3\cos q_5 - L_{ms}\dot{q}_2\dot{q}_4\sin q_5$$

$$= -L_{ms}[(\dot{q}_1\dot{q}_3 + \dot{q}_2\dot{q}_4)\sin q_5 + (\dot{q}_1\dot{q}_4 - \dot{q}_2\dot{q}_3)\cos q_5],$$

$$\frac{\partial\Gamma}{\partial\dot{q}_5} = J\dot{q}_5,$$

$$\frac{\partial\Pi}{\partial q_1} = 0, \quad \frac{\partial\Pi}{\partial q_2} = 0, \quad \frac{\partial\Pi}{\partial q_3} = 0, \quad \frac{\partial\Pi}{\partial q_4} = 0, \quad \frac{\partial\Pi}{\partial q_5} = 0,$$

$$\frac{\partial D}{\partial\dot{q}_1} = r_s\dot{q}_1, \quad \frac{\partial D}{\partial\dot{q}_2} = r_s\dot{q}_2, \quad \frac{\partial D}{\partial\dot{q}_3} = r'_r\dot{q}_3, \quad \frac{\partial D}{\partial\dot{q}_4} = r'_r\dot{q}_4, \quad \frac{\partial D}{\partial\dot{q}_5} = B_m\dot{q}_5.$$

Once we have found the derivative terms of five Lagrange equations (5.12), using the generalized coordinates $(\dot{q}_1 = i_{as}, \dot{q}_2 = i_{bs}, \dot{q}_3 = i'_{ar}, \dot{q}_4 = i'_{br}, \dot{q}_5 = \omega_r)$ and generalized forces $(Q_1 = u_{as}, Q_2 = u_{bs}, Q_3 = u'_{ar}, Q_4 = u'_{ar}, Q_5 = -T_L)$, one obtains the following differential equations

$$L_{ss}\frac{di_{as}}{dt} + L_{ms}\frac{d(i'_{ar}\cos\theta_r)}{dt} - L_{ms}\frac{d(i'_{br}\sin\theta_r)}{dt} + r_s i_{as} = u_{as},$$

$$L_{ss}\frac{di_{bs}}{dt} + L_{ms}\frac{d(i'_{ar}\sin\theta_r)}{dt} + L_{ms}\frac{d(i'_{br}\cos\theta_r)}{dt} + r_s i_{bs} = u_{bs},$$

$$L_{ms}\frac{d(i_{as}\cos\theta_r)}{dt} + L_{ms}\frac{d(i_{bs}\sin\theta_r)}{dt} + L'_{rr}\frac{di'_{ar}}{dt} + r'_r i'_{ar} = u'_{ar},$$

$$-L_{ms}\frac{d(i_{as}\sin\theta_r)}{dt} + L_{ms}\frac{d(i_{bs}\cos\theta_r)}{dt} + L'_{rr}\frac{di'_{br}}{dt} + r'_r i'_{br} = u'_{br},$$

$$J\frac{d^2\theta_r}{dt^2} + L_{ms}[(i_{as}i'_{ar} + i_{bs}i'_{br})\sin\theta_r + (i_{as}i'_{br} - i_{bs}i'_{ar})\cos\theta_r] + B_m\frac{d\theta_r}{dt} = -T_L. \qquad (5.13)$$

From (5.13), for *P*-pole induction motors, by using

$$\frac{d\theta_r}{dt} = \omega_r,$$

six differential equations (5.5) and (5.9) result, which were found applying other physical laws discussed in Section 5.1.2. The differential equations in the Cauchy form (5.10) result. The advantage of the Lagrange concept is that Kirchhoff's, Newton's, Faraday's, Lorenz's, coenergy, or other laws are not used to derive the resulting models. The Lagrange equations of motion provide a general and coherent procedure. Furthermore, one may derive the *emf* terms, electromagnetic torque, and so on. For example, from equation

$$\frac{d}{dt}\left(\frac{\partial\Gamma}{\partial\dot{q}_5}\right) - \frac{\partial\Gamma}{\partial q_5} + \frac{\partial D}{\partial\dot{q}_5} + \frac{\partial\Pi}{\partial q_5} = Q_5,$$

one concludes that the electromagnetic torque is

$$T_e = \frac{\partial\Gamma}{\partial q_5} = \frac{\partial\Gamma}{\partial\theta_r} = -L_{ms}[(\dot{q}_1\dot{q}_3 + \dot{q}_2\dot{q}_4)\sin q_5 + (\dot{q}_1\dot{q}_4 - \dot{q}_2\dot{q}_3)\cos q_5]$$

$$= -L_{ms}[(i_{as}i'_{ar} + i_{bs}i'_{br})\sin\theta_r + (i_{as}i'_{br} - i_{bs}i'_{ar})\cos\theta_r]$$

The same expression, as given by (5.8), was derived by applying the coenergy.

5.1.4   Torque-Speed Characteristics and Control of Induction Motors

The angular velocity of induction motors must be controlled, and the torque-speed curves $\omega_r = \Omega_T(T_e)$ are studied. The electromagnetic torque developed by induction motors is a function of the stator and rotor currents as well as rotor displacement. Induction motors are controlled by changing the frequency f and magnitude u_M of the voltages supplied to the phase windings. The magnitude of the voltages applied to the stator windings cannot exceed the rated voltage u_{Mmax}, and $u_{Mmin} \leq u_M \leq u_{Mmax}$. The angular frequency of the applied phase voltages ω_f is also bounded, and $\omega_f = 2\pi f, f_{min} \leq f \leq f_{max}, f_{min} > 0$, and $f_{max} > 0$.

The synchronous angular velocity ω_e of induction machines, as a function of f, is

$$\omega_e = \frac{4\pi f}{P}$$

The electrical angular velocity of induction motors ω_r is less or equal (at no load and no friction) to ω_e. Hence, for induction motors, $\omega_r \leq \omega_e$. In contrast, synchronous motors rotate at ω_e, and $\omega_r = \omega_e$.

The steady-state curves $\omega_r = \Omega_T(T_e)$ are found by plotting the angular velocity versus the electromagnetic torque. The National Electric Manufacturers Association (NEMA) in the USA and the International Electromechanical Commission (IEC) in Europe have defined four basic classes of induction machines to be A, B, C, and D. For different classes, typical steady-state torque-speed characteristic curves in the motor region are depicted in Figure 5.4a, where the *slip* is given as

$$slip = \frac{\omega_e - \omega_r}{\omega_e} \quad \text{and} \quad \omega_r = (1 - slip)\omega_e.$$

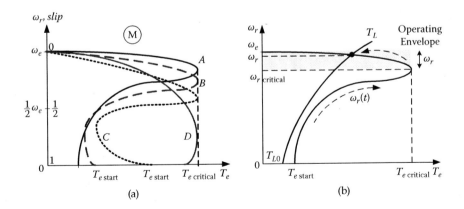

FIGURE 5.4
(a) Typical torque-speed characteristics of induction motors; (b) torque-speed and load curves.

These torque-speed characteristics can be found using the experimental data by measuring the torque and angular velocity which evolve in time, for example, one applies $T_e(t)$ and $\omega_r(t)$. The measured experimental dynamic characteristics $\omega_r(t) = \Omega_T[T_e(t)]$ will be different as compared to the steady-state $\omega_r = \Omega_T(T_e)$. In fact, $\omega_r = \Omega_T(T_e)$ "average" $\omega_r(t) = \Omega_T[T_e(t)]$. However, $\omega_r(t) = \Omega_T[T_e(t)]$ are much more general and descriptive. To derive experimental dynamic characteristics $\omega_r(t) = \Omega_T[T_e(t)]$, if $T_e(t)$ is not directly measurable, one also may measure (or observe) the phase currents to obtain $T_e(t)$.

The steady-state angular velocity with which induction motor rotates is found as the intersection of the torque-speed $\omega_r = \Omega_T(T_e)$ and load $T_L(\omega_r)$ characteristics as illustrated in Figure 5.4b. From the Newton second law, neglecting the friction torque (which can be considered as a part of the load torque T_L), we have

$$\frac{d\omega_r}{dt} = \frac{1}{J}(T_e - T_L).$$

One concludes that $\omega_r = $ const it $T_e = T_L$. Furthermore, assuming that $T_e > T_L$ is guaranteed and $T_{estart} > T_{L0}$, the motor accelerates until ω_r is reached when $T_e = T_L$. The critical angular velocity $\omega_{rcritical}$ is documented in Figure 5.4b. To ensure the acceleration and rotation, one must guarantee $T_e > T_L$ in the full operating envelope for all T_L, and the rated torque may be specified to be $T_{emax} = T_{ecritical}$. The induction motor torque-speed operating envelope $[\omega_r \, T_e]$ is defined by $\omega_r \in [\omega_{rcritical} \, \omega_e]$ and $T_e \in [0 \, T_{ecritical}]$, $T_e > T_L$, $\forall \, T_L$.

The industrial induction motors are usually designed to be ether the A- or B-class machines. These motors have normal starting torque and low slip, which is usually ~0.05. In contrast, the C-class induction motors have higher starting torque because of double-rotor design, and the slip is greater than 0.05. The D-class induction motors have high rotor resistance, and high starting torque results. Typically, the *slip* of the D-class induction motors is in the range from 0.5 to 0.9. The E- and F-class induction motors have very low starting torque, and the rotor bars are deeply buried resulting in high leakage inductances. Figure 5.5 illustrates the torque-speed characteristic of A-class induction machines in the motor, generator, and braking regions.

The electromagnetic torque developed by two-phase induction motors is given by (5.8)

$$T_e = -\frac{P}{2}L_{ms}[(i_{as}i'_{ar} + i_{bs}i'_{br})\sin\theta_r + (i_{as}i'_{br} - i_{bs}i'_{ar})\cos\theta_r],$$

whereas for three-phase induction motors it will be found that

$$T_e = -\frac{P}{2}L_{ms}\left\{ \left[i_{as}\left(i'_{ar} - \frac{1}{2}i'_{br} - \frac{1}{2}i'_{cr} \right) + i_{bs}\left(i'_{br} - \frac{1}{2}i'_{ar} - \frac{1}{2}i'_{cr} \right) \right. \right.$$
$$\left. \left. + i_{cs}\left(i'_{cr} - \frac{1}{2}i'_{br} - \frac{1}{2}i'_{ar} \right) \right]\sin\theta_r + \frac{\sqrt{3}}{2}\left[i_{as}(i'_{br} - i'_{cr}) + i_{bs}(i'_{cr} - i'_{ar}) + i_{cs}(i'_{ar} - i'_{br}) \right]\cos\theta_r \right\}.$$

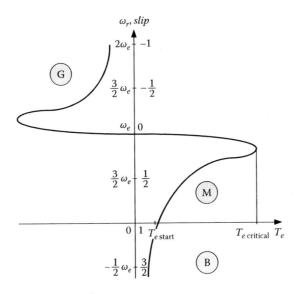

FIGURE 5.5
Torque-speed characteristic curves in the motor (M), generator (G), and braking (B) regions.

To guarantee the balanced operation of two-phase induction motors, one supplies the following phase voltages to the stator windings

$$u_{as}(t) = \sqrt{2}u_M \cos(\omega_f t), \quad u_{bs}(t) = \sqrt{2}u_M \sin(\omega_f t),$$

and the sinusoidal steady-state phase currents are

$$i_{as}(t) = \sqrt{2}i_M \cos(\omega_f t - \varphi_i), \quad i_{bs}(t) = \sqrt{2}i_M \sin(\omega_f t - \varphi_i),$$

where u_M and i_M are the magnitude of the *as* and *bs* stator voltages and currents; ω_f is the angular frequency of the supplied phase voltages, $\omega_f = 2\pi f$; f is the frequency of the supplied voltage; φ_i is the phase difference.

For three-phase induction motors, one supplies the following phase voltages

$$u_{as}(t) = \sqrt{2}u_M \cos(\omega_f t), \quad u_{bs}(t) = \sqrt{2}u_M \cos\left(\omega_f t - \frac{2}{3}\pi\right) \text{ and }$$

$$u_{cs}(t) = \sqrt{2}u_M \cos\left(\omega_f t + \frac{2}{3}\pi\right).$$

The applied voltage to the motor windings cannot exceed the rated voltage u_{Mmax}, for example, $u_{Mmin} \leq u_M \leq u_{Mmax}$. The synchronous angular velocity ω_e is found by using the number of poles P and the frequency f as

$$\omega_e = \frac{4\pi f}{P}.$$

The frequency is constrained as $f_{min} \leq f \leq f_{max}$ due to the power electronics limits (for f_{min}) and mechanical limits on the maximum angular velocity which define f_{max}. The synchronous and electrical (mechanical) angular velocities ω_e and ω_r can be regulated by changing the frequency f. To vary ω_r, one can change both the magnitude of the applied voltages u_M and f. The torque-speed curves of induction motors can be studied using the equivalent circuits. Alternatively, from transient dynamics, as found from the experimental results or solving the derived differential equations, one can find the experimental and analytical evolution of $\omega_r = \Omega_T(T_e)$ by plotting the angular velocity ω_r versus the electromagnetic torque T_e.

The following principles are used to control the angular velocity of squirrel-cage induction motors [1–8].

Voltage Control

By changing the magnitude u_M of the supplied phase voltages to the stator windings, the angular velocity is regulated in the stable operating region (see Figure 5.6a). It was emphasized that $u_{Mmin} \leq u_M \leq u_{Mmax}$, where u_{Mmax} is the maximum allowed (rated) voltage. By reducing u_M, one reduces T_{estart} and $T_{ecritical}$. The operating envelope for the angular velocity is $\omega_r \in [\omega_{rcritical}\ \omega_e]$. For example, for A-, B-, and C-class induction motors, one is not able to effectively regulate the angular velocity. The voltage control is applicable only for low-efficiency D-class induction motors in order to avoid the use of power converters.

Frequency Control

The magnitude of the supplied phase voltages is constant $u_M = $ const, and the angular velocity is regulated by changing the synchronous angular velocity ω_e by varying the frequency of the supplied voltages f, $f_{min} \leq f \leq f_{max}$ (usually f_{min} is ~2 Hz). This concept is justified using the formula

$$\omega_e = \frac{4\pi f}{P}.$$

The angular frequency ω_f of the supplied voltages is proportional to the frequency f, $\omega_f = 2\pi f$, $f_{min} \leq f \leq f_{max}$. The torque-speed characteristics for different values of the frequency are shown in Figure 5.6b.

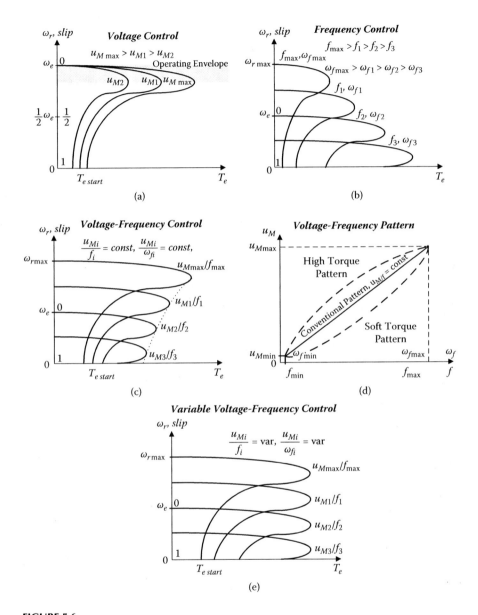

FIGURE 5.6

Torque-speed characteristics $\omega_r = \Omega_T(T_e)$: (a) Voltage control; (b) frequency control; (c) voltage-frequency control: *constant volts per hertz* control; (d) voltage-frequency patterns; (e) variable voltage-frequency control.

Voltage-Frequency Control

To minimize losses, the voltage magnitude u_M is regulated if the frequency f is changed. In particular, u_M can be decreased linearly while reducing f. To guarantee the so-called *constant volts per hertz* control, one maintains the following relationship

$$\frac{u_{Mi}}{f_i} = \text{const} \quad \text{or} \quad \frac{u_{Mi}}{\omega_{fi}} = \text{const.}$$

The corresponding torque-speed characteristics are documented in Figure 5.6c. Regulating the voltage-frequency patterns, one varies the torque-speed curves. For example, one may apply

$$\sqrt{\frac{u_{Mi}}{f_i}} = \text{const}$$

to adjust the magnitude u_M and frequency f. To attain the specified acceleration, settling time, overshoot, rise time, and other specifications, the general purpose (conventional), soft- and high-starting torque patterns are implemented based upon the requirements imposed and operating envelope. The conventional (*constant volts per hertz* control), soft- and high-torque patterns are illustrated in Figure 5.6d. That is, assigning $\omega_f = \varphi(u_M)$ with domain $u_{Mmin} \leq u_M \leq u_{Mmax}$ and range $\omega_{fmin} \leq \omega_f \leq \omega_{fmax}$ ($f_{min} \leq f \leq f_{max}$), one maintains

$$\frac{u_{Mi}}{f_i} = \text{var} \quad \text{or} \quad \frac{u_{Mi}}{\omega_{fi}} = \text{var.}$$

The desired torque-speed characteristics, as documented in Figure 5.6e, can be guaranteed.

The steady-state torque-speed characteristics of induction motors can be obtained by using the equivalent circuits. It will be demonstrated in Examples 5.2, 5.3, 5.6, and 5.7 that solving the differential equations and analyzing the dynamics for $\omega_r(t)$ and evolution of $T_e(t)$, one obtains $\omega_r(t) = \Omega_T[T_e(t)]$. These dynamic torque-speed characteristics $\omega_r(t) = \Omega_T[T_e(t)]$, which can be obtained experimentally, result in a coherent performance evaluations. To approach the preliminary analysis, the following formula can be applied to obtain the torque-speed characteristics for two- and three-phase induction motors

$$T_e = \frac{3(u_M \frac{X_M}{X_s+X_M})^2 \frac{r_r'}{slip}}{\omega_e \left[(r_s(\frac{X_M}{X_s+X_M})^2 + \frac{r_r'}{slip})^2 + (X_s + X_r')^2 \right]}, \quad slip = \frac{\omega_e - \omega_r}{\omega_e}, \quad \omega_e = \frac{4\pi f}{P},$$

$$(5.14)$$

where X_s and X'_r are the stator and rotor reactances; X_M is the magnetizing reactance.

The torque-speed characteristics $\omega_r = \Omega_T(T_e)$ are found assigning different values of the magnitude u_M and frequency f of the phase voltage supplied.

Example 5.1

Calculate and plot the torque-speed characteristic for a four-pole induction motor. The motor parameters are: $r_s = 24.5$ ohm, $r'_r = 23$ ohm, $X_s = 10$ ohm, $X'_r = 40$ ohm, and $X_M = 25$ ohm. The rated voltage is $u_{Mmax} = 110$ V. The maximum frequency of the supplied phase voltages is $f_{max} = 60$ Hz. Assume the phase voltages are supplied with frequencies 20, 40, and 60 Hz.

For each value of the assigned frequency, we calculate the synchronous angular velocity by applying the equation

$$\omega_e = \frac{4\pi f}{P}.$$

Then, using (5.14), where

$$slip = \frac{\omega_e - \omega_r}{\omega_e}$$

the torque-speed characteristics are found assigning different values for the angular velocity. The following MATLAB file is developed to calculate and plot $\omega_r = \Omega_T(T_e)$

```
clear all
% parameters of an induction motor
rs=24.5; rr=23; Xs=10; Xr=40; Xm=25;
uM=110; f=60; P=4; we=4*pi*f/P;
% calculation of a torque-speed characteristic
for wr=[1:1:4*pi*f/P]; % angular velocity
slip=(we-wr)/we; % slip
Te=3*(uM*Xm/(Xs+Xm))^2*(rr/slip)/(we*((rs*(Xm/(Xs+Xm))^2+rr.../slip)^2+(Xs+Xr)^2));
plot(Te,wr,'o'); title('Torque-Speed Characteristics','FontSize',14);
xlabel('Electromagnetic Torque T _ e, N-m','FontSize',14);
ylabel('Angular velocity \omega _ r, rad/sec','FontSize',14);
hold on;
end;
```

The resulting torque-speed characteristics for 20, 40, and 60 Hz are documented in Figure 5.7. Using the plots $\omega_r = \Omega_T(T_e)$ reported, one assesses the control features. The frequency-centered control is a baseline principle in control of high-performance electric drives with induction motors. The starting T_{estart} is maximum at the minimum frequency f_{min}, and T_{estart} is ~1.8 N-m. □

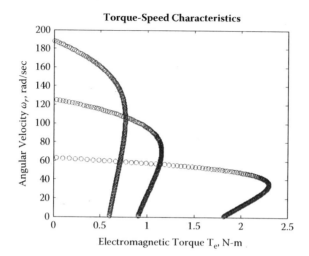

FIGURE 5.7
Torque-speed characteristics of a three-phase induction motor, f is 20, 40, and 60 Hz.

Example 5.2

Our goal is to perform simulations and analyze the performance of two two-phase 115 V (*rms*), 60 Hz, four-pole ($P = 4$) induction motors. The dynamics is described by differential equations (5.10). The ultimate objective is to examine the performance of inductions motors analyzing the acceleration capabilities, settling time, efficiency, and so on. The torque-speed characteristics will be studied by solving the differential equations in order to compare the dynamic and steady-state torque-speed characteristics $\omega_r(t) = \Omega_T[T_e(t)]$ and $\omega_r = \Omega_T(T_e)$. The parameters of the A- and D-class motors under our consideration are:

- A-class induction motor: $r_s = 1.2$ ohm, $r'_r = 1.5$ ohm, $L_{ms} = 0.16$ H, $L_{ls} = 0.02$ H, $L_{ss} = L_{ls} + L_{ms}$, $L'_{lr} = 0.02$ H, $L'_{rr} = L'_{lr} + L'_{ms}$, $B_m = 1 \times 10^{-6}$ N-m-sec/rad, and $J = 0.005$ kg-m^2.

- D-class induction motor: $r_s = 24.5$ ohm, $r'_r = 23$ ohm, $L_{ms} = 0.27$ H, $L_{ls} = 0.027$ H, $L_{ss} = L_{ls} + L_{ms}$, $L'_{lr} = 0.027$ H, $L'_{rr} = L'_{lr} + L'_{ms}$, $B_m = 1 \times 10^{-6}$ N-m-sec/rad, and $J = 0.001$ kg-m^2.

To guarantee the balanced operation, the supplied phase voltages are

$$u_{as}(t) = \sqrt{2}u_M \cos(\omega_f t) \quad \text{and} \quad u_{bs}(t) = \sqrt{2}u_M \sin(\omega_f t).$$

For the rated voltage, we have

$$u_{as}(t) = \sqrt{2}\,115\cos(377t) \quad \text{and} \quad u_{bs}(t) = \sqrt{2}\,115\sin(377t).$$

No load and loaded conditions are examined assigning the load torque to be 0 and 0.5 N-m. The simulations are performed, and transient dynamics of the stator and rotor currents in the *as, bs, ar,* and *br* windings $i_{as}(t)$, $i_{bs}(t)$, $i'_{ar}(t)$, and $i'_{br}(t)$, as well as the mechanical angular velocity $\omega_{rm}(t)$, are plotted in Figures 5.8 and 5.9. Figure 5.8 illustrates the transient dynamics of the A-class motor. The motor accelerates from stall, that is, ($\omega_{rm0} = 0$ rad/sec). Figures 5.8.a and 5.8.b depict the motor dynamics if $T_L = 0$ and $T_L = 0.5$ N-m (applied at $t = 0$ sec), respectively. The dynamics and acceleration capabilities of the D-class motor are documented in Figures 5.9a and 5.9b.

The A-class induction motor reaches the steady-state angular velocity within 0.8 sec (with no load), whereas the settling time is 1.4 sec if $T_L = 0.5$ N-m is applied at $t = 0$ sec. Better acceleration capabilities are observed for the D-class motor (0.5 and 0.8 sec for no load and loaded conditions). Analyzing the torque-speed characteristics, it was emphasized that the D-class induction motors may develop higher starting electromagnetic torque as compared to the A- and B-class motors (see Figure 5.4.a). However, the D-class motors may not necessarily possess higher T_{estart} and $T_{ecritical}$ as compared to the A-class motor. In fact, as illustrated in Figures 5.8 and 5.9, A-class induction motors possess higher T_{estart} and $T_{ecritical}$ as well as higher ratio $T_{ecritical}/T_{estart}$. For the A- and D-class induction motors studied, the $T_{ecritical}$ are ~3 and 1.1 N-m, respectively. Furthermore, the efficiency of D-class induction motors is low as a result of high r_r.

The moment of inertial significantly affects the acceleration capabilities. Using equation

$$\frac{d\omega_{rm}}{dt} = \frac{1}{J}(T_e - B_m\omega_{rm} - T_L),$$

we conclude that the A-class induction motor possesses better acceleration capabilities for the moments of inertia used (J are 0.005 and 0.001 kg-m^2). Figures 5.8 and 5.9 illustrate that the electromagnetic torque for A- and D-class motors reaches 4.1 and 1.8 N-m, respectively. Hence, A-class induction motor ensures better performance. In the A-class motor, the magnitude of the stator phase currents are higher (the input power is higher, but the losses are lower because of low phase resistances), whereas L_{ms} is lower, as compared to the D-class motor.

This analysis is confirmed by assessing the torque-speed characteristics. The dynamics of $\omega_{rm}(t)$ and evolutions of $T_e(t)$ are documented in Figures 5.8 and 5.9. The resulting characteristics $\omega_{rm}(t) = \Omega_T[T_e(t)]$ are obtained by plotting the mechanical angular velocity versus the electromagnetic torque evolution. The torque-speed characteristics are derived by using the simulation results carried out studying the motor dynamics. Thus, the steady-state-centered analysis can be accomplished by examining the transient dynamics, but not vice versa. Figure 5.8 documents the torque-speed characteristics of the A-class motor, whereas Figure 5.9 illustrates $\omega_{rm}(t) = \Omega_T[T_e(t)]$ for the D-class motor.

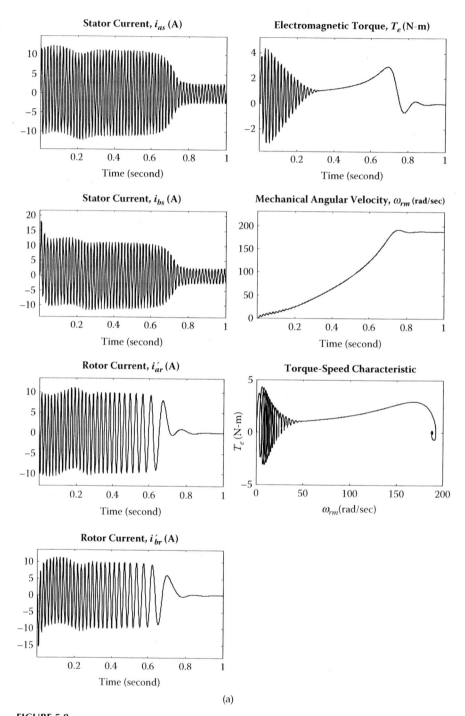

(a)

FIGURE 5.8
Dynamics and torque-speed characteristics of the A-class induction motor:
(a) $T_L = 0$ N-m.

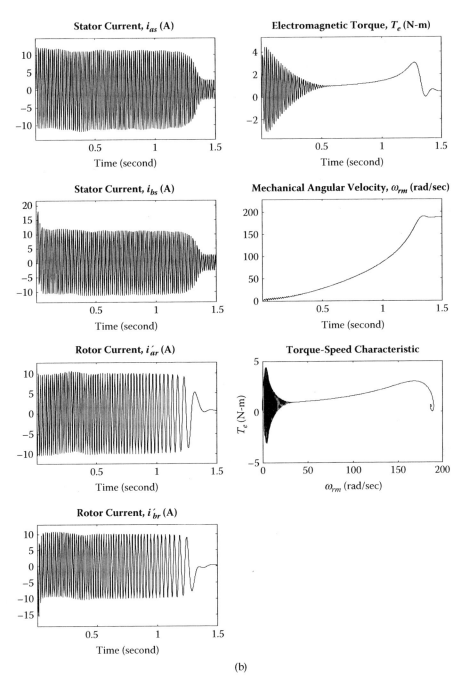

(b)

FIGURE 5.8
(Continued). (b) $T_L = 0.5$ N-m, $u_{as}(t) = \sqrt{2}\,115\cos(377t)$ and $u_{bs}(t) = \sqrt{2}\,115\sin(377t)$.

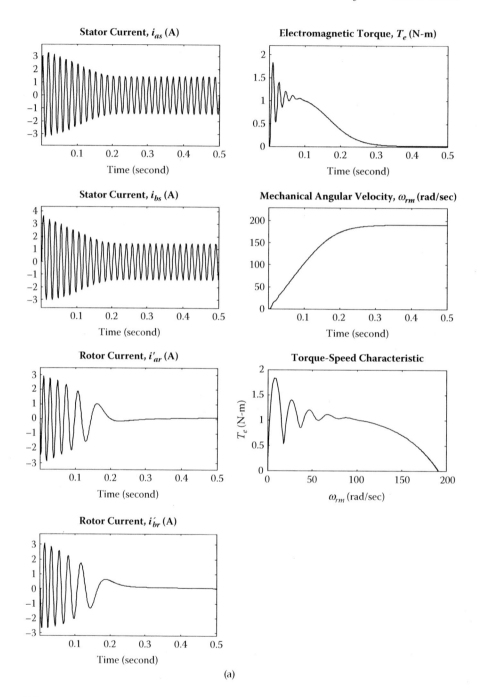

FIGURE 5.9
Dynamics and torque-speed characteristics of the D-class induction motor:
(a) $T_L = 0$ N-m.

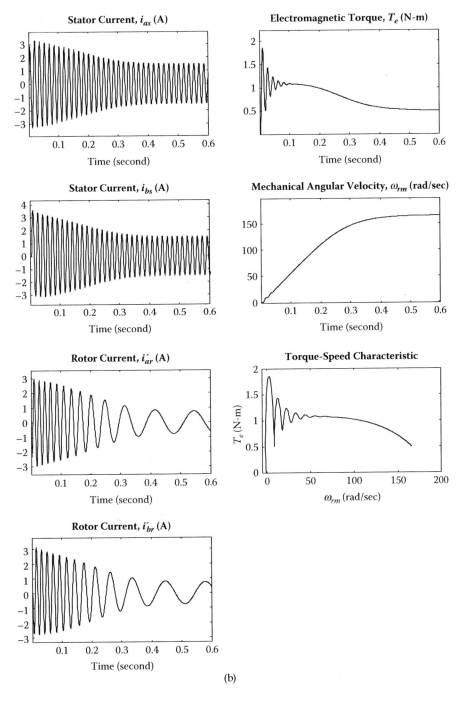

FIGURE 5.9

(Continued). (b) $T_L = 0.5$ N-m, $u_{as}(t) = \sqrt{2}\,115\cos(377t)$ and $u_{bs}(t) = \sqrt{2}\,115\sin(377t)$.

One may conclude that the steady-state torque-speed characteristic curves, as plotted in Figures 5.4 through 5.7, may not result in coherent analysis of the evolution for $T_e(t)$ as induction motors are in transients (acceleration, disacceleration, loading, disturbances, etc.). Hence, the steady-state analysis can be used mainly in preliminary studies.   □

5.1.5 Advanced Topics in Analysis of Induction Machines

The analysis was performed assuming optimal design of induction machines, linearity of magnetic system, and so on. The designer can achieve near-optimal design in the specified operating envelope. However, the undesired effects can significantly degrade the machine and system performance and capabilities. In this section, we document how to perform the advanced studies examining near-optimal stator-rotor magnetic coupling. The magnetic coupling between stator and rotor windings may not obey the expressions

$$L_{asar} = L_{aras} = L_{sr}\cos\theta_r, \quad L_{asbr} = L_{bras} = -L_{sr}\sin\theta_r,$$
$$L_{bsar} = L_{arbs} = L_{sr}\sin\theta_r \quad \text{and} \quad L_{bsbr} = L_{brbs} = L_{sr}\cos\theta_r.$$

One may not expect that in the full operating envelope, an optimal design will result in this ideal pure sinusoidal distribution which eliminates the torque ripple, current chattering, overheating, and so on. Depending on the induction machine overall design, in the full operating envelope, one may have

$$L_{asar} = L_{aras} = \sum_{n=1}^{\infty} L_{sr\,n} \cos^{2n-1}\theta_r, \quad L_{asbr} = L_{bras} = -\sum_{n=1}^{\infty} L_{sr\,n} \sin^{2n-1}\theta_r,$$

$$L_{bsar} = L_{arbs} = \sum_{n=1}^{\infty} L_{sr\,n} \sin^{2n-1}\theta_r, \quad L_{bsbr} = L_{brbs} = \sum_{n=1}^{\infty} L_{sr\,n} \cos^{2n-1}\theta_r. \qquad (5.15)$$

Modeling, simulations, and evaluation can be achieved as the machine design is accomplished and experimental studies are conducted to finalize motor parameters, inductance mapping, and other descriptive features used in the analysis tasks. As reported, using the circuitry-electromagnetic equations of motion (5.4) and *torsional-mechanical* dynamics (5.7), the transient behavior can be described with the ultimate objective to examine machine performance and capabilities. The motor parameters, induced *emf*, torque, magnetic field, and other qualitative and quantitative quantities can be experimentally obtained in the full operating envelope.

Having obtained (5.15), one finds the inductance mapping $\mathbf{L}'_{sr}(\theta_r)$. Using (5.4)

$$\mathbf{u}_{abs} = \mathbf{r}_s \mathbf{i}_{abs} + \frac{d\boldsymbol{\psi}_{abs}}{dt}, \quad \mathbf{u}'_{abr} = \mathbf{r}'_r \mathbf{i}'_{abr} + \frac{d\boldsymbol{\psi}'_{abr}}{dt},$$

and applying the expression for the flux linkages

$$\begin{bmatrix} \boldsymbol{\psi}_{abs} \\ \boldsymbol{\psi}'_{abr} \end{bmatrix} = \begin{bmatrix} \mathbf{L}_s & \mathbf{L}'_{sr}(\theta_r) \\ \mathbf{L}'^T_{sr}(\theta_r) & \mathbf{L}'_r \end{bmatrix} \begin{bmatrix} \mathbf{i}_{abs} \\ \mathbf{i}'_{abr} \end{bmatrix}$$

to find the total derivative terms, the circuitry-electromagnetic equations of motion results. The electromagnetic torque is found using the coenergy

$$T_e = \frac{P}{2} \frac{\partial W_c(\mathbf{i}_{abs}, \mathbf{i}'_{abr}, \theta_r)}{\partial \theta_r} = \frac{P}{2} \mathbf{i}^T_{abs} \frac{\partial \mathbf{L}'_{sr}(\theta_r)}{\partial \theta_r} \mathbf{i}'_{abr},$$

and the *torsional-mechanical* dynamics (5.7) is integrated.

Assume that for an induction machine

$$L_{asar} = L_{aras} = L_{sr1}\cos\theta_r + L_{sr2}\cos^3\theta_r, \quad L_{asbr} = L_{bras} = -L_{sr1}\sin\theta_r - L_{sr2}\sin^3\theta_r$$

and

$$L_{bsar} = L_{arbs} = L_{sr1}\sin\theta_r + L_{sr2}sin^3\theta_r, \quad L_{bsbr} = L_{brbs} = L_{sr1}\cos\theta_r + L_{sr2}\cos^3\theta_r$$

which corresponds to (5.15).

Taking note of the turn ratio, the flux linkages are

$$\begin{bmatrix} \boldsymbol{\psi}_{abs} \\ \boldsymbol{\psi}'_{abs} \end{bmatrix} = \begin{bmatrix} \mathbf{L}_s & \mathbf{L}'_{sr}(\theta_r) \\ \mathbf{L}'^T_{sr}(\theta_r) & \mathbf{L}'_r \end{bmatrix} \begin{bmatrix} \mathbf{i}_{abs} \\ \mathbf{i}'_{abr} \end{bmatrix}, \quad \mathbf{L}_s = \begin{bmatrix} L_{ss} & 0 \\ 0 & L_{ss} \end{bmatrix}, \quad \mathbf{L} = \begin{bmatrix} L'_{rr} & 0 \\ 0 & L'_{rr} \end{bmatrix},$$

$$\mathbf{L}'_{sr}(\theta_r) = \left(\frac{N_s}{N_r}\right) \mathbf{L}_{sr}(\theta_r) = \begin{bmatrix} L_{ms1}\cos\theta_r + L_{ms2}\cos^3\theta_r & -L_{ms1}\sin\theta - L_{ms2}\sin^3\theta_r \\ L_{ms1}\sin\theta_r + L_{ms2}\sin^3\theta_r & L_{ms1}\cos\theta_r + L_{ms2}\cos^3\theta_r \end{bmatrix},$$

or

$$\begin{bmatrix} \psi_{as} \\ \psi_{bs} \\ \psi'_{ar} \\ \psi'_{br} \end{bmatrix} = \begin{bmatrix} L_{ss} & 0 & L_{ms1}\cos\theta_r + L_{ms2}\cos^3\theta_r & -L_{ms1}\sin\theta_r - L_{ms2}\sin^3\theta_r \\ 0 & L_{ss} & L_{ms1}\sin\theta_r + L_{ms2}\sin^3\theta_r & L_{ms1}\cos\theta_r + L_{ms2}\cos^3\theta_r \\ L_{ms1}\cos\theta_r + L_{ms2}\cos^3\theta_r & L_{ms1}\sin\theta_r + L_{ms2}\sin^3\theta_r & L'_{rr} & 0 \\ -L_{ms1}\sin\theta_r - L_{ms2}\sin^3\theta_r & L_{ms1}\cos\theta_r + L_{ms2}\cos^3\theta_r & 0 & L'_{rr} \end{bmatrix} \begin{bmatrix} i_{as} \\ i_{bs} \\ i'_{ar} \\ i'_{br} \end{bmatrix}.$$

(5.16)

Using the differential equations (5.4)

$$\mathbf{u}_{abs} = \mathbf{r}_s \mathbf{i}_{abs} + \frac{d\boldsymbol{\psi}_{abs}}{dt} \quad \text{and} \quad \mathbf{u}'_{abr} = \mathbf{r}'_r \mathbf{i}'_{abr} + \frac{d\boldsymbol{\psi}'_{abr}}{dt}$$

with (5.16), results in

$$
L_{ss}\frac{di_{as}}{dt} + L_{ms1}\frac{d(i'_{ar}\cos\theta_r)}{dt} + L_{ms2}\frac{d(i'_{ar}\cos^3\theta_r)}{dt}
$$

$$
- L_{ms1}\frac{d(i'_{br}\sin\theta_r)}{dt} - L_{ms2}\frac{d(i'_{br}\sin^3\theta_r)}{dt} = -r_s i_{as} + u_{as},
$$

$$
L_{ss}\frac{di_{bs}}{dt} + L_{ms1}\frac{d(i'_{ar}\sin\theta_r)}{dt} + L_{ms2}\frac{d(i'_{ar}\sin^3\theta_r)}{dt}
$$

$$
+ L_{ms1}\frac{d(i'_{br}\cos\theta_r)}{dt} + L_{ms2}\frac{d(i'_{br}\cos^3\theta_r)}{dt} = -r_s i_{bs} + u_{bs},
$$

$$
L_{ms1}\frac{d(i_{as}\cos\theta_r)}{dt} + L_{ms2}\frac{d(i_{as}\cos^3\theta_r)}{dt} + L_{ms1}\frac{d(i_{bs}\sin\theta_r)}{dt}
$$

$$
+ L_{ms2}\frac{d(i_{bs}\sin^3\theta_r)}{dt} + L'_{rr}\frac{di'_{ar}}{dt} = -r'_r i'_{ar} + u'_{ar},
$$

$$
-L_{ms1}\frac{d(i_{as}\sin\theta_r)}{dt} - L_{ms2}\frac{d(i_{as}\sin^3\theta_r)}{dt} + L_{ms1}\frac{d(i_{bs}\cos\theta_r)}{dt}
$$

$$
+ L_{ms2}\frac{d(i_{bs}\cos^3\theta_r)}{dt} + L'_{rr}\frac{di'_{br}}{dt} = -r'_r i'_{br} + u'_{br}. \tag{5.17}
$$

The rotor motional *emf* terms in the steady-state operation are

$$
emf_{ar\omega} = (L_{ms1}i_{as}\sin\theta_r + 3L_{ms2}i_{as}\sin\theta_r\cos^2\theta_r - L_{ms1}i_{bs}\cos\theta_r
$$

$$
- 3L_{ms2}i_{bs}\cos\theta_r\sin^2\theta_r)\omega_r,
$$

$$
emf_{br\omega} = (L_{ms1}i_{as}\cos\theta_r + 3L_{ms2}i_{as}\cos\theta_r\sin^2\theta_r + L_{ms1}i_{bs}\sin\theta_r
$$

$$
+ 3L_{ms2}i_{bs}\sin\theta_r\cos^2\theta_r)\omega_r.
$$

The *torsional-mechanical* equations are used, and the expression for T_e is of a great importance. The electromagnetic torque developed is found as

$$
T_e = \frac{P}{2}\frac{\partial W_c(\mathbf{i}_{abs}, \mathbf{i}'_{abr}, \theta_r)}{\partial\theta_r} = \frac{P}{2}\mathbf{i}^T_{abs}\frac{\partial \mathbf{L}'_{sr}(\theta_r)}{\partial\theta_r}\mathbf{i}'_{abr}
$$

$$
= \frac{P}{2}\begin{bmatrix} i_{as} & i_{bs} \end{bmatrix}\begin{bmatrix} -L_{ms1}\sin\theta_r - 3L_{ms2}\sin\theta_r\cos^2\theta_r & -L_{ms1}\cos\theta_r - 3L_{ms2}\cos\theta_r\sin^2\theta_r \\ L_{ms1}\cos\theta_r + 3L_{ms2}\cos\theta_r\sin^2\theta_r & -L_{ms1}\sin\theta_r - 3L_{ms2}\sin\theta_r\cos^2\theta_r \end{bmatrix}\begin{bmatrix} i'_{ar} \\ i'_{br} \end{bmatrix}
$$

$$
= -\frac{P}{2}\{L_{ms1}[(i_{as}i'_{ar} + i_{bs}i'_{br})\sin\theta_r + (i_{as}i'_{br} - i_{bs}i'_{ar})\cos\theta_r]
$$

$$
+ 3L_{ms2}[(i_{as}i'_{ar} + i_{bs}i'_{br})\sin\theta_r\cos^2\theta_r + (i_{as}i'_{br} - i_{bs}i'_{ar})\cos\theta_r\sin^2\theta_r]\}. \tag{5.18}
$$

Using the Newtonian dynamics (5.7) and the derived expression for T_e (5.18), the *torsional-mechanical* equations are found to be

$$\frac{d\omega_r}{dt} = -\frac{P^2}{4J}\{L_{ms1}[(i_{as}i'_{ar} + i_{bs}i'_{br})\sin\theta_r + (i_{as}i'_{br} - i_{bs}i'_{ar})\cos\theta_r]$$

$$+ 3L_{ms2}[(i_{as}i'_{ar} + i_{bs}i'_{br})\sin\theta_r\cos^2\theta_r + (i_{as}i'_{br} - i_{bs}i'_{ar})\cos\theta_r\sin^2\theta_r]\}$$

$$-\frac{B_m}{J}\omega_r - \frac{P}{2J}T_L,$$

$$\frac{d\theta_r}{dt} = \omega_r. \tag{5.19}$$

The resulting equations of motion are found by using (5.17) and (5.19). We have

$$\frac{di_{as}}{dt} = \frac{1}{L_{ss}}\left[-r_s i_{as} - L_{ms1}\frac{d(i'_{ar}\cos\theta_r)}{dt} - L_{ms2}\frac{d(i'_{ar}\cos^3\theta_r)}{dt}\right.$$

$$\left. + L_{ms1}\frac{d(i'_{br}\sin\theta_r)}{dt} + L_{ms2}\frac{d(i'_{br}\sin^3\theta_r)}{dt} + u_{as}\right],$$

$$\frac{di_{bs}}{dt} = \frac{1}{L_{ss}}\left[-r_s i_{bs} - L_{ms1}\frac{d(i'_{ar}\sin\theta_r)}{dt} - L_{ms2}\frac{d(i'_{ar}\sin^3\theta_r)}{dt}\right.$$

$$\left. - L_{ms1}\frac{d(i'_{br}\cos\theta_r)}{dt} - L_{ms2}\frac{d(i'_{br}\cos^3\theta_r)}{dt} + u_{bs}\right],$$

$$\frac{di'_{ar}}{dt} = \frac{1}{L'_{rr}}\left[-r'_r i'_{ar} - L_{ms1}\frac{d(i_{as}\cos\theta_r)}{dt} - L_{ms2}\frac{d(i_{as}\cos^3\theta_r)}{dt}\right.$$

$$\left. - L_{ms1}\frac{d(i_{bs}\sin\theta_r)}{dt} - L_{ms2}\frac{d(i_{bs}\sin^3\theta_r)}{dt} + u'_{ar}\right],$$

$$\frac{di'_{br}}{dt} = \frac{1}{L'_{rr}}\left[-r'_r i'_{br} + L_{ms1}\frac{d(i_{as}\sin\theta_r)}{dt} + L_{ms2}\frac{d(i_{as}\sin^3\theta_r)}{dt}\right.$$

$$\left. - L_{ms1}\frac{d(i_{bs}\cos\theta_r)}{dt} - L_{ms2}\frac{d(i_{bs}\cos^3\theta_r)}{dt} + u'_{br}\right],$$

$$\frac{d\omega_r}{dt} = -\frac{P^2}{4J}\{L_{ms1}[(i_{as}i'_{ar} + i_{bs}i'_{br})\sin\theta_r + (i_{as}i'_{br} - i_{bs}i'_{ar})\cos\theta_r]$$

$$+ 3L_{ms2}[(i_{as}i'_{ar} + i_{bs}i'_{br})\sin\theta_r\cos^2\theta_r + (i_{as}i'_{br} - i_{bs}i'_{ar})\cos\theta_r\sin^2\theta_r]\}$$

$$-\frac{B_m}{J}\omega_r - \frac{P}{2J}T_L,$$

$$\frac{d\theta_r}{dt} = \omega_r. \tag{5.20}$$

Cauchy's form of differential equations are obtained by applying (5.17) or (5.20). One finds

$$
\frac{di_{as}}{dt} = \frac{1}{L_\Sigma}\Big(-L'_{rr}r_s i_{as} - 3L^2_{ms2}i_{bs}\omega_r \sin^4\theta_r + L_{ms1}r'_r i'_{ar}\cos\theta_r - L_{ms1}u'_{ar}\cos\theta_r
$$

$$
+ L_{ms2}u'_{br}\sin^3\theta_r + L_{ms1}u'_{br}\sin\theta_r + L^2_{ms1}i_{bs}\omega_r + 3L_{ms2}L'_{rr}i'_{ar}\omega_r\sin\theta_r
$$

$$
+ L_{ms2}u'_{ar}\cos\theta_r\sin^2\theta_r + L_{ms2}r'_r i'_{ar}\cos\theta_r + 3L^2_{ms2}i_{bs}\omega_r\sin^2\theta_r
$$

$$
+ 4L_{ms1}L_{ms2}i_{bs}\omega_r\sin^2\theta_r + L'_{rr}u_{as} + 8L_{ms1}L_{ms2}i_{as}\omega_r\cos\theta_r
$$

$$
- 4L_{ms1}L_{ms2}i_{as}\omega_r\cos\theta_r\sin\theta_r + 3L_{ms2}L'_{rr}i'_{ar}\omega_r\cos\theta_r\sin^2\theta_r + L_{ms1}L'_{rr}i'_{ar}\omega_r\sin\theta_r
$$

$$
- L_{ms2}u'_{ar}\cos\theta_r - L_{ms1}r'_r i'_{br}\sin\theta_r - 3L_{ms2}L_{rr}i'_{ar}\omega_r\sin^3\theta_r
$$

$$
+ 6L^2_{ms2}i_{as}\omega_r\cos\theta_r\sin^3\theta_r - 4L_{ms1}L_{ms2}i_{bs}\omega_r\sin^4\theta_r
$$

$$
+ L_{ms1}L_{ms2}i_{bs}\omega_r - L_{ms2}r'_r i'_{ar}\cos\theta_r\sin^2\theta_r
$$

$$
- 3L^2_{ms2}i_{as}\omega_r\cos\theta_r\sin\theta_r - L_{ms2}r'_r i'_{br}\sin^3\theta_r + L_{ms1}L'_{rr}i'_{br}\omega_r\cos\theta_r\Big),
$$

$$
\frac{di_{bs}}{dt} = \frac{1}{L_\Sigma}\Big(-L'_{rr}r_s i_{bs} - L_{ms1}u'_{ar}\sin\theta_r - L_{ms2}u'_{br}\cos\theta_r - 3L^2_{ms2}i_{bs}\cos\theta_r\sin\theta_r
$$

$$
+ L_{ms1}L'_{rr}i'_{br}\omega_r\sin\theta_r - 3L_{ms2}L'_{rr}i'_{br}\omega_r\sin^3\theta_r + 3L_{ms1}L'_{rr}i'_{br}\omega_r\sin\theta_r
$$

$$
+ 4L_{ms1}L_{ms2}i_{as}\omega_r\sin^4\theta_r - L_{ms1}L_{ms2}i_{as}\omega_r + 8L_{ms1}L_{ms2}i_{bs}\omega_r\cos\theta_r\sin^3\theta_r
$$

$$
+ L_{ms1}r'_r i'_{br}\cos\theta_r + L_{ms2}u'_{br}\cos\theta_r\sin^2\theta_r + L_{ms2}r'_r i'_{br}\cos\theta_r + L_{ms1}r'_r i'_{ar}\sin\theta_r
$$

$$
- 3L^2_{ms2}i_{as}\omega_r\sin^2\theta_r - 4L_{ms1}L_{ms2}i_{as}\omega_r\sin^2\theta_r - L_{ms2}r'_r i'_{br}\cos\theta_r\sin^2\theta_r
$$

$$
+ 6L^2_{ms2}i_{bs}\omega_r\cos\theta_r\sin^3\theta_r - 4L_{ms1}L_{ms2}i_{bs}\omega_r\cos\theta_r\sin\theta_r - L_{ms1}L'_{rr}i'_{ar}\omega_r\cos\theta_r
$$

$$
- 3L_{ms2}L'_{rr}i'_{ar}\omega_r\cos\theta_r\sin^2\theta_r + 3L^2_{ms2}i_{as}\omega_r\sin^4\theta_r + L_{ms2}r'_r i'_{ar}\sin^3\theta_r
$$

$$
- L^2_{ms1}i_{as}\omega_r - L_{ms1}u'_{br}\cos\theta_r - L_{ms2}u'_{ar}\sin^3\theta_r + L'_{rr}u_{bs}\Big),
$$

$$
\frac{di'_{ar}}{dt} = \frac{1}{L_\Sigma}\Big(-L_{ss}r'_r i'_{ar} - L_{ms2}r_s i_{as}\cos\theta_r\sin^2\theta_r + L_{ms2}u_{as}\cos\theta_r\sin^2\theta_r
$$

$$
+ L_{ms2}r_s i_{as}\cos\theta_r + L_{ms1}r_s i_{as}\cos\theta_r - L^2_{ms1}i'_{br}\omega_r - L_{ms2}u_{as}\cos\theta_r
$$

$$
- 4L_{ms1}L_{ms2}i'_{br}\omega_r\sin^2\theta_r + 4L_{ms1}L_{ms2}i'_{br}\sin^4\theta_r - L_{ms2}u_{bs}\sin^3\theta_r
$$

$$
+ 8L_{ms1}L_{ms2}i'_{ar}\omega_r\cos\theta_r\sin^3\theta_r - 3L^2_{ms2}i'_{ar}\omega_r\cos\theta_r\sin\theta_r
$$

$$
- 4L_{ms1}L_{ms2}i'_{ar}\omega_r\cos\theta_r\sin\theta_r - 3L_{ms2}L_{ss}\omega_r\sin^3\theta_r - 3L^2_{ms2}i'_{br}\omega_r\sin^2\theta_r
$$

$$
- 3L_{ms2}L_{ss}i_{bs}\omega_r\cos\theta_r\sin^2\theta_r + 3L_{ms2}L_{ss}i_{as}\omega_r\sin\theta_r + L_{ms1}L_{ss}i_{bs}\omega_r\sin\theta_r
$$

$$
- 3L^2_{ms2}i'_{br}\omega_r\sin^4\theta_r + L_{ms1}r_s i_{bs}\sin\theta_r - L_{ms1}L_{ms2}i'_{br}\omega_r + L_{ms2}r_s i_{bs}\sin^3\theta_r
$$

$$
- L_{ms1}u_{as}\cos\theta_r - L_{ms1}u_{bs}\sin\theta_r + 6L^2_{ms2}i'_{ar}\omega_r\cos\theta_r\sin^3\theta_r
$$

$$
- L_{ms1}L_{ss}i_{bs}\omega_r\cos\theta_r + L_{ss}u'_{ar}\Big),
$$

$$\frac{di'_{br}}{dt} = \frac{1}{L_\Sigma}\left(-L_{ss}r'_r i'_{br} + u'_{br}L_{ss} + L_{ms1}L_{ms2}i'_{ar}\omega_r - 4L_{ms1}L_{ms2}i'_{br}\omega_r \cos\theta_r \sin\theta_r\right.$$

$$+3L_{ms2}L_{ss}i_{as}\omega_r \cos\theta_r \sin^2\theta_r - L_{ms1}r_s i_{bs}\cos\theta_r - 4L_{ms1}L_{ms2}i'_{ar}\omega_r \sin^4\theta_r$$

$$-L_{ms2}r_s i_{bs}\cos\theta_r \sin^2\theta_r + L_{ms1}L_{ss}i_{bs}\omega_r \sin\theta_r + L_{ms2}u_{bs}\cos\theta_r \sin^2\theta_r$$

$$+L_{ms1}L_{ss}i_{as}\omega_r \cos\theta_r + 6L_{ms2}^2 i'_{br}\omega_r \cos\theta_r \sin^3\theta_r - 3L_{ms2}^2 i'_{br}\omega_r \cos\theta_r \sin\theta_r$$

$$+8L_{ms1}L_{ms2}i'_{br}\omega_r \cos\theta_r \sin^3\theta_r - 3L_{ms2}L_{ss}i_{bs}\omega_r \sin^3\theta_r + 3L_{ms2}L_{ss}i_{bs}\omega_r \sin\theta_r$$

$$+L_{ms1}u_{as}\omega_r \sin\theta_r + L_{ms2}u_{as}\sin^3\theta_r - L_{ms1}u_{bs}\cos\theta_r - L_{ms2}u_{bs}\cos\theta_r + L_{ms1}^2 i'_{ar}\omega_r$$

$$+3L_{ms2}^2 i'_{ar}\omega_r \sin^2\theta_r - 3L_{ms2}^2 i'_{ar}\omega_r \sin^4\theta_r - L_{ms2}r_s i_{as}\sin^3\theta_r - L_{ms2}r_s i_{bs}\cos\theta_r$$

$$\left. +L_{ms1}r_s i_{as}\sin\theta_r + 4L_{ms1}L_{ms2}i'_{ar}\omega_r \sin^2\theta_r\right),$$

$$\frac{d\omega_r}{dt} = -\frac{P^2}{4J}\left\{L_{ms1}[(i_{as}i'_{ar} + i_{bs}i'_{br})\sin\theta_r + (i_{as}i'_{br} - i_{bs}i'_{ar})\cos\theta_r]\right.$$

$$\left. +3L_{ms2}[(i_{as}i'_{ar} + i_{bs}i'_{br})\sin\theta_r \cos^2\theta_r + (i_{as}i'_{br} - i_{bs}i'_{ar})\cos\theta_r \sin^2\theta_r]\right\}$$

$$-\frac{B_m}{J}\omega_r - \frac{P}{2J}T_L,$$

$$\frac{d\theta_r}{dt} = \omega_r, \tag{5.21}$$

where

$$L_\Sigma = L_{ss}L'_{rr} - L_{ms2}^2 - L_{ms1}^2 + 3L_{ms2}^2 \sin^2\theta_r - 3L_{ms2}^2 \sin^4\theta_r - 2L_{ms1}L_{ms2}$$
$$+ 4L_{ms1}L_{ms2}\sin^2\theta_r - 4L_{ms1}L_{ms2}\sin^4\theta_r.$$

The analysis and simulation can be carried out using the nonlinear differential equations, which are in Cauchy's form or not in Cauchy's form.

Example 5.3

We perform simulations for an A-class two-phase, 115 V (*rms*), 60 Hz, four-pole ($P = 4$) induction motor by using the differential equations derived. To ensure the descriptive features, the motor parameters are in correspondence as used in Example 5.2, except $L_{ms} = L_{ms1} + 3L_{ms2}$. We have: $r_s = 1.2$ ohm, $r'_r = 1.5$ ohm, $L_{ms1} = 0.145$ H, $L_{ms2} = 0.005$ H, $L_{ls} = 0.02$ H, $L_{ss} = L_{ls} + L_{ms1} + 3L_{ms2}$, $L'_{lr} = 0.02$ H, $L'_{rr} = L'_{lr} + L_{ms1} + 3L_{ms2}$, $B_m = 1 \times 10^{-6}$ N-m-sec/rad, and $J = 0.005$ kg-m^2. The supplied phase voltages are

$$u_{as}(t) = \sqrt{2}115\cos(377t) \quad \text{and} \quad u_{bs}(t) = \sqrt{2}115\sin(377t).$$

Using the derived equations of motion in non-Cauchy's and Cauchy's forms, as given by (5.19) and (5.20), the Simulink models (diagrams) are developed. For differential equations (5.19), the Simulink model is illustrated in Figure 5.10. The parameters are uploaded by the MATLAB file as reported below.

```
% Parameters of a Two-Phase Induction Motor
Rs = 1.2; Rr = 1.5;
Lms1 = 0.16; Lms2 = 0.0; % Ideal magnetic coupling
Lms1 = 0.145; Lms2 = 0.005; % Near-optimal coupling
Lls = 0.02; Llr = 0.02; % Leakage inductances
Lss = Lls+Lms1+3*Lms2;
Lrr = Llr+Lms1+3*Lms2;
Bm = .000001; % Friction coefficient
J = 0.005; % Moment of inertia
P = 4; % Number of poles
T _ L=0; % Load torque
% Simulate the Simulink Model
sim('ch5 _ TwoPhaseInductionMotor',2.5);
```

The simulations are performed. The transient behavior and dynamic torque-speed characteristics for a pure sinusoidal magnetic coupling are reported in Figure 5.11a. For the realistic (near-optimal) coupling, the results are documented in Figures 5.11b and 5.11c for unloaded and loaded ($T_L = 0.5$ N-m at $t = 1.5$ sec) motor as it accelerates from the stall. One can assess the (1) acceleration capabilities; (2) transient dynamics for all state variables $i_{as}(t)$, $i_{bs}(t)$, $i_{ar}(t)$, $i_{br}(t)$, $\omega_r(t)$, and $\theta_r(t)$; (3) evolution of the electromagnetic torque T_e; (4) dynamic torque-speed characteristics $\omega_{rm}(t) = \Omega_r[T_e(t)]$; (5) efficiency and losses; (6) thermodynamics; (7) motional $emf_{ar\omega}$ and $emf_{br\omega}$ induced in the rotor windings; and so on. The results indicate that even a very small deviation from the ideal pure sinusoidal stator-rotor magnetic coupling results in a significant degradation of the motor and system performance and capabilities. We found a significant reduction of the efficiency, degradation of acceleration capabilities, torque ripple (which results in vibration, noise, mechanical wearing, etc.), and other undesirable effects. The analysis performed supports the need for a coherent structural design, accurate modeling, and nonlinear simulations with minimum level of simplifications and assumptions. The realistic motor design must be integrated with sound studies of motion devices and systems in the behavior domain. □

5.1.6 Three-Phase Induction Motors in the Machine Variables

We examined two-phase induction motors using the *machine* variables utilizing the *as*, *bs*, *ar*, and *br* physical quantities (voltage, current, and flux linkages). The large majority of industrial induction machines are three-phase motors. Our goal is to develop the equations of motion in order to accomplish various analysis tasks for three-phase induction motors, as shown in Figure 5.12.

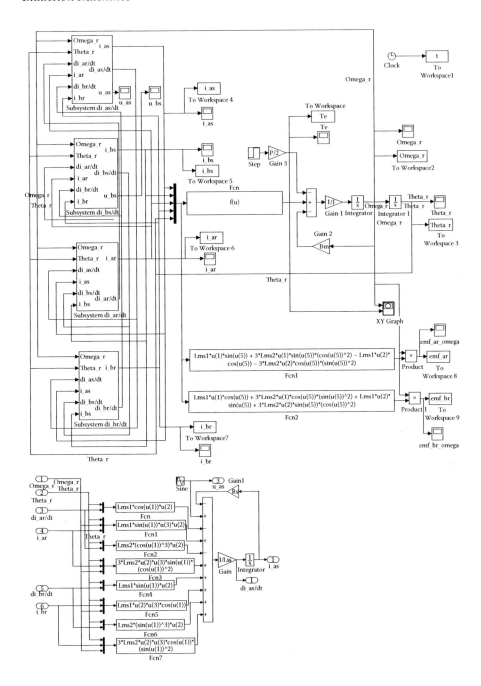

FIGURE 5.10
Simulink model to simulate induction motor dynamics.

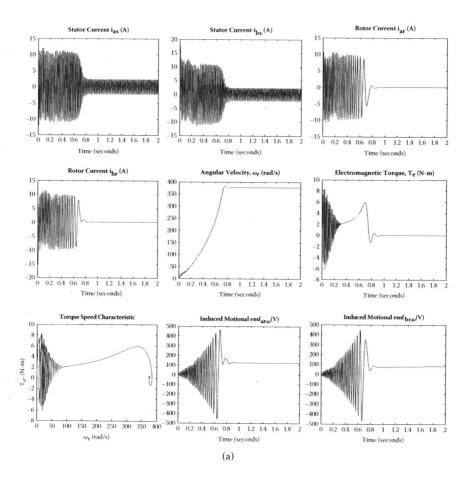

(a)

FIGURE 5.11
Dynamics of the A-class induction motor: (a) $T_L = 0$ N-m, $L_{asar} = L_{aras} = L_{sr}\cos\theta_r$, $L_{asbr} = L_{bras} = -L_{sr}\sin\theta_r$, $L_{bsar} = L_{arbs} = L_{sr}\sin\theta_r$, and $L_{bsbr} = L_{brbs} = L_{sr}\cos\theta_r$.

Kirchhoff's voltage law gives the equations for the voltages, supplied to the *abc* stator and rotor windings, the *abc* stator and rotor currents, and flux linkages. We have

$$u_{as} = r_s i_{as} + \frac{d\psi_{as}}{dt}, \quad u_{bs} = r_s i_{bs} + \frac{d\psi_{bs}}{dt}, \quad u_{cs} = r_s i_{cs} + \frac{d\psi_{cs}}{dt},$$

$$u_{ar} = r_r i_{ar} + \frac{d\psi_{ar}}{dt}, \quad u_{br} = r_r i_{br} + \frac{d\psi_{br}}{dt}, \quad u_{cr} = r_r i_{cr} + \frac{d\psi_{cr}}{dt}. \quad (5.22)$$

where u_{as}, u_{bs} and u_{cs} are the phase voltages supplied to the *as*, *bs*, and *cs* stator windings; u_{ar}, u_{br} and u_{cr} are the phase voltages supplied to the *ar*, *br*, and *cr*

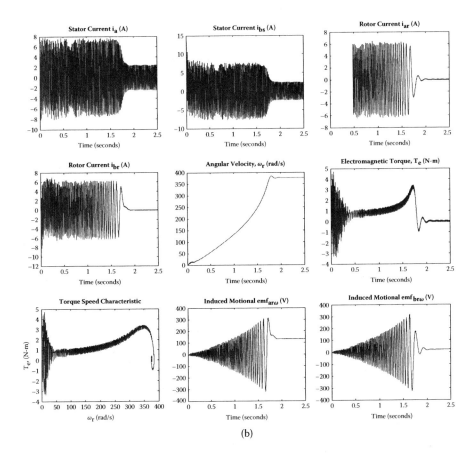

(b)

FIGURE 5.11

(Continued) (b) $T_L = 0$ N-m, $L_{asar} = L_{aras} = L_{sr1}\cos\theta_r + L_{sr2}\cos^3\theta_r$, $L_{asbr} = L_{bras} = -L_{sr1}\sin\theta_r - L_{sr2}\sin^3\theta_r$, $L_{bsar} = L_{arbs} = L_{sr1}\sin\theta_r + L_{sr2}\sin^3\theta_r$, and $L_{bsbr} = L_{brbs} = L_{sr1}\cos\theta_r + L_{sr2}\cos^3\theta_r$.

rotor windings (for squirrel-cage motors $u_{ar} = 0$, $u_{br} = 0$, and $u_{cr} = 0$); i_{as}, i_{bs}, and i_{cs} are the phase currents in the stator windings; i_{ar}, i_{br} and i_{cr} are the phase currents in the rotor windings; ψ_{as}, ψ_{bs} and ψ_{cs} are the stator flux linkages; ψ_{ar}, ψ_{br} and ψ_{cr} are the rotor flux linkages.

The *abc* stator and rotor voltages, currents and flux linkages are used as the variables. From (5.22), the application of vector notations yields a set of two differential equations

$$\mathbf{u}_{abcs} = \mathbf{r}_s \mathbf{i}_{abcs} + \frac{d\psi_{abcs}}{dt},$$

$$\mathbf{u}_{abcr} = \mathbf{r}_r \mathbf{i}_{abcr} + \frac{d\psi_{abcr}}{dt}, \tag{5.23}$$

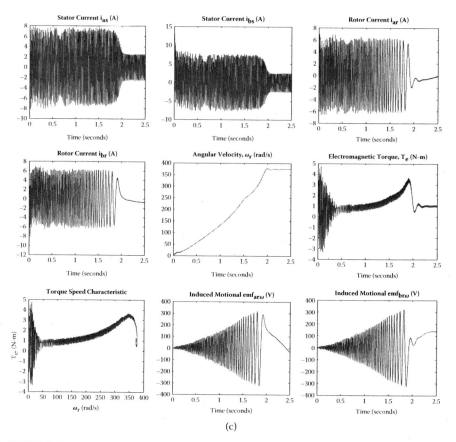

FIGURE 5.11
(Continued) (c) $T_L = 0.5$ N-m (at $t = 1.5$ sec), $L_{asar} = L_{aras} = L_{sr1}\cos\theta_r + L_{sr2}\cos^3\theta_r$, $L_{asbr} = L_{bras} = -L_{sr1}\sin\theta_r - L_{sr2}\sin^3\theta_r$, $L_{bsar} = L_{arbs} = L_{sr1}\sin\theta_r + L_{sr2}\sin^3\theta_r$, and $L_{bsbr} = L_{brbs} = L_{sr1}\cos\theta_r + L_{sr2}\cos^3\theta_r$.

where the *abc* stator and rotor voltages, currents, and flux linkages vectors are given as

$$\mathbf{u}_{abcs} = \begin{bmatrix} u_{as} \\ u_{bs} \\ u_{cs} \end{bmatrix}, \mathbf{u}_{abcr} = \begin{bmatrix} u_{ar} \\ u_{br} \\ u_{cr} \end{bmatrix}, \mathbf{i}_{abcs} = \begin{bmatrix} i_{as} \\ i_{bs} \\ i_{cs} \end{bmatrix}, \mathbf{i}_{abcr} = \begin{bmatrix} i_{ar} \\ i_{br} \\ i_{cr} \end{bmatrix}, \boldsymbol{\psi}_{abcs} = \begin{bmatrix} \psi_{as} \\ \psi_{bs} \\ \psi_{cs} \end{bmatrix}, \text{ and } \boldsymbol{\psi}_{abcr} = \begin{bmatrix} \psi_{ar} \\ \psi_{br} \\ \psi_{cr} \end{bmatrix}.$$

In (5.23), the diagonal matrices of the stator and rotor resistances are

$$\mathbf{r}_s = \begin{bmatrix} r_s & 0 & 0 \\ 0 & r_s & 0 \\ 0 & 0 & r_s \end{bmatrix} \text{ and } \mathbf{r}_r = \begin{bmatrix} r_r & 0 & 0 \\ 0 & r_r & 0 \\ 0 & 0 & r_r \end{bmatrix}.$$

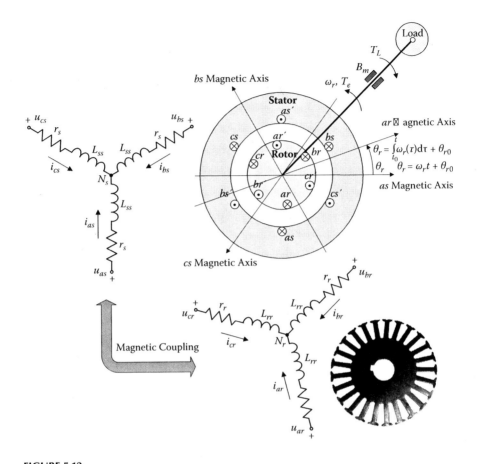

FIGURE 5.12
Three-phase symmetric induction motor. Rotor windings are placed in the slots in the laminated rotor.

The flux linkages are found as functions of the corresponding currents in the stator and rotor windings using the self- and mutual inductances. The analysis of the stator and rotor magnetically coupled system, accomplished referencing to Figure 5.12, gives the following equations

$$\psi_{as} = L_{asas}i_{as} + L_{asbs}i_{bs} + L_{ascs}i_{cs} + L_{asar}i_{ar} + L_{asbr}i_{br} + L_{ascr}i_{cr},$$

$$\psi_{bs} = L_{bsas}i_{as} + L_{bsbs}i_{bs} + L_{bscs}i_{cs} + L_{bsar}i_{ar} + L_{bsbr}i_{br} + L_{bscr}i_{cr},$$

$$\psi_{cs} = L_{csas}i_{as} + L_{csbs}i_{bs} + L_{cscs}i_{cs} + L_{csar}i_{ar} + L_{csbr}i_{br} + L_{cscr}i_{cr},$$

$$\psi_{ar} = L_{aras}i_{as} + L_{arbs}i_{bs} + L_{arcs}i_{cs} + L_{arar}i_{ar} + L_{arbr}i_{br} + L_{arcr}i_{cr},$$

$$\psi_{br} = L_{bras}i_{as} + L_{brbs}i_{bs} + L_{brcs}i_{cs} + L_{brar}i_{ar} + L_{brbr}i_{br} + L_{brcr}i_{cr},$$

$$\psi_{cr} = L_{cras}i_{as} + L_{crbs}i_{bs} + L_{crcs}i_{cs} + L_{crar}i_{ar} + L_{crbr}i_{br} + L_{crcr}i_{cr},$$

where L_{asas}, L_{bsbs}, L_{cscs}, L_{arar}, L_{brbr} and l_{crcr} are the stator and rotor self-inductances; L_{asbs}, L_{ascs}, L_{asar}, L_{asbr}, L_{ascr} ..., L_{cras}, L_{crbs}, L_{crcs}, L_{crar} and L_{crbr} are the mutual inductances between stator-stator, stator-rotor, and rotor-rotor windings.

The stator and rotor *abc* windings are identical and displaced magnetically by $2/3\pi$.

Hence, there exists a coupling between the *abc* stator and rotor windings. The mutual inductances between the stator windings are equal, and

$$L_{asbs} = L_{ascs} = L_{bscs} = L_{ms}\cos\left(\frac{2}{3}\pi\right) = -\frac{1}{2}L_{ms}, \quad L_{ms} = \frac{N_s^2}{\Re_m}.$$

The rotor windings are displaced by 120 electrical degrees. One finds the following equations for the mutual inductances between the rotor windings

$$L_{arbr} = L_{arcr} = L_{brcr} = L_{mr}\cos\left(\frac{2}{3}\pi\right) = -\frac{1}{2}L_{mr}, \quad L_{mr} = \frac{N_r^2}{N_s^2}L_{ms}.$$

The stator and rotor self-inductances are

$$L_{ss} = L_{ls} + L_{ms} \quad \text{and} \quad L_{rr} = L_{lr} + L_{mr}.$$

The matrices of self- and mutual inductances $\mathbf{L}_s$ and $\mathbf{L}_r$ are

$$\mathbf{L}_s = \begin{bmatrix} L_{ls} + L_{ms} & -\frac{1}{2}L_{ms} & -\frac{1}{2}L_{ms} \\ -\frac{1}{2}L_{ms} & L_{ls} + L_{ms} & -\frac{1}{2}L_{ms} \\ -\frac{1}{2}L_{ms} & -\frac{1}{2}L_{ms} & L_{ls} + L_{ms} \end{bmatrix} \quad \text{and} \quad \mathbf{L}_r = \begin{bmatrix} L_{lr} + L_{mr} & -\frac{1}{2}L_{mr} & -\frac{1}{2}L_{mr} \\ -\frac{1}{2}L_{mr} & L_{lr} + L_{mr} & -\frac{1}{2}L_{mr} \\ -\frac{1}{2}L_{mr} & -\frac{1}{2}L_{mr} & L_{lr} + L_{mr} \end{bmatrix}.$$

The mutual inductances between the stator and rotor windings are periodic functions of the electrical angular displacement θ_r, and the period is 2π. Assume that the mutual inductances are sinusoidal functions such that

$$L_{asar} = L_{aras} = L_{sr}\cos\theta_r, \quad L_{asbr} = L_{bras} = L_{sr}\cos\left(\theta_r + \frac{2}{3}\pi\right),$$

$$L_{ascr} = L_{cras} = L_{sr}\cos\left(\theta_r - \frac{2}{3}\pi\right),$$

$$L_{bsar} = L_{arbs} = L_{sr}\cos\left(\theta_r - \frac{2}{3}\pi\right), \quad L_{bsbr} = L_{brbs} = L_{sr}\cos\theta_r,$$

$$L_{bscr} = L_{crbs} = L_{sr}\cos\left(\theta_r + \frac{2}{3}\pi\right),$$

$$L_{csar} = L_{arcs} = L_{sr}\cos\left(\theta_r + \frac{2}{3}\pi\right), \quad L_{csbr} = L_{brcs} = L_{sr}\cos\left(\theta_r - \frac{2}{3}\pi\right),$$

$$L_{cscr} = L_{crcs} = L_{sr}\cos\theta_r,$$

where

$$L_{sr} = \frac{N_s N_r}{\Re_m}.$$

We have the following stator-rotor mutual inductance mapping

$$\mathbf{L}_{sr}(\theta_r) = L_{sr} \begin{bmatrix} \cos\theta_r & \cos(\theta_r + \frac{2}{3}\pi) & \cos(\theta_r - \frac{2}{3}\pi) \\ \cos(\theta_r - \frac{2}{3}\pi) & \cos\theta_r & \cos(\theta_r + \frac{2}{3}\pi) \\ \cos(\theta_r + \frac{2}{3}\pi) & \cos(\theta_r - \frac{2}{3}\pi) & \cos\theta_r \end{bmatrix}.$$

One obtains

$$\begin{bmatrix} \boldsymbol{\psi}_{abcs} \\ \boldsymbol{\psi}_{abcr} \end{bmatrix} = \begin{bmatrix} \mathbf{L}_s & \mathbf{L}_{sr}(\theta_r) \\ \mathbf{L}_{sr}^T(\theta_r) & \mathbf{L}_r \end{bmatrix} \begin{bmatrix} \mathbf{i}_{abcs} \\ \mathbf{i}_{abcr} \end{bmatrix},$$

$$\boldsymbol{\psi}_{abcs} = \mathbf{L}_s \mathbf{i}_{abcs} + \mathbf{L}_{sr}(\theta_r)\mathbf{i}_{abcr}, \quad \boldsymbol{\psi}_{abcr} = \mathbf{L}_{sr}^T(\theta_r)\mathbf{i}_{abcs} + \mathbf{L}_r \mathbf{i}_{abcr}. \tag{5.24}$$

Using the number of turns N_s and N_r, we have

$$\mathbf{u}'_{abcr} = \frac{N_s}{N_r}\mathbf{u}_{abcr}, \quad \mathbf{i}'_{abcr} = \frac{N_r}{N_s}\mathbf{i}_{abcr} \quad \text{and} \quad \boldsymbol{\psi}'_{abcr} = \frac{N_s}{N_r}\boldsymbol{\psi}_{abcr}.$$

The inductances are

$$L_{ms} = \frac{N_s}{N_r}L_{sr}, \quad L_{sr} = \frac{N_s N_r}{\Re_m} \quad \text{and} \quad L_{ms} = \frac{N_s^2}{\Re_m}.$$

Hence,

$$\mathbf{L}'_{sr}(\theta_r) = \frac{N_s}{N_r}\mathbf{L}_{sr}(\theta_r) = L_{ms} \begin{bmatrix} \cos\theta_r & \cos(\theta_r + \frac{2}{3}\pi) & \cos(\theta_r - \frac{2}{3}\pi) \\ \cos(\theta_r - \frac{2}{3}\pi) & \cos\theta_r & \cos(\theta_r + \frac{2}{3}\pi) \\ \cos(\theta_r + \frac{2}{3}\pi) & \cos(\theta_r - \frac{2}{3}\pi) & \cos\theta_r \end{bmatrix},$$

$$\mathbf{L}'_r = \frac{N_s^2}{N_r^2}\mathbf{L}_r = \begin{bmatrix} L'_{lr} + L_{ms} & -\frac{1}{2}L_{ms} & -\frac{1}{2}L_{ms} \\ -\frac{1}{2}L_{ms} & L'_{lr} + L_{ms} & -\frac{1}{2}L_{ms} \\ -\frac{1}{2}L_{ms} & -\frac{1}{2}L_{ms} & L'_{lr} + L_{ms} \end{bmatrix},$$

where

$$L'_{lr} = \frac{N_s^2}{N_r^2}L_{lr}.$$

The matrix equation (5.24) for the flux linkages is rewritten as

$$\begin{bmatrix} \boldsymbol{\psi}_{abcs} \\ \boldsymbol{\psi}'_{abcr} \end{bmatrix} = \begin{bmatrix} \mathbf{L}_s & \mathbf{L}'_{sr}(\theta_r) \\ \mathbf{L}'^{T}_{sr}(\theta_r) & \mathbf{L}'_r \end{bmatrix} \begin{bmatrix} \mathbf{i}_{abcs} \\ \mathbf{i}'_{abcr} \end{bmatrix}. \tag{5.25}$$

Substituting the expressions for $\mathbf{L}_s$, $\mathbf{L}'_{sr}(\theta_r)$, and $\mathbf{L}'_r$ in (5.25), we have

$$\begin{bmatrix} \psi_{as} \\ \psi_{bs} \\ \psi_{cs} \\ \psi'_{ar} \\ \psi'_{br} \\ \psi'_{cr} \end{bmatrix} = \begin{bmatrix} L_{ls}+L_{ms} & -\frac{1}{2}L_{ms} & -\frac{1}{2}L_{ms} & L_{ms}\cos\theta_r & L_{ms}\cos(\theta_r+\frac{2}{3}\pi) & L_{ms}\cos(\theta_r-\frac{2}{3}\pi) \\ -\frac{1}{2}L_{ms} & L_{ls}+L_{ms} & -\frac{1}{2}L_{ms} & L_{ms}\cos(\theta_r-\frac{2}{3}\pi) & L_{ms}\cos\theta_r & L_{ms}\cos(\theta_r+\frac{2}{3}\pi) \\ -\frac{1}{2}L_{ms} & -\frac{1}{2}L_{ms} & L_{ls}+L_{ms} & L_{ms}\cos(\theta_r+\frac{2}{3}\pi) & L_{ms}\cos(\theta_r-\frac{2}{3}\pi) & L_{ms}\cos\theta_r \\ L_{ms}\cos\theta_r & L_{ms}\cos(\theta_r-\frac{2}{3}\pi) & L_{ms}\cos(\theta_r+\frac{2}{3}\pi) & L'_{lr}+L_{ms} & -\frac{1}{2}L_{ms} & -\frac{1}{2}L_{ms} \\ L_{ms}\cos(\theta_r+\frac{2}{3}\pi) & L_{ms}\cos\theta_r & L_{ms}\cos(\theta_r-\frac{2}{3}\pi) & -\frac{1}{2}L_{ms} & L'_{lr}+L_{ms} & -\frac{1}{2}L_{ms} \\ L_{ms}\cos(\theta_r-\frac{2}{3}\pi) & L_{ms}\cos(\theta_r+\frac{2}{3}\pi) & L_{ms}\cos\theta_r & -\frac{1}{2}L_{ms} & -\frac{1}{2}L_{ms} & L'_{lr}+L_{ms} \end{bmatrix} \begin{bmatrix} i_{as} \\ i_{bs} \\ i_{cs} \\ i'_{ar} \\ i'_{br} \\ i'_{cr} \end{bmatrix}.$$

In expanded form the expressions for the flux linkages are

$$\psi_{as} = (L_{ls}+L_{ms})i_{as} - \frac{1}{2}L_{ms}i_{bs} - \frac{1}{2}L_{ms}i_{cs} + L_{ms}\cos\theta_r\, i'_{ar} + L_{ms}\cos\left(\theta_r+\frac{2}{3}\pi\right)i'_{br}$$

$$+ L_{ms}\cos\left(\theta_r-\frac{2}{3}\pi\right)i'_{cr},$$

$$\psi_{bs} = -\frac{1}{2}L_{ms}i_{as} + (L_{ls}+L_{ms})i_{bs} - \frac{1}{2}L_{ms}i_{cs} + L_{ms}\cos\left(\theta_r-\frac{2}{3}\pi\right)i'_{ar} + L_{ms}\cos\theta_r\, i'_{br}$$

$$+ L_{ms}\cos\left(\theta_r+\frac{2}{3}\pi\right)i'_{cr},$$

$$\psi_{cs} = -\frac{1}{2}L_{ms}i_{as} - \frac{1}{2}L_{ms}i_{bs} + (L_{ls}+L_{ms})i_{cs} + L_{ms}\cos\left(\theta_r+\frac{2}{3}\pi\right)i'_{ar}$$

$$+ L_{ms}\cos\left(\theta_r-\frac{2}{3}\pi\right)i'_{br} + L_{ms}\cos\theta_r\, i'_{cr},$$

$$\psi'_{ar} = L_{ms}\cos\theta_r\, i_{as} + L_{ms}\cos\left(\theta_r-\frac{2}{3}\pi\right)i_{bs} + L_{ms}\cos\left(\theta_r+\frac{2}{3}\pi\right)i_{cs}$$

$$+ (L'_{lr}+L_{ms})i'_{ar} - \frac{1}{2}L_{ms}i'_{br} - \frac{1}{2}L_{ms}i'_{cr},$$

$$\psi'_{br} = L_{ms}\cos\left(\theta_r+\frac{2}{3}\pi\right)i_{as} + L_{ms}\cos\theta_r\, i_{bs} + L_{ms}\cos\left(\theta_r-\frac{2}{3}\pi\right)i_{cs} - \frac{1}{2}L_{ms}i'_{ar}$$

$$+ (L'_{lr}+L_{ms})i'_{br} - \frac{1}{2}L_{ms}i'_{cr},$$

$$\psi'_{cr} = L_{ms}\cos\left(\theta_r-\frac{2}{3}\pi\right)i_{as} + L_{ms}\cos\left(\theta_r+\frac{2}{3}\pi\right)i_{bs} + L_{ms}\cos\theta_r\, i_{cs} - \frac{1}{2}L_{ms}i'_{ar}$$

$$- \frac{1}{2}L_{ms}i'_{br} + (L'_{lr}+L_{ms})i'_{cr}. \tag{5.26}$$

From (5.26), one finds the flux linkages total derivatives

$$\frac{d\psi_{abcs}}{dt} \text{ and } \frac{d\psi'_{abcr}}{dt}$$

which provide the expressions for *emf* terms. Using (5.23) and (5.25), we obtain

$$\mathbf{u}_{abcs} = \mathbf{r}_s \mathbf{i}_{abcs} + \frac{d\psi_{abcs}}{dt} = \mathbf{r}_s \mathbf{i}_{abcs} + \mathbf{L}_s \frac{d\mathbf{i}_{abcs}}{dt} + \frac{d(\mathbf{L}'_{sr}(\theta_r)\mathbf{i}'_{abcr})}{dt},$$

$$\mathbf{u}'_{abcr} = \mathbf{r}'_r \mathbf{i}'_{abcr} + \frac{d\psi'_{abcr}}{dt} = \mathbf{r}'_r \mathbf{i}'_{abcr} + \mathbf{L}'_r \frac{d\mathbf{i}'_{abcr}}{dt} + \frac{d(\mathbf{L}'^T_{sr}(\theta_r)\mathbf{i}_{abcs})}{dt}, \qquad (5.27)$$

where

$$\mathbf{r}'_r = \frac{N_s^2}{N_r^2}\mathbf{r}_r.$$

Equations (5.27) in expanded form, using (5.26), are given as

$$u_{as} = r_s i_{as} + (L_{ls} + L_{ms})\frac{di_{as}}{dt} - \frac{1}{2}L_{ms}\frac{di_{bs}}{dt} - \frac{1}{2}L_{ms}\frac{di_{cs}}{dt} + L_{ms}\frac{d(i'_{ar}\cos\theta_r)}{dt}$$

$$+ L_{ms}\frac{d(i'_{br}\cos(\theta_r + \frac{2\pi}{3}))}{dt} + L_{ms}\frac{d(i'_{cr}\cos(\theta_r - \frac{2\pi}{3}))}{dt},$$

$$u_{bs} = r_s i_{bs} - \frac{1}{2}L_{ms}\frac{di_{as}}{dt} + (L_{ls} + L_{ms})\frac{di_{bs}}{dt} - \frac{1}{2}L_{ms}\frac{di_{cs}}{dt} + L_{ms}\frac{d(i'_{ar}\cos(\theta_r - \frac{2\pi}{3}))}{dt}$$

$$+ L_{ms}\frac{d(i'_{br}\cos\theta_r)}{dt} + L_{ms}\frac{d(i'_{cr}\cos(\theta_r + \frac{2\pi}{3}))}{dt},$$

$$u_{cs} = r_s i_{cs} - \frac{1}{2}L_{ms}\frac{di_{as}}{dt} - \frac{1}{2}L_{ms}\frac{di_{bs}}{dt} + (L_{ls} + L_{ms})\frac{di_{cs}}{dt} + L_{ms}\frac{d(i'_{ar}\cos(\theta_r + \frac{2\pi}{3}))}{dt}$$

$$+ L_{ms}\frac{d(i'_{br}\cos(\theta_r - \frac{2\pi}{3}))}{dt} + L_{ms}\frac{d(i'_{cr}\cos\theta_r)}{dt},$$

$$u'_{ar} = r'_r i'_{ar} + L_{ms}\frac{d(i_{as}\cos\theta_r)}{dt} + L_{ms}\frac{d(i_{bs}\cos(\theta_r - \frac{2\pi}{3}))}{dt} + L_{ms}\frac{d(i_{cs}\cos(\theta_r + \frac{2\pi}{3}))}{dt}$$

$$+ (L'_{lr} + L_{ms})\frac{di'_{ar}}{dt} - \frac{1}{2}L_{ms}\frac{di'_{br}}{dt} - \frac{1}{2}L_{ms}\frac{di'_{cr}}{dt},$$

$$u'_{br} = r'_r i'_{br} + L_{ms}\frac{d(i_{as}\cos(\theta_r + \frac{2\pi}{3}))}{dt} + L_{ms}\frac{d(i_{bs}\cos\theta_r)}{dt} + L_{ms}\frac{d(i_{cs}\cos(\theta_r - \frac{2\pi}{3}))}{dt}$$

$$- \frac{1}{2}L_{ms}\frac{di'_{ar}}{dt} + (L'_{lr} + L_{ms})\frac{di'_{br}}{dt} - \frac{1}{2}L_{ms}\frac{di'_{cr}}{dt},$$

$$u'_{cr} = r'_r i'_{cr} + L_{ms}\frac{d(i_{as}\cos(\theta_r - \frac{2\pi}{3}))}{dt} + L_{ms}\frac{d(i_{bs}\cos(\theta_r + \frac{2\pi}{3}))}{dt} + L_{ms}\frac{d(i_{cs}\cos\theta_r)}{dt}$$

$$- \frac{1}{2}L_{ms}\frac{di'_{ar}}{dt} - \frac{1}{2}L_{ms}\frac{di'_{br}}{dt} + (L'_{lr} + L_{ms})\frac{di'_{cr}}{dt}.$$

We obtain the following set of equations that describe the circuitry-electromagnetic dynamics of three-phase induction motors

$$u_{as} = r_s i_{as} + (L_{ls} + L_{ms}) \frac{di_{as}}{dt} - \frac{1}{2} L_{ms} \frac{di_{bs}}{dt} - \frac{1}{2} L_{ms} \frac{di_{cs}}{dt} + L_{ms} \cos\theta_r \frac{di'_{ar}}{dt}$$

$$+ L_{ms} \cos\left(\theta_r + \frac{2\pi}{3}\right) \frac{di'_{br}}{dt} + L_{ms} \cos\left(\theta_r - \frac{2\pi}{3}\right) \frac{di'_{cr}}{dt}$$

$$- L_{ms} \left[i'_{ar} \sin\theta_r + i'_{br} \sin\left(\theta_r + \frac{2\pi}{3}\right) + i'_{cr} \sin\left(\theta_r - \frac{2\pi}{3}\right) \right] \omega_r,$$

$$u_{bs} = r_s i_{bs} - \frac{1}{2} L_{ms} \frac{di_{as}}{dt} + (L_{ls} + L_{ms}) \frac{di_{bs}}{dt} - \frac{1}{2} L_{ms} \frac{di_{cs}}{dt}$$

$$+ L_{ms} \cos\left(\theta_r + \frac{2\pi}{3}\right) \frac{di'_{ar}}{dt} + L_{ms} \cos\theta_r \frac{di'_{br}}{dt} + L_{ms} \cos\left(\theta_r + \frac{2\pi}{3}\right) \frac{di'_{cr}}{dt}$$

$$- L_{ms} \left[i'_{ar} \sin\left(\theta_r + \frac{2\pi}{3}\right) + i'_{br} \sin\theta_r + i'_{cr} \sin\left(\theta_r + \frac{2\pi}{3}\right) \right] \omega_r,$$

$$u_{cs} = r_s i_{cs} - \frac{1}{2} L_{ms} \frac{di_{as}}{dt} - \frac{1}{2} L_{ms} \frac{di_{bs}}{dt} + (L_{ls} + L_{ms}) \frac{di_{cs}}{dt} + L_{ms} \cos\left(\theta_r + \frac{2\pi}{3}\right) \frac{di'_{ar}}{dt}$$

$$+ L_{ms} \cos\left(\theta_r - \frac{2\pi}{3}\right) \frac{di'_{br}}{dt} + L_{ms} \cos\theta_r \frac{di'_{cr}}{dt}$$

$$- L_{ms} \left[i'_{ar} \sin\left(\theta_r + \frac{2\pi}{3}\right) + i'_{br} \sin\left(\theta_r - \frac{2\pi}{3}\right) + i'_{cr} \sin\theta_r \right] \omega_r,$$

$$u'_{ar} = r'_r i'_{ar} + L_{ms} \cos\theta_r \frac{di_{as}}{dt} + L_{ms} \cos\left(\theta_r - \frac{2\pi}{3}\right) \frac{di_{bs}}{dt} + L_{ms} \cos\left(\theta_r + \frac{2\pi}{3}\right) \frac{di_{cs}}{dt}$$

$$+ (L'_{lr} + L_{ms}) \frac{di'_{ar}}{dt} - \frac{1}{2} L_{ms} \frac{di'_{br}}{dt} - \frac{1}{2} L_{ms} \frac{di'_{cr}}{dt}$$

$$- L_{ms} \left[i_{as} \sin\theta_r + i_{bs} \sin\left(\theta_r - \frac{2\pi}{3}\right) + i_{cs} \sin\left(\theta_r + \frac{2\pi}{3}\right) \right] \omega_r,$$

$$u'_{br} = r'_r i'_{br} + L_{ms} \cos\left(\theta_r + \frac{2\pi}{3}\right) \frac{di_{as}}{dt} + L_{ms} \cos\theta_r \frac{di_{bs}}{dt}$$

$$+ L_{ms} \cos\left(\theta_r - \frac{2\pi}{3}\right) \frac{di_{cs}}{dt} - \frac{1}{2} L_{ms} \frac{di'_{ar}}{dt} + (L'_{lr} + L_{ms}) \frac{di'_{br}}{dt} - \frac{1}{2} L_{ms} \frac{di'_{cr}}{dt}$$

$$- L_{ms} \left[i_{as} \sin\left(\theta_r + \frac{2\pi}{3}\right) + i_{bs} \sin\theta_r + i_{cs} \sin\left(\theta_r - \frac{2\pi}{3}\right) \right] \omega_r,$$

$$u'_{br} = r'_r i'_{cr} + L_{ms} \cos\left(\theta_r - \frac{2\pi}{3}\right) \frac{di_{as}}{dt} + L_{ms} \cos\left(\theta_r + \frac{2\pi}{3}\right) \frac{di_{bs}}{dt}$$

$$+ L_{ms} \cos\theta_r \frac{di_{cs}}{dt} - \frac{1}{2} L_{ms} \frac{di'_{ar}}{dt} - \frac{1}{2} L_{ms} \frac{di'_{br}}{dt} + (L'_{lr} + L_{ms}) \frac{di'_{cr}}{dt}$$

$$- L_{ms} \left[i_{as} \sin\left(\theta_r - \frac{2\pi}{3}\right) + i_{bs} \sin\left(\theta_r + \frac{2\pi}{3}\right) + i_{cs} \sin\theta_r \right] \omega_r. \tag{5.28}$$

Equation (5.28) yields the differential equations in Cauchy's form as

$$
\begin{bmatrix}
\dfrac{di_{as}}{dt} \\[4pt]
\dfrac{di_{bs}}{dt} \\[4pt]
\dfrac{di_{cs}}{dt} \\[4pt]
\dfrac{di'_{ar}}{dt} \\[4pt]
\dfrac{di'_{br}}{dt} \\[4pt]
\dfrac{di'_{cr}}{dt}
\end{bmatrix}
= \frac{1}{L_{\Sigma L}}
\begin{bmatrix}
-r_s L_{\Sigma m} & -\frac{1}{2} r_s L_{ms} & -\frac{1}{2} r_s L_{ms} & 0 & 0 & 0 \\[4pt]
-\frac{1}{2} r_s L_{ms} & -r_s L_{\Sigma m} & -\frac{1}{2} r_s L_{ms} & 0 & 0 & 0 \\[4pt]
-\frac{1}{2} r_s L_{ms} & -\frac{1}{2} r_s L_{ms} & -r_s L_{\Sigma m} & 0 & 0 & 0 \\[4pt]
0 & 0 & 0 & -r_r L_{\Sigma m} & -\frac{1}{2} r_r L_{ms} & -\frac{1}{2} r_r L_{ms} \\[4pt]
0 & 0 & 0 & -\frac{1}{2} r_r L_{ms} & -r_r L_{\Sigma m} & -\frac{1}{2} r_r L_{ms} \\[4pt]
0 & 0 & 0 & -\frac{1}{2} r_r L_{ms} & -\frac{1}{2} r_r L_{ms} & -r_r L_{\Sigma m}
\end{bmatrix}
\begin{bmatrix}
i_{as} \\ i_{bs} \\ i_{cs} \\ i'_{ar} \\ i'_{br} \\ i'_{cr}
\end{bmatrix}
$$

$$
+ \frac{1}{L_{\Sigma L}}
\begin{bmatrix}
0 & 0 & 0 & r_r L_{ms}\cos\theta_r & r_r L_{ms}\cos(\theta_r+\frac{2}{3}\pi) & r_r L_{ms}\cos(\theta_r-\frac{2}{3}\pi) \\[4pt]
0 & 0 & 0 & r_r L_{ms}\cos(\theta_r-\frac{2}{3}\pi) & r_r L_{ms}\cos\theta_r & r_r L_{ms}\cos(\theta_r+\frac{2}{3}\pi) \\[4pt]
0 & 0 & 0 & r_r L_{ms}\cos(\theta_r+\frac{2}{3}\pi) & r_r L_{ms}\cos(\theta_r-\frac{2}{3}\pi) & r_r L_{ms}\cos\theta_r \\[4pt]
r_s L_{ms}\cos\theta_r & r_s L_{ms}\cos(\theta_r-\frac{2}{3}\pi) & r_s L_{ms}\cos(\theta_r+\frac{2}{3}\pi) & 0 & 0 & 0 \\[4pt]
r_s L_{ms}\cos(\theta_r+\frac{2}{3}\pi) & r_s L_{ms}\cos\theta_r & r_s L_{ms}\cos(\theta_r-\frac{2}{3}\pi) & 0 & 0 & 0 \\[4pt]
r_s L_{ms}\cos(\theta_r-\frac{2}{3}\pi) & r_s L_{ms}\cos(\theta_r+\frac{2}{3}\pi) & r_s L_{ms}\cos\theta_r & 0 & 0 & 0
\end{bmatrix}
\times
\begin{bmatrix}
i_{as} \\ i_{bs} \\ i_{cs} \\ i'_{ar} \\ i'_{br} \\ i'_{cr}
\end{bmatrix}
$$

$$
+ \frac{1}{L_{\Sigma L}}
\begin{bmatrix}
0 & 1.299 L_{ms}^2 \omega_r & -1.299 L_{ms}^2 \omega_r & L_{\Sigma m}\omega_r\sin\theta_r & L_{\Sigma m}\omega_r\sin(\theta_r+\frac{2}{3}\pi) & L_{\Sigma m}\omega_r\sin(\theta_r-\frac{2}{3}\pi) \\[4pt]
-1.299 L_{ms}^2 \omega_r & 0 & 1.299 L_{ms}^2 \omega_r & L_{\Sigma m}\omega_r\sin(\theta_r-\frac{2}{3}\pi) & L_{\Sigma m}\omega_r\sin\theta_r & L_{\Sigma m}\omega_r\sin(\theta_r+\frac{2}{3}\pi) \\[4pt]
1.299 L_{ms}^2 \omega_r & -1.299 L_{ms}^2 \omega_r & 0 & L_{\Sigma m}\omega_r\sin(\theta_r+\frac{2}{3}\pi) & L_{\Sigma m}\omega_r\sin(\theta_r-\frac{2}{3}\pi) & L_{\Sigma m}\omega_r\sin\theta_r \\[4pt]
L_{\Sigma m}\omega_r\sin\theta_r & L_{\Sigma m}\omega_r\sin(\theta_r-\frac{2}{3}\pi) & L_{\Sigma m}\omega_r\sin(\theta_r+\frac{2}{3}\pi) & 0 & -1.299 L_{ms}^2 \omega_r & 1.299 L_{ms}^2 \omega_r \\[4pt]
L_{\Sigma m}\omega_r\sin(\theta_r+\frac{2}{3}\pi) & L_{\Sigma m}\omega_r\sin\theta_r & L_{\Sigma m}\omega_r\sin(\theta_r-\frac{2}{3}\pi) & 1.299 L_{ms}^2 \omega_r & 0 & -1.299 L_{ms}^2 \omega_r \\[4pt]
L_{\Sigma m}\omega_r\sin(\theta_r-\frac{2}{3}\pi) & L_{\Sigma m}\omega_r\sin(\theta_r+\frac{2}{3}\pi) & L_{\Sigma m}\omega_r\sin\theta_r & -1.299 L_{ms}^2 \omega_r & 1.299 L_{ms}^2 \omega_r & 0
\end{bmatrix}
\times
\begin{bmatrix}
i_{as} \\ i_{bs} \\ i_{cs} \\ i'_{ar} \\ i'_{br} \\ i'_{cr}
\end{bmatrix}
$$

$$+\frac{1}{L_{\Sigma L}}\begin{bmatrix} 2L_{ms}+L'_{lr} & \frac{1}{2}L_{ms} & \frac{1}{2}L_{ms} & -L_{ms}\cos\theta_r & -L_{ms}\cos(\theta_r+\frac{2}{3}\pi) & -L_{ms}\cos(\theta_r-\frac{2}{3}\pi) \\ \frac{1}{2}L_{ms} & 2L_{ms}+L'_{lr} & \frac{1}{2}L_{ms} & -L_{ms}\cos(\theta_r-\frac{2}{3}\pi) & -L_{ms}\cos\theta_r & -L_{ms}\cos(\theta_r+\frac{2}{3}\pi) \\ \frac{1}{2}L_{ms} & \frac{1}{2}L_{ms} & 2L_{ms}+L'_{lr} & -L_{ms}\cos(\theta_r+\frac{2}{3}\pi) & -L_{ms}\cos(\theta_r-\frac{2}{3}\pi) & -L_{ms}\cos\theta_r \\ -L_{ms}\cos\theta_r & -L_{ms}\cos(\theta_r-\frac{2}{3}\pi) & -L_{ms}\cos(\theta_r+\frac{2}{3}\pi) & 2L_{ms}+L'_{lr} & \frac{1}{2}L_{ms} & \frac{1}{2}L_{ms} \\ -L_{ms}\cos(\theta_r+\frac{2}{3}\pi) & -L_{ms}\cos\theta_r & -L_{ms}\cos(\theta_r-\frac{2}{3}\pi) & \frac{1}{2}L_{ms} & 2L_{ms}+L'_{lr} & \frac{1}{2}L_{ms} \\ -L_{ms}\cos(\theta_r-\frac{2}{3}\pi) & -L_{ms}\cos(\theta_r+\frac{2}{3}\pi) & -L_{ms}\cos\theta_r & \frac{1}{2}L_{ms} & \frac{1}{2}L_{ms} & 2L_{ms}+L'_{lr} \end{bmatrix}$$

$$\times\begin{bmatrix} u_{as} \\ u_{bs} \\ u_{cs} \\ u'_{ar} \\ u'_{br} \\ u'_{cr} \end{bmatrix}. \tag{5.29}$$

Here,

$$L_{\Sigma L}=(3L_{ms}+L'_{lr})L'_{lr}, \quad L_{\Sigma m}=2L_{ms}+L'_{lr} \quad \text{and} \quad L_{\Sigma ms}=\frac{3}{2}L_{ms}^2+L_{ms}L'_{lr}.$$

Newton's second law of motion is applied. The expression for the electromagnetic torque developed by induction motors is obtained using the coenergy $W_c(\mathbf{i}_{abcs}, \mathbf{i}'_{abcr}, \theta_r)$. For P-pole three-phase induction machines, we have

$$T_e=\frac{P}{2}\frac{\partial W_c(\mathbf{i}_{abcs}, \mathbf{i}'_{abcr}, \theta_r)}{\partial\theta_r}.$$

The coenergy is given by the following expression

$$W_c=W_f=\frac{1}{2}\mathbf{i}_{abcs}^T(\mathbf{L}_s-L_{ls}\mathbf{I})\mathbf{i}_{abcs}+\mathbf{i}_{abcs}^T\mathbf{L}'_{sr}(\theta_r)\mathbf{i}'_{abcr}+\frac{1}{2}\mathbf{i}_{abcr}^{\prime T}(\mathbf{L}'_r-L'_{lr}\mathbf{I})\mathbf{i}'_{abcr}.$$

In W_c, matrices $\mathbf{L}_s$, $L_{ls}\mathbf{I}$, $\mathbf{L}'_r$, and $L'_{lr}\mathbf{I}$ are not functions of the electrical angular displacement θ_r. Using

$$\mathbf{L}'_{sr}(\theta_r)=L_{ms}\begin{bmatrix} \cos\theta_r & \cos(\theta_r+\frac{2}{3}\pi) & \cos(\theta_r-\frac{2}{3}\pi) \\ \cos(\theta_r-\frac{2}{3}\pi) & \cos\theta_r & \cos(\theta_r+\frac{2}{3}\pi) \\ \cos(\theta_r+\frac{2}{3}\pi) & \cos(\theta_r-\frac{2}{3}\pi) & \cos\theta_r \end{bmatrix},$$

the expression for the electromagnetic torque is found to be

$$
T_e = \frac{P}{2} \mathbf{i}_{abcs}^{T} \frac{\partial \mathbf{L}_{sr}'(\theta_r)}{\partial \theta_r} \mathbf{i}_{abcr}'
$$

$$
= -\frac{P}{2} L_{ms} \begin{bmatrix} i_{as} & i_{bs} & i_{cs} \end{bmatrix} \begin{bmatrix} \sin\theta_r & \sin(\theta_r + \frac{2}{3}\pi) & \sin(\theta_r - \frac{2}{3}\pi) \\ \sin(\theta_r - \frac{2}{3}\pi) & \sin\theta_r & \sin(\theta_r + \frac{2}{3}\pi) \\ \sin(\theta_r + \frac{2}{3}\pi) & \sin(\theta_r - \frac{2}{3}\pi) & \sin\theta_r \end{bmatrix} \begin{bmatrix} i_{ar}' \\ i_{br}' \\ i_{cr}' \end{bmatrix}
$$

$$
= -\frac{P}{2} L_{ms} \Big[(i_{as} i_{ar}' + i_{bs} i_{br}' + i_{cs} i_{cr}') \sin\theta_r + (i_{as} i_{cr}' + i_{bs} i_{ar}' + i_{cs} i_{br}') \sin\left(\theta_r - \frac{2}{3}\pi\right)
$$

$$
+ (i_{as} i_{br}' + i_{bs} i_{cr}' + i_{cs} i_{ar}') \sin\left(\theta_r + \frac{2}{3}\pi\right) \Big]
$$

$$
= -\frac{P}{2} L_{ms} \Bigg\{ \left[i_{as}\left(i_{ar}' - \frac{1}{2} i_{br}' - \frac{1}{2} i_{cr}' \right) + i_{bs}\left(i_{br}' - \frac{1}{2} i_{ar}' - \frac{1}{2} i_{cr}' \right) \right.
$$

$$
\left. + i_{cs}\left(i_{cr}' - \frac{1}{2} i_{br}' - \frac{1}{2} i_{ar}' \right) \right] \sin\theta_r + \frac{\sqrt{3}}{2} [i_{as}(i_{br}' - i_{cr}') + i_{bs}(i_{cr}' - i_{ar}') + i_{cs}(i_{ar}' - i_{br}')] \cos\theta_r \Bigg\}.
$$

$$(5.30)$$

Newton's second law of rotational motion leads to the *torsional-mechanical* equations, where T_e is given by (5.30). We have

$$
\frac{d\omega_r}{dt} = \frac{P}{2J} T_e - \frac{B_m}{J} \omega_r - \frac{P}{2J} T_L
$$

$$
= -\frac{P^2}{4J} L_{ms} \Bigg[(i_{as} i_{ar}' + i_{bs} i_{br}' + i_{cs} i_{cr}') \sin\theta_r + (i_{as} i_{cr}' + i_{bs} i_{ar}' + i_{cs} i_{br}') \sin\left(\theta_r - \frac{2}{3}\pi\right)
$$

$$
+ (i_{as} i_{br}' + i_{bs} i_{cr}' + i_{cs} i_{ar}') \sin(\theta_r + \tfrac{2}{3}\pi) \Bigg] - \frac{B_m}{J} \omega_r - \frac{P}{2J} T_L
$$

$$
= -\frac{P^2}{4J} L_{ms} \Bigg\{ \left[i_{as}\left(i_{ar}' - \frac{1}{2} i_{br}' - \frac{1}{2} i_{cr}' \right) + i_{bs}\left(i_{br}' - \frac{1}{2} i_{ar}' - \frac{1}{2} i_{cr}' \right) \right.
$$

$$
\left. + i_{cs}\left(i_{cr}' - \frac{1}{2} i_{br}' - \frac{1}{2} i_{ar}' \right) \right] \sin\theta_r + \frac{\sqrt{3}}{2} [i_{as}(i_{br}' - i_{cr}') + i_{bs}(i_{cr}' - i_{ar}') + i_{cs}(i_{ar}' - i_{br}')] \cos\theta_r \Bigg\}
$$

$$
- \frac{B_m}{J} \omega_r - \frac{P}{2J} T_L,
$$

$$
\frac{d\theta_r}{dt} = \omega_r.
$$

$$(5.31)$$

Combining differential equations (5.29) and (5.31), the resulting model for three-phase induction motors in the *machine* variables results.

5.2 Dynamics and Analysis of Induction Motors Using the *Quadrature* and *Direct* Variables

5.2.1 *Arbitrary,* Stationary, Rotor, and Synchronous Reference Frames

The differential equations to examine the steady-state and dynamic performance of induction machines were developed in the *machine* variables by using Kirchhoff's second law, Newton's law of motion as well as the Lagrange concept. We described the motor steady-state and transient behavior applying the well-defined *abc* stator and rotor variables and quantities. The *quadrature* and *direct* (*qd*) quantities can be applied to reduce the complexity of the resulting differential equations which describe the dynamics of electric machines. For three-phase induction and synchronous machines, the transformations of the *machine* variables (stator and rotor voltages, currents and flux linkages) to the *quadrature-, direct-,* and *zero*-axis components of stator and rotor voltages, currents and flux linkages are performed using the Park and other transformations. In the most general case, the *arbitrary* reference frame is applied. The reference frames can be fixed with the rotor and/or stator, and the frame "rotates." In the *arbitrary* reference frame, the frame angular velocity ω is not specified. Assigning the frame angular velocity to be $\omega = 0$, $\omega = \omega_r$, and $\omega = \omega_e$, the models in three reference frames can be found. In particular, the stationary ($\omega = 0$), rotor ($\omega = \omega_r$), and synchronous ($\omega = \omega_e$) reference frames are used. Table 5.1 documents the transformations of the *machine* variables to the *quadrature, direct,* and *zero* (*qd0*) quantities, and vice versa. Here, ω and θ are the angular velocity and displacement of the reference frame.

Example 5.4

Assume the following ac voltages are supplied to the *as, bs,* and *cs* windings

$$u_{as}(t) = 100\cos(377t) \text{ V}, \ u_{bs}(t) = 100\cos\left(377t - \frac{2}{3}\pi\right) \text{ V} \quad \text{and}$$

$$u_{cs}(t) = 100\cos\left(377t + \frac{2}{3}\pi\right) \text{ V}.$$

Assigning the frame angular velocities to be $\omega = 377$ rad/sec and $\omega = 0$ rad/sec (the synchronous and stationary reference frames, respectively), one can obtain the corresponding *qd0* voltage components.

We obtain the *quadrature-, direct-,* and *zero*-axis components of voltages as given in the *arbitrary* reference frame. As given in Table 5.1., in the *arbitrary* reference frame, the *direct* Park transformation is

$$\mathbf{u}_{qd0s} = \mathbf{K}_s \mathbf{u}_{abcs},$$

TABLE 5.1

Transformations of the *Machine* and *qd0* Variables

Stator, Rotor, Quadrature, and Direct Magnetic Axes	Transformation of Variables Using Transformation Matrices

Arbitrary reference frame ω unspecified

Direct transformation (stator variables)

$$\mathbf{u}_{qdos} = \mathbf{K}_s \mathbf{u}_{abcs}, \mathbf{i}_{qdos} = \mathbf{K}_s \mathbf{i}_{abcs}, \boldsymbol{\psi}_{qdos}, = \mathbf{K}_s \boldsymbol{\psi}_{abcs}$$

Inverse transformation (stator variables)

$$\mathbf{u}_{abcs} = \mathbf{K}_s^{-1} \mathbf{u}_{qdos}, \mathbf{i}_{abcs} = \mathbf{K}_s^{-1} \mathbf{i}_{qdos}, \boldsymbol{\psi}_{abcs} = \mathbf{K}_s^{-1} \boldsymbol{\psi}_{qdos}$$

Stator transformation matrices

$$\mathbf{K}_s = \frac{2}{3} \begin{bmatrix} \cos\theta & \cos(\theta - \frac{2}{3}\pi) & \cos(\theta + \frac{2}{3}\pi) \\ \sin\theta & \sin(\theta - \frac{2}{3}\pi) & \sin(\theta + \frac{2}{3}\pi) \\ \frac{1}{2} & \frac{1}{2} & \frac{1}{2} \end{bmatrix},$$

$$\mathbf{K}_s^{-1} = \begin{bmatrix} \cos\theta & \sin\theta & 1 \\ \cos(\theta - \frac{2}{3}\pi) & \sin(\theta - \frac{2}{3}\pi) & 1 \\ \cos(\theta + \frac{2}{3}\pi) & \sin(\theta + \frac{2}{3}\pi) & 1 \end{bmatrix}$$

Direct transformation (rotor variables)

$$\mathbf{u}_{qdor} = \mathbf{K}_r \mathbf{u}_{abcr}, \mathbf{i}_{qdor} = \mathbf{K}_r \mathbf{i}_{abcr}, \boldsymbol{\psi}_{qdor} = \mathbf{K}_r \boldsymbol{\psi}_{abcr}$$

Inverse transformation (rotor variables)

$$\mathbf{u}_{abcr} = \mathbf{K}_r^{-1} \mathbf{u}_{qdor}, \mathbf{i}_{abcr} = \mathbf{K}_r^{-1} \mathbf{i}_{qdor}, \boldsymbol{\psi}_{abcr} = \mathbf{K}_r^{-1} \boldsymbol{\psi}_{qdor}$$

Rotor transformation matrices

$$\mathbf{K}_r = \frac{2}{3} \begin{bmatrix} \cos(\theta - \theta_r) & \cos(\theta - \theta_r - \frac{2}{3}\pi) & \cos(\theta - \theta_r + \frac{2}{3}\pi) \\ \sin(\theta - \theta_r) & \sin(\theta - \theta_r - \frac{2}{3}\pi) & \sin(\theta - \theta_r + \frac{2}{3}\pi) \\ \frac{1}{2} & \frac{1}{2} & \frac{1}{2} \end{bmatrix},$$

$$\mathbf{K}_r^{-1} = \begin{bmatrix} \cos(\theta - \theta_r) & \sin(\theta - \theta_r) & 1 \\ \cos(\theta - \theta_r - \frac{2}{3}\pi) & \sin(\theta - \theta_r - \frac{2}{3}\pi) & 1 \\ \cos(\theta - \theta_r + \frac{2}{3}\pi) & \sin(\theta - \theta_r + \frac{2}{3}\pi) & 1 \end{bmatrix}$$

Stationary reference frame, $\omega = 0$

Direct transformation

$$\mathbf{u}_{qdos}^s = \mathbf{K}_s^s \mathbf{u}_{abcs}, \mathbf{i}_{qdos}^s = \mathbf{K}_s^s \mathbf{i}_{abcs}, \boldsymbol{\psi}_{qdos}^s = \mathbf{K}_s^s \boldsymbol{\psi}_{abcs}$$

Inverse transformation

$$\mathbf{u}_{abcs} = \mathbf{K}_s^{s-1} \mathbf{u}_{qdos}^s, \mathbf{i}_{abcs} = \mathbf{K}_s^{s-1} \mathbf{i}_{qdos}^s, \boldsymbol{\psi}_{abcs} = \mathbf{K}_s^{s-1} \boldsymbol{\psi}_{qdos}^s$$

Direct transformation

$$\mathbf{u}_{qdor}^s = \mathbf{K}_r^s \mathbf{u}_{abcr}, \mathbf{i}_{qdor}^s = \mathbf{K}_r^s \mathbf{i}_{abcr}, \boldsymbol{\psi}_{qdor}^s = \mathbf{K}_r^s \boldsymbol{\psi}_{abcr}$$

Inverse transformation

$$\mathbf{u}_{abcr} = \mathbf{K}_r^{s-1} \mathbf{u}_{qdor}^s, \mathbf{i}_{abcr} = \mathbf{K}_r^{s-1} \mathbf{i}_{qdor}^s, \boldsymbol{\psi}_{abcr} = \mathbf{K}_r^{s-1} \boldsymbol{\psi}_{qdor}^s$$

(Continued)

TABLE 5.1

Transformations of the *Machine* and *qd0* Variables (Continued)

Stator, Rotor, Quadrature, and Direct Magnetic Axes	Transformation of Variables using Transformation Matrices
Rotor reference frame, $\omega = \omega_r$	*Direct transformation* $$\mathbf{u}^r_{qdos} = \mathbf{K}^r_s\mathbf{u}_{abcs}\,,\mathbf{i}^r_{qdos} = \mathbf{K}^r_s\mathbf{i}_{abcs}\,,\boldsymbol{\psi}^r_{qdos} = \mathbf{K}^r_s\boldsymbol{\psi}_{abcs}$$ *Inverse transformation* $$\mathbf{u}_{abcs} = \mathbf{K}^{r-1}_s\mathbf{u}^r_{qdos}\,,\mathbf{i}_{abcs} = \mathbf{K}^{r-1}_s\mathbf{i}^r_{qdos}\,,\boldsymbol{\psi}_{abcs} = \mathbf{K}^{r-1}_s\boldsymbol{\psi}^r_{qdos}$$ *Direct transformation* $$\mathbf{u}^r_{qdor} = \mathbf{K}^r_r\mathbf{u}_{abcr}\,,\mathbf{i}^r_{qdor} = \mathbf{K}^r_r\mathbf{i}_{abcr}\,,\boldsymbol{\psi}^r_{qdor} = \mathbf{K}^r_r\boldsymbol{\psi}_{abcr}$$ *Inverse transformation* $$\mathbf{u}_{abcr} = \mathbf{K}^{r-1}_r\mathbf{u}^r_{qdor}\,,\mathbf{i}_{abcr} = \mathbf{K}^{r-1}_r\mathbf{i}^r_{qdor}\,,\boldsymbol{\psi}_{abcr} = \mathbf{K}^{r-1}_r\boldsymbol{\psi}^r_{qdor}$$
Synchronous reference frame, $\omega = \omega_r$	*Direct transformation* $$\mathbf{u}^e_{qdos} = \mathbf{K}^e_s\mathbf{u}_{abcs}\,,\mathbf{i}^e_{qdos} = \mathbf{K}^e_s\mathbf{i}_{abcs}\,,\boldsymbol{\psi}^e_{qdos} = \mathbf{K}^e_s\boldsymbol{\psi}_{abcs}$$ *Inverse transformation* $$\mathbf{u}_{abcs} = \mathbf{K}^{e-1}_s\mathbf{u}^e_{qdos}\,,\mathbf{i}_{abcs} = \mathbf{K}^{e-1}_s\mathbf{i}^e_{qdos}\,,\boldsymbol{\psi}_{abcs} = \mathbf{K}^{e-1}_s\boldsymbol{\psi}^e_{qdos}$$ *Direct transformation* $$\mathbf{u}^e_{qdor} = \mathbf{K}^e_r\mathbf{u}_{abcr}\,,\mathbf{i}^e_{qdor} = \mathbf{K}^e_r\mathbf{i}_{abcr}\,,\boldsymbol{\psi}^e_{qdor} = \mathbf{K}^e_r\boldsymbol{\psi}_{abcr}$$ *Inverse transformation* $$\mathbf{u}_{abcr} = \mathbf{K}^{e-1}_r\mathbf{u}^e_{qdor}\,,\mathbf{i}_{abcr} = \mathbf{K}^{e-1}_r\mathbf{i}^e_{qdor}\,,\boldsymbol{\psi}_{abcr} = \mathbf{K}^{e-1}_r\boldsymbol{\psi}^e_{qdor}$$

where $\mathbf{K}_s$ is the transformation matrix,

$$\mathbf{K}_s = \frac{2}{3}\begin{bmatrix} \cos\theta & \cos(\theta - \frac{2}{3}\pi) & \cos(\theta + \frac{2}{3}\pi) \\ \sin\theta & \sin(\theta - \frac{2}{3}\pi) & \sin(\theta + \frac{2}{3}\pi) \\ \frac{1}{2} & \frac{1}{2} & \frac{1}{2} \end{bmatrix};$$

θ is the angular displacement of the reference frame.
Hence, from $\mathbf{u}_{qdos} = \mathbf{K}_s\mathbf{u}_{abcs}$, we have

$$\begin{bmatrix} u_{qs} \\ u_{ds} \\ u_{os} \end{bmatrix} = \frac{2}{3}\begin{bmatrix} \cos\theta & \cos(\theta - \frac{2}{3}\pi) & \cos(\theta + \frac{2}{3}\pi) \\ \sin\theta & \sin(\theta - \frac{2}{3}\pi) & \sin(\theta + \frac{2}{3}\pi) \\ \frac{1}{2} & \frac{1}{2} & \frac{1}{2} \end{bmatrix}\begin{bmatrix} u_{as} \\ u_{bs} \\ u_{cs} \end{bmatrix}.$$

In the *arbitrary* reference frame the *quadrature-*, *direct-*, and *zero-*axis components of voltages are

$$u_{qs}(t) = \frac{2}{3}\left(\cos\theta u_{as}(t) + \cos\left(\theta - \frac{2}{3}\pi\right)u_{bs}(t) + \cos\left(\theta + \frac{2}{3}\pi\right)u_{cs}(t) \right),$$

$$u_{ds}(t) = \frac{2}{3}\left(\sin\theta u_{as}(t) + \sin\left(\theta - \frac{2}{3}\pi\right)u_{bs}(t) + \sin\left(\theta + \frac{2}{3}\pi\right)u_{cs}(t) \right),$$

$$u_{0s}(t) = \frac{1}{3}(u_{as}(t) + u_{bs}(t) + u_{cs}(t)).$$

Using

$$u_{as}(t) = 100\cos(377t),\, u_{bs}(t) = 100\cos\left(377t - \frac{2}{3}\pi\right),\, u_{cs}(t) = 100\cos\left(377t + \frac{2}{3}\pi\right),$$

one obtains the equations for the *quadrature-*, *direct-*, and *zero*-axis components of voltages in the *arbitrary* reference frame. In particular,

$$u_{qs}(t) = \frac{200}{3}\left(\cos\theta\cos(377t) + \cos\left(\theta - \frac{2}{3}\pi\right)\cos\left(377t - \frac{2}{3}\pi\right) \right.$$
$$\left. + \cos\left(\theta + \frac{2}{3}\pi\right)\cos\left(377t + \frac{2}{3}\pi\right) \right),$$

$$u_{ds}(t) = \frac{200}{3}\left(\sin\theta\cos(377t) + \sin\left(\theta - \frac{2}{3}\pi\right)\cos\left(377t - \frac{2}{3}\pi\right) \right.$$
$$\left. + \sin\left(\theta + \frac{2}{3}\pi\right)\cos\left(377t + \frac{2}{3}\pi\right) \right),$$

$$u_{0s}(t) = \frac{200}{3}\left(\cos(377t) + \cos\left(377t - \frac{2}{3}\pi\right) + \cos\left(377t + \frac{2}{3}\pi\right) \right).$$

For $f = 60$ Hz, one concludes that the synchronous reference frame is in place because the angular frequency of the supplied voltages is 377 rad/sec. Assuming that $\theta_0 = 0$, we have $\theta_e = \omega_e t$. Using the trigonometric identities, one finds

$$u_{qs}^e(t) = \frac{200}{3}\left(\cos^2(377t) + \cos^2\left(377t - \frac{2}{3}\pi\right) + \cos^2\left(377t + \frac{2}{3}\pi\right) \right)$$
$$= \frac{200}{3}\frac{3}{2} = 100 \text{ V},$$

$$u_{ds}^e(t) = \frac{200}{3}\left(\sin(377t)\cos(377t) + \sin\left(377t - \frac{2}{3}\pi\right)\cos\left(377t - \frac{2}{3}\pi\right) \right.$$
$$\left. + \sin\left(377t + \frac{2}{3}\pi\right)\cos\left(377t + \frac{2}{3}\pi\right) \right) = 0 \text{ V},$$

$$u_{0s}^e(t) = \frac{200}{3}\left(\cos(377t) + \cos\left(377t - \frac{2}{3}\pi\right) + \cos\left(377t + \frac{2}{3}\pi\right) \right) = 0 \text{ V}.$$

For the balanced three-phase voltage set (a set of equal-amplitude sinusoidal voltages displaced by 120 degrees) as

$$u_{as}(t) = 100\cos(377t),\ u_{bs}(t) = 100\sin\left(377t - \frac{2}{3}\pi\right),\text{ and}$$

$$u_{cs}(t) = 100\sin\left(377t + \frac{2}{3}\pi\right),$$

the resulting $qd0$ components

$$u^e_{qs}(t),\quad u^e_{ds}(t)\quad\text{and}\quad u^e_{0s}(t)$$

are found to be dc voltages. Furthermore,

$$u^e_{0s}(t) = 0\quad\text{and}\quad u^e_{0s}(t) = 0$$

In the stationary reference frame, one assigns $\omega = 0$. Thus, $\theta = 0$. Using

$$\mathbf{K}_s = \frac{2}{3}\begin{bmatrix} \cos\theta & \cos(\theta - \frac{2}{3}\pi) & \cos(\theta + \frac{2}{3}\pi) \\ \sin\theta & \sin(\theta - \frac{2}{3}\pi) & \sin(\theta + \frac{2}{3}\pi) \\ \frac{1}{2} & \frac{1}{2} & \frac{1}{2} \end{bmatrix},$$

the matrix needed to be applied in the stationary reference frame is

$$\mathbf{K}^s_s = \frac{2}{3}\begin{bmatrix} \cos\theta & \cos(\theta - \frac{2}{3}\pi) & \cos(\theta + \frac{2}{3}\pi) \\ \sin\theta & \sin(\theta - \frac{2}{3}\pi) & \sin(\theta + \frac{2}{3}\pi) \\ \frac{1}{2} & \frac{1}{2} & \frac{1}{2} \end{bmatrix}_{\theta=0}$$

$$= \frac{2}{3}\begin{bmatrix} 1 & -\frac{1}{2} & -\frac{1}{2} \\ 0 & -\frac{\sqrt{3}}{2} & \frac{\sqrt{3}}{2} \\ \frac{1}{2} & \frac{1}{2} & \frac{1}{2} \end{bmatrix} = \begin{bmatrix} \frac{2}{3} & -\frac{1}{3} & -\frac{1}{3} \\ 0 & -\frac{1}{\sqrt{3}} & \frac{1}{\sqrt{3}} \\ \frac{1}{3} & \frac{1}{3} & \frac{1}{3} \end{bmatrix}.$$

From

$$\mathbf{u}^s_{qdos} = \mathbf{K}^s_s\mathbf{u}_{abcs}\begin{bmatrix} u^s_{qs} \\ u^s_{ds} \\ u^s_{os} \end{bmatrix} = \begin{bmatrix} \frac{2}{3} & -\frac{1}{3} & -\frac{1}{3} \\ 0 & -\frac{1}{\sqrt{3}} & \frac{1}{\sqrt{3}} \\ \frac{1}{3} & \frac{1}{3} & \frac{1}{3} \end{bmatrix}\begin{bmatrix} u_{as} \\ u_{bs} \\ u_{cs} \end{bmatrix},$$

one finds

$$u^s_{qs}(t) = \frac{2}{3} u_{as}(t) - \frac{1}{3} u_{bs}(t) - \frac{1}{3} u_{cs}(t),$$

$$u^s_{ds}(t) = -\frac{1}{\sqrt{3}} u_{bs}(t) + \frac{1}{\sqrt{3}} u_{cs}(t),$$

$$u^s_{os}(t) = \frac{1}{3} u_{as}(t) + \frac{1}{3} u_{bs}(t) + \frac{1}{3} u_{cs}(t).$$

Therefore, the ac voltages result as

$$u^s_{qs}(t) = \frac{200}{3} \cos(377t) - \frac{100}{3} \cos\left(377t - \frac{2}{3}\pi\right) - \frac{100}{3} \cos\left(377t + \frac{2}{3}\pi\right)$$

$$= 100 \cos(377t),$$

$$u^s_{ds}(t) = -\frac{100}{\sqrt{3}} \cos\left(377t - \frac{2}{3}\pi\right) + \frac{100}{\sqrt{3}} \cos\left(377t + \frac{2}{3}\pi\right) = -100\sin(377t),$$

$$u^s_{os}(t) = \frac{100}{3} \cos(377t) + \frac{100}{3} \cos\left(377t - \frac{2}{3}\pi\right) + \frac{100}{3} \cos\left(377t + \frac{2}{3}\pi\right) = 0. \quad \square$$

5.2.2 Induction Motors in the *Arbitrary* Reference Frame

To derive the most general results, we will derive the equations of motion for three-phase induction machines in the *arbitrary* reference frame. One recalls that in the *arbitrary* reference frame, the frame angular velocity ω is not specified. Assigning the frame angular velocities ($\omega = 0$, $\omega = \omega_r$, $\omega = \omega_e$, or other), the models in the stationary ($\omega = 0$), rotor ($\omega = \omega_r$), and synchronous ($\omega = \omega_e$) reference frames result.

Consider three-phase induction motors with *quadrature* and *direct* magnetic axes as illustrated in Figure 5.13.

The *abc* stator and rotor variables must be transformed to the *quadrature, direct,* and *zero* quantities. To transform the *machine (abc)* stator voltages, currents, and flux linkages to the *quadrature-, direct-,* and *zero-*axis components of stator voltages, currents, and flux linkages, the *direct Park transformation* is used. Using the transformations and results reported Table 5.1, we have

$$\mathbf{u}_{qdos} = \mathbf{K}_s \mathbf{u}_{abcs}, \quad \mathbf{i}_{qdos} = \mathbf{K}_s \mathbf{i}_{abcs}, \quad \boldsymbol{\psi}_{qdos} = \mathbf{K}_s \boldsymbol{\psi}_{abcs}, \tag{5.32}$$

where the stator transformation matrix $\mathbf{K}_s$ is

$$\mathbf{K}_s = \frac{2}{3} \begin{bmatrix} \cos\theta & \cos(\theta - \frac{2}{3}\pi) & \cos(\theta + \frac{2}{3}\pi) \\ \sin\theta & \sin(\theta - \frac{2}{3}\pi) & \sin(\theta + \frac{2}{3}\pi) \\ \frac{1}{2} & \frac{1}{2} & \frac{1}{2} \end{bmatrix}. \tag{5.33}$$

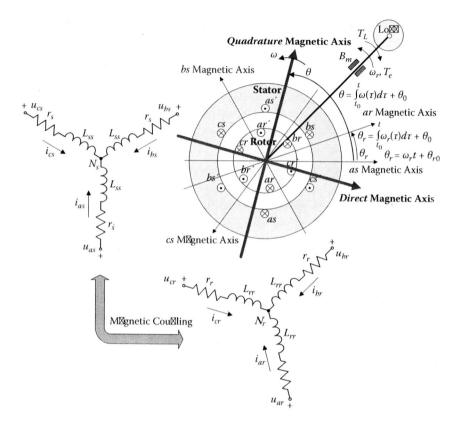

FIGURE 5.13
Three-phase symmetric induction motor with the rotating reference frame.

The angular displacement of the reference frame is

$$\theta = \int_{t_0}^{t} \omega(\tau)d\tau + \theta_0.$$

Using the rotor transformations matrix $\mathbf{K}_r$, the *quadrature-, direct-,* and *zero-axis* components of rotor voltages, currents, and flux linkages are found by using the *abc* rotor voltages, currents, and flux linkages as

$$\mathbf{u}'_{qdor} = \mathbf{K}_r \mathbf{u}'_{abcr}, \quad \mathbf{i}'_{qdor} = \mathbf{K}_r \mathbf{i}'_{abcr}, \quad \psi'_{qdor} = \mathbf{K}_r \psi'_{abcs}, \tag{5.34}$$

where the rotor transformation matrix $\mathbf{K}_r$ is

$$\mathbf{K}_r = \frac{2}{3}\begin{bmatrix} \cos(\theta - \theta_r) & \cos(\theta - \theta_r - \tfrac{2}{3}\pi) & \cos(\theta - \theta_r + \tfrac{2}{3}\pi) \\ \sin(\theta - \theta_r) & \sin(\theta - \theta_r - \tfrac{2}{3}\pi) & \sin(\theta - \theta_r + \tfrac{2}{3}\pi) \\ \tfrac{1}{2} & \tfrac{1}{2} & \tfrac{1}{2} \end{bmatrix}. \tag{5.35}$$

From differential equations (5.27)

$$\mathbf{u}_{abcs} = \mathbf{r}_s \mathbf{i}_{abcs} + \frac{d\mathbf{\psi}_{abcs}}{dt},$$

$$\mathbf{u}'_{abcr} = \mathbf{r}'_r \mathbf{i}'_{abcr} + \frac{d\mathbf{\psi}'_{abcr}}{dt},$$

using the *inverse* Park transformation matrices $\mathbf{K}_s^{-1}$ and $\mathbf{K}_r^{-1}$ we have

$$\mathbf{K}_s^{-1} \mathbf{u}_{qdos} = \mathbf{r}_s \mathbf{K}_s^{-1} \mathbf{i}_{qdos} + \frac{d(\mathbf{K}_s^{-1} \mathbf{\psi}_{qdos})}{dt},$$

$$\mathbf{K}_r^{-1} \mathbf{u}'_{qdor} = \mathbf{r}'_r \mathbf{K}_r^{-1} \mathbf{i}'_{qdor} + \frac{d(\mathbf{K}_r^{-1} \mathbf{\psi}'_{qdor})}{dt}. \qquad (5.36)$$

Using (5.33) and (5.35), one finds inverse matrices $\mathbf{K}_s^{-1}$ and $\mathbf{K}_r^{-1}$ as

$$\mathbf{K}_s^{-1} = \begin{bmatrix} \cos\theta & \sin\theta & 1 \\ \cos(\theta - \frac{2}{3}\pi) & \sin(\theta - \frac{2}{3}\pi) & 1 \\ \cos(\theta + \frac{2}{3}\pi) & \sin(\theta + \frac{2}{3}\pi) & 1 \end{bmatrix},$$

$$\mathbf{K}_r^{-1} = \begin{bmatrix} \cos(\theta - \theta_r) & \sin(\theta - \theta_r) & 1 \\ \cos(\theta - \theta_r - \frac{2}{3}\pi) & \sin(\theta - \theta_r - \frac{2}{3}\pi) & 1 \\ \cos(\theta - \theta_r + \frac{2}{3}\pi) & \sin(\theta - \theta_r + \frac{2}{3}\pi) & 1 \end{bmatrix}.$$

Multiplying left and right sides of equations (5.36) by $\mathbf{K}_s$ and $\mathbf{K}_r$, yield

$$\mathbf{u}_{qdos} = \mathbf{K}_s \mathbf{r}_s \mathbf{K}_s^{-1} \mathbf{i}_{qdos} + \mathbf{K}_s \frac{d\mathbf{K}_s^{-1}}{dt} \mathbf{\psi}_{qdos} + \mathbf{K}_s \mathbf{K}_s^{-1} \frac{d\mathbf{\psi}_{qdos}}{dt},$$

$$\mathbf{u}'_{qdor} = \mathbf{K}_r \mathbf{r}'_r \mathbf{K}_r^{-1} \mathbf{i}'_{qdor} + \mathbf{K}_r \frac{d\mathbf{K}_r^{-1}}{dt} \mathbf{\psi}'_{qdor} + \mathbf{K}_r \mathbf{K}_r^{-1} \frac{d\mathbf{\psi}'_{qdor}}{dt}. \qquad (5.37)$$

The matrices of the stator and rotor resistances $\mathbf{r}_s$ and $\mathbf{r}'_s$ are diagonal. Hence,

$$\mathbf{K}_s \mathbf{r}_s \mathbf{K}_s^{-1} = \mathbf{r}_s \quad \text{and} \quad \mathbf{K}_r \mathbf{r}'_r \mathbf{K}_r^{-1} = \mathbf{r}'_r.$$

The differentiation of terms in (5.37) gives

$$\frac{d\mathbf{K}_s^{-1}}{dt} = \omega \begin{bmatrix} -\sin\theta & \cos\theta & 0 \\ -\sin(\theta - \frac{2}{3}\pi) & \cos(\theta - \frac{2}{3}\pi) & 0 \\ -\sin(\theta + \frac{2}{3}\pi) & \cos(\theta + \frac{2}{3}\pi) & 0 \end{bmatrix},$$

$$\frac{d\mathbf{K}_r^{-1}}{dt} = (\omega - \omega_r) \begin{bmatrix} -\sin(\theta - \theta_r) & \cos(\theta - \theta_r) & 0 \\ -\sin(\theta - \theta_r - \frac{2}{3}\pi) & \cos(\theta - \theta_r - \frac{2}{3}\pi) & 0 \\ -\sin(\theta - \theta_r + \frac{2}{3}\pi) & \cos(\theta - \theta_r + \frac{2}{3}\pi) & 0 \end{bmatrix}.$$

Therefore,

$$\mathbf{K}_s \frac{d\mathbf{K}_s^{-1}}{dt} = \omega \begin{bmatrix} 0 & 1 & 0 \\ -1 & 0 & 0 \\ 0 & 0 & 0 \end{bmatrix}, \quad \mathbf{K}_r \frac{d\mathbf{K}_r^{-1}}{dt} = (\omega - \omega_r) \begin{bmatrix} 0 & 1 & 0 \\ -1 & 0 & 0 \\ 0 & 0 & 0 \end{bmatrix}.$$

From (5.37), one obtains the following equations for stator and rotor circuits in the *arbitrary* reference frame when the angular velocity of the reference frame ω is not specified

$$\mathbf{u}_{qdos} = \mathbf{r}_s \mathbf{i}_{qdos} + \begin{bmatrix} 0 & \omega & 0 \\ -\omega & 0 & 0 \\ 0 & 0 & 0 \end{bmatrix} \boldsymbol{\psi}_{qdos} + \frac{d\boldsymbol{\psi}_{qdos}}{dt},$$

$$\mathbf{u}'_{qdor} = \mathbf{r}'_r \mathbf{i}'_{qdor} + \begin{bmatrix} 0 & \omega - \omega_r & 0 \\ -\omega + \omega_r & 0 & 0 \\ 0 & 0 & 0 \end{bmatrix} \boldsymbol{\psi}'_{qdor} + \frac{d\boldsymbol{\psi}'_{qdor}}{dt}. \tag{5.38}$$

Using (5.38), six differential equations derived are

$$u_{qs} = r_s i_{qs} + \omega \psi_{ds} + \frac{d\psi_{qs}}{dt}, \quad u_{ds} = r_s i_{ds} - \omega \psi_{qs} + \frac{d\psi_{ds}}{dt},$$

$$u_{os} = r_s i_{os} + \frac{d\psi_{os}}{dt}, \quad u'_{qr} = r'_r i'_{qr} + (\omega - \omega_r) \psi'_{dr} + \frac{d\psi'_{qr}}{dt},$$

$$u'_{dr} = r'_r i'_{dr} - (\omega - \omega_r) \psi'_{qr} + \frac{d\psi'_{dr}}{dt}, \quad u'_{or} = r'_r i'_{or} + \frac{d\psi'_{or}}{dt}. \tag{5.39}$$

From

$$\begin{bmatrix} \boldsymbol{\psi}_{abcs} \\ \boldsymbol{\psi}'_{abcr} \end{bmatrix} = \begin{bmatrix} \mathbf{L}_s & \mathbf{L}'_{sr}(\theta_r) \\ \mathbf{L}'_{sr}{}^T(\theta_r) & \mathbf{L}'_r \end{bmatrix} \begin{bmatrix} \mathbf{i}_{abcs} \\ \mathbf{i}'_{abcr} \end{bmatrix},$$

we have

$$\boldsymbol{\psi}_{abcs} = \mathbf{L}_s \mathbf{i}_{abcs} + \mathbf{L}'_{sr}(\theta_r) \mathbf{i}'_{abcr}, \quad \boldsymbol{\psi}'_{abcr} = \mathbf{L}'_{sr}{}^T(\theta_r) \mathbf{i}_{abcs} + \mathbf{L}'_r \mathbf{i}'_{abcr}.$$

These *machine* flux linkages should be transformed to the *quadrature, direct,* and *zero* quantities by employing the Park transformation matrices. One finds

$$\mathbf{K}_s^{-1}\boldsymbol{\psi}_{qdos} = \mathbf{L}_s\mathbf{K}_s^{-1}\mathbf{i}_{qdos} + \mathbf{L}'_{sr}(\theta_r)\mathbf{K}_r^{-1}\mathbf{i}'_{qdor},$$

$$\mathbf{K}_r^{-1}\boldsymbol{\psi}'_{qdor} = \mathbf{L}'^T_{sr}(\theta_r)\mathbf{K}_s^{-1}\mathbf{i}_{qdos} + \mathbf{L}'_r\mathbf{K}_r^{-1}\mathbf{i}'_{abcr}.$$

Thus,

$$\boldsymbol{\psi}_{qdos} = \mathbf{K}_s\mathbf{L}_s\mathbf{K}_s^{-1}\mathbf{i}_{qdos} + \mathbf{K}_s\mathbf{L}'_{sr}(\theta_r)\mathbf{K}_r^{-1}\mathbf{i}'_{qdor},$$

$$\boldsymbol{\psi}'_{qdor} = \mathbf{K}_r\mathbf{L}'^T_{sr}(\theta_r)\mathbf{K}_s^{-1}\mathbf{i}_{qdos} + \mathbf{K}_r\mathbf{L}'_r\mathbf{K}_r^{-1}\mathbf{i}'_{abcr}. \tag{5.40}$$

Using the transformation matrices and applying the derived expressions for $\mathbf{L}_s$, $\mathbf{L}'_{sr}(\theta_r)$ and $\mathbf{L}'_r$, the matrices multiplication gives

$$\mathbf{K}_s\mathbf{L}_s\mathbf{K}_s^{-1} = \begin{bmatrix} L_{ls} + M & 0 & 0 \\ 0 & L_{ls} + M & 0 \\ 0 & 0 & L_{ls} \end{bmatrix},$$

$$\mathbf{K}_s\mathbf{L}'_{sr}(\theta_r)\mathbf{K}_r^{-1} = \mathbf{K}_r\mathbf{L}'^T_{sr}(\theta_r)\mathbf{K}_s^{-1} = \begin{bmatrix} M & 0 & 0 \\ 0 & M & 0 \\ 0 & 0 & 0 \end{bmatrix},$$

$$\mathbf{K}_r\mathbf{L}'_r\mathbf{K}_r^{-1} = \begin{bmatrix} L'_{lr} + M & 0 & 0 \\ 0 & L'_{lr} + M & 0 \\ 0 & 0 & L'_{lr} \end{bmatrix},$$

where

$$M = \frac{3}{2}L_{ms}.$$

In the expanded form, the flux linkage equations (5.40) are

$$\psi_{qs} = L_{ls}i_{qs} + Mi_{qs} + Mi'_{qr}, \quad \psi_{ds} = L_{ls}i_{ds} + Mi_{ds} + Mi'_{dr}, \quad \psi_{os} = L_{ls}i_{os},$$

$$\psi'_{qr} = L'_{lr}i'_{qr} + Mi_{qs} + Mi'_{qr}, \quad \psi'_{dr} = L'_{lr}i'_{dr} + Mi_{ds} + Mi'_{dr}, \quad \psi'_{or} = L'_{lr}i'_{or}. \tag{5.41}$$

Substituting (5.41) in (5.39), the following differential equations result

$$u_{qs} = r_s i_{qs} + \omega(L_{ls} i_{ds} + M i_{ds} + M i'_{dr}) + \frac{d(L_{ls} i_{qs} + M i_{qs} + M i'_{qr})}{dt},$$

$$u_{ds} = r_s i_{ds} - \omega(L_{ls} i_{qs} + M i_{qs} + M i'_{qr}) + \frac{d(L_{ls} i_{ds} + M i_{ds} + M i'_{dr})}{dt},$$

$$u_{os} = r_s i_{os} + \frac{d(L_{ls} i_{os})}{dt},$$

$$u'_{qr} = r'_r i'_{qr} + (\omega - \omega_r)(L'_{lr} i'_{dr} + M i_{ds} + M i'_{dr}) + \frac{d(L'_{lr} i'_{qr} + M i_{qs} + M i'_{qr})}{dt},$$

$$u'_{dr} = r'_r i'_{dr} - (\omega - \omega_r)(L'_{lr} i'_{qr} + M i_{qs} + M i'_{qr}) + \frac{d(L'_{lr} i'_{dr} + M i_{ds} + M i'_{dr})}{dt},$$

$$u'_{or} = r'_r i'_{or} + \frac{d(L'_{lr} i'_{or})}{dt}.$$

Cauchy's form of the differential equations is

$$\frac{di_{qs}}{dt} = \frac{1}{L_{SM} L_{RM} - M^2}[-L_{RM} r_s i_{qs} - (L_{SM} L_{RM} - M^2)\omega i_{ds} + M r'_r i'_{qr}$$
$$- M(M i_{ds} + L_{RM} i'_{dr})\omega_r + L_{RM} u_{qs} - M u'_{qr}],$$

$$\frac{di_{ds}}{dt} = \frac{1}{L_{SM} L_{RM} - M^2}[(L_{SM} L_{RM} - M^2)\omega i_{qs} - L_{RM} r_s i_{ds} + M r'_r i'_{dr}$$
$$+ M(M i_{qs} + L_{RM} i'_{qr})\omega_r + L_{RM} u_{ds} - M u'_{dr}],$$

$$\frac{di_{os}}{dt} = \frac{1}{L_{ls}}(-r_s i_{os} + u_{os})$$

$$\frac{di'_{qr}}{dt} = \frac{1}{L_{SM} L_{RM} - M^2}[M r_s i_{qs} - L_{SM} r'_r i'_{qr} - (L_{SM} L_{RM} - M^2)\omega i'_{dr}$$
$$+ L_{SM}(M i_{ds} + L_{RM} i'_{dr})\omega_r - M u_{qs} + L_{SM} u'_{qr}],$$

$$\frac{di'_{dr}}{dt} = \frac{1}{L_{SM} L_{RM} - M^2}[M r_s i_{ds} + (L_{SM} L_{RM} - M^2)\omega i'_{qr} - L_{SM} r'_r i'_{dr}$$
$$- L_{SM}(M i_{qs} + L_{RM} i'_{qr})\omega_r - M u_{ds} + L_{SM} u'_{dr}],$$

$$\frac{di'_{or}}{dt} = \frac{1}{L'_{lr}}(-r'_r i'_{or} + u'_{or}),$$ (5.42)

where

$$L_{SM} = L_{ls} + M = L_{ls} + \frac{3}{2}L_{ms}, \quad L_{RM} = L'_{lr} + M = L'_{lr} + \frac{3}{2}L_{ms}.$$

The torsional-mechanical equations are

$$T_e - B_m \omega_{rm} - T_L = J\frac{d\omega_{rm}}{dt}, \quad \frac{d\theta_{rm}}{dt} = \omega_{rm}. \tag{5.43}$$

The expression for T_e should be obtained in terms of the *quadrature-* and *direct*-axis components of stator and rotor currents. Using

$$W_c = \frac{1}{2}\mathbf{i}_{abcs}^T(\mathbf{L}_s - L_{ls}\mathbf{I})\mathbf{i}_{abcs} + \mathbf{i}_{abcs}^T\mathbf{L}'_{sr}(\theta_r)\mathbf{i}'_{abcr} + \frac{1}{2}\mathbf{i}_{abcr}^{\prime T}(\mathbf{L}'_r - L'_{lr}\mathbf{I})\mathbf{i}'_{abcr},$$

one finds

$$T_e = \frac{P}{2}\frac{\partial W_c(\mathbf{i}_{abcs}, \mathbf{i}'_{abcr}, \theta_r)}{\partial \theta_r} = \frac{P}{2}\mathbf{i}_{abcs}^T\frac{\partial \mathbf{L}'_{cr}(\theta_r)}{\partial \theta_r}\mathbf{i}'_{abcr}.$$

We have

$$T_e = \frac{P}{2}(\mathbf{K}_s^{-1}\mathbf{i}_{qdos})^T\frac{\partial \mathbf{L}'_{sr}(\theta_r)}{\partial \theta_r}\mathbf{K}_r^{-1}\mathbf{i}'_{qdor} = \frac{P}{2}\mathbf{i}_{qdos}^T\mathbf{K}_s^{-1^T}\frac{\partial \mathbf{L}'_{sr}(\theta_r)}{\partial \theta_r}\mathbf{K}_r^{-1}\mathbf{i}'_{qdor}.$$

By performing the multiplication of matrices, the following formula for T_e results

$$T_e = \frac{3P}{4}M(i_{qs}i'_{dr} - i_{ds}i'_{qr}). \tag{5.44}$$

From (5.43) and (5.44), one has

$$\frac{d\omega_r}{dt} = \frac{3P^2}{8J}M(i_{qs}i'_{dr} - i_{ds}i'_{qr}) - \frac{B_m}{J}\omega_r - \frac{P}{2J}T_L,$$

$$\frac{d\theta_r}{dt} = \omega_r. \tag{5.45}$$

Combining the circuitry-electromagnetic and *torsional-mechanical* dynamics, as given by (5.42) and (5.45), the model for three-phase induction motors in the *arbitrary* reference frame is given as a set of eight nonlinear differential equations

$$\frac{di_{qs}}{dt} = \frac{1}{L_{SM}L_{RM} - M^2}[-L_{RM}r_s i_{qs} - (L_{SM}L_{RM} - M^2)\omega i_{ds} + Mr'_r i'_{qr}$$

$$- M(Mi_{ds} + L_{RM}i'_{dr})\omega_r + L_{RM}u_{qs} - Mu'_{qr}],$$

$$\frac{di_{ds}}{dt} = \frac{1}{L_{SM}L_{RM} - M^2}[(L_{SM}L_{RM} - M^2)\omega i_{qs} - L_{RM}r_s i_{ds} + Mr'_r i'_{dr}$$

$$+ M(Mi_{qs} + L_{RM}i'_{qr})\omega_r + L_{RM}u_{ds} - Mu'_{dr}],$$

$$\frac{di_{os}}{dt} = \frac{1}{L_{ls}}(-r_s i_{os} + u_{os}),$$

$$\frac{di'_{qr}}{dt} = \frac{1}{L_{SM}L_{RM} - M^2}[Mr_s i_{qs} - L_{SM}r'_r i'_{qr} - (L_{SM}L_{RM} - M^2)\omega i'_{dr}$$

$$+ L_{SM}(Mi_{ds} + L_{RM}i'_{dr})\omega_r - Mu_{qs} + L_{SM}u'_{qr}],$$

$$\frac{di'_{dr}}{dt} = \frac{1}{L_{SM}L_{RM} - M^2}[Mr_s i_{ds} + (L_{SM}L_{RM} - M^2)\omega i'_{qr} - L_{SM}r'_r i'_{dr}$$

$$- L_{SM}(Mi_{qs} + L_{RM}i'_{qr})\omega_r - Mu_{ds} + L_{SM}u'_{dr}],$$

$$\frac{di'_{or}}{dt} = \frac{1}{L'_{lr}}(-r'_r i'_{or} + u'_{or}),$$

$$\frac{d\omega_r}{dt} = \frac{3P^2}{8J}M(i_{qs}i'_{dr} - i_{ds}i'_{qr}) - \frac{B_m}{J}\omega_r - \frac{P}{2J}T_L,$$

$$\frac{d\theta_r}{dt} = \omega_r. \tag{5.46}$$

The last differential equation

$$\frac{d\theta_r}{dt} = \omega_r$$

in (5.46) can be omitted in the analysis and simulations if induction motors are used in drives. That is, for electric drives one finds the following state-space equation

$$
\begin{bmatrix}
\dfrac{di_{qs}}{dt} \\[2mm]
\dfrac{di_{ds}}{dt} \\[2mm]
\dfrac{di_{os}}{dt} \\[2mm]
\dfrac{di'_{qr}}{dt} \\[2mm]
\dfrac{di'_{dr}}{dt} \\[2mm]
\dfrac{di'_{or}}{dt} \\[2mm]
\dfrac{d\omega_r}{dt}
\end{bmatrix}
=
\begin{bmatrix}
-\dfrac{L_{RM}r_s}{L_{SM}L_{RM}-M^2} & -\omega & 0 & \dfrac{Mr'_r}{L_{SM}L_{RM}-M^2} & 0 & 0 & 0 \\[3mm]
\omega & -\dfrac{L_{RM}r_s}{L_{SM}L_{RM}-M^2} & 0 & 0 & \dfrac{Mr'_r}{L_{SM}L_{RM}-M^2} & 0 & 0 \\[3mm]
0 & 0 & -\dfrac{r_s}{L_{ls}} & 0 & 0 & 0 & 0 \\[3mm]
\dfrac{Mr_s}{L_{SM}L_{RM}-M^2} & 0 & 0 & -\dfrac{L_{SM}r'_r}{L_{SM}L_{RM}-M^2} & -\omega & 0 & 0 \\[3mm]
0 & \dfrac{Mr_s}{L_{SM}L_{RM}-M^2} & 0 & \omega & -\dfrac{L_{SM}r'_r}{L_{SM}L_{RM}-M^2} & 0 & 0 \\[3mm]
0 & 0 & 0 & 0 & 0 & -\dfrac{r'_r}{L'_{lr}} & 0 \\[3mm]
0 & 0 & 0 & 0 & 0 & 0 & -\dfrac{B_m}{J}
\end{bmatrix}
$$

$$
\times
\begin{bmatrix}
i_{qs} \\ i_{ds} \\ i_{os} \\ i'_{qr} \\ i'_{dr} \\ i'_{or} \\ \omega_r
\end{bmatrix}
+
\begin{bmatrix}
-\dfrac{M(Mi_{ds}+L_{RM}i'_{dr})\omega_r}{L_{SM}L_{RM}-M^2} \\[3mm]
\dfrac{M(Mi_{qs}+L_{RM}i'_{qr})\omega_r}{L_{SM}L_{RM}-M^2} \\[3mm]
0 \\[3mm]
\dfrac{L_{SM}(Mi_{ds}+L_{RM}i'_{dr})\omega_r}{L_{SM}L_{RM}-M^2} \\[3mm]
-\dfrac{L_{SM}(Mi_{qs}+L_{RM}i'_{qr})\omega_r}{L_{SM}L_{RM}-M^2} \\[3mm]
0 \\[3mm]
\dfrac{3P^2}{8J}M\left(i_{qs}i'_{dr}-i_{ds}i'_{qr}\right)
\end{bmatrix}
$$

$$
+
\begin{bmatrix}
\dfrac{L_{RM}}{L_{SM}L_{RM}-M^2} & 0 & 0 & -\dfrac{M}{L_{SM}L_{RM}-M^2} & 0 & 0 \\[3mm]
0 & \dfrac{L_{RM}}{L_{SM}L_{RM}-M^2} & 0 & 0 & -\dfrac{M}{L_{SM}L_{RM}-M^2} & 0 \\[3mm]
0 & 0 & \dfrac{1}{L_{ls}} & 0 & 0 & 0 \\[3mm]
-\dfrac{M}{L_{SM}L_{RM}-M^2} & 0 & 0 & \dfrac{L_{SM}}{L_{SM}L_{RM}-M^2} & 0 & 0 \\[3mm]
0 & -\dfrac{M}{L_{SM}L_{RM}-M^2} & 0 & 0 & \dfrac{L_{SM}}{L_{SM}L_{RM}-M^2} & 0 \\[3mm]
0 & 0 & 0 & 0 & 0 & \dfrac{1}{L'_{lr}} \\[3mm]
0 & 0 & 0 & 0 & 0 & 0
\end{bmatrix}
\begin{bmatrix}
u_{qs} \\ u_{ds} \\ u_{os} \\ u'_{qr} \\ u'_{dr} \\ u'_{or}
\end{bmatrix}
-
\begin{bmatrix}
0 \\ 0 \\ 0 \\ 0 \\ 0 \\ 0 \\ \dfrac{P}{2J}
\end{bmatrix}
T_L. \quad (5.47)
$$

The corresponding s-domain diagram for three-phase induction motors in the *arbitrary* reference frame is developed using (5.46), as depicted in Figure 5.14.

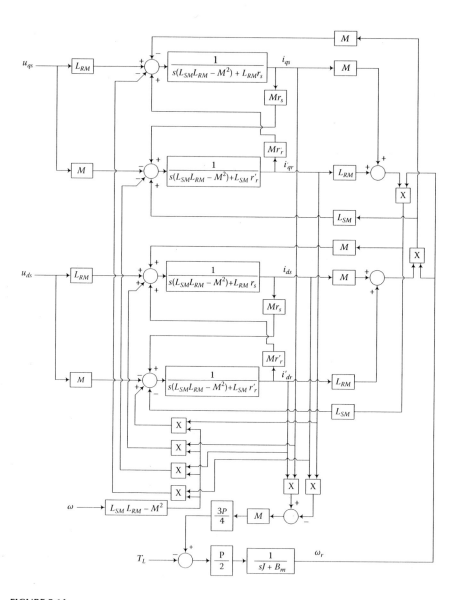

FIGURE 5.14
The s-domain diagram of three-phase squirrel-cage induction motors in the *arbitrary* reference frame.

In squirrel-cage motors, the rotor windings are short-circuited. To guarantee the balanced operating conditions, one supplies the following phase voltages

$$u_{as}(t) = \sqrt{2}u_M \cos(\omega_f t), \quad u_{bs}(t) = \sqrt{2}u_M \cos(\omega_f t - \frac{2}{3}\pi) \quad \text{and}$$

$$u_{cs}(t) = \sqrt{2}u_M \cos\left(\omega_f t + \frac{2}{3}\pi\right).$$

The *quadrature-*, *direct-*, and *zero*-axis components of stator voltages are obtained by using the stator Park transformation matrix. One applies

$$\mathbf{u}_{qdos} = \mathbf{K}_s \mathbf{u}_{abcs}, \ \mathbf{K}_s = \frac{2}{3} \begin{bmatrix} \cos\theta & \cos(\theta - \frac{2}{3}\pi) & \cos(\theta + \frac{2}{3}\pi) \\ \sin\theta & \sin(\theta - \frac{2}{3}\pi) & \sin(\theta + \frac{2}{3}\pi) \\ \frac{1}{2} & \frac{1}{2} & \frac{1}{2} \end{bmatrix}.$$

The stationary, rotor, and synchronous reference frames are commonly used. For the above-mentioned reference frames, the frame angular velocities are $\omega = 0$, $\omega = \omega_r$, and $\omega = \omega_e$. The corresponding angular displacement θ results. Letting $\theta_0 = 0$, for the stationary, rotor, and synchronous reference frames, one finds $\theta = 0$, $\theta = \theta_r$, and $\theta = \theta_e$. Hence, the *quadrature-*, *direct-*, and *zero*-axis components of voltages can be obtained to guarantee the balance operation of induction motors as illustrated in Section 5.2.3 and Examples 5.4 and 5.5.

5.2.3 Induction Motors in the Synchronous Reference Frame

The synchronous reference frame is most commonly used to study induction and synchronous machines. The differential equations, obtained for the *arbitrary* reference frame, as given by (5.46), are modified using the specified $\omega = \omega_e$. From (5.46), we have

$$\frac{di_{qs}^e}{dt} = \frac{1}{L_{SM}L_{RM} - M^2}\left[-L_{RM}r_s i_{qs}^e - (L_{SM}L_{RM} - M^2)\omega_e i_{ds}^e + Mr_r' i_{qr}'^e\right.$$

$$\left. - M(Mi_{ds}^e + L_{RM}i_{dr}'^e)\omega_r + L_{RM}u_{qs}^e - Mu_{qr}'^e\right],$$

$$\frac{di_{ds}^e}{dt} = \frac{1}{L_{SM}L_{RM} - M^2}\left[(L_{SM}L_{RM} - M^2)\omega_e i_{qs}^e - L_{RM}r_s i_{ds}^e + Mr_r' i_{dr}'^e\right.$$

$$\left. + M(Mi_{qs}^e + L_{RM}i_{qr}'^e)\omega_r + L_{RM}u_{ds}^e - Mu_{dr}'^e\right],$$

$$\frac{di_{os}^e}{dt} = \frac{1}{L_{ls}}(-r_s i_{os}^e + u_{os}^e),$$

$$\frac{di_{qr}'^e}{dt} = \frac{1}{L_{SM}L_{RM} - M^2}\left[Mr_s i_{qs}^e - L_{SM}r_r' i_{qr}'^e - (L_{SM}L_{RM} - M^2)\omega_e i_{dr}'^e \right.$$

$$\left. + L_{SM}(Mi_{ds}^e + L_{RM}i_{dr}'^e)\omega_r - Mu_{qs}^e + L_{SM}u_{qr}'^e \right],$$

$$\frac{di_{dr}'^e}{dt} = \frac{1}{L_{SM}L_{RM} - M^2}\left[Mr_s i_{ds}^e + (L_{SM}L_{RM} - M^2)\omega_e i_{qr}'^e - L_{SM}r_r' i_{dr}'^e \right.$$

$$\left. - L_{SM}(Mi_{qs}^e + L_{RM}i_{qr}'^e)\omega_r - Mu_{ds}^e + L_{SM}u_{dr}'^e \right],$$

$$\frac{di_{or}'^e}{dt} = \frac{1}{L_{lr}'}(-r_r' i_{or}'^e + u_{or}'^e),$$

$$\frac{d\omega_r}{dt} = \frac{3P^2}{8J}M(i_{qs}^e i_{dr}'^e - i_{ds}^e i_{or}'^e) - \frac{B_m}{J}\omega_r - \frac{P}{2J}T_L,$$

$$\frac{d\theta_r}{dt} = \omega_r. \tag{5.48}$$

The *quadrature, direct,* and *zero* components of stator voltages u_{qs}^e, u_{ds}^e and u_{os}^e to guarantee the balanced operation of induction motors are found by using $\mathbf{u}_{qdos}^e = \mathbf{K}_s^e \mathbf{u}_{abcs}$.

From

$$\mathbf{K}_s = \frac{2}{3}\begin{bmatrix} \cos\theta & \cos(\theta - \frac{2}{3}\pi) & \cos(\theta + \frac{2}{3}\pi) \\ \sin\theta & \sin(\theta - \frac{2}{3}\pi) & \sin(\theta + \frac{2}{3}\pi) \\ \frac{1}{2} & \frac{1}{2} & \frac{1}{2} \end{bmatrix},$$

letting $\theta = \theta_e$, we have

$$\mathbf{K}_s^e = \frac{2}{3}\begin{bmatrix} \cos\theta_e & \cos(\theta_e - \frac{2}{3}\pi) & \cos(\theta_e + \frac{2}{3}\pi) \\ \sin\theta_e & \sin(\theta_e - \frac{2}{3}\pi) & \sin(\theta_e + \frac{2}{3}\pi) \\ \frac{1}{2} & \frac{1}{2} & \frac{1}{2} \end{bmatrix}.$$

Using

$$\begin{bmatrix} u_{qs}^e \\ u_{ds}^e \\ u_{os}^e \end{bmatrix} = \frac{2}{3}\begin{bmatrix} \cos\theta_e & \cos(\theta_e - \frac{2}{3}\pi) & \cos(\theta_e + \frac{2}{3}\pi) \\ \sin\theta_e & \sin(\theta_e - \frac{2}{3}\pi) & \sin(\theta_e + \frac{2}{3}\pi) \\ \frac{1}{2} & \frac{1}{2} & \frac{1}{2} \end{bmatrix}\begin{bmatrix} u_{as} \\ u_{bs} \\ u_{cs} \end{bmatrix},$$

one obtains

$$u_{qs}^e(t) = \frac{2}{3}\left[u_{as} \cos\theta_e + u_{bs} \cos\left(\theta_e - \frac{2}{3}\pi\right) + u_{cs} \cos\left(\theta_e + \frac{2}{3}\pi\right) \right],$$

$$u_{ds}^e(t) = \frac{2}{3}\left[u_{as} \sin\theta_e + u_{bs} \sin\left(\theta_e - \frac{2}{3}\pi\right) + u_{cs} \sin\left(\theta_e + \frac{2}{3}\pi\right) \right],$$

$$u_{0s}^e(t) = \frac{1}{3}(u_{as} + u_{bs} + u_{cs}).$$

The three-phase balanced voltage set is

$$u_{as}(t) = \sqrt{2}u_M \cos(\omega_f t), \quad u_{bs}(t) = \sqrt{2}u_M \cos\left(\omega_f t - \frac{2}{3}\pi\right) \quad \text{and}$$

$$u_{cs}(t) = \sqrt{2}u_M \cos\left(\omega_f t + \frac{2}{3}\pi\right).$$

Assume that the initial displacement of the *quadrature* magnetic axis is zero. From $\theta_e = \omega_f t$, we conclude that the following *quadrature*, *direct*, and *zero* components of stator voltage must be utilized to guarantee the balance operation

$$u_{qs}^e(t) = \sqrt{2}u_M, \quad u_{ds}^e(t) = 0, \quad u_{0s}^e(t) = 0. \tag{5.49}$$

The deviations and corresponding trigonometric identities were reported in Example 5.4.

The only advantage of the stationary, rotor, and synchronous reference frames is the mathematical simplicity of equations of motion. The existing analysis and simulation software tools allow one to straightforwardly use the most sophisticated induction machine models in the *machine* variable, as documented in Sections 5.1 and 5.3. There are significant disadvantages of the *arbitrary* (stationary, rotor, and synchronous) reference frame from the implementation standpoints, as discussed in Example 5.5.

Example 5.5   Practicality of Synchronous Reference Frame: Vector Control of Induction Motors, Hardware-Software Complexity, and Performance Assessments

It was found that, in the synchronous reference frame, the *quadrature-*, *direct-*, and *zero*-axis components of stator and rotor ac voltages, currents and flux linkages have a dc form. To control induction motors, one may conclude that the dc *quadrature* voltage $u_{qs}^e(t)$ should be regulated because $u_{ds}^e(t) = 0$ and $u_{0s}^e(t) = 0$. This assessment results from the mathematical studies performed. Although the deviations are mathematically elegant and accurate, the results may have a limited practicality. We obtained the *qd0* quantities that mathematically (formally) correspond to the physical *machine* variables as the transformations are performed. For example, one measures $i_{as}(t)$, $i_{bs}(t)$, and $i_{cs}(t)$, but $i^e_{qs}(t)$, $i^e_{ds}(t)$, and $i^e_{os}(t)$, are not measured. Most important, the ac phase voltages u_{as}, u_{bs}, and u_{cs} must be supplied to the stator windings. One does not (and cannot not) supply $u^e_{qs}(t)$, $u^e_{ds}(t)$, and $u^e_{os}(t)$ to the phase windings.

One may feel that the mathematical simplicity and/or control ease are ensured by using the stationary, rotor, and/or synchronous reference frames. However, ac electric machines variables have ac quantities (voltage, current, flux linkages, etc.). One does not directly measure or observe the *quadrature-*, *direct-*, and *zero*-axis components. To rotate induction motors, **the ac phase voltages are supplied to the phase windings**. Therefore, the *qd0* voltages, thought can be viewed as controls, are not applied to the phase windings. If one decided to control induction machines in the *qd0* frames deriving $\mathbf{u}^e_{qdos}$, one should calculate (in real-time)

$$\mathbf{u}_{abcs} \text{ as } \mathbf{u}_{abcs} = \mathbf{K}^{r\,\text{-1}}_s \mathbf{u}^e_{qdos}.$$

The frequency *f* (which affects ω_e and θ_e, $\theta_e = \omega_f t$) is varied by the power converter to control the angular velocity. Thus, as one may derive $\mathbf{u}^e_{qdos}$, the phase voltages u_{as}, u_{bs}, and u_{cs} should be calculated in real-time by using the *inverse* Park transformation

$$\mathbf{u}_{abcs} = \mathbf{K}^{r\,\text{-1}}_s \mathbf{u}^e_{qdos}.$$

The most advanced DSPs are required if the designer decided to apply the synchronous (as well as stationary and rotor) reference frame. For example, the angular displacement θ_r must be measured (or observed) and utilized as one applies the rotor reference frame. The calculations of the phase voltages $\mathbf{u}_{abcs}$ must be accomplished in real-time resulting in the need for advanced DSPs significantly complicating software and hardware solutions. Although the so-called *vector* control of induction motors can be applied, quite limited practical benefits may emerge because the stationary, rotor, or synchronous reference frames imply the application of the *qd0* quantities.

The variable voltage-frequency control

$$u_{Mi}/f_i = \text{var}, \quad u_{M\min} \leq u_M \leq u_{M\max}, \quad f_{\min} \leq f \leq f_{\max}, \quad \omega_{f\min} \leq \omega_f \leq \omega_{f\max},$$

as discussed in Section 5.1.4, guarantees the high-torque patterns that surpass (or at least match) the capabilities of the *vector* control or other *qd0*-centered concepts. The highest T_{estart} and $T_{ecritical}$ are developed using the frequency control, and $f_{\min} \leq f \leq f_{\max}$. The $T_{estart\,max}$ corresponds to $f_{\min}$. This $f_{\min}$ is defined by the converter topologies, solid-state devices, driving ICs, efficiency, and other hardware specifications. Therefore, the analysis of induction machines in the *machine* variables and the use of sound concepts may be prioritized. □

5.3 Simulation and Analysis of Induction Motors in the MATLAB Environment

Using the differential equations derived, which should be derived for the electric machine under consideration, the simulation and analysis can be performed in the MATLAB environment [9, 10]. Cauchy's and not Cauchy's forms of differential equations can be used. Two-phase induction motors were simulated and analyzed in Examples 5.2 and 5.3. As the differential equations of three-phase induction motors in the *machine* variables were derived in Section 5.1.6, Section 5.2 documented that the use of the *arbitrary* reference frame may simplify the modeling tasks. However, practicality, affordability, performance, capabilities, hardware complexity, and other issues were emphasized in Example 5.5 as the *qd0*-centered concepts are applied. The analysis and control of electric machines in the *machine* variables should be prioritized from engineering, practical, and fundamental standpoints. Using advanced software, the nonlinear heterogeneous simulation and data-intensive analysis of motion devices in the *machine* variables are a sound and straightforward task. We perform and report simulations and analysis of three-phase induction motors.

Example 5.6 Simulation of Three-Phase Induction Motors Using Simulink

One may use Simulink to simulate three-phase induction motors modeled not in Cauchy's form. Using the circuitry-electromagnetic equations (5.28) and the *torsional-mechanical* equations of motion (5.31), we have

$$u_{as} = r_s i_{as} + (L_{ls} + L_{ms})\frac{di_{as}}{dt} - \frac{1}{2}L_{ms}\frac{di_{bs}}{dt} - \frac{1}{2}L_{ms}\frac{di_{cs}}{dt} + L_{ms}\frac{d(i'_{ar}\cos\theta_r)}{dt}$$

$$+ L_{ms}\frac{d(i'_{br}\cos(\theta_r + \frac{2\pi}{3}))}{dt} + L_{ms}\frac{d(i'_{cr}\cos(\theta_r - \frac{2\pi}{3}))}{dt},$$

$$u_{bs} = r_s i_{bs} - \frac{1}{2}L_{ms}\frac{di_{as}}{dt} + (L_{ls} + L_{ms})\frac{di_{bs}}{dt} - \frac{1}{2}L_{ms}\frac{di_{cs}}{dt} + L_{ms}\frac{d(i'_{ar}\cos(\theta_r - \frac{2\pi}{3}))}{dt}$$

$$+ L_{ms}\frac{d(i'_{br}\cos\theta_r)}{dt} + L_{ms}\frac{d(i'_{cr}\cos(\theta_r + \frac{2\pi}{3}))}{dt},$$

$$u_{cs} = r_s i_{cs} - \frac{1}{2} L_{ms} \frac{di_{as}}{dt} - \frac{1}{2} L_{ms} \frac{di_{bs}}{dt} + (L_{ls} + L_{ms}) \frac{di_{cs}}{dt} + L_{ms} \frac{d(i'_{ar} \cos(\theta_r + \frac{2\pi}{3}))}{dt}$$

$$+ L_{ms} \frac{d(i'_{br} \cos(\theta_r - \frac{2\pi}{3}))}{dt} + L_{ms} \frac{d(i'_{cr} \cos\theta_r)}{dt},$$

$$u'_{ar} = r'_r i'_{ar} + L_{ms} \frac{d(i_{as} \cos\theta_r)}{dt} + L_{ms} \frac{d(i_{bs} \cos(\theta_r - \frac{2\pi}{3}))}{dt} + L_{ms} \frac{d(i_{cs} \cos(\theta_r + \frac{2\pi}{3}))}{dt}$$

$$+ \left(L'_{lr} + L_{ms}\right) \frac{di'_{ar}}{dt} - \frac{1}{2} L_{ms} \frac{di'_{br}}{dt} - \frac{1}{2} L_{ms} \frac{di'_{cr}}{dt},$$

$$u'_{br} = r'_r i'_{br} + L_{ms} \frac{d(i_{as} \cos(\theta_r + \frac{2\pi}{3}))}{dt} + L_{ms} \frac{d(i_{bs} \cos\theta_r)}{dt} + L_{ms} \frac{d(i_{cs} \cos(\theta_r - \frac{2\pi}{3}))}{dt}$$

$$- \frac{1}{2} L_{ms} \frac{di'_{ar}}{dt} + \left(L'_{lr} + L_{ms}\right) \frac{di'_{br}}{dt} - \frac{1}{2} L_{ms} \frac{di'_{cr}}{dt},$$

$$u'_{cr} = r'_r i'_{cr} + L_{ms} \frac{d(i_{as} \cos(\theta_r - \frac{2\pi}{3}))}{dt} + L_{ms} \frac{d(i_{bs} \cos(\theta_r + \frac{2\pi}{3}))}{dt} + L_{ms} \frac{d(i_{cs} \cos\theta_r)}{dt}$$

$$- \frac{1}{2} L_{ms} \frac{di'_{ar}}{dt} - \frac{1}{2} L_{ms} \frac{di'_{br}}{dt} + \left(L'_{lr} + L_{ms}\right) \frac{di'_{cr}}{dt},$$

$$\frac{d\omega_r}{dt} = -\frac{P^2}{4J} L_{ms} \left\{ \left[i_{as} \left(i'_{ar} - \frac{1}{2} i'_{br} - \frac{1}{2} i'_{cr} \right) + i_{bs} \left(i'_{br} - \frac{1}{2} i'_{ar} - \frac{1}{2} i'_{cr} \right) \right.\right.$$

$$\left. + i_{cs} \left(i'_{cr} - \frac{1}{2} i'_{br} - \frac{1}{2} i'_{ar} \right) \right] \sin\theta_r$$

$$\left. + \frac{\sqrt{3}}{2} \left[i_{as}(i'_{br} - i'_{cr}) + i_{bs}(i'_{cr} - i'_{ar}) + i_{cs}(i'_{ar} - i'_{br}) \right] \cos\theta_r \right\} - \frac{B_m}{J} \omega_r - \frac{P}{2J} T_L,$$

$$\frac{d\theta_r}{dt} = \omega_r.$$

The differential equations are rewritten as

$$\frac{di_{as}}{dt} = \frac{1}{L_{ls} + L_{ms}} \left[-r_s i_{as} + \frac{1}{2} L_{ms} \frac{di_{bs}}{dt} + \frac{1}{2} L_{ms} \frac{di_{cs}}{dt} - L_{ms} \frac{d(i'_{ar} \cos\theta_r)}{dt} \right.$$

$$\left. - L_{ms} \frac{d(i'_{br} \cos(\theta_r + \frac{2\pi}{3}))}{dt} - L_{ms} \frac{d(i'_{cr} \cos(\theta_r - \frac{2\pi}{3}))}{dt} + u_{as} \right],$$

$$\frac{di_{bs}}{dt} = \frac{1}{L_{ls} + L_{ms}}\left[-r_s i_{bs} + \frac{1}{2}L_{ms}\frac{di_{as}}{dt} + \frac{1}{2}L_{ms}\frac{di_{cs}}{dt} - L_{ms}\frac{d(i'_{ar}\cos(\theta_r - \frac{2\pi}{3}))}{dt}\right.$$

$$\left. - L_{ms}\frac{d(i'_{br}\cos\theta_r)}{dt} - L_{ms}\frac{d(i'_{cr}\cos(\theta_r + \frac{2\pi}{3}))}{dt} + u_{bs}\right],$$

$$\frac{di_{cs}}{dt} = \frac{1}{L_{ls} + L_{ms}}\left[-r_s i_{cs} + \frac{1}{2}L_{ms}\frac{di_{as}}{dt} + \frac{1}{2}L_{ms}\frac{di_{bs}}{dt} - L_{ms}\frac{d(i'_{ar}\cos(\theta_r + \frac{2\pi}{3}))}{dt}\right.$$

$$\left. - L_{ms}\frac{d(i'_{br}\cos(\theta_r - \frac{2\pi}{3}))}{dt} - L_{ms}\frac{d(i'_{cr}\cos\theta_r)}{dt} + u_{cs}\right],$$

$$\frac{di'_{ar}}{dt} = \frac{1}{L'_{lr} + L_{ms}}\left[-r'_r i'_{ar} - L_{ms}\frac{d(i_{as}\cos\theta_r)}{dt} - L_{ms}\frac{d(i_{bs}\cos(\theta_r - \frac{2\pi}{3}))}{dt}\right.$$

$$\left. - L_{ms}\frac{d(i_{cs}\cos(\theta_r + \frac{2\pi}{3}))}{dt} + \frac{1}{2}L_{ms}\frac{di'_{br}}{dt} + \frac{1}{2}L_{ms}\frac{di'_{cr}}{dt} + u'_{ar}\right],$$

$$\frac{di'_{br}}{dt} = \frac{1}{L'_{lr} + L_{ms}}\left[-r'_r i'_{br} - L_{ms}\frac{d(i_{as}\cos(\theta_r + \frac{2\pi}{3}))}{dt} - L_{ms}\frac{d(i_{bs}\cos\theta_r)}{dt}\right.$$

$$\left. - L_{ms}\frac{d(i_{cs}\cos(\theta_r - \frac{2\pi}{3}))}{dt} + \frac{1}{2}L_{ms}\frac{di'_{ar}}{dt} + \frac{1}{2}L_{ms}\frac{di'_{cr}}{dt} + u'_{br}\right],$$

$$\frac{di'_{cr}}{dt} = \frac{1}{L'_{lr} + L_{ms}}\left[-r'_r i'_{cr} - L_{ms}\frac{d(i_{as}\cos(\theta_r - \frac{2\pi}{3}))}{dl} - L_{ms}\frac{d(i_{bs}\cos(\theta_r + \frac{2\pi}{3}))}{dt}\right.$$

$$\left. - L_{ms}\frac{d(i_{cs}\cos\theta_r)}{dt} + \frac{1}{2}L_{ms}\frac{di'_{ar}}{dt} + \frac{1}{2}L_{ms}\frac{di'_{br}}{dt} + u'_{cr}\right],$$

$$\frac{d\omega_r}{dt} = -\frac{P^2}{4J}L_{ms}\left\{\left[i_{as}\left(i'_{ar} - \frac{1}{2}i'_{br} - \frac{1}{2}i'_{cr}\right) + i_{bs}\left(i'_{br} - \frac{1}{2}i'_{ar} - \frac{1}{2}i'_{cr}\right)\right.\right.$$

$$\left.\left. + i_{cs}\left(i'_{cr} - \frac{1}{2}i'_{br} - \frac{1}{2}i'_{ar}\right)\right]\sin\theta_r + \frac{\sqrt{3}}{2}\left[i_{as}(i'_{br} - i'_{cr}) + i_{bs}(i'_{cr} - i'_{ar}) + i_{cs}(i'_{ar} - i'_{br})\right]\cos\theta_r\right\}$$

$$- \frac{B_m}{J}\omega_r - \frac{P}{2J}T_L,$$

$$\frac{d\theta_r}{dt} = \omega_r. \tag{5.50}$$

The Simulink diagram is developed as illustrated in Figure 5.15. The subsystem to represent the first differential equation in (5.50)

$$
\frac{di_{as}}{dt} = \frac{1}{L_{ls} + L_{ms}} \left[-r_s i_{as} + \frac{1}{2} L_{ms} \frac{di_{bs}}{dt} + \frac{1}{2} L_{ms} \frac{di_{cs}}{dt} - L_{ms} \frac{d(i'_{ar} \cos \theta_r)}{dt} \right.
$$

$$
\left. - L_{ms} \frac{d(i'_{br} \cos(\theta_r + \frac{2\pi}{3}))}{dt} - L_{ms} \frac{d(i'_{cr} \cos(\theta_r - \frac{2\pi}{3}))}{dt} + u_{as} \right]
$$

which results in

$$
\frac{di_{as}}{dt} = \frac{1}{L_{ls} + L_{ms}} \left\{ -r_s i_{as} + \frac{1}{2} L_{ms} \frac{di_{bs}}{dt} + \frac{1}{2} L_{ms} \frac{di_{cs}}{dt} - L_{ms} \cos \theta_r \frac{di'_{ar}}{dt} \right.
$$

$$
- L_{ms} \cos \left(\theta_r + \frac{2\pi}{3} \right) \frac{di'_{br}}{dt} - L_{ms} \cos \left(\theta_r - \frac{2\pi}{3} \right) \frac{di'_{cr}}{dt}
$$

$$
\left. + L_{ms} \left[i'_{ar} \sin \theta_r + i'_{br} \sin \left(\theta_r + \frac{2\pi}{3} \right) + i'_{cr} \sin \left(\theta_r - \frac{2\pi}{3} \right) \right] \omega_r + u_{as} \right\},
$$

is also reported in Figure 5.15. In the developed Simulink diagram, different blocks from the Simulink Library Browser are used, including the XY Graph to depict the torque-speed characteristic using the dynamics of $\omega_r(t)$ and evolution for $T_e(t)$.

We perform numerical simulations for a 220 V, 60 Hz, two-pole induction motor. The parameters are: $r_s = 0.8$ ohm, $r_r = 1$ ohm, $L_{ms} = 0.1$ H, $L_{ls} = 0.01$ H, $L_{lr} = 0.01$ H, $B_m = 4 \times 10^{-4}$ N-m-sec/rad, and $J = 0.002$ kg-m². These parameters are uploaded in the Command Window as

```
Um=220; P=2; Rs=0.8; Rr=1; Lms=0.1; Lls=0.01; Llr=0.01; Bm=0.0004; J=0.002;
```

The balanced three-phase voltage set is

$$
u_{as}(t) = \sqrt{2} u_M \cos(\omega_f t), \quad u_{bs}(t) = \sqrt{2} u_M \cos \left(\omega_f t - \frac{2}{3} \pi \right),
$$

$$
u_{cs}(t) = \sqrt{2} u_M \cos \left(\omega_f t + \frac{2}{3} \pi \right), \quad u_M = 220 \, \text{V}.
$$

The frequency of the supplied voltage is 60 Hz. Hence, $\omega_e = 4\pi f/P = 377$ rad/sec. If $T_L = 0$ and the friction torque is negligible, the steady-state value of angular velocity should approach ω_e. Hence, in the steady-state with no loads, $\omega_r = \omega_e$.

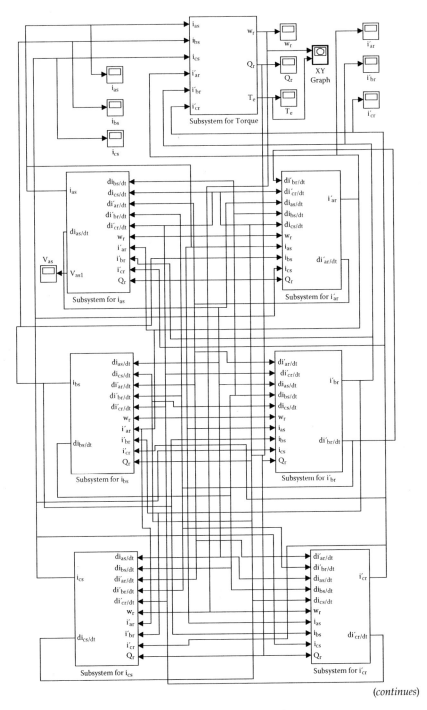

FIGURE 5.15

Simulink diagram to simulate squirrel-cage induction motors (ch5 _ 01.mdl).

(continues)

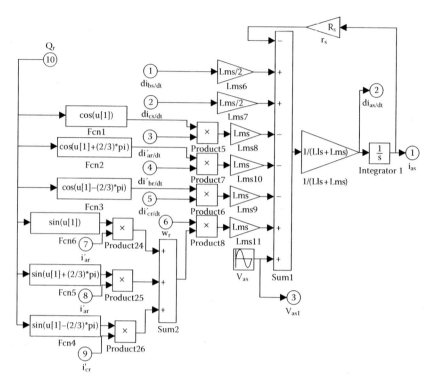

FIGURE 5.15
(Continued).

Using the Simulink model developed, as reported in Figure 5.15, nonlinear simulations are performed. The transient dynamics of the stator and rotor currents

$$i_{as}(t), \; i_{bs}(t), \; i_{cs}(t), \; i'_{ar}(t), \; i'_{br}(t), \; i'_{cr}(t),$$

as well as the angular velocity $\omega_r(t)$, can be viewed in the scopes. Figure 5.16 illustrates the transient dynamics for the angular velocity. To depict $\omega_r(t)$, using the saved scope data, one may use the following statement

```
plot(wr(:,1),wr(:,2)); xlabel('Time [seconds]','FontSize',14);
ylabel('\omega _ r','FontSize',14);
title('Angular Velocity Dynamics, \omega _ r [rad/sec]','FontSize',14);
```

The dynamic torque-speed characteristic $\omega_r(t) = \Omega_T[T_e(t)]$ is found using the $\omega_r(t)$ dynamics and $T_e(t)$ evolution. Figure 5.17 documents $\omega_r(t) = \Omega_T[T_e(t)]$ from the XY Graph. The results are also depicted using the statement

```
plot(wr(:,2),Te(:,2));
xlabel('Angular Velocity, \omega _ r [rad/sec]','FontSize',14);
```

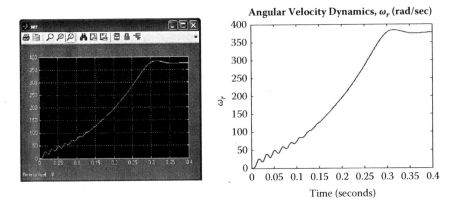

FIGURE 5.16
Transient dynamics for the angular velocity $\omega_r(t)$.

```
ylabel('Electromagnetic Torque, T _ e [N-m]','FontSize',14);
title('Torque-Speed Characteristic','FontSize',14);
```

The dynamics of $\omega_r(t)$ and $\omega_r(t) = \Omega_T[T_e(t)]$ if the load torque $T_L = 4$ N-m is applied at 0.4 sec (the motor rotates at ~377 rad/sec at no load) are plotted in Figure 5.18. One observes that the angular velocity decreases as T_L is applied. One refers to Figure 5.4b to visualize that the steady-state value of ω_r at which rotor rotates decreases if T_L is applied. The critical angular velocity from the plot $\omega_r(t) = \Omega_T[T_e(t)]$ is ~315 rad/sec. If ~4.5 N-m or higher load would be applied, $T_{emax} < T_L$, and the motor will disaccelerate to the stall. The induction motor cannot start if $T_L > ~1.5$ N-m is applied when the motor

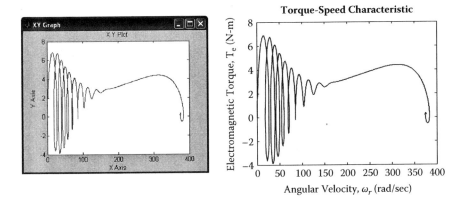

FIGURE 5.17
Dynamic torque-speed characteristic $\omega_r(t) = \Omega_T[T_e(t)]$.

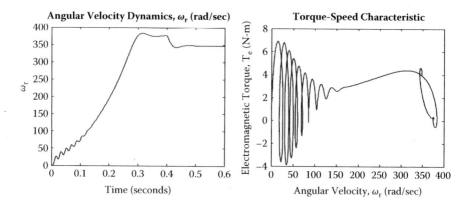

FIGURE 5.18
Transient dynamics for the angular velocity $\omega_r(t)$ and evolution of $\omega_r(t) = \Omega_T[T_e(t)]$.

at stall ($\omega_r = 0$ rad/sec). Correspondingly, one uses the voltage-frequency or frequency control to start the motor at high load (T_{Lmax} is ~4.5 N-m) as well as operate the motor at the desired angular velocity.   □

Example 5.7   Simulation of a Three-Phase Squirrel-Cage Induction Motor

Consider a 220 V, 60 Hz, two-pole induction motor. The parameters are: $r_s = 0.3$ ohm, $r'_r = 0.2$ ohm, $L_{ms} = 0.035$ H, $L_{ls} = 0.003$ H, $L'_{lr} = 0.003$ H, $B_m = 1 \times 10^{-3}$ N-m-sec/rad, and $J = 0.02$ kg-m^2.

The three-phase balanced voltage set is

$$u_{as}(t) = \sqrt{2}u_M \cos(\omega_f t),$$

$$u_{bs}(t) = \sqrt{2}u_M \cos\left(\omega_f t - \frac{2}{3}\right) \quad \text{and} \quad u_{cs}(t) = \sqrt{2}u_M \cos\left(\omega_f t + \frac{2}{3}\right),$$

where $u_M = 220$ V.

The frequency of the supplied voltage is 60 Hz, yielding $\omega_e = 4\pi f/P = 377$ rad/sec. If $T_L = 0$ and the friction torque is negligibly small, the steady-state value of angular velocity should approach ω_e. For example, at the steady-state with no loads, $\omega_r = \omega_e$. We apply the load torque 40 N-m at 0.7 sec.

The simulations are performed using Cauchy's form differential equations as given by (5.29) and (5.31). Two MATLAB files are developed. In particular, the first MATLAB file (ch5 _ 03.m) is

```
% Simulation of Three-Phase Induction Motors in Machine Variables
function yprime = motor(t,y);
global mag freq P J Rs Rr L Lms Bm TL0
% The Load Torque is Applied at 0.5 sec
if t(1,:) < 0.7
TL=0;
else TL=TL0;
end
```

```
% Squirrel-Cage Induction Motor: Rotor Windings are Shorted
UAR=0; UBR=0; UCR=0;
% Balanced Voltage Set
UAS=sqrt(2)*mag*cos(freq*2*pi*t);
UBS=sqrt(2)*mag*cos(freq*2*pi*t-2*pi/3);
UCS=sqrt(2)*mag*cos(freq*2*pi*t+2*pi/3);
theta=y(8,:);
A=cos(theta); B=cos(theta+2*pi/3); C=cos(theta-2*pi/3);
S1=sin(theta); S2=sin(theta+2*pi/3); S3=sin(theta-2*pi/3);
IAS=y(1,:); IBS=y(2,:); ICS=y(3,:);
IAR=y(4,:); IBR=y(5,:); ICR=y(6,:);
W=y(7,:);
TE=-0.5*P*Lms.*((IAS.*IAR+IBS.*IBR+ICS.*ICR).*S1+...(IAS.*IBR+IBS.*ICR+ICS.*IAR).*S2+...
(IAS.*ICR+IBS.*IAR+ICS.*IBR).*S3);
LS1=1/(L*(L+3*Lms));
% Differential Equations
yprime=[LS1*(-Rs*IAS*(2*... Lms+L)-0.5*Rs*Lms*(IBS+ICS)+Rr*Lms*(A*IAR+B*IBR+C*ICR)+...
1.299*(Lms^2)*W*(IBS-ICS)+(L*Lms+1.5*Lms^2)*W*(S1*IAR+S2*IBR+S3*ICR)+...
(2*Lms+L)*UAS+0.5*Lms*(UBS+UCS)-Lms*(A*UAR+B*UBR+C*UCR));...
LS1*(-Rs*IBS*(2*Lms+L)-0.5*Rs*Lms*(IAS+ICS)+Rr*Lms*(C*IAR+A*IBR+B*ICR)+...
1.299*(Lms^2)*W*(ICS-IAS)+(L*Lms+1.5*Lms^2)*W*(S3*IAR+S1*IBR+S2*ICR)+...
(2*Lms+L)*UBS+0.5*Lms*(UAS+UCS)-Lms*(C*UAR+A*UBR+B*UCR));...
LS1*(-Rs*ICS*(2*Lms+L)-0.5*Rs*Lms*(IAS+IBS)+Rr*Lms*(B*IAR+C*IBR+A*ICR)+...
1.299*(Lms^2)*W*(IAS-IBS)+(L*Lms+1.5*Lms^2)*W*(S2*IAR+S3*IBR+S1*ICR)+...
(2*Lms+L)*UCS+0.5*Lms*(UAS+UBS)-Lms*(B*UAR+C*UBR+A*UCR));...
LS1*(-Rr*IAR*(2*Lms+L)-0.5*Rr*Lms*(ICR+IBR)+Rs*Lms*(A*IAS+C*IBS+B*ICS)...
+1.299*(Lms^2)*W*(ICR-IBR)+(L*Lms+1.5*Lms^2)*W*(S1*IAS+S3*IBS+S2*ICS)...
+(2*Lms+L)*UAR+0.5*Lms*(UBR+UCR)-Lms*(A*UAS+C*UBS+B*UCS));...
LS1*(-Rr*IBR*(2*Lms+L)-0.5*Rr*Lms*(IAR+ICR)+Rs*Lms*(B*IAS+A*IBS+C*ICS)...
+1.299*(Lms^2)*W*(IAR-ICR)+(L*Lms+1.5*Lms^2)*W*(S2*IAS+S1*IBS+S3*ICS)+...
(2*Lms+L)*UBR+0.5*Lms*(UAR+UCR)-Lms*(B*UAS+A*UBS+C*UCS));...
LS1*(-Rr*ICR*(2*Lms+L)-0.5*Rr*Lms*(IAR+IBR)+Rs*Lms*(C*IAS+B*IBS+A*ICS)+...
1.299*(Lms^2)*W*(IBR-IAR)+(L*Lms+1.5*Lms^2)*W*(S3*IAS+S2*IBS+S1*ICS)+...
(2*Lms+L)*UCR+0.5*Lms*(UBR+UAR)-Lms*(C*UAS+B*UBS+A*UCS));...
(P/(2*J)*(TE-TL))-Bm*W/J;...
W];
```

The MATLAB file (ch5_04.m) is

```
% Simulation of Three-Phase Induction Motors in Machine Variables
echo on; clc; clear all;
global mag freq P J Rs Rr L Lms Bm TL0
% *********** Motor Parameters *************************
P=2; % Number of Poles
Rs=0.3; % Stator Winding Resistance
Rr=0.2; % Rotor Winding Resistance
Lms=0.035; % Mutual Stator-Rotor Inductance
L=0.003; % Leakage Inductance
Bm=0.001; % Viscous Friction Coefficient
J=0.02; % Moment of Inertia
TL0=40; % Load Torque Applied
time=1; % Final Time for Simulations
mag=220; % Applied Voltage Magnitude to the abc Windings
freq=60; % Frequency of the Applied Voltage
% ************************************************************
tspan=[0 time];
y0=[0 0 0 0 0 0 0 0]; % initial conditions
options=odeset('RelTol',1e-4,'AbsTol',[1e-4 1e-4 1e-4 1e-4 1e-4 1e-4 1e-4 1e-4]);
[t,y]=ode45('ch5_03',tspan,y0,options);
UAS=sqrt(2)*mag*cos(freq*2*pi*t);
UBS=sqrt(2)*mag*cos(freq*2*pi*t-2*pi/3);
UCS=sqrt(2)*mag*cos(freq*2*pi*t+2*pi/3);
```

```
theta=y(:,8);
S1=sin(theta); S2=sin(theta+2*pi/3); S3=sin(theta-2*pi/3);
IAS=y(:,1); IBS=y(:,2); ICS=y(:,3);
IAR=y(:,4); IBR=y(:,5); ICR=y(:,6);
W=y(:,7);
TE=-0.5*P*Lms.*(S1.*(IAR.*IAS+ICS.*ICR+IBR.*IBS)+S2.*(ICS.*IAR+IBR.*IAS+IBS.*ICR)+...
S3.*(IAR.*IBS+ICS.*IBR+ICR.*IAS));


% *********** Plots ************************
plot(t,UAS,t,UBS,t,UCS);
title('Stator Phase Voltages Applied, u _ a _ s, u _ b _ s and u _ c _ s [V]','FontSize',14);
axis([0 0.1 -sqrt(2)*225 sqrt(2)*225]);
xlabel('Time [seconds]','FontSize',14); ylabel('u _ a _ s, u _ b _ s, u _ c _ s','FontSize',14);
grid; pause;
plot(t,y(:,1));
title('Stator Current, i _ a _ s [A]','FontSize',14);
xlabel('Time [seconds]','FontSize',14); ylabel('i _ a _ s','FontSize',14);
grid; pause;
plot(t,y(:,2));
title('Stator Current, i _ b _ s [A]','FontSize',14);
xlabel('Time [seconds]','FontSize',14); ylabel('i _ b _ s','FontSize',14);
grid; pause;
plot(t,y(:,3));
title('Stator Current, i _ c _ s [A]','FontSize',14);
xlabel('Time [seconds]','FontSize',14); ylabel('i _ c _ s','FontSize',14);
grid; pause;
plot(t,y(:,4));
title('Rotor Current, i _ a _ r [A]','FontSize',14);
xlabel('Time [seconds]','FontSize',14); ylabel('i _ a _ r','FontSize',14);
grid; pause;
plot(t,y(:,5));
title('Rotor Current, i _ b _ r [A]','FontSize',14);
xlabel('Time [seconds]','FontSize',14); ylabel('i _ b _ r','FontSize',14);
grid; pause;
plot(t,y(:,6));
title('Rotor Current, i _ c _ r [A]','FontSize',14);
xlabel('Time [seconds]','FontSize',14); ylabel('i _ c _ r','FontSize',14);
grid; pause;
plot(t,y(:,7));
title('Angular Velocity, \omega _ r [rad/sec]','FontSize',14);
xlabel('Time [seconds]','FontSize',14); ylabel('\omega _ r','FontSize',14);
grid; pause;
Te(:,1)=(-P*M/2)*((y(:,1).*(y(:,4)-0.5*y(:,5)-0.5*y(:,6))+y(:,2).*(y(:,5)-0.5*y(:,4)...
-0.5*y(:,6))+y(:,3).*(y(:,6)-0.5*y(:,5)-...
0.5*y(:,4))).*sin(y(:,8))+0.865*(y(:,1).*(y(:,5)-y(:,6))+y(:,2).*(y(:,6)-...
y(:,4))+y(:,3).*(y(:,5)-y(:,4))).*cos(y(:,8)));
plot(t,TE);
title('Electromagnetic Torque, T _ e [N-m]','FontSize',14);
xlabel('Time [seconds]','FontSize',14); ylabel('T _ e [N-m]','FontSize',14);
grid; pause;
plot(W,TE); title('Torque-Speed Characteristic','FontSize',14);
xlabel('Angular Velocity, \omega _ r [rad/sec]','FontSize',14);
ylabel('Electromagnetic Torque, T _ e [N-m]','FontSize',14);
```

Using the `ode45` differential equations solver, we numerically solve a set of eight differential equations (5.29) and (5.31) with describe the induction motor dynamics in the *machine* variables. The transients of the stator and rotor currents

$$i_{as}(t),\ i_{bs}(t),\ i_{cs}(t)\ ,\ i'_{ar}(t),\ i'_{br}(t),\ i'_{cr}(t),$$

as well as the angular velocity $\omega_r(t)$, are plotted using the statements embedded in the second file. Figure 5.19 illustrates the phase voltages applied, phase currents dynamics, angular velocity behavior, evolution of the electromagnetic torque $T_e(t)$, as well as the torque-speed characteristic. One can easily examine the changes occuring at 0.7 sec when $T_L = 40$ N-m is applied. The analysis of induction motor dynamics, performance and capabilities are straightforwardly performed. In particular, the settling time, acceleration, losses, and other performance characteristics are readily accessible. □

5.4 Power Converters

The angular velocity of squirrel-cage induction motors is regulated by changing the magnitude and frequency of the phase voltages applied to the stator windings using power converters. The basic components of variable-frequency converters are rectifier, filter, and inverter. The simplest rectifiers are the single-phase half- and full-wave rectifiers. To control medium- and high-power induction motors, *polyphase* rectifiers are used. *Polyphase* rectifiers contain several ac sources, and the rectified voltage is summated at the output. The rectified voltage is filtered to reduce the harmonic content of the rectifier output voltage. Passive and active harmonic reduction, harmonic elimination, and harmonic cancellation can be attained by using passive and active filters. To control the frequency, inverters are used. Voltage- and current-fed inverters convert the dc voltage or current, respectively. Pulse-width modulation (PWM) reduces the *total harmonic distortion*. The PWM concept requires control circuitry to drive the switches with high frequency, however, the filtering requirements are significantly relaxed. Power converters provide an interface between the input energy source and the induction motor needed to be controlled (Figure 5.20).

Power converters produce sinusoidal voltages, supplied to the induction motor windings. The dc voltage is obtained by rectifying and filtering the line voltage, and the magnitude of the voltage can be controlled. The sinusoidal ac voltage or current with the regulated frequency, fed to the phase windings of induction motors, are obtained by using DC-to-AC inverters. Design and deployments of high-performance electric drives have been directly related to the availability of power semiconductor devices. High-frequency switching power transistors, which need to be used in power converters for light-, medium-, and heavy-power applications, exist. The specialized design is

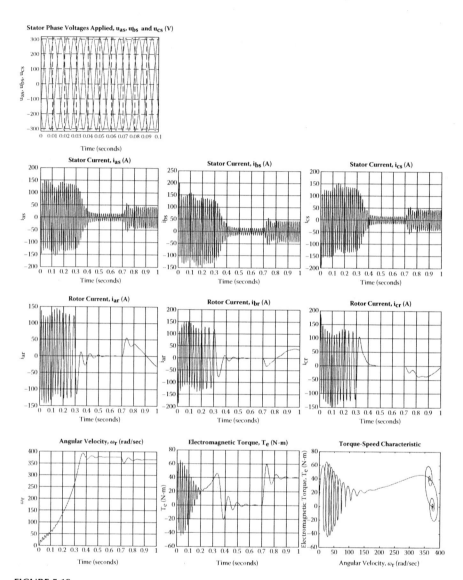

FIGURE 5.19

Transient dynamic for the phase currents, angular velocity $\omega_r(t)$, evolution of $T_e(t)$, and the dynamic torque-speed characteristic $\omega_r(t) = \Omega_T[T_e(t)]$.

available, and, for example, 3000 V, 1000 A IGBTs are integrated with diodes in the same package. The ~200 kVA soft switching resonant-link inverters have the switching frequency ~70 kHz. The development of the gate turn-off (GTO) thyristor was the key to extending the power rating of electric drives with induction machines to the megawatt range. Power converters with GTO have found a widespread application in traction drives (electric drive trains in ships and locomotives). Gate turn-off thyristors are current-controlled devices that

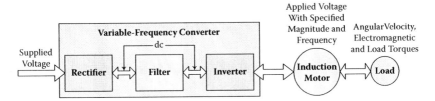

FIGURE 5.20
Variable-frequency power converter and loaded induction motor.

require large gate current to enable turn-off in the anode current. Hence, large snubbers are needed to ensure turn-off without failures.

There are two basic types of inverters. The voltage source inverters supply induction motor windings with variable frequency phase voltages. In contrast, the variable frequency phase currents are fed to the induction motor windings by current source inverters. Figure 5.21 illustrates high-level

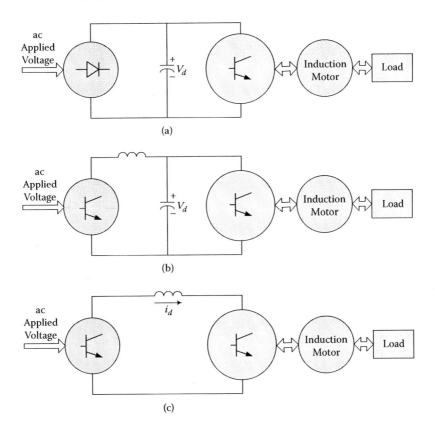

FIGURE 5.21
Variable-frequency power converters: (a) Pulse-width-modulated voltage source inverter with an unregulated rectifier; (b) squire-wave voltage source inverter with a regulated rectifier; (c) current source inverter with a regulated rectifier.

diagrams of power converters, which include PWM voltage source inverter
with an unregulated rectifies, squire-wave voltage source inverter with a
regulated rectifier, and current source inverter with a regulated rectifier.

Typical PWM power converter configurations consist three legs, one for
each phase, to control the frequency and the magnitude of the phase voltages
applied to the motor windings. Figure 5.22 document a representative schematics. The inverter converts the dc bus voltage into a *polyphase* ac voltage at
the desired frequency to attain the specified torque-speed characteristics, efficiency, starting capabilities, acceleration, and so on. Switching stresses, losses,
high electromagnetic interference, extended operating areas, and some other
drawbacks of hard-switching inverters, lead to the implementation of soft-
switching technology, as illustrated in Figure 5.22b. Soft-switching by using
resonant-linked converters allows one to attain zero voltage across (current
through) the switching device. That is, the semiconductor device is switched
when the voltage across it, or the current through it, is zero. Hence, low losses
and electromagnetic interference, high efficiency, and switching capabilities
result compared with the hard-switching inverters shown in Figure 5.22a.

For hard- and soft-switching inverter configurations, shown in Figure 5.22,
the phase voltage waveform, supplied to the motor winding, is illustrated in
Figures 5.23a and 5.23b, respectively.

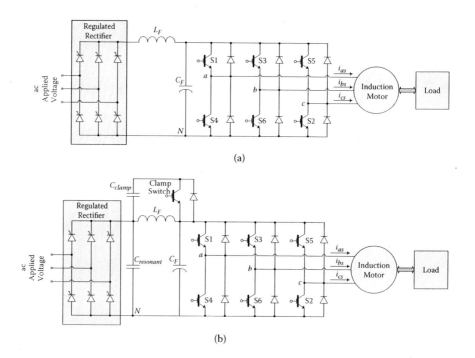

(a)

(b)

FIGURE 5.22
(a) Power converter with three-phase hard-switching inverter; (b) power converter with three-
phase soft-switching inverter.

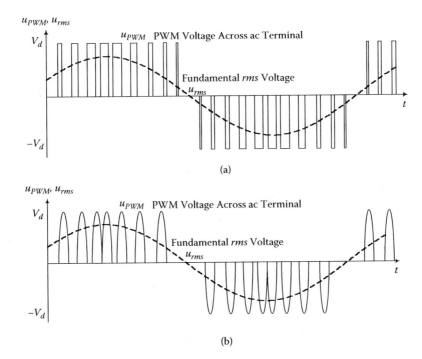

FIGURE 5.23
(a) Phase voltage waveforms in the hard-switching PWM inverter; (b) phase voltage waveforms in the soft-switching PWM inverter.

Transistors and thyristors are current-controlled solid-state devices. Transistors require continuous drive signals, whereas thyristors need a momentary gate current to turn on and turn off. For example, a base current must be regulated to maintain the BJT in the conducting state, and the turn-on and turn-off times depend on how rapidly the charge needed to be supplied (to turn on) or removed (to turn off) can be delivered to the base region. The turn-off switching speed is decreased by initially applying a spike of base current, and, then, reducing the current to the magnitude needed. In contrast, the turn-off switching speed is decreased by initially applying a spike of negative base current. Control ICs should drive high-frequency transistors with the overall objective to vary the magnitude and frequency of phase voltages. The closed-loop systems for induction motor with power converter are designed.

Consider the hard-switching inverter as shown in Figure 5.22a with three switch pairs. To obtain three-phase balanced output voltages using PWM concept, a triangular signal-level voltage is compared with three sinusoidal specified frequency control signals, and these control signals are shifted by 120°C, (see Figure 5.24a). High-frequency switches S1 and S4, S3 and S6, and S5 and S2 close and open opposite of each other. That is, switches in each pair

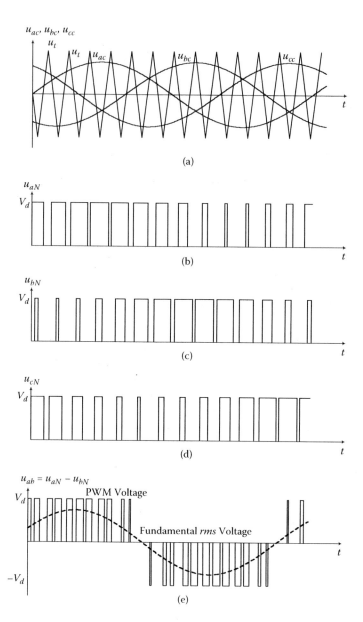

FIGURE 5.24
Voltage waveforms in three-phase hard-switching inverters.

are turned on and off simultaneously. If, for example, S1 and S4 are closed at the same instant, the circuit is short-circuited across the source. The instantaneous voltages u_{aN}, u_{bN}, and u_{cN} are either equal to V_d or 0. The signal level voltages u_{ac}, u_{bc}, and u_{cc} are compared with the triangular signal u_t. If, for example, u_{ac} is greater than u_t, then S1 is closed, whereas S4 is open. If the signal-level voltage

u_{ac} is less than u_t, then S4 is closed, whereas S1 is open. The resulting waveform for the phase voltage u_{aN} is shown in Figure 5.24b. In the similar manner, the phase voltages u_{bN} and u_{cN} are defined by comparing the signal-level voltages u_{bc} and u_{cc} with u_t to open or close switches S3-S6 and S5-S2. The resulting voltages u_{bN} and u_{cN} possess the same pattern as the aN voltage, except that u_{bN} and u_{cN} are shifted by 120° and 240° as illustrated in Figures 5.24c and 5.24d. The voltages u_{aN}, u_{bN}, and u_{cN} are measured with respect to the negative dc bus. These dc components are canceled as one considers the line-to-line voltage, which is plotted in Figure 5.24e. The line-to-line voltage u_{ab} is found by subtracting voltage u_{bN} from u_{aN}. One can analyze the waveforms of the instantaneous and *rms* voltages, shown in Figures 5.23 and 5.24e.

The square-wave voltage source inverters, known as six-step inverters, are commonly used. The three-phase square-wave voltage source inverter bridge is shown in Figure 5.22. The rectifier rectifies the three-phase ac applied voltage, and a large electrolytic capacitor C_F maintains a near-constant dc voltage as well as provides a path for the rapidly changing currents drawn by the inverter. The inductor L_F attenuates current spikes. Assume that the inverter consists of six ideal switches. We consider the basic operation of the square-wave voltage inverters. Each switch is closed for 180° and is opened for the remaining 180° in a cyclic pattern. Furthermore, S3 is closed 120° after S1, S5 is closed 120° after S3, S4 is closed 180° after S1, S6 is closed 180° after S3, and S2 is closed 180° after S5 (see Figure 5.25). The result of this switching operation is that a

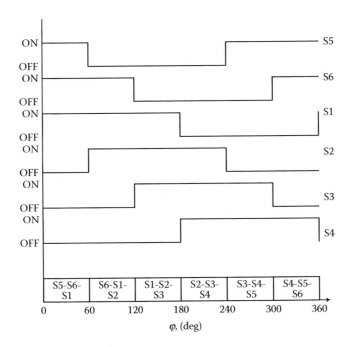

FIGURE 5.25
Switching pattern for three-phase six-step inverters.

Electromechanical Systems and Devices

combination of three switches are closed simultaneously for every 60° duration as shown in Figure 5.25. That is, in three-phase sex-step inverters, the
switching appears every 60° interval (1/6 T time interval).

To determine the voltage waveforms, applied to the *abc* windings, consider
the six-step inverter and motor circuitry as illustrated in Figure 5.22. During the interval from 0° to 60°, where switches S5, S6, and S1 are closed, the
phase *a* is in parallel with *c*, and they are connected to the phase *b* in series,
which is connected to the source via S6. The voltage waveforms, as shown in
Figure 5.26, result. In particular,

$$u_{aN} = u_{cN} = V_d \quad \text{and} \quad u_{bN} = 0.$$

Hence,

$$u_{ab} = V_d, \ u_{bc} = -V_d \quad \text{and} \quad u_{ca} = 0.$$

Because phases *a* and *c* are connected in parallel, the apparent impedance, seen
from the neutral of the motor (depicted in Figure 5.26 as a point N'), is halved.
Hence, the voltage drop across the phases *as* and *cs* is

$$\frac{2}{3}V_d,$$

whereas voltage drop across the phase *bs* is

$$\frac{1}{3}V_d.$$

That is,

$$u_{as} = \frac{1}{3}V_d, \ u_{bs} = -\frac{2}{3}V_d, \quad \text{and} \quad u_{cs} = \frac{1}{3}V_d.$$

Hence, the voltage drop across the phase is always

$$\frac{1}{3}V_d \quad \text{or} \quad \frac{2}{3}V_d$$

depending on the connection of the phases (series or parallel). The waveforms for u_{as}, u_{bs}, and u_{cs} are shown in Figure 5.26.

Compared with the voltage source inverters, the power converters if the
current source inverters are implemented, are different. In particular, the
current source inverter is fed from a constant current source which is generated by a controlled rectifier with a large dc link inductor L to smooth the
current. The representative schematics of a current source inverter is shown
in Figure 5.27.

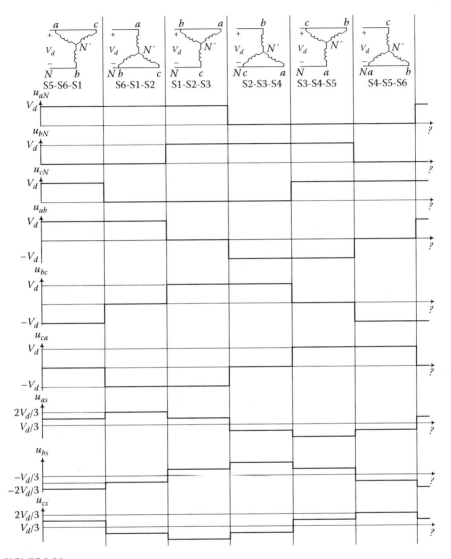

FIGURE 5.26
Voltage waveforms at the terminals aN, bN, and cN, line-to-line, and line-to-neutral voltages applied to the induction motor phase windings.

At any time, only two thyristors conduct. In particular, one of the thyristors is connected to the positive dc link, and the other is connected to the negative dc link. The current is switched sequentially into one of the phases of a three-phase induction motor by the top half of the inverter and returns from another phase to the dc link by the bottom half of the inverter. Because the current is constant, there will be a constant voltage drop across the stator winding of the motor and zero voltage drop across the self inductance of the

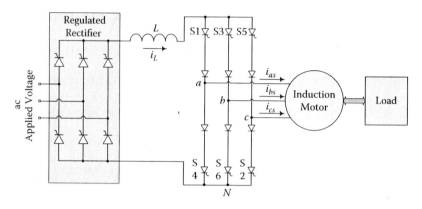

FIGURE 5.27
Power converter with a thyristor current source inverter.

winding. Hence, the motor terminal voltage is not set by the inverter but by the resistance of the stator winding of the motor. Because the motor is wound with sinusoidally distributed windings, the voltages that appear on the terminals of the motor are sinusoidal. The current waveforms of the motor connected to a current source inverter are shown in Figure 5.28.

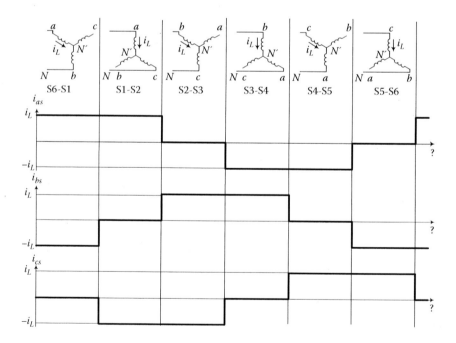

FIGURE 5.28
Phase currents for the induction motors fed by current source inverter.

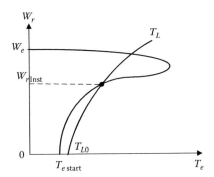

FIGURE 5.29
Torque-speed and load characteristic.

Homework Problems

Problem 5.1

Consider a four-phase, 60 Hz, two-pole induction motor.

5.1.1. Derive the expression for ψ_{ar} assuming that the stator-rotor mutual inductances obey a $\cos^3\theta_r$ distribution. Hence, the stator-rotor mutual inductances become to be functions of $\cos^3(\theta_r + phase)$.

5.1.2. Derive the expression for the emf_{ar} induced in the rotor winding ar. Justify why the studied ac motor is called the induction motor.

5.1.3. The induction motor accelerates, and the motor "instantaneous" angular velocity is denoted as ω_{rInst} in Figure 5.29. What is the final angular velocity with which the motor will operate (report the numerical value of ω_r).

Problem 5.2

Consider an A-class two-phase, 115 V (*rms*), 60 Hz, four-pole ($P = 4$) induction motor. Assume

$$L_{asar} = L_{aras} = L_{ms1}\cos\theta_r + L_{ms3}\cos^5\theta_r, \; L_{asbr} = L_{bras} = -L_{ms1}\sin\theta_r - L_{ms3}\sin^5\theta_r,$$
$$L_{bsar} = L_{arbs} = L_{ms1}\sin\theta_r + L_{ms3}\sin^5\theta_r, \text{ and } L_{bsbr} = L_{brbs} = L_{ms1}\cos\theta_r + L_{ms3}\cos^5\theta_r.$$

The motor parameters are: $r_s = 1.2$ ohm, $r'_r = 1.5$ ohm, $L_{ms1} = 0.14$ H, $L_{ms2} = 0.004$ H, $L_{ls} = 0.02$ H, $L_{ss} = L_{ls} + L_{ms1} + 5L_{ms3}$, $L'_{lr} = 0.02$ H, $L'_{rr} = L'_{lr} + L_{ms1} + 5L_{ms3}$, $B_m = 1 \times 10^{-6}$ N-m-sec/rad, and $J = 0.005$ kg-m². The phase voltages supplied are

$$u_{as}(t) = \sqrt{2}115\cos(377t) \quad \text{and} \quad u_{bs}(t) = \sqrt{2}115\sin(377t).$$

5.2.1. In MATLAB, calculate, plot and compare

$$L_{asar} = L_{ms}\cos\theta_r \ (L_{ms} = 0.16 \ \text{H}) \quad \text{and} \quad L_{asar} = L_{ms1}\cos\theta_r + L_{ms3}\cos^5\theta_r;$$

5.2.2. Derive the circuitry-electromagnetic equations of motion;

5.2.3. Report emf_{as} and emf_{ar};

5.2.4. Find the expression for the electromagnetic torque;

5.2.5. Using Newton's second law, obtain the *torsional-mechanical* model;

5.2.6. Report the equations of motion in non-Cauchy's form;

5.2.7. Develop the Simulink mdl model or MATLAB file using ode45 differential equations solver to simulate induction motor dynamics.

5.2.8. Plot the transient dynamics for all state variables, in particular, report $i_{as}(t)$, $i_{bs}(t)$, $i_{ar}(t)$, $i_{br}(t)$, $\omega_r(t)$, and $\theta_r(t)$;

5.2.9. Plot the motional $emf_{ar\omega}$ and $emf_{br\omega}$ induced in the rotor windings;

5.2.10. Plot the torque-speed characteristics $\omega_r = \Omega_T(T_e)$ using the simulation results;

5.2.11. Analyze the induction motor performance (acceleration, settling time, load attenuation, etc.).

Problem 5.3

Consider the *arbitrary* reference frame for two-phase electric machines and electromechanical motion devices. Using the Park transformations, the *machine (ab)* variables can be transformed to the *quadrature* and *direct (qd)* quantities. Figure 5.30 illustrates the *machine* stator and rotor magnetic axes, as well as the *quadrature* and *direct* magnetic axes. Letting the angular velocity of the reference frame be ω (not specified), $\omega = 0$, $\omega = \omega_r$, and $\omega = \omega_e$, one deals with the *arbitrary*, stationary, rotor, and synchronous reference frames. The *direct* and *inverse* transformations, as well as corresponding matrices for different reference frames, are reported in Table 5.2.

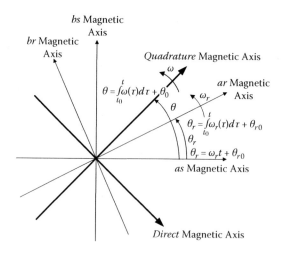

FIGURE 5.30
Magnetic axes.

TABLE 5.2

Transformations of the *Machine* and *qd* Variables

Stator, Rotor, and Reference Frame Magnetic Axes	Transformation Matrices	Transformation of Variables
Arbitrary reference frame ω unspecified	Stator transformation matrices $$\mathbf{K}_s = \begin{bmatrix} \cos\theta & \sin\theta \\ \sin\theta & -\cos\theta \end{bmatrix}$$ $$\mathbf{K}_s^{-1} = \begin{bmatrix} \cos\theta & \sin\theta \\ \sin\theta & -\cos\theta \end{bmatrix}$$ Rotor transformation matrices $$\mathbf{K}_r = \begin{bmatrix} \cos(\theta-\theta_r) & \sin(\theta-\theta_r) \\ \sin(\theta-\theta_r) & -\cos(\theta-\theta_r) \end{bmatrix}$$ $$\mathbf{K}_r^{-1} = \begin{bmatrix} \cos(\theta-\theta_r) & \sin(\theta-\theta_r) \\ \sin(\theta-\theta_r) & -\cos(\theta-\theta_r) \end{bmatrix}$$	*Direct* transformation $$\mathbf{u}_{qds} = \mathbf{K}_s \mathbf{u}_{abs}, \mathbf{i}_{qds} = \mathbf{K}_s \mathbf{i}_{abs}$$ $$\boldsymbol{\psi}_{qds} = \mathbf{K}_s \boldsymbol{\psi}_{abs}$$ *Inverse* transformation $$\mathbf{u}_{abs} = \mathbf{K}_s^{-1} \mathbf{u}_{qds}, \mathbf{i}_{abs} = \mathbf{K}_s^{-1} \mathbf{i}_{qds}$$ $$\boldsymbol{\psi}_{abs} = \mathbf{K}_s^{-1} \boldsymbol{\psi}_{qds}$$ *Direct* transformation $$\mathbf{u}_{qdr} = \mathbf{K}_r \mathbf{u}_{abr}, \mathbf{i}_{qdr} = \mathbf{K}_r \mathbf{i}_{abr}$$ $$\boldsymbol{\psi}_{qdr} = \mathbf{K}_r \boldsymbol{\psi}_{abr}$$ *Inverse* transformation $$\mathbf{u}_{abr} = \mathbf{K}_r^{-1} \mathbf{u}_{qdr}, \mathbf{i}_{abr} = \mathbf{K}_r^{-1} \mathbf{i}_{qdr}$$ $$\boldsymbol{\psi}_{abr} = \mathbf{K}_r^{-1} \boldsymbol{\psi}_{qdr}$$
Stationary reference frame, $\omega = 0$	Stator transformation matrices $$\mathbf{K}_s^s = \begin{bmatrix} 1 & 0 \\ 0 & -1 \end{bmatrix},$$ $$\mathbf{K}_s^{s-1} = \begin{bmatrix} 1 & 0 \\ 0 & -1 \end{bmatrix}$$ Rotor transformation matrices $$\mathbf{K}_r^s = \begin{bmatrix} \cos\theta_r & \sin\theta_r \\ \sin\theta_r & -\cos\theta_r \end{bmatrix},$$ $$\mathbf{K}_r^{r-1} = \begin{bmatrix} \cos\theta_r & \sin\theta_r \\ \sin\theta_r & -\cos\theta_r \end{bmatrix}$$	*Direct* transformation $$\mathbf{u}_{qds}^s = \mathbf{K}_s^s \mathbf{u}_{abs}, \mathbf{i}_{qds}^s = \mathbf{K}_s^s \mathbf{i}_{abs}$$ $$\boldsymbol{\psi}_{qds}^s = \mathbf{K}_s^s \boldsymbol{\psi}_{abs}$$ *Inverse* transformation $$\mathbf{u}_{abs} = \mathbf{K}_s^{s-1} \mathbf{u}_{qds}^s, \mathbf{i}_{abs} = \mathbf{K}_s^{s-1} \mathbf{i}_{qds}^s$$ $$\boldsymbol{\psi}_{abs} = \mathbf{K}_s^{s-1} \boldsymbol{\psi}_{qds}^s$$ *Direct* transformation $$\mathbf{u}_{qdr}^s = \mathbf{K}_r^s \mathbf{u}_{abr}, \mathbf{i}_{qdr}^s = \mathbf{K}_r^s \mathbf{i}_{abr}$$ $$\boldsymbol{\psi}_{qdr}^s = \mathbf{K}_r^s \boldsymbol{\psi}_{abr}$$ *Inverse* transformation $$\mathbf{u}_{abr} = \mathbf{K}_r^{s-1} \mathbf{u}_{qdr}^s, \mathbf{i}_{abr} = \mathbf{K}_r^{s-1} \mathbf{i}_{qdr}^s$$ $$\boldsymbol{\psi}_{abr} = \mathbf{K}_r^{s-1} \boldsymbol{\psi}_{qdr}^s$$

(Continued)

TABLE 5.2

Transformations of the *Machine* and *qd* Variables (Continued)

Stator, Rotor, and Reference frame Magnetic Axes	Transformation Matrices	Transformation of Variables
Rotor reference frame, $\omega = \omega_r$	Stator transformation matrices $$\mathbf{K}_s^r = \begin{bmatrix} \cos\theta_r & \sin\theta_r \\ \sin\theta_r & -\cos\theta_r \end{bmatrix}$$ $$\mathbf{K}_s^{r-1} = \begin{bmatrix} \cos\theta_r & \sin\theta_r \\ \sin\theta_r & -\cos\theta_r \end{bmatrix}$$ Rotor transformation matrices $$\mathbf{K}_r^r = \begin{bmatrix} 1 & 0 \\ 0 & -1 \end{bmatrix}$$ $$\mathbf{K}_r^{r-1} = \begin{bmatrix} 1 & 0 \\ 0 & -1 \end{bmatrix}$$	*Direct* transformation $$\mathbf{u}_{qds}^r = \mathbf{K}_s^r \mathbf{u}_{abs}, \mathbf{i}_{qds}^r = \mathbf{K}_s^r \mathbf{i}_{abs}$$ $$\psi_{qds}^r = \mathbf{K}_s^r \psi_{abs}$$ *Inverse* transformation $$\mathbf{u}_{abs} = \mathbf{K}_s^{r-1} \mathbf{u}_{qds}^r, \mathbf{i}_{abs} = \mathbf{K}_s^{r-1} \mathbf{i}_{qds}^r$$ $$\psi_{abs} = \mathbf{K}_s^{r-1} \psi_{qds}^r$$ *Direct* transformation $$\mathbf{u}_{qdr}^r = \mathbf{K}_r^r \mathbf{u}_{abr}, \mathbf{i}_{qdr}^r = \mathbf{K}_r^r \mathbf{i}_{abr}$$ $$\psi_{qdr}^r = \mathbf{K}_r^r \psi_{abr}$$ *Inverse* transformation $$\mathbf{u}_{abr} = \mathbf{K}_r^{r-1} \mathbf{u}_{qdr}^r, \mathbf{i}_{abr} = \mathbf{K}_r^{r-1} \mathbf{i}_{qdr}^r$$ $$\psi_{abr} = \mathbf{K}_r^{r-1} \psi_{qdr}^r$$
Synchronous reference frame, $\omega = \omega_e$	Stator transformation matrices $$\mathbf{K}_s^e = \begin{bmatrix} \cos\theta_e & \sin\theta_e \\ \sin\theta_e & -\cos\theta_e \end{bmatrix}$$ $$\mathbf{K}_s^{e-1} = \begin{bmatrix} \cos\theta_e & \sin\theta_e \\ \sin\theta_e & -\cos\theta_e \end{bmatrix}$$ Rotor transformation matrices $$\mathbf{K}_r^e = \begin{bmatrix} \cos(\theta_e-\theta_r) & \sin(\theta_e-\theta_r) \\ \sin(\theta_e-\theta_r) & -\cos(\theta_e-\theta_r) \end{bmatrix}$$ $$\mathbf{K}_r^{e-1} = \begin{bmatrix} \cos(\theta_e-\theta_r) & \sin(\theta_e-\theta_r) \\ \sin(\theta_e-\theta_r) & -\cos(\theta_e-\theta_r) \end{bmatrix}$$	*Direct* transformation $$\mathbf{u}_{qds}^e = \mathbf{K}_s^e \mathbf{u}_{abs}, \mathbf{i}_{qds}^e = \mathbf{K}_s^e \mathbf{i}_{abs}$$ $$\psi_{qds}^e = \mathbf{K}_s^e \psi_{abs}$$ *Inverse* transformation $$\mathbf{u}_{abs} = \mathbf{K}_s^{e-1} \mathbf{u}_{qds}^e, \mathbf{i}_{abs} = \mathbf{K}_s^{e-1} \mathbf{i}_{qds}^e$$ $$\psi_{abs} = \mathbf{K}_s^{e-1} \psi_{qds}^e$$ *Direct* transformation $$\mathbf{u}_{qdr}^e = \mathbf{K}_r^e \mathbf{u}_{abr}, \mathbf{i}_{qdr}^e = \mathbf{K}_r^e \mathbf{i}_{abr}$$ $$\psi_{qdr}^e = \mathbf{K}_r^e \psi_{abr}$$ *Inverse* transformation $$\mathbf{u}_{abr} = \mathbf{K}_r^{e-1} \mathbf{u}_{qdr}^e, \mathbf{i}_{abr} = \mathbf{K}_r^{e-1} \mathbf{i}_{qdr}^e$$ $$\psi_{abr} = \mathbf{K}_r^{e-1} \psi_{qdr}^e$$

5.3.1. Find the *quadrature-* and *direct*-axis components of stator voltages in the *arbitrary* reference frame if the voltages

$$u_{as}(t) = 100\cos(377t) \quad \text{and} \quad u_{as}(t) = 100\sin(377t)$$

are supplied to the *as* and *bs* windings.

5.3.2. Find

$$u_{qs}^e(t), \ u_{ds}^e(t), \text{ and } u_{qs}^s(t), \ u_{ds}^s(t)$$

in the synchronous and stationary reference frame assigning

$$\omega = 377 \, \frac{rad}{\sec} \text{ and } \omega = 0.$$

5.3.3. Prove that $u_{qs}(t)$ and $u_{qs}(t)$ are the dc or ac voltage components in the synchronous and stationary reference frames, respectively.

References

1. S.J. Chapman, *Electric Machinery Fundamentals*, McGraw-Hill, New York, 1999.
2. A.E. Fitzgerald, C. Kingsley, and S.D. Umans, *Electric Machinery*, McGraw-Hill, New York, 1990.
3. P.C. Krause and O. Wasynczuk, *Electromechanical Motion Devices*, McGraw-Hill, New York, 1989.
4. P.C. Krause, O. Wasynczuk, and S.D. Sudhoff, *Analysis of Electric Machinery*, IEEE Press, New York, 1995.
5. W. Leonhard, *Control of Electrical Drives*, Springer, Berlin, 1996.
6. S.E. Lyshevski, *Electromechanical Systems, Electric Machines, and Applied Mechatronics*, CRC Press, Boca Raton, FL, 1999.
7. D.W. Novotny and T.A. Lipo, *Vector Control and Dynamics of AC Drives*, Clarendon Press, Oxford, 1996.
8. G.R. Slemon, *Electric Machines and Drives*, Addison-Wesley Publishing Company, Reading, MA, 1992.
9. S.E. Lyshevski, *Engineering and Scientific Computations Using MATLAB®*, Wiley-Interscience, Hoboken, NJ, 2003.
10. *MATLAB R2006b*, CD-ROM, MathWorks, Inc., 2007.

6

Synchronous Machines

6.1 Introduction to Synchronous Machines

Permanent-magnet synchronous machines are the high-performance electromechanical motion devices which possess superior performance and capabilities surpassing any other electric machines such as permanent-magnet DC electric machines, induction motors, and so on. In addition, conventional synchronous machines are utilized. In synchronous motors, the electromagnetic torque results as a result of the interaction of time-varying magnetic field established by the stator windings and stationary magnetic field produced by the windings or magnets on the rotor. In addition, there exist synchronous reluctance machines that exhibit, in general, low performance. In synchronous reluctance motors, the electromagnetic torque is produced to align the minimum-reluctance path of the rotor with the time-varying rotating airgap *mmf*.

In high-performance drives, servos, and power generation systems (up to ~200 kW), three-phase permanent-magnet synchronous machines (motors and generators) are the preferable choice. In high-power (from hundreds of kW to hundreds of MW range) generation systems, conventional three-phase synchronous generators are utilized. In this chapter, we consider radial and axial topology synchronous machines which can be classified as: (1) synchronous reluctance; (2) permanent-magnet synchronous machines (two- and three-phase); (3) conventional synchronous machines. There exist translational (linear) and rotational synchronous machines [1-6]. Because a great majority of electromechanical systems utilize rotational electric machines, we concentrate on the rotational motion devices.

Our objective is to examine the energy conversion, torque production, control, and other important issues. The angular velocity of synchronous motors is fixed with the frequency of the supplied phase voltages to the stator windings. It will be illustrated that the phase voltages must be applied as functions of the rotor angular displacement θ_r. The steady-state torque-speed characteristics can be represented as a family of horizontal lines in Figure 6.1. The electrical angular velocity is equal to the synchronous angular velocity $\omega_e = 4\pi f/P$. The designer examines the maximum load torque $T_{L\,max}$ to satisfy $T_{e\,rated} > T_{L\,max}$. For a short period of time (~1 min for majority of permanent-magnet synchronous motors), one may significantly overload

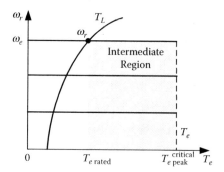

FIGURE 6.1
Torque-speed characteristic curves of synchronous motors.

motors, and the ratio $T_{e\,peak}/T_{e\,rated}$ could be from ~2 to ~10. If $T_{e\,critical} < T_L$, the rotor magnetic field is no longer locked to the stator magnetic field (motor cannot produce the needed torque due to the constraints imposed including rated voltage and current of power converters). For $T_{e\,critical} < T_L$, the rotor magnetic field slips behind the stator field. Because of the loss of synchronization, the electromagnetic torque developed by the motor reverses (surges). Therefore, the condition $T_{e\,critical} > T_L$ must be always guaranteed, and the peak power converter capabilities (peak voltage and current) must be studied. Although the permanent magnet synchronous machines can be overloaded by the factor of 10, the maximum current overloading capabilities of PWM amplifiers is up to ~2, whereas the peak voltage cannot exceed the bus voltage.

Radial and axial topology permanent-magnet synchronous machines are widely used as actuators (motors) and generators. The motors are controlled by power converters (PWM amplifiers). The implications of microelectronics to motion devices have received meticulous consideration as technologies to fabricate various advanced machines have becoming developed and widely deployed. Mini- and micromachines have been fabricated utilizing CMOS-centered and micromachining technologies and processes. The images of 2 and 4 mm in diameter permanent-magnet synchronous machines are shown in Figures 6.2a and 6.2b. These permanent-magnet synchronous machines are smaller than the operational amplifier or ICs-centered electronics to control them. However, to guarantee rotation and actuation, the condition $T_e > T_L$ and $F_e > F_L$ must be satisfied for any electromechanical motion devices. Correspondingly, the operating envelope (torque/force, load, load profile, angular velocity, etc.) defines T_e thereby resulting in the motor dimensionality. The acceleration capability, which defines the settling time and repositioning, depends on the ratio $(T_e - T_L)/J$. The torque and power densities, rated angular velocity, and other characteristics are defined by the machine design, dimensionality, materials, technologies, and other factors. In the preliminary power estimate, one may assume that the power density ~1 W/cm³ can be reached. Figure 6.2c documents the images of a permanent-magnet synchronous

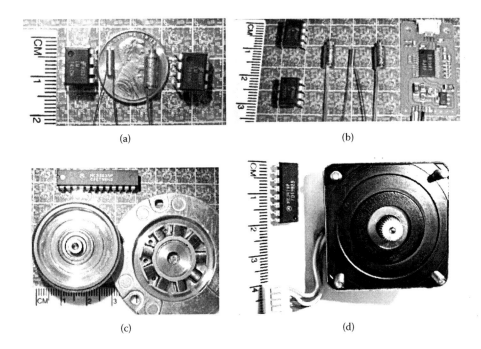

(a)

(b)

(c)

(d)

FIGURE 6.2

(a) Images of permanent-magnet synchronous machines and operational amplifiers on a silicon not diced wafer with ICs on it; (b) permanent-magnet synchronous motors, operational amplifiers (on left) and power electronics board (on right) to control them; (c) permanent-magnet synchronous motor to drive a computer hard drive and a monolithic driver; (d) two-phase permanent-magnet synchronous motor (stepper motor) and a controller/driver.

motor in a computer hard drive and VHS with the monolithic ICs driver. A two-phase permanent magnet synchronous motor (usually called a stepper motor) with the monolithic controller/driver is shown in Figure 6.2d.

In this chapter, we address and solve a spectrum of problems in analysis, modeling, and control of various synchronous machines. The electromechanical motion devices must be synthesized in the structural domain optimizing their electromagnetic and mechanical features through a coherent design. Using a sound device physics, which is electromagnetically defined, various machine topologies (radial and axial, rotational and translational, hybrid, etc.) have been devised and researched. The operating principles, energy conversion, control, and other issues contribute to sequential tasks in analysis and design. Rotational and translational ac electromechanical motion devices can be classified as synchronous and induction. A step-by-step procedure in the structural synthesis and behavioral design is:

1. Devise (or examine existing) electromechanical motion devices researching device physics, operating principles, topologies, geometry, electromagnetic systems, and other features.

2. Study electromechanical energy conversion and control mechanisms.
3. Specify application and environmental requirements.
4. Define specifications and requirements imposed against reachable performance and capabilities.
5. Perform electromagnetic, energy conversion, mechanical, thermal, vibroacoustic, and sizing/dimension (stator, rotor, magnets, air gap, winding, etc.) estimates.
6. Define technologies, processes, and materials to fabricate structures (stator, rotor, windings, magnets, bearing, shaft, etc.) and assemble electromechanical motion devices.
7. Perform coherent electromagnetic, mechanical, and thermodynamic design, optimization, and analysis assessing electromechanical motion device performance and capabilities.
8. Modify and refine the design in order to ensure optimal *achievable* performance and capabilities.

Assuming an optimal or near-optimal structural design of electric machines (which is not within the objective of this book), we concentrate on modeling and simulation, which results in coherent analysis, sound control, and optimal design in the behavior domain. Examining complex electromagnetic, electromechanical, and vibroacoustic phenomena in electromechanical motion devices, data-intensive analysis can be performed. These tasks result in deriving sound control concepts to ensure *achievable* performance and capabilities. For example, maximum efficiency, minimal losses, maximum torque and power densities, vibration, and noise minimization, as well as other critical improvements, can be achieved.

6.2   Radial Topology Synchronous Reluctance Motors

6.2.1   Single-Phase Synchronous Reluctance Motors

The single-phase synchronous reluctance motor is illustrated in Figure 6.3. We examine radial topology single-phase reluctance motors to study the operation, analyze important features, research the torque production, as well as evaluate and define sound control concepts.

The *quadrature* and *direct* magnetic axes are fixed with the rotor, which rotates with angular velocity ω_r. These magnetic axes rotate with the angular velocity ω. Under normal operation, the angular velocity of synchronous machines is equal to the synchronous angular velocity ω_e. Hence, $\omega_r = \omega_e$ and $\omega = \omega_r = \omega_e$. The angular displacements of the rotor θ_r and the angular displacement of the *quadrature* magnetic axis θ are equal. Assuming that the initial conditions are zero, we have

$$\theta_r = \theta = \int_{t_0}^{t} \omega_r(\tau)d\tau = \int_{t_0}^{t} \omega(\tau)d\tau.$$

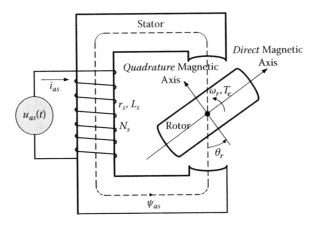

FIGURE 6.3
Radial topology reluctance motor with a variable reluctance path.

The magnetizing reluctance $\mathfrak{R}_m$ is a function of the rotor angular displacement θ_r. Using the number of turns N_s, the magnetizing inductance is

$$L_m(\theta_r) = N_s^2 / \mathfrak{R}_m(\theta_r).$$

This magnetizing inductance varies twice per one revolution of the rotor and has minimum and maximum values. We have

$$L_{m\,\min} = \left.\frac{N_s^2}{\mathfrak{R}_{m\,\max}(\theta_r)}\right|_{\theta_r = 0,\pi,2\pi,\dots} \quad \text{and} \quad L_{m\,\max} = \left.\frac{N_s^2}{\mathfrak{R}_{m\,\min}(\theta_r)}\right|_{\theta_r = \frac{1}{2}\pi,\frac{3}{2}\pi,\frac{5}{2}\pi,\dots}.$$

Assume that the inductance variation is an ideally sinusoidal. Emphasizing that the magnetizing inductance is a function of the rotor angular displacement, we have

$$L_m(\theta_r) = \bar{L}_m - L_{\Delta m} \cos 2\theta_r,$$

where $\bar{L}_m$ is the average value of the magnetizing inductance; $L_{\Delta m}$ is the half of amplitude of the sinusoidal variation of the magnetizing inductance.

The plot for $L_m(\theta_r)$ is documented in Figure 6.4.

The electromagnetic torque, developed by single-phase reluctance motors, is found using the expression for the coenergy $W_c(i_{as}, \theta_r)$. From

$$W_c(i_{as}, \theta_r) = \frac{1}{2}(L_{ls} + \bar{L}_m - L_{\Delta m} \cos 2\theta_r) i_{as}^2,$$

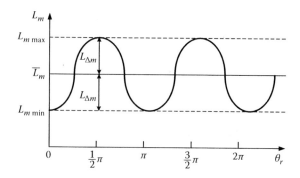

FIGURE 6.4
Magnetizing inductance $L_m(\theta_r)$.

one finds the electromagnetic torque

$$T_e = \frac{\partial W_c(i_{as}, \theta_r)}{\partial \theta_r} = \frac{\partial \left(\frac{1}{2} i_{as}^2 (L_{ls} + \bar{L}_m - L_{\Delta m} \cos 2\theta_r) \right)}{\partial \theta_r} = L_{\Delta m} i_{as}^2 \sin 2\theta_r.$$

The electromagnetic torque is not developed by synchronous reluctance motors if one feeds the dc current or voltage to the winding. Hence, sound control concepts, which are based on electromagnetic features, must be studied. The average value of T_e is not equal to zero if the current is a function of θ_r. As an illustration, let the phase current in the *as* winding is

$$i_{as} = i_M \operatorname{Re}(\sqrt{\sin 2\theta_r}).$$

The electromagnetic torque is

$$T_e = L_{\Delta m} i_{as}^2 \sin 2\theta_r = L_{\Delta m} i_M^2 \left(\operatorname{Re} \sqrt{\sin 2\theta_r} \right)^2 \sin 2\theta_r \neq 0,$$

and

$$T_{eav} = \frac{1}{\pi} \int_0^\pi L_{\Delta m} i_{as}^2 \sin 2\theta_r d\theta_r = \frac{1}{4} L_{\Delta m} i_M^2.$$

From $T_e = L_{\Delta m} i_{as}^2 \sin 2\theta_r$, one formally finds the phase current to maximize T_e and eliminate the torque ripple to be

$$i_{as} = i_M \frac{1}{\sqrt{\sin 2\theta_r}}.$$

This phase current theoretically leads to $T_e = L_{\Delta m} i_M^2$. However, it is impossible to implement

$$i_{as} = i_M \frac{1}{\sqrt{\sin 2\theta_r}}$$

because of the constraints on the maximum current, singularity, denominator becomes complex for negative values of $\sin 2\theta_r$, and so on. Therefore, the derived expression for the phase current $i_{as}(\theta_r)$ should be modified.

The mathematical model of the single-phase reluctance motor is found by using Kirchhoff's and Newton's second laws

$$u_{as} = r_s i_{as} + \frac{d\psi_{as}}{dt} \quad \text{(circuitry-electromagnetic equation of motion)},$$

$$J \frac{d^2\theta_r}{dt^2} = T_e - B_m \omega_r - T_L \quad \text{(torsional-mechanical equation of motion)}.$$

From

$$\psi_{as} = (L_{ls} + \overline{L}_m - L_{\Delta m} \cos 2\theta_r) i_{as},$$

one obtains a set of three first-order nonlinear differential equations, which describes single-phase reluctance motor dynamic and steady-state behavior. In particular, we have

$$\frac{di_{as}}{dt} = -\frac{r_s}{L_{ls} + \overline{L}_m - L_{\Delta m} \cos 2\theta_r} i_{as} - \frac{2L_{\Delta m}}{L_{ls} + \overline{L}_m - L_{\Delta m} \cos 2\theta_r} i_{as} \omega_r \sin 2\theta_r$$

$$+ \frac{1}{L_{ls} + \overline{L}_m - L_{\Delta m} \cos 2\theta_r} u_{as},$$

$$\frac{d\omega_r}{dt} = \frac{1}{J} \left(L_{\Delta m} i_{as}^2 \sin 2\theta_r - B_m \omega_r - T_L \right),$$

$$\frac{d\theta_r}{dt} = \omega_r. \tag{6.1}$$

One can simulate and analyze reluctance machines using the equations of motion. For example, the MATLAB environment [7,8] can be used to solve the differential equations and perform the analysis.

Example 6.1

The parameters and other quantities of interest, for example the variation of $L_m(\theta_r)$, of electromechanical motion devices can be estimated as machine is designed. The coefficients and quantities can be measured as the machine is fabricated. For a single-phase synchronous reluctance motor, as illustrated in Figure 6.3, the parameters are: $r_s = 1$ ohm, $L_{md} = 0.4$ H, $L_{mq} = 0.04$ H, $L_{ls} = 0.05$ H, $J = 0.00001$ kg-m^2 and $B_m = 0.00005$ N-m-sec/rad.

The voltage applied to the stator winding is

$$u_{as} = u_M \text{Re}\left(\sqrt{\sin 2(\theta_r - 0.2)} \right), \quad u_M = 50 \text{ V}.$$

For no-load condition, the motor parameters are uploaded as

```
% Synchronous reluctance motor parameters
P=2; rs=1; Lmd=0.4; Lmq=0.04; Lls=0.05; J=0.00001; Bm=0.00005;
```

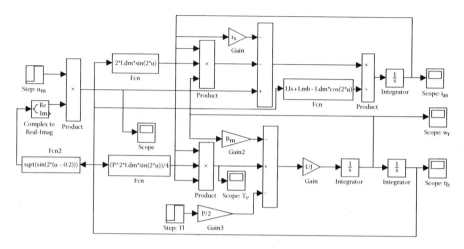

FIGURE 6.5
Simulink block diagram to simulate and analyze single-phase synchronous reluctance motors
(ch6 _ 01.mdl).

```
Lmb=(Lmq+Lmd)/3; Ldm=(Lmd-Lmq)/3;
um=50; % rms value of the applied voltage
Tl=0; % load torque
```

One uploads the parameters by running the m-file or typing the parameter values in the Command Window. The developed Simulink block diagram, which corresponds the differential equations (6.1), is documented in
Figure 6.5.

The transient of the angular velocity $\omega_r(t)$ is plotted in Figure 6.6. The motor
reaches the steady-state 280 rad/sec within 0.75 sec. The plotting statement is

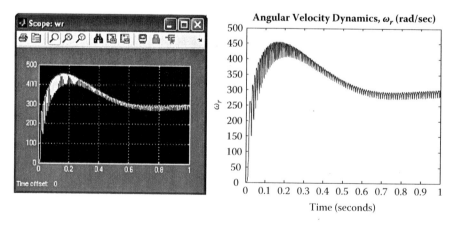

FIGURE 6.6
Motor behavior (transient response) for the angular velocity.

```
plot(wr(:,1),wr(:,2)); xlabel('Time [seconds]','FontSize',14);
ylabel('\omega _ r','FontSize',14);
title('Angular Velocity Dynamics, \omega _ r [rad/sec]','FontSize',14);
```

One observes the electromagnetic torque ripple and phase current chattering. These effects lead to low efficiency, heating, vibration, noise, mechanical wearing, and other undesired features. Therefore, single-phase synchronous motors are not used because of the low performance and inadequate capabilities. This electromechanical motion device was studied as a *primer* to introduce advanced topics. It is found that the phase voltage must be supplied as a function of the rotor angular displacement θ_r, in order to develop the electromagnetic torque. □

6.2.2 Three-Phase Synchronous Reluctance Motors

The goal of this section is to expand the results to three-phase synchronous reluctance motors. We will solve a spectrum of problems in analysis, modeling, and control. The machine electromagnetics is studied before attempting to control it. The electromagnetic features define the control concepts to be applied. The motor is controlled (angular velocity or displacement are regulated) by changing the phase voltages applied or phase currents fed to the windings. We study synchronous reluctance motors in the *machine (abc)* and in the *quadrature-direct-zero (qd0)* variables. The circuitry-electromagnetic equations of motion are found by using the Kirchhoff voltage law, whereas the Newtonian mechanics is applied to derive the *torsional-mechanical* dynamics. The three-phase synchronous reluctance motor is illustrated in Figure 6.7.

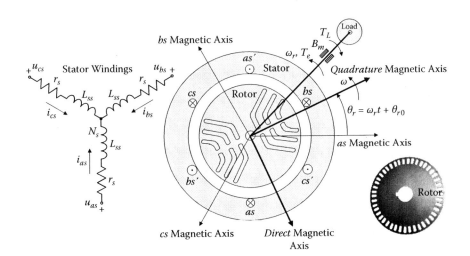

FIGURE 6.7
Three-phase synchronous reluctance motor.

The machine parameters are the stator resistance r_s (it is assumed that the phase resistances are equal), the magnetizing inductances in the *quadrature* and *direct* axes L_{mq} and L_{md}, the average magnetizing inductance $\bar{L}_m$, the leakage inductance L_{ls}, the moment of inertia J, and the viscous friction coefficient B_m.

The circuitry-electromagnetic dynamics is found from the equation

$$\mathbf{u}_{abcs} = \mathbf{r}_s \mathbf{i}_{abcs} + \frac{d\psi_{abcs}}{dt}, \tag{6.2}$$

where u_{as}, u_{bs}, and u_{cs} are the phase voltages supplied to the *as*, *bs*, and *cs* stator windings; i_{as}, i_{bs}, and i_{cs} are the phase currents; ψ_{as}, ψ_{bs}, and ψ_{cs} are the flux linkages; $\psi_{abcs} = \mathbf{L}_s(\theta_r)\mathbf{i}_{abcs}$.

In (6.2), the resistance matrix is

$$\mathbf{r}_s = \begin{bmatrix} r_s & 0 & 0 \\ 0 & r_s & 0 \\ 0 & 0 & r_s \end{bmatrix},$$

whereas the inductance mapping $\mathbf{L}_s(\theta_r)$ is

$$\mathbf{L}_s(\theta_r) = \begin{bmatrix} L_{ls} + \bar{L}_m - L_{\Delta m}\cos 2(\theta_r) & -\frac{1}{2}\bar{L}_m - L_{\Delta m}\cos 2\left(\theta_r - \frac{1}{3}\pi\right) & -\frac{1}{2}\bar{L}_m - L_{\Delta m}\cos 2\left(\theta_r + \frac{1}{3}\pi\right) \\ -\frac{1}{2}\bar{L}_m - L_{\Delta m}\cos 2\left(\theta_r - \frac{1}{3}\pi\right) & L_{ls} + \bar{L}_m - L_{\Delta m}\cos 2\left(\theta_r - \frac{2}{3}\pi\right) & -\frac{1}{2}\bar{L}_m - L_{\Delta m}\cos 2\left(\theta_r + \pi\right) \\ -\frac{1}{2}\bar{L}_m - L_{\Delta m}\cos 2\left(\theta_r + \frac{1}{3}\pi\right) & -\frac{1}{2}\bar{L}_m - L_{\Delta m}\cos 2\left(\theta_r + \pi\right) & L_{ls} + \bar{L}_m - L_{\Delta m}\cos 2\left(\theta_r + \frac{2}{3}\pi\right) \end{bmatrix},$$

where

$$\bar{L}_m = \frac{1}{3}(L_{mq} + L_{md}) \quad \text{and} \quad L_{\Delta m} = \frac{1}{3}(L_{md} - L_{mq}).$$

The expressions for inductances are nonlinear functions of the electrical angular displacement θ_r. Hence, the *torsional-mechanical* dynamics must be used. From Newton's second law of rotational motion, and using the electrical angular velocity and displacement as the state variables (ω_r and θ_r can be considered to be the mechanical variables, and the mechanical angular velocity is $\omega_{rm} = 2\omega_r/P$), one obtains

$$J\frac{2}{P}\frac{d\omega_r}{dt} = T_e - B_m\frac{2}{P}\omega_r - T_L,$$

$$\frac{d\theta_r}{dt} = \omega_r. \tag{6.3}$$

Using the coenergy, the electromagnetic torque T_e should be found. From

$$W_c = \frac{1}{2}[i_{as} \ i_{bs} \ i_{cs}] \, \mathbf{L}_s \begin{bmatrix} i_{as} \\ i_{bs} \\ i_{cs} \end{bmatrix}$$

$$= \frac{1}{2}[i_{as} \ i_{bs} \ i_{cs}] \begin{bmatrix} L_{ls} + \bar{L}_m - L_{\Delta m}\cos 2\theta_r & -\frac{1}{2}\bar{L}_m - L_{\Delta m}\cos 2\left(\theta_r - \frac{1}{3}\pi\right) & -\frac{1}{2}\bar{L}_m - L_{\Delta m}\cos 2\left(\theta_r + \frac{1}{3}\pi\right) \\ -\frac{1}{2}\bar{L}_m - L_{\Delta m}\cos 2\left(\theta_r - \frac{1}{3}\pi\right) & L_{ls} + \bar{L}_m - L_{\Delta m}\cos 2\left(\theta_r - \frac{2}{3}\pi\right) & -\frac{1}{2}\bar{L}_m - L_{\Delta m}\cos 2\theta_r \\ -\frac{1}{2}\bar{L}_m - L_{\Delta m}\cos 2\left(\theta_r + \frac{1}{3}\pi\right) & -\frac{1}{2}\bar{L}_m - L_{\Delta m}\cos 2\theta_r & L_{ls} + \bar{L}_m - L_{\Delta m}\cos 2\left(\theta_r + \frac{1}{3}\pi\right) \end{bmatrix} \begin{bmatrix} i_{as} \\ i_{bs} \\ i_{cs} \end{bmatrix},$$

we have

$$T_r = \frac{P}{2}\frac{\partial W_c}{\partial \theta_r} = \frac{P}{2}\frac{1}{2}[i_{as} \ i_{bs} \ i_{cs}] \begin{bmatrix} 2L_{\Delta m}\sin 2\theta_r & 2L_{\Delta m}\sin 2(\theta_r - \frac{1}{3}\pi) & 2L_{\Delta m}\sin 2(\theta_r + \frac{1}{3}\pi) \\ 2L_{\Delta m}\sin 2(\theta_r - \frac{1}{3}\pi) & 2L_{\Delta m}\sin 2(\theta_r - \frac{2}{3}\pi) & 2L_{\Delta m}\sin 2\theta_r \\ 2L_{\Delta m}\sin 2(\theta_r + \frac{1}{3}\pi) & 2L_{\Delta m}\sin 2\theta_r & 2L_{\Delta m}\sin 2(\theta_r + \frac{2}{3}\pi) \end{bmatrix} \begin{bmatrix} i_{as} \\ i_{bs} \\ i_{cs} \end{bmatrix}.$$

One obtains

$$T_e = \frac{P}{2}L_{\Delta m}\left[i_{as}^2 \sin 2\theta_r + 2i_{as}i_{bs}\sin 2\left(\theta_r - \frac{1}{3}\pi\right) + 2i_{as}i_{cs}\sin 2\left(\theta_r + \frac{1}{3}\pi\right)\right.$$

$$\left. + i_{bs}^2 \sin 2\left(\theta_r - \frac{2}{3}\pi\right) + 2i_{bs}i_{cs}\sin 2\theta_r + i_{cs}^2 \sin 2\left(\theta_r + \frac{2}{3}\pi\right)\right]. \tag{6.4}$$

Hence, T_e is a nonlinear function of the motor variables (phase currents and electrical angular displacement θ_r) and motor parameters (number of poles P and inductance $L_{\Delta m}$). For three-phase synchronous reluctance motors,

$$L_{mq} = \frac{3}{2}(\bar{L}_m - L_{\Delta m}) \quad \text{and} \quad L_{md} = \frac{3}{2}(\bar{L}_m + L_{\Delta m}).$$

Therefore,

$$\bar{L}_m = \frac{1}{3}(L_{mq} + L_{md}) \quad \text{and} \quad L_{\Delta m} = \frac{1}{3}(L_{md} - L_{mq}).$$

Using the trigonometric identities, from

$$L_{\Delta m} = \frac{1}{3}(L_{md} - L_{mq}),$$

one finds the following formula for the electromagnetic torque

$$T_e = \frac{P(L_{md} - L_{mq})}{6}\left[\left(i_{as}^2 - \frac{1}{2}i_{bs}^2 - \frac{1}{2}i_{cs}^2 - i_{as}i_{bs} - i_{as}i_{cs} + 2i_{bs}i_{cs}\right)\sin 2\theta_r\right.$$

$$\left. + \frac{\sqrt{3}}{2}\left(i_{bs}^2 - i_{cs}^2 - 2i_{as}i_{bs} + 2i_{as}i_{cs}\right)\cos 2\theta_r\right].$$

To control the angular velocity, the electromagnetic torque must be regulated. To maximize the electromagnetic torque, one must feed the following phase currents as functions of the angular displacement θ_r measured or observed (sensorless control)

$$i_{as} = \sqrt{2}i_M \sin\left(\theta_r + \frac{1}{3}\varphi_i\pi\right), \quad i_{bs} = \sqrt{2}i_M \sin\left(\theta_r - \frac{1}{3}(2-\varphi_i)\pi\right) \text{ and}$$

$$i_{cs} = \sqrt{2}i_M \sin\left(\theta_r + \frac{1}{3}(2+\varphi_i)\pi\right).$$

For $\varphi_i = 0.3245$, one obtains

$$T_e = \sqrt{2}PL_{\Delta m}i_M^2.$$

That is, T_e is maximized, and T_e is controlled by changing the magnitude of the phase currents i_M. The angular rotor displacement θ_r must be measured to control synchronous motors. For an ideal design, using the equations derived, one may not expect the current chattering and torque ripple. In practice, using the experimental results, one finds that there are current chattering and torque ripple as a result of the nonlinear magnetic system, cogging, fringing effects, eccentricity, electromagnetic field nonuniformity, nonuniformity of magnetic materials, pulse-width-modulation, applied voltage waveforms, and other phenomena. In general, even the high-fidelity models, derived using three-dimensional Maxwell's equations, and heterogeneous simulations do not allow one to ensure an absolute consistency. However, very important features were found by applying sound assumptions. We establish a control concept to ensure a near-optimal operation of synchronous reluctance motors. Power amplifiers usually control the phase voltages u_{as}, u_{bs}, and u_{cs}. The three-phase balanced voltage set is

$$u_{as} = \sqrt{2}u_M \sin\left(\theta_r + \frac{1}{3}\varphi_i\pi\right), \quad u_{bs} = \sqrt{2}u_M \sin\left(\theta_r - \frac{1}{3}(2-\varphi_i)\pi\right) \text{ and}$$

$$u_{cs} = \sqrt{2}u_M \sin\left(\theta_r + \frac{1}{3}(2+\varphi_i)\pi\right),$$

where u_M is the magnitude of the supplied voltages.

Example 6.2

Calculate and plot the electromagnetic torque if the balanced current set

$$i_{as} = \sqrt{2}i_M \sin\left(\theta_r + \frac{1}{3}\varphi_i\pi\right), \quad i_{bs} = \sqrt{2}i_M \sin\left(\theta_r - \frac{1}{3}(2-\varphi_i)\pi\right),$$

$$i_{cs} = \sqrt{2}i_M \sin\left(\theta_r + \frac{1}{3}(2+\varphi_i)\pi\right), \quad \varphi_i = 0.3245$$

is applied. Let $i_M = 10$ A, $P = 4$, and $L_{\Delta m} = 0.05$ H.

The expression for the electromagnetic torque (6.4) leads to

$$T_e = \sqrt{2}\ 10\ \text{N-m}.$$

By using the balanced three-phase current set, the MATLAB m-file is developed as documented

```
% Calculation of the Developed Electromagnetic Torque
th=0:0.01:4*pi; % angular rotor displacement
phi=0.3245; % phase current angle
IM=10; P=4; LDm=0.05;
% Balanced three-phase current set
Ias=IM*sin(th+phi*pi/3); % current in the as winding
Ibs=IM*sin(th-(2-phi)*pi/3); % current in the bs winding
Ics=IM*sin(th+(2+phi)*pi/3); % current in the cs winding
% Calculation of the electromagnetic torque developed
Te=P*LDm*(Ias.*(sin(2*th).*Ias+2*sin(2*th-...
2*pi/3).*Ibs+2*sin(2*th+2*pi/3).*Ics)...
+Ibs.*(sin(2*th...-4*pi/3).*Ibs+2*sin(2*th).*Ics)+Ics.*sin(2*th+4*pi/3).*Ics)/2;
% Plot the currents applied to abc windings
plot(th,Ias,'-',th,Ibs,'--',th,Ics,'-.'); axis([0,4*pi,-10,10]);
xlabel('Angular Displacement, \theta_r [rad]','FontSize',14);
ylabel('Phase Currents','FontSize',14);
title('Phase Currents, i_a_s, i_b_s and i_c_s [A]','FontSize',14);
pause;
% Plot of the torque developed versus the angular displacement
plot(th,Te); axis([0,4*pi,0,15]);
xlabel('Angular Displacement, \theta_r [rad]','FontSize',14);
ylabel('Electromagnetic Torque','FontSize',14);
title('Electromagnetic Torque, T_e [N-m]','FontSize',14);
```

The steady-state T_e was calculated as a function of the currents fed and θ_r. The evolutions of the phase currents and T_e are reported in Figure 6.8. The analysis shows that the maximum electromagnetic torque is developed by three-phase synchronous reluctance motors when the balanced three-phase current set is fed to the windings. There is no torque ripple. From

$$T_e = \sqrt{2}\,PL_{\Delta m}i_M^2,$$

one obtains

$$T_e = \sqrt{2}\,10\,\text{N-m}.\qquad\square$$

Mathematical Models of Three-Phase Synchronous Reluctance Motors

The equations of motion for synchronous reluctance motors are derived in non-Cauchy's and Cauchy's forms. These models in the *abc* (*machine*) variables are found using Kirchhoff's second law (6.2) with $\psi_{abcs} = \mathbf{L}_s(\theta_r)\mathbf{i}_{abcs}$,

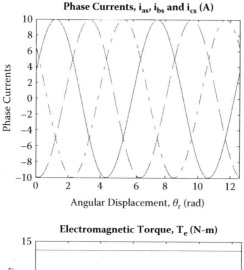

FIGURE 6.8
Phase currents and electromagnetic torque.

and Newton's second law of motions (6.3). The electromagnetic torque is expressed by the equation (6.4). Using the inductance mapping $\mathbf{L}_s(\theta_r)$ derived

$$\mathbf{L}_s(\theta_r) = \begin{bmatrix} L_{ls} + \bar{L}_m - L_{\Delta m}\cos 2\theta_r & -\tfrac{1}{2}\bar{L}_m - L_{\Delta m}\cos 2\left(\theta_r - \tfrac{1}{3}\pi\right) & -\tfrac{1}{2}\bar{L}_m - L_{\Delta m}\cos 2\left(\theta_r + \tfrac{1}{3}\pi\right) \\ -\tfrac{1}{2}\bar{L}_m - L_{\Delta m}\cos 2\left(\theta_r - \tfrac{1}{3}\pi\right) & L_{ls} + \bar{L}_m - L_{\Delta m}\cos 2\left(\theta_r - \tfrac{2}{3}\pi\right) & -\tfrac{1}{2}\bar{L}_m - L_{\Delta m}\cos 2\left(\theta_r + \pi\right) \\ -\tfrac{1}{2}\bar{L}_m - L_{\Delta m}\cos 2\left(\theta_r + \tfrac{1}{3}\pi\right) & -\tfrac{1}{2}\bar{L}_m - L_{\Delta m}\cos 2(\theta_r + \pi) & L_{ls} + \bar{L}_m - L_{\Delta m}\cos 2\left(\theta_r + \tfrac{2}{3}\pi\right) \end{bmatrix},$$

from (6.2) we have the resulting differential equations, which describe the dynamics of three-phase synchronous motors in non-Cauchy's form

$$\frac{di_{as}}{dt} = \frac{1}{L_{ls} + \bar{L}_m - L_{\Delta m}\cos 2\theta_r}\left[-r_s i_{as} + u_{as} + \left(\frac{1}{2}\bar{L}_m + L_{\Delta m}\cos 2\left(\theta_r - \frac{1}{3}\pi\right)\right)\frac{di_{bs}}{dt}\right.$$

$$+ \left(\frac{1}{2}\bar{L}_m + L_{\Delta m}\cos 2\left(\theta_r + \frac{1}{3}\pi\right)\right)\frac{di_{cs}}{dt}$$

$$\left. -2L_{\Delta m}\omega_r\left(i_{as}\sin 2\theta_r + i_{bs}\sin 2\left(\theta_r - \frac{1}{3}\pi\right) + i_{cs}\sin 2\left(\theta_r + \frac{1}{3}\pi\right)\right)\right],$$

$$\frac{di_{bs}}{dt} = \frac{1}{L_{ls} + \bar{L}_m - L_{\Delta m}\cos 2\left(\theta_r - \frac{2}{3}\pi\right)}\left[-r_s i_{bs} + u_{bs}\right.$$

$$+ \left(\frac{1}{2}\bar{L}_m + L_{\Delta m}\cos 2\left(\theta_r - \frac{1}{3}\pi\right)\right)\frac{di_{as}}{dt} + \left(\frac{1}{2}\bar{L}_m + L_{\Delta m}\cos 2\theta_r\right)\frac{di_{cs}}{dt}$$

$$\left. -2L_{\Delta m}\omega_r\left(i_{as}\sin 2\left(\theta_r - \frac{1}{3}\pi\right) + i_{bs}\sin 2\left(\theta_r - \frac{2}{3}\pi\right) + i_{cs}\sin 2\theta_r\right)\right],$$

$$\frac{di_{cs}}{dt} = \frac{1}{L_{ls} + \bar{L}_m - L_{\Delta m}\cos 2\left(\theta_r + \frac{2}{3}\pi\right)}\left[-r_s i_{cs} + u_{cs}\right.$$

$$+ \left(\frac{1}{2}\bar{L}_m + L_{\Delta m}\cos 2\left(\theta_r + \frac{1}{3}\pi\right)\right)\frac{di_{as}}{dt} + \left(\frac{1}{2}\bar{L}_m + L_{\Delta m}\cos 2\theta_r\right)\frac{di_{bs}}{dt}$$

$$\left. -2L_{\Delta m}\omega_r\left(i_{as}\sin 2\left(\theta_r + \frac{1}{3}\pi\right) + i_{bs}\sin 2\theta_r + i_{cs}\sin 2\left(\theta_r + \frac{2}{3}\pi\right)\right)\right],$$

$$\frac{d\omega_r}{dt} = \frac{P^2}{4J}L_{\Delta m}\left[i_{as}^2\sin 2\theta_r + 2i_{as}i_{bs}\sin 2\left(\theta_r - \frac{1}{3}\pi\right) + 2i_{as}i_{cs}\sin 2\left(\theta_r + \frac{1}{3}\pi\right)\right.$$

$$\left. + i_{bs}^2\sin 2\left(\theta_r - \frac{2}{3}\pi\right) + 2i_{bs}i_{cs}\sin 2\theta_r + i_{cs}^2\sin 2\left(\theta_r + \frac{2}{3}\pi\right)\right] - \frac{B_m}{J}\omega_r - \frac{P}{2J}T_L,$$

$$\frac{d\theta_r}{dt} = \omega_r. \tag{6.5}$$

Using the MATLAB Symbolic Math Toolbox, an m-file was developed to find the differential equations for nonlinear circuitry-electromagnetic dynamics in Cauchy's form. We use the following notations

$$Lbm = \bar{L}_m, \quad Ldm = L_{\Delta m}, \quad Lls = L_{ls}, \quad rs = r_s, \quad Bm = B_m,$$

$$S1 = \sin 2\theta_r, \quad S2 = \sin 2\left(\theta_r - \frac{1}{3}\pi\right), \quad S3 = \sin 2\left(\theta_r + \frac{1}{3}\pi\right),$$

$$S4 = \sin 2\left(\theta_r - \frac{2}{3}\pi\right), \quad S5 = \sin 2(\theta_r + \pi), \quad S6 = \sin 2\left(\theta_r + \frac{2}{3}\pi\right)$$

and

$$C1 = \cos 2\theta_r, \quad C2 = \cos 2\left(\theta_r - \frac{1}{3}\pi\right), \quad C3 = \cos 2\left(\theta_r + \frac{1}{3}\pi\right),$$

$$C4 = \cos 2\left(\theta_r - \frac{2}{3}\pi\right), \quad C5 = \cos 2(\theta_r + \pi), \quad C6 = \cos 2\left(\theta_r + \frac{2}{3}\pi\right).$$

The analytic results are derived by running the following file

```
L=sym('[Lls+Lbm-Ldm*C1,-Lbm/2-Ldm*C2,-Lbm/2-Ldm*C3,0;-Lbm/2-Ldm*C2,Lls+Lbm-Ldm*C4,
  -Lbm/2-Ldm*C5,0;-Lbm/2-Ldm*C3,-Lbm/2-Ldm*C5,Lls+Lbm-Ldm*C6,0;0,0,0,2*J/P] ');
R=sym('[-rs,0,0,0;0,-rs,0,0;0,0,-rs,0;0,0,0,-2*Bm/P]');
I=sym('[Ias; Ibs; Ics; Wr] ');
V=sym('[vas;vbs;vcs;-TL] ');
K=sym('[Ldm*2*Wr*(S1*Ias+S2*Ibs+S3*Ics); Ldm*2*Wr*(S2*Ias+S4Ibs+S5*Ics);
Ldm*2*Wr*(S3*Ias+S5*Ibs+S6*Ics);Te] ');
L1=inv(L); L2=simplify(L1);
FS1=L2*R*I; FS2=simplify(FS1)
FS3=L2*V; FS4=simplify(FS3)
FS5=L2*K; FS6=simplify(FS5)
FS7=FS2+FS4-FS6; FS=simplify(FS7)
```

and utilizing the trigonometric identities. The resulting nonlinear differential equations obtained are

$$\frac{di_{as}}{dt} = \frac{1}{L_D}\left[\left(r_s i_{as} - u_{as}\right)\left(4L_{ls}^2 + 3\bar{L}_m^2 - 3L_{\Delta m}^2 + 8\bar{L}_m L_{ls} - 4L_{ls}L_{\Delta m}\cos 2\theta_r\right)\right.$$

$$+ \left(r_s i_{bs} - u_{bs}\right)\left(3\bar{L}_m^2 - 3L_{\Delta m}^2 + 2\bar{L}_m L_{ls} + 4L_{ls}L_{\Delta m}\cos 2\left(\theta_r - \frac{1}{3}\pi\right)\right)$$

$$+ \left(r_s i_{cs} - u_{cs}\right)\left(3\bar{L}_m^2 - 3L_{\Delta m}^2 + 2\bar{L}_m L_{ls} + 4L_{ls}L_{\Delta m}\cos 2\left(\theta_r + \frac{1}{3}\pi\right)\right)$$

$$+ 6\sqrt{3}L_{\Delta m}^2 L_{ls}\omega_r\left(i_{cs} - i_{bs}\right)$$

$$+ \left(8L_{\Delta m}L_{ls}^2\omega_r + 12L_{\Delta m}\bar{L}_m L_{ls}\omega_r\right)\left(\sin 2\theta_r i_{as} + \sin 2\left(\theta_r - \frac{1}{3}\pi\right)i_{bs}\right.$$

$$\left. + \sin 2\left(\theta_r + \frac{1}{3}\pi\right)i_{cs}\right],$$

$$\frac{di_{bs}}{dt} = \frac{1}{L_D}\left[(r_s i_{as} - u_{as})\left(3\bar{L}_m^2 - 3L_{\Delta m}^2 + 2\bar{L}_m L_{ls} + 4L_{ls}L_{\Delta m}\cos 2\left(\theta_r - \frac{1}{3}\pi\right)\right)\right.$$

$$+ (r_s i_{bs} - u_{bs})\left(4L_{ls}^2 + 3\bar{L}_m^2 - 3L_{\Delta m}^2 + 8\bar{L}_m L_{ls} - 4L_{ls}L_{\Delta m}\cos 2\left(\theta_r + \frac{1}{3}\pi\right)\right)$$

$$+ (r_s i_{cs} - u_{cs})\left(3\bar{L}_m^2 - 3L_{\Delta m}^2 + 2\bar{L}_m L_{ls} + 4L_{ls}L_{\Delta m}\cos 2\theta_r\right)$$

$$+ 6\sqrt{3}L_{\Delta m}^2 L_{ls}\omega_r(i_{as} - i_{cs}) + \left(8L_{\Delta m}L_{ls}^2\omega_r + 12L_{\Delta m}\bar{L}_m L_{ls}\omega_r\right)\left(\sin 2\left(\theta_r - \frac{1}{3}\pi\right)i_{as}\right.$$

$$+ \left.\left.\sin 2\left(\theta_r + \frac{1}{3}\pi\right)i_{bs} + \sin 2\theta_r i_{cs}\right)\right],$$

$$\frac{di_{cs}}{dt} = \frac{1}{L_D}\left[(r_s i_{as} - u_{as})\left(3\bar{L}_m^2 - 3L_{\Delta m}^2 + 2\bar{L}_m L_{ls} + 4L_{ls}L_{\Delta m}\cos 2\left(\theta_r + \frac{1}{3}\pi\right)\right)\right.$$

$$+ (r_s i_{bs} - u_{bs})\left(3\bar{L}_m^2 - 3L_{\Delta m}^2 + 2\bar{L}_m L_{ls} + 4L_{ls}L_{\Delta m}\cos 2\theta_r\right)$$

$$+ (r_s i_{cs} - u_{cs})\left(4L_{ls}^2 + 3\bar{L}_m^2 - 3L_{\Delta m}^2 + 8\bar{L}_m L_{ls} - 4L_{ls}L_{\Delta m}\cos 2\left(\theta_r - \frac{1}{3}\pi\right)\right)$$

$$+ 6\sqrt{3}L_{\Delta m}^2 L_{ls}\omega_r(i_{bs} - i_{as}) + \left(8L_{\Delta m}L_{ls}^2\omega_r + 12L_{\Delta m}\bar{L}_m L_{ls}\omega_r\right)\left(\sin 2\left(\theta_r + \frac{1}{3}\pi\right)i_{as}\right.$$

$$+ \left.\left.\sin 2\theta_r i_{bs} + \sin 2\left(\theta_r - \frac{1}{3}\pi\right)i_{cs}\right)\right],$$

$$\frac{d\omega_r}{dt} = \frac{P^2}{4J}L_{\Delta m}\left(i_{as}^2 \sin 2\theta_r + 2i_{as}i_{bs}\sin 2\left(\theta_r - \frac{1}{3}\pi\right) + 2i_{as}i_{cs}\sin 2\left(\theta_r + \frac{1}{3}\pi\right)\right.$$

$$+ \left.i_{bs}^2 \sin 2\left(\theta_r - \frac{2}{3}\pi\right) + 2i_{bs}i_{cs}\sin 2\theta_r + i_{cs}^2 \sin 2\left(\theta_r + \frac{2}{3}\pi\right)\right) - \frac{B_m}{J}\omega_r - \frac{P}{2J}T_L,$$

$$\frac{d\theta_r}{dt} = \omega_r. \tag{6.6}$$

In these differential equations, the following notations are used

$$\bar{L}_m = \frac{1}{3}(L_{mq} + L_{md}), \; L_{\Delta m} = \frac{1}{3}(L_{md} - L_{mq}) \text{ and } L_D = L_{ls}\left(9L_{\Delta m}^2 - 4L_{ls}^2 - 12\bar{L}_m L_{ls} - 9\bar{L}_m^2\right).$$

Highly nonlinear differential equations (6.5) and (6.6) describe the dynamics of synchronous reluctance motors in the *machine* (*abc*) variables. The analytic solution of these equations is virtually impossible. The nonlinear simulations and analysis can be performed in MATLAB.

Synchronous reluctance synchronous machines can be described in the
qd0 variables. In the rotor reference frame, we apply the Park transformation.
From Table 5.1, we have

$$\mathbf{u}^r_{qd0s} = \mathbf{K}^r_s \mathbf{u}_{abcs}, \quad \mathbf{i}^r_{qd0s} = \mathbf{K}^r_s \mathbf{i}_{abcs}, \quad \boldsymbol{\psi}^r_{qd0s} = \mathbf{K}^r_s \boldsymbol{\psi}_{abcs},$$

$$\mathbf{K}^r_s = \frac{2}{3}\begin{bmatrix} \cos\theta_r & \cos\left(\theta_r - \frac{2}{3}\pi\right) & \cos\left(\theta_r + \frac{2}{3}\pi\right) \\ \sin\theta_r & \sin\left(\theta_r - \frac{2}{3}\pi\right) & \sin\left(\theta_r + \frac{2}{3}\pi\right) \\ \frac{1}{2} & \frac{1}{2} & \frac{1}{2} \end{bmatrix},$$

where $u_{qs}, u_{ds}, u_{0s}, i_{qs}, i_{ds}, i_{0s}$ and $\psi_{qs}, \psi_{ds}, \psi_{0s}$ are the *qd0* components of voltages, currents, and flux linkages.

Using the circuitry-electromagnetic (6.2) and *torsional-mechanical* (6.3)
dynamics, as well as the expression for T_e (6.4), one finds the following
nonlinear differential equations to describe synchronous reluctance motors
in the rotor reference frame

$$\frac{di^r_{qs}}{dt} = -\frac{r_s}{L_{ls}+L_{mq}}i^r_{qs} - \frac{L_{ls}+L_{md}}{L_{ls}+L_{mq}}i^r_{ds}\omega_r + \frac{1}{L_{ls}+L_{mq}}u^r_{qs},$$

$$\frac{di^r_{ds}}{dt} = -\frac{r_s}{L_{ls}+L_{md}}i^r_{ds} + \frac{L_{ls}+L_{mq}}{L_{ls}+L_{md}}i^r_{qs}\omega_r + \frac{1}{L_{ls}+L_{md}}u^r_{ds},$$

$$\frac{di^r_{0s}}{dt} = -\frac{r_s}{L_{ls}}i^r_{0s} + \frac{1}{L_{ls}}u^r_{0s},$$

$$\frac{d\omega_r}{dt} = \frac{3P^2}{8J}(L_{md}-L_{mq})i^r_{qs}i^r_{ds} - \frac{B_m}{J}\omega_r - \frac{P}{2J}T_L,$$

$$\frac{d\theta_r}{dt} = \omega_r. \tag{6.7}$$

The models in the rotor and synchronous reference frames are identical
because $\omega_r = \omega_e$.

To attain the balanced operation, the *quadrature*- and *direct*-axis components of currents and voltages must be derived. This problem is solved by
applying the *direct* Park transformations

$$\mathbf{i}^r_{qd0s} = \mathbf{K}^r_s \mathbf{i}_{abcs} \quad \text{and} \quad \mathbf{u}^r_{qd0s} = \mathbf{K}^r_s \mathbf{u}_{abcs}.$$

By using the three-phase balanced voltage set, one finds

$$u^r_{qs} = \sqrt{2}u_M, \quad u^r_{ds} = 0, \quad \text{and} \quad u^r_{0s} = 0.$$

We derived the mathematical models of three-phase synchronous reluctance motors. Using these resulting differential equations, nonlinear analysis can be performed. The phase currents and voltages which guarantee the balance operation were found using the electromagnetic features. As was reported for induction motors, ac electromechanical motion devices are controlled by applying the phase voltages u_{as}, u_{bs} and u_{cs}. Therefore, the $qd0$-centered modeling and control concepts should be applied with a complete understanding. Control of ac machines using the *arbitrary* (stationary, rotor, and synchronous) reference frame results in the need for advanced DSP to perform the Park transformations in real-time. Those issues were discussed in Example 5.5.

Example 6.3

The nonlinear simulations and analysis are performed by using the differential equations of three-phase synchronous reluctance motors in the *machine* variables (6.6). The four-pole, 110 V, 400 rad/sec, 40 kW motor parameters are: $r_s = 0.01$ ohm, $L_{md} = 0.0012$ H, $L_{mq} = 0.0002$ H, $J = 0.6$ kg-m², and $B_m = 0.003$ N-m-sec/rad.

The three-phase voltage set

$$u_{as} = \sqrt{2}u_M \sin\left(\theta_r + \frac{1}{3}\varphi_i\pi\right), \quad u_{bs} = \sqrt{2}u_M \sin\left(\theta_r - \frac{1}{3}(2 - \varphi_i)\pi\right),$$

$$u_{cs} = \sqrt{2}u_M \sin\left(\theta_r + \frac{1}{3}(2 + \varphi_i)\pi\right),$$

is supplied. The magnitude of the phase voltages is $u_M = 110$ V and $\varphi_u = 0.3882$. Figure 6.9 illustrates the transient dynamics for the angular velocity if the

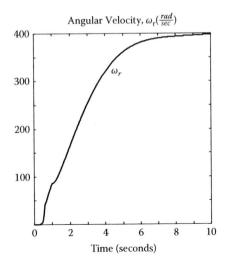

FIGURE 6.9
Angular velocity dynamics.

motor accelerates from the stall and the load torque 10 N-m is applied at 1 sec. The settling time is 10 sec and the steady-state angular velocity is 400 rad/sec.

□

6.3 Radial Topology Permanent-Magnet Synchronous Machines

Consider the radial topology permanent-magnet synchronous machines. These motion devices surpass other electric machines. High efficiency, high power and torque densities, superior overloading, robustness, expanded operating envelope, and other features contribute to superior performance and excellent capabilities. The studied permanent-magnet synchronous motion devices are brushless electric machines because the excitation flux is produced by permanent magnets placed on the rotor. The images of permanent-magnet synchronous machines used in computer hard drives and VHS devices are illustrated in Figure 6.10. The permanent-magnet synchronous motors are frequently called the "brushless DC motors" perhaps because of the similarity of the torque-speed characteristics. However, the basic electromagnetic and operating principles of permanent-magnet DC and synchronous machines are fundamentally distinct. The terminology "brushless DC motor" could be very deceptive; therefore it is not used in this book.

6.3.1 Two-Phase Permanent-Magnet Synchronous Motors and Stepper Motors

Consider a radial topology two-phase permanent-magnet synchronous motor. Using Kirchhoff's voltage law, we have the following two equations

$$u_{as} = r_s i_{as} + \frac{d\psi_{as}}{dt}, \quad u_{bs} = r_s i_{bs} + \frac{d\psi_{bs}}{dt}, \tag{6.8}$$

FIGURE 6.10
Permanent-magnet synchronous machines.

where the flux linkages are expressed as

$$\psi_{as} = L_{asas}i_{as} + L_{asbs}i_{bs} + \psi_{asm} \quad \text{and} \quad \psi_{bs} = L_{bsas}i_{as} + L_{bsbs}i_{bs} + \psi_{bsm}.$$

Here, u_{as} and u_{bs} are the phase voltages supplied to the stator windings *as* and *bs*; i_{as} and i_{bs} are the phase currents; ψ_{as} and ψ_{bs} are the stator flux linkages; r_s is the resistances of the stator windings; L_{asas} and L_{bsbs} are the self-inductances; L_{asbs} and L_{bsas} are the mutual inductances.

The flux linkages are periodic functions of the angular displacement (rotor position). Let

$$\psi_{asm} = \psi_m \sin\theta_r \quad \text{and} \quad \psi_{bsm} = -\psi_m \cos\theta_r.$$

The self-inductances of the stator windings are

$$L_{ss} = L_{asas} = L_{bsbs} = L_{ls} + \bar{L}_m.$$

The stator windings are displaced by 90 electrical degrees. Hence, the mutual inductances between the stator windings are $L_{asbs} = L_{bsas} = 0$. We have

$$\psi_{as} = L_{ss}i_{as} + \psi_m \sin\theta_r \quad \text{and} \quad \psi_{bs} = L_{ss}i_{bs} - \psi_m \cos\theta_r.$$

One finds

$$u_{as} = r_s i_{as} + \frac{d(L_{ss}i_{as} + \psi_m \sin\theta_r)}{dt} = r_s i_{as} + L_{ss}\frac{di_{as}}{dt} + \psi_m \omega_r \cos\theta_r$$

$$u_{bs} = r_s i_{bs} + \frac{d(L_{ss}i_{bs} - \psi_m \cos\theta_r)}{dt} = r_s i_{bs} + L_{ss}\frac{di_{bs}}{dt} - \psi_m \omega_r \sin\theta_r. \tag{6.9}$$

Using Newton's second law

$$J\frac{d^2\theta_{rm}}{dt^2} = T_e - B_m\omega_{rm} - T_L,$$

we have

$$\frac{d\omega_{rm}}{dt} = \frac{1}{J}(T_e - B_m\omega_{rm} - T_L),$$

$$\frac{d\theta_{rm}}{dt} = \omega_{rm}. \tag{6.10}$$

The expression for the electromagnetic torque developed by permanent-magnet motors is obtained by using the coenergy

$$W_c = \frac{1}{2}\left(L_{ss}i_{as}^2 + L_{ss}i_{bs}^2\right) + \psi_m i_{as} \sin\theta_r - \psi_m i_{bs} \cos\theta_r + W_{PM}.$$

From

$$T_e = \frac{\partial W_c}{\partial \theta_r}$$

one finds

$$T_e = \frac{P\psi_m}{2}(i_{as}\cos\theta_r + i_{bs}\sin\theta_r) \tag{6.11}$$

Integrating the circuitry-electromagnetic equations (6.9) with the *torsional-mechanical* dynamics (6.10), and using (6.11), one finds

$$\frac{di_{as}}{dt} = -\frac{r_s}{L_{ss}}i_{as} - \frac{\psi_m}{L_{ss}}\omega_r\cos\theta_r + \frac{1}{L_{ss}}u_{as},$$

$$\frac{di_{bs}}{dt} = -\frac{r_s}{L_{ss}}i_{bs} + \frac{\psi_m}{L_{ss}}\omega_r\sin\theta_r + \frac{1}{L_{ss}}u_{bs},$$

$$\frac{d\omega_r}{dt} = \frac{P^2\psi_m}{4J}(i_{as}\cos\theta_r + i_{bs}\sin\theta_r) - \frac{B_m}{J}\omega_r - \frac{P}{2J}T_L,$$

$$\frac{d\theta_r}{dt} = \omega_r. \tag{6.12}$$

For two-phase permanent-magnet synchronous motors (assuming the sinusoidal winding distributions and the sinusoidal *mmf* waveforms), the electromagnetic torque was found in (6.11). Hence, to guarantee the balanced operation, one feeds

$$i_{as} = \sqrt{2}i_M\cos\theta_r \quad \text{and} \quad i_{bs} = \sqrt{2}i_M\sin\theta_r.$$

The phase voltages to be applied are

$$u_{as} = \sqrt{2}u_M\cos\theta_r \quad \text{and} \quad u_{bs} = \sqrt{2}u_M\sin\theta_r.$$

We maximize the electromagnetic torque obtaining

$$T_e = \frac{P\psi_m}{2}\sqrt{2}i_M\left(\cos^2\theta_r + \sin^2\theta_r\right) = \frac{P\psi_m}{\sqrt{2}}i_M.$$

For two-phase permanent-magnet synchronous motors, one may use the mechanical angular displacement (rotor displacement) θ_{rm}, which is related to the electrical angular displacement as $\theta_{rm} = 2\theta_r/P$. The equations can be easily refined using θ_{rm}. One may use θ_{rm} because two-phase permanent-magnet synchronous motors are commonly utilized as the stepper motors which usually operate in the open-loop configuration. One rotates the stepper motor step-by-step by properly "energizing" the windings by supplying u_{as} and u_{bs}. The full-, half-, and mini-step operations can be achieved.

The stepper motors are designed with a high P. For example, if $P = 50$, one easily achieves 3.6 degree rotation using the full-step operation.

The permanent-magnet synchronous motors with high P develop high electromagnetic torque, whereas the mechanical angular velocity is relatively low because $\omega_{rm} = 2\omega_r/P$. These motors are effectively utilized as direct drives and servos. This direct connection of motors without matching mechanical coupling allows one to achieve a remarkable level of efficiency, reliability, and performance.

Permanent-magnet synchronous motors are controlled to ensure the desired dynamics, steady-state operation, stability, precision, disturbance attenuation, and so on. For stepper motors, illustrated in Figures 6.2d and 6.11, one "energizes" the stator windings supplying the proper sequence of u_{as} and u_{bs}, and rotor rotates in the *counterclockwise* or *clockwise* direction because of the T_e developed. By supplying u_{as} and u_{bs}, one achieves the angular increment rotor displacement equal to full- or half-step. The rotor repositioning (instead of using angular displacement or velocity, usually, the number of "steps per second" is used) is regulated by changing the frequency of the voltages supplied to the phase windings. Because of the possibility to operate stepper motors in the open-loop mode sequentially supplying u_{as} and u_{bs}, some warnings should be stated. The stepper motor can *miss* the step (or steps) if: (1) the instantaneous $T_{e\,inst}$ is not sufficient; (2) $T_{e\,inst} < T_L$; (3) u_{as} and u_{bs} are supplied at high frequency with respect to the dynamic features defined by L_{ss}, J, and other parameters; and so on. Furthermore, the stepper motor can *pass* the step (or steps) if: (1) $T_e \gg T_L$; (2) high kinetic energy is stored in the moving rotor because of high J; and so on. Other factors contribute to *missing* or *passing* steps such as varying J, varying and potentially bidirectional T_L, disturbances, parameter variations, and so on. Therefore, open-loop electromechanical systems with stepper motors are utilized if $T_L \approx$ const and $J \approx$ const deriving the phase voltage switching by coherently examining the system dynamics, disturbances and parameter variations. The images of the stepper motor are reported in Figure 6.11.

For stepper motors, the electrical angular velocity and displacement are found using the number of rotor tooth RT, for example, $\omega_r = RT\omega_{rm}$ and $\theta_r = RT\theta_{rm}$. The flux linkages are the functions of the number of the rotor tooth and displacement. Let

$$\psi_{asm} = \psi_m \cos(RT\theta_{rm}) \quad \text{and} \quad \psi_{bsm} = \psi_m \sin(RT\theta_{rm}).$$

We have

$$\psi_{as} = L_{ss}i_{as} + \psi_m \cos(RT\theta_{rm}) \quad \text{and} \quad \psi_{bs} = L_{ss}i_{bs} + \psi_m \sin(RT\theta_{rm}).$$

Equation (6.8) yields

$$u_{as} = r_s i_{as} + \frac{d(L_{ss}i_{as} + \psi_m \cos(RT\theta_{rm}))}{dt} = r_s i_{as} + L_{ss}\frac{di_{as}}{dt} - RT\psi_m\omega_{rm}\sin(RT\theta_{rm}),$$

$$u_{bs} = r_s i_{bs} + \frac{d(L_{ss}i_{bs} + \psi_m \sin(RT\theta_{rm}))}{dt} = r_s i_{bs} + L_{ss}\frac{di_{bs}}{dt} + RT\psi_m\omega_{rm}\cos(RT\theta_{rm}).$$

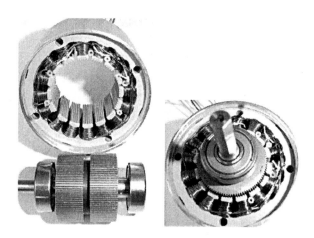

FIGURE 6.11
Stepper motor.

Therefore,

$$\frac{di_{as}}{dt} = -\frac{r_s}{L_{ss}} i_{as} + \frac{RT\psi_m}{L_{ss}} \omega_{rm} \sin(RT\theta_{rm}) + \frac{1}{L_{ss}} u_{as},$$

$$\frac{di_{bs}}{dt} = -\frac{r_s}{L_{ss}} i_{bs} - \frac{RT\psi_m}{L_{ss}} \omega_{rm} \cos(RT\theta_{rm}) + \frac{1}{L_{ss}} u_{bs}. \tag{6.13}$$

The expression for T_e is found using the coenergy

$$W_c = \frac{1}{2}\left(L_{ss} i_{as}^2 + L_{ss} i_{bs}^2\right) + \psi_m i_{as} \cos(RT\theta_{rm}) + \psi_m i_{bs} \sin(RT\theta_{rm}) + W_{PM}.$$

We have

$$T_e = \frac{\partial W_c}{\partial \theta_{rm}} = RT\psi_m [-i_{as} \sin(RT\theta_{rm}) + i_{bs} \cos(RT0_{rm})]. \tag{6.14}$$

Using Newton's second law (6.10) and (6.14), the behavior of the rotor angular velocity and displacement are described by

$$\frac{d\omega_{rm}}{dt} = \frac{RT\psi_m}{J} [-i_{as} \sin(RT\theta_{rm}) + i_{bs} \cos(RT\theta_{rm})] - \frac{B_m}{J} \omega_{rm} - \frac{1}{J} T_L,$$

$$\frac{d\theta_{rm}}{dt} = \omega_{rm}. \tag{6.15}$$

From (6.13) and (6.15), one has

$$\frac{di_{as}}{dt} = -\frac{r_s}{L_{ss}}i_{as} + \frac{RT\psi_m}{L_{ss}}\omega_{rm}\sin(RT\theta_{rm}) + \frac{1}{L_{ss}}u_{as},$$

$$\frac{di_{bs}}{dt} = -\frac{r_s}{L_{ss}}i_{bs} - \frac{RT\psi_m}{L_{ss}}\omega_{rm}\cos(RT\theta_{rm}) + \frac{1}{L_{ss}}u_{bs},$$

$$\frac{d\omega_{rm}}{dt} = \frac{RT\psi_m}{J}[-i_{as}\sin(RT\theta_{rm}) + i_{bs}\cos(RT\theta_{rm})] - \frac{B_m}{J}\omega_{rm} - \frac{1}{J}T_L,$$

$$\frac{d\theta_{rm}}{dt} = \omega_{rm}. \tag{6.16}$$

Using (6.16), an s-domain diagram is developed and illustrated in Figure 6.12.

The analysis of (6.14) leads one to the expressions for a balanced two-phase current sinusoidal set

$$i_{as} = -\sqrt{2}i_M\sin(RT\theta_{rm}) \quad \text{and} \quad i_{bs} = \sqrt{2}i_M\cos(RT\theta_{rm}),$$

which results in

$$T_e = \sqrt{2}RT\psi_m i_M.$$

The phase voltages u_{as} and u_{bs} should be supplied as functions of the rotor angular displacement

$$u_{as} = -\sqrt{2}u_M\sin(RT\theta_{rm}) \quad \text{and} \quad u_{bs} = \sqrt{2}u_M\cos(RT\theta_{rm})$$

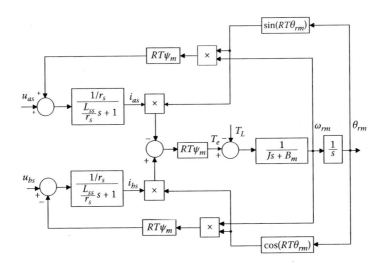

FIGURE 6.12
The s-domain diagram of permanent-magnet stepper motors.

to maximize the electromagnetic torque. However, to implement current or voltage sets, one needs to measure (or observe) the angular displacement using sensors (or observers which use the induced *emf* or phase current measurements). The Hall sensors are commonly used to measure the rotor angular displacement in permanent-magnet synchronous motors. Alternatively, to eliminate the need for sensors or observers (sensorless control), stepper motors can operate in the open-loop configuration by supplying the phase voltages with the allowable frequency ω_a. For example, without measuring θ_{rm}, one, for example, may supply the pulses

$$u_{as} = -\sqrt{2}u_M \, \text{sgn}(\sin(\omega_a t)) \text{ and } u_{as} = \sqrt{2}u_M \, \text{sgn}\left(\sin\left(\omega_a t - \frac{1}{2}\pi\right)\right).$$

Having examined two-phase permanent-magnet synchronous motors, including stepper motors, in the *machine* variable, we study stepper motors in the *qd* quantities. The rotor and synchronous reference frames are among the most frequently applied. The results of using rotor and synchronous reference frames are the same because $\omega_e = \omega_r$. In the *machine* variables, Kirchhoff's voltage law results in two nonlinear differential equations (6.13). We apply the *direct* Park formation as reported in Table 5.2, Homework Problem 5.3. In the rotor reference frame, one has

$$\begin{bmatrix} u_{qs}^r \\ u_{ds}^r \end{bmatrix} = \begin{bmatrix} -\sin(RT\theta_{rm}) & \cos(RT\theta_{rm}) \\ \cos(RT\theta_{rm}) & \sin(RT\theta_{rm}) \end{bmatrix} \begin{bmatrix} u_{as} \\ u_{bs} \end{bmatrix},$$

$$\begin{bmatrix} i_{qs}^r \\ i_{ds}^r \end{bmatrix} = \begin{bmatrix} -\sin(RT\theta_{rm}) & \cos(RT\theta_{rm}) \\ \cos(RT\theta_{rm}) & \sin(RT\theta_{rm}) \end{bmatrix} \begin{bmatrix} i_{as} \\ i_{bs} \end{bmatrix}.$$

The following differential equations in the *qd* quantities result

$$u_{qs}^r = r_s i_{qs}^r + L_{ss}\frac{di_{qs}^r}{dt} + RT\psi_m\omega_{rm} + RTL_{ss}i_{ds}^r\omega_{rm},$$

$$u_{ds}^r = r_s i_{ds}^r + L_{ss}\frac{di_{ds}^r}{dt} - RTL_{ss}i_{qs}^r\omega_{rm}.$$

Hence,

$$\frac{di_{qs}^r}{dt} = -\frac{r_s}{L_{ss}}i_{qs}^r - \frac{RT\psi_m}{L_{ss}}\omega_{rm} - RTi_{ds}^r\omega_{rm} + \frac{1}{L_{ss}}u_{qs}^r,$$

$$\frac{di_{ds}^r}{dt} = -\frac{r_s}{L_{ss}}i_{ds}^r + RTi_{qs}^r\omega_{rm} + \frac{1}{L_{ss}}u_{ds}^r. \tag{6.17}$$

From the derived T_e in (6.14), using the *inverse* Park transformation

$$\begin{bmatrix} i_{as} \\ i_{bs} \end{bmatrix} = \begin{bmatrix} -\sin(RT\theta_{rm}) & \cos(RT\theta_{rm}) \\ \cos(RT\theta_{rm}) & \sin(RT\theta_{rm}) \end{bmatrix} \begin{bmatrix} i_{qs}^r \\ i_{ds}^r \end{bmatrix},$$

we have $T_e = RT\psi_m i_{qs}^r$.

From Newton's second law of motions (6.15), one finds

$$\frac{d\omega_{rm}}{dt} = \frac{RT\psi_m}{J} i_{qs}^r - \frac{B_m}{J} \omega_{rm} - \frac{1}{J} T_L,$$

$$\frac{d\theta_{rm}}{dt} = \omega_{rm}. \tag{6.18}$$

Differential equations (6.17) and (6.18) yield the model of permanent-magnet synchronous motors in the rotor reference frame as

$$\frac{di_{qs}^r}{dt} = -\frac{r_s}{L_{ss}} i_{qs}^r - \frac{RT\psi_m}{L_{ss}} \omega_{rm} - RT i_{ds}^r \omega_{rm} + \frac{1}{L_{ss}} u_{qs}^r,$$

$$\frac{di_{ds}^r}{dt} = -\frac{r_s}{L_{ss}} i_{ds}^r + RT i_{qs}^r \omega_{rm} + \frac{1}{L_{ss}} u_{ds}^r,$$

$$\frac{d\omega_{rm}}{dt} = \frac{RT\psi_m}{J} i_{qs}^r - \frac{B_m}{J} \omega_{rm} - \frac{1}{J} T_L,$$

$$\frac{d\theta_{rm}}{dt} = \omega_{rm}. \tag{6.19}$$

The phase voltages supplied to the *as* and *bs* windings must be found. Recalling

$$i_{as} = -\sqrt{2} i_M \sin(RT\theta_{rm}), \quad i_{bs} = \sqrt{2} i_M \cos(RT\theta_{rm}),$$

and

$$u_{as} = -\sqrt{2} u_M \sin(RT\theta_{rm}), \quad u_{bs} = \sqrt{2} u_M \cos(RT\theta_{rm}),$$

we apply the Park transformations

$$\begin{bmatrix} i_{qs}^r \\ i_{ds}^r \end{bmatrix} = \begin{bmatrix} -\sin(RT\theta_{rm}) & \cos(RT\theta_{rm}) \\ \cos(RT\theta_{rm}) & \sin(RT\theta_{rm}) \end{bmatrix} \begin{bmatrix} i_{as} \\ i_{bs} \end{bmatrix}$$

and

$$\begin{bmatrix} u^r_{qs} \\ u^r_{ds} \end{bmatrix} = \begin{bmatrix} -\sin(RT\theta_{rm}) & \cos(RT\theta_{rm}) \\ \cos(RT\theta_{rm}) & \sin(RT\theta_{rm}) \end{bmatrix} \begin{bmatrix} u_{as} \\ u_{bs} \end{bmatrix}.$$

From

$$i^r_{qs} = -i_{as}\sin(RT\theta_{rm}) + i_{bs}\cos(RT\theta_{rm}), \quad i^r_{ds} = i_{as}\cos(RT\theta_{rm}) + i_{bs}\sin(RT\theta_{rm}),$$

one finds

$$i^r_{qs} = \sqrt{2}i_M \sin^2(RT\theta_{rm}) + \sqrt{2}i_M \cos^2(RT\theta_{rm}) = \sqrt{2}i_M$$

and

$$i^r_{ds} = -\sqrt{2}i_M \sin(RT\theta_{rm})\cos(RT\theta_{rm}) + \sqrt{2}i_M \sin(RT\theta_{rm})\cos(RT\theta_{rm}) = 0.$$

Hence,

$$i^r_{qs} = \sqrt{2}i_M \quad \text{and} \quad i^r_{ds} = 0.$$

Similarly, the *quadrature* and *direct* voltage components are

$$u^r_{qs} = \sqrt{2}u_M \quad \text{and} \quad u^r_{ds} = 0.$$

To the author's best knowledge, stepper motors never have been controlled using the qd voltage or current components because the phase voltages u_{as} and u_{bs} must be supplied. The derived u^r_{qs} and u^r_{ds} illustrate a very limited practicality and obscurity of the application of the qd quantities.

To control the stepper motor output (angular velocity or displacement), the power amplifiers "energize" the as and bs windings ensuring the full step or other operations. As an example, the Motorola monolithic MC3479 Stepper Motor Driver (6 V and 0.35 A) can be used [5]. This driver is designed to drive a two-phase stepper motor bidirectionally. The pin connection for the plastic 648C case, representative block diagram, circuitry, and timing/output diagram are shown in Figure 6.13.

This MC3479 driver is designed to drive a stepper minimotors in various applications such as positioning tables, disk drives, small robots, and so on. As illustrated in Figure 6.13, the H-bridge topology power stages supply the phase voltages u_{as} and u_{bs} to the motor windings (only one coil with terminals L1 and L2 is shown in Figure 6.13). The applied voltage polarity depends on which transistor (Q_H or Q_L) is *on*, and these transistors are driven by the signals from the decoding circuitry. The maximum sink current is a function of the resistor between Pin 6 and ground. When the outputs are in a high impedance state, both transistors (Q_H or Q_L) are *off*. The pin V_D provides a current path for the winding (coil) current during transients (switching) in order to attenuate the *back emf* (voltage) spikes. Pin V_D is normally connected to V_M (Pin 16) through a diode, resistor, or directly. The peaks instantaneous

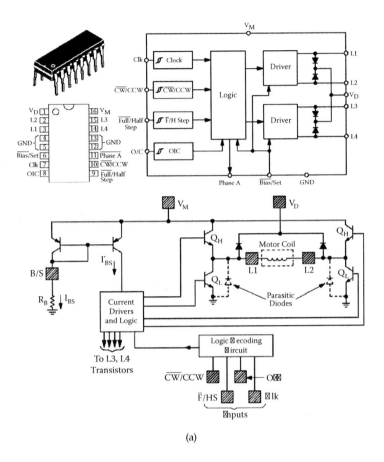

(a)

FIGURE 6.13
Motorola monolithic MC3479 Stepper Motor Driver (copyright of Motorola, used with permission) [5]. (Continues on following page.)

voltage at the outputs must not exceed the value V_M, which is 6 V. The parasitic diodes across Q_L of each output results in a circuit path for the switched current. When the input is at a Logic "0" (less than 0.8 V), the outputs correspond to a full step operation with each clock cycle. The direction depends on the CW/CCW input. There are four switching phases for each cycle of the sequencing logic. Phase voltages are supplied, and currents i_{as} and i_{bs} flow in the motor windings. For a Logic "1" (more than 2 V), the outputs change a half step during each clock cycle. Eight switching phases result for each complete cycle of the sequencing logic. The output sequences and timing diagrams are shown in Figure 6.13. A complete description, application notes, and important details are available from Motorola. We very briefly covered the MC3479 Stepper Motor Driver; and other specialized ICs and motor controllers/drivers exist.

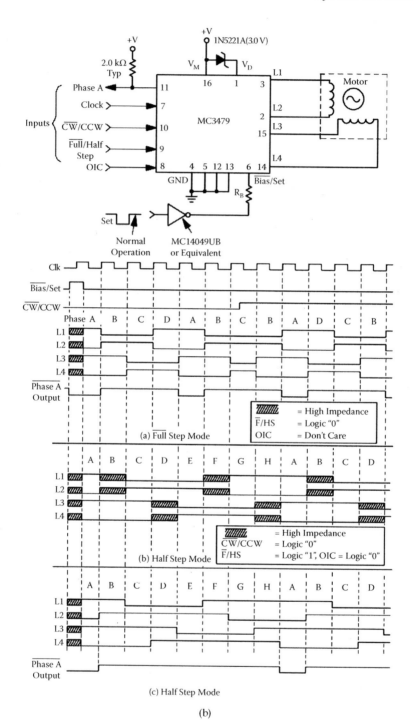

FIGURE 6.13
(Continued).

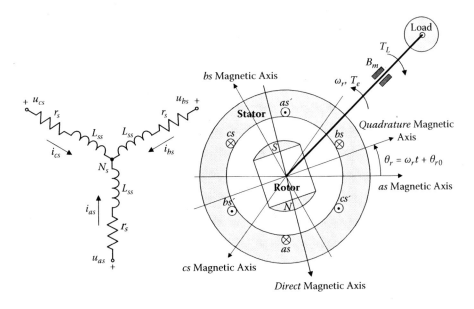

FIGURE 6.14
Two-pole permanent-magnet synchronous motor.

6.3.2 Radial Topology Three-Phase Permanent-Magnet Synchronous Machines

Three-phase two-pole permanent-magnet synchronous machines (motors and generators) are illustrated in Figures 6.14 and 6.15.

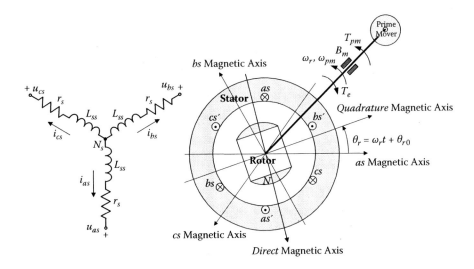

FIGURE 6.15
Three-phase wye-connected synchronous generator.

From Kirchhoff's second law, one obtains three differential equations for the *as*, *bs*, and *cs* stator windings as

$$u_{as} = r_s i_{as} + \frac{d\psi_{as}}{dt},$$

$$u_{bs} = r_s i_{bs} + \frac{d\psi_{bs}}{dt},$$

$$u_{cs} = r_s i_{cs} + \frac{d\psi_{cs}}{dt}. \tag{6.20}$$

From (6.20), one yields

$$\mathbf{u}_{abcs} = \mathbf{r}_s \mathbf{i}_{abcs} + \frac{d\mathbf{\psi}_{abcs}}{dt}, \quad \begin{bmatrix} u_{as} \\ u_{bs} \\ u_{cs} \end{bmatrix} = \begin{bmatrix} r_s & 0 & 0 \\ 0 & r_s & 0 \\ 0 & 0 & r_s \end{bmatrix} \begin{bmatrix} i_{as} \\ i_{bs} \\ i_{cs} \end{bmatrix} + \begin{bmatrix} \dfrac{d\psi_{as}}{dt} \\ \dfrac{d\psi_{bs}}{dt} \\ \dfrac{d\psi_{cs}}{dt} \end{bmatrix}.$$

The flux linkages are

$$\psi_{as} = L_{asas} i_{as} + L_{asbs} i_{bs} + L_{ascs} i_{cs} + \psi_{asm},$$

$$\psi_{bs} = L_{bsas} i_{as} + L_{bsbs} i_{bs} + L_{bscs} i_{cs} + \psi_{bsm},$$

$$\psi_{cs} = L_{csas} i_{as} + L_{csbs} i_{bs} + L_{cscs} i_{cs} + \psi_{csm}.$$

The flux linkages ψ_{asm}, ψ_{bsm}, and ψ_{csm}, established by the permanent magnet, are periodic functions of θ_r. The stator windings are displaced by 120 electrical degrees. Denoting the magnitude of the flux linkages established by the permanent magnet as ψ_m, assume that ψ_{asm}, ψ_{bsm}, and ψ_{csm} vary as

$$\psi_{asm} = \psi_m \sin\theta_r, \quad \psi_{bsm} = \psi_m \sin\left(\theta_r - \tfrac{2}{3}\pi\right), \quad \text{and} \quad \psi_{csm} = \psi_m \sin\left(\theta_r + \frac{2}{3}\pi\right).$$

Self- and mutual inductances for three-phase permanent-magnet synchronous machines can be derived. In particular, the equations for the magnetizing *quadrature* and *direct* inductances are

$$L_{mq} = \frac{N_s^2}{\Re_{mq}} \quad \text{and} \quad L_{md} = \frac{N_s^2}{\Re_{md}}.$$

The *quadrature* and *direct* magnetizing reluctances can be different, and $\Re_{mq} > \Re_{md}$. Hence, we have $L_{mq} < L_{md}$.

The minimum value of L_{asas} occurs periodically at $\theta_r = 0,\ \pi,\ 2\pi,\ldots$, while the maximum value of L_{asas} occurs at $\theta_r = \frac{1}{2}\pi,\ \frac{3}{2}\pi,\ \frac{5}{2}\pi,\ldots$. One concludes that $L_{asas}(\theta_r)$, which is $L_{ls} + L_{mq} \leq L_{asas} \leq L_{ls} + L_{md}$, is a periodic function of θ_r. Hence, $L_{asas}(\theta_r)$ varies as a sine function and has a constant component. We let

$$L_{asas} = L_{ls} + \bar{L}_m - L_{\Delta m}\cos 2\theta_r,$$

where $\bar{L}_m$ is the average value of the magnetizing inductance; $L_{\Delta m}$ is half the amplitude of the sinusoidal variation of the magnetizing inductance.

The relationships between L_{mq}, L_{md} and $\bar{L}_m, L_{\Delta m}$ are

$$L_{mq} = \frac{3}{2}(\bar{L}_m - L_{\Delta m}) \quad \text{and} \quad L_{md} = \frac{3}{2}(\bar{L}_m + L_{\Delta m}).$$

Therefore,

$$\bar{L}_m = \frac{1}{3}(L_{mq} + L_{md}) \quad \text{and} \quad L_{\Delta m} = \frac{1}{3}(L_{md} - L_{mq}).$$

Using the expressions for L_{mq} and L_{md}, we have

$$\bar{L}_m = \frac{1}{3}\left(\frac{N_s^2}{\mathfrak{R}_{mq}} + \frac{N_s^2}{\mathfrak{R}_{md}}\right) \quad \text{and} \quad L_{\Delta m} = \frac{1}{3}\left(\frac{N_s^2}{\mathfrak{R}_{md}} - \frac{N_s^2}{\mathfrak{R}_{mq}}\right).$$

The following equations for the flux linkages result

$$\psi_{as} = (L_{ls} + \bar{L}_m - L_{\Delta m}\cos 2\theta_r)i_{as} + \left(-\frac{1}{2}\bar{L}_m - L_{\Delta m}\cos 2\left(\theta_r - \frac{1}{3}\pi\right)\right)i_{bs}$$

$$+ \left(-\frac{1}{2}\bar{L}_m - L_{\Delta m}\cos 2\left(\theta_r + \frac{1}{3}\pi\right)\right)i_{cs} + \psi_m \sin\theta_r,$$

$$\psi_{bs} = \left(-\frac{1}{2}\bar{L}_m - L_{\Delta m}\cos 2\left(\theta_r - \frac{1}{3}\pi\right)\right)i_{as} + \left(L_{ls} + \bar{L}_m - L_{\Delta m}\cos 2\left(\theta_r - \frac{2}{3}\pi\right)\right)i_{bs}$$

$$+ \left(-\frac{1}{2}\bar{L}_m - L_{\Delta m}\cos 2\theta_r\right)i_{cs} + \psi_m \sin\left(\theta_r - \frac{2}{3}\pi\right),$$

$$\psi_{cs} = \left(-\frac{1}{2}\bar{L}_m - L_{\Delta m}\cos 2\left(\theta_r + \frac{1}{3}\pi\right)\right)i_{as} + \left(-\frac{1}{2}\bar{L}_m - L_{\Delta m}\cos 2\theta_r\right)i_{bs}$$

$$+ \left(L_{ls} + \bar{L}_m - L_{\Delta m}\cos 2\left(\theta_r + \frac{2}{3}\pi\right)\right)i_{cs} + \psi_m \sin\left(\theta_r + \frac{2}{3}\pi\right). \tag{6.21}$$

From (6.21), one has

$$\boldsymbol{\psi}_{abcs} = \mathbf{L}_s \mathbf{i}_{abcs} + \boldsymbol{\psi}_m$$

$$
= \begin{bmatrix}
L_{ls} + \bar{L}_m - L_{\Delta m} \cos 2\theta_r & -\frac{1}{2}\bar{L}_m - L_{\Delta m} \cos 2\left(\theta_r - \frac{1}{3}\pi\right) & -\frac{1}{2}\bar{L}_m - L_{\Delta m} \cos 2\left(\theta_r + \frac{1}{3}\pi\right) \\
-\frac{1}{2}\bar{L}_m - L_{\Delta m} \cos 2\left(\theta_r - \frac{1}{3}\pi\right) & L_{ls} + \bar{L}_m - L_{\Delta m} \cos 2\left(\theta_r - \frac{2}{3}\pi\right) & -\frac{1}{2}\bar{L}_m - L_{\Delta m} \cos 2\theta_r \\
-\frac{1}{2}\bar{L}_m - L_{\Delta m} \cos 2\left(\theta_r + \frac{1}{3}\pi\right) & -\frac{1}{2}\bar{L}_m - L_{\Delta m} \cos 2\theta_r & L_{ls} + \bar{L}_m - L_{\Delta m} \cos 2\left(\theta_r + \frac{2}{3}\pi\right)
\end{bmatrix}
$$

$$
\times \begin{bmatrix} i_{as} \\ i_{bs} \\ i_{cs} \end{bmatrix} + \psi_m \begin{bmatrix} \sin\theta_r \\ \sin\left(\theta_r - \frac{2}{3}\pi\right) \\ \sin\left(\theta_r + \frac{2}{3}\pi\right) \end{bmatrix}.
$$

The inductance mapping $\mathbf{L}_s(\theta_r)$ is

$$
\mathbf{L}_s = \begin{bmatrix}
L_{ls} + \bar{L}_m - L_{\Delta m} \cos 2\theta_r & -\frac{1}{2}\bar{L}_m - L_{\Delta m} \cos 2\left(\theta_r - \frac{1}{3}\pi\right) & -\frac{1}{2}\bar{L}_m - L_{\Delta m} \cos 2\left(\theta_r + \frac{1}{3}\pi\right) \\
-\frac{1}{2}\bar{L}_m - L_{\Delta m} \cos 2\left(\theta_r - \frac{1}{3}\pi\right) & L_{ls} + \bar{L}_m - L_{\Delta m} \cos 2\left(\theta_r - \frac{2}{3}\pi\right) & -\frac{1}{2}\bar{L}_m - L_{\Delta m} \cos 2\theta_r \\
-\frac{1}{2}\bar{L}_m - L_{\Delta m} \cos 2\left(\theta_r + \frac{1}{3}\pi\right) & -\frac{1}{2}\bar{L}_m - L_{\Delta m} \cos 2\theta_r & L_{ls} + \bar{L}_m - L_{\Delta m} \cos 2\left(\theta_r + \frac{2}{3}\pi\right)
\end{bmatrix}.
$$

It was shown that

$$\bar{L}_m = \frac{1}{3}\left(\frac{N_s^2}{\Re_{mq}} + \frac{N_s^2}{\Re_{md}}\right) \quad \text{and} \quad L_{\Delta m} = \frac{1}{3}\left(\frac{N_s^2}{\Re_{md}} - \frac{N_s^2}{\Re_{mq}}\right).$$

Permanent-magnet synchronous motion devices are round-rotor electric machines (the magnetic paths in the *quadrature* and *direct* magnetic axes are identical, yielding $\Re_{mq} = \Re_{md}$). Thus,

$$\bar{L}_m = \frac{2N_s^2}{3\Re_{mq}} = \frac{2N_s^2}{3\Re_{md}}, \quad L_{\Delta m} = 0 \quad \text{and} \quad \bar{L}_m = L_{ss} - L_{ls}.$$

From the inductance mapping $\mathbf{L}_s(\theta_r)$, one obtains the inductance matrix as

$$
\mathbf{L}_s = \begin{bmatrix}
L_{ls} + \bar{L}_m & -\frac{1}{2}\bar{L}_m & -\frac{1}{2}\bar{L}_m \\
-\frac{1}{2}\bar{L}_m & L_{ls} + \bar{L}_m & -\frac{1}{2}\bar{L}_m \\
-\frac{1}{2}\bar{L}_m & -\frac{1}{2}\bar{L}_m & L_{ls} + \bar{L}_m
\end{bmatrix}.
$$

The expressions for the flux linkages (6.21) are simplified to

$$\psi_{as} = (L_{ls} + \bar{L}_m)i_{as} - \frac{1}{2}\bar{L}_m i_{bs} - \frac{1}{2}\bar{L}_m i_{cs} + \psi_m \sin\theta_r,$$

$$\psi_{bs} = -\frac{1}{2}\bar{L}_m i_{as} + (L_{ls} + \bar{L}_m)i_{bs} - \frac{1}{2}\bar{L}_m i_{cs} + \psi_m \sin\left(\theta_r - \frac{2}{3}\pi\right),$$

$$\psi_{cs} = -\frac{1}{2}\bar{L}_m i_{as} - \frac{1}{2}\bar{L}_m i_{bs} + (L_{ls} + \bar{L}_m)i_{cs} + \psi_m \sin\left(\theta_r + \frac{2}{3}\pi\right), \qquad (6.22)$$

or in matrix form

$$\psi_{abcs} = \mathbf{L}_s \mathbf{i}_{abcs} + \psi_m = \begin{bmatrix} L_{ls} + \bar{L}_m & -\frac{1}{2}\bar{L}_m & -\frac{1}{2}\bar{L}_m \\ -\frac{1}{2}\bar{L}_m & L_{ls} + \bar{L}_m & -\frac{1}{2}\bar{L}_m \\ -\frac{1}{2}\bar{L}_m & -\frac{1}{2}\bar{L}_m & L_{ls} + \bar{L}_m \end{bmatrix} \begin{bmatrix} i_{as} \\ i_{bs} \\ i_{cs} \end{bmatrix} + \psi_m \begin{bmatrix} \sin\theta_r \\ \sin\left(\theta_r - \frac{2}{3}\pi\right) \\ \sin\left(\theta_r + \frac{2}{3}\pi\right) \end{bmatrix}.$$

Using (6.20) and (6.22), we have

$$\mathbf{u}_{abcs} = \mathbf{r}_s \mathbf{i}_{abcs} + \frac{d\psi_{abcs}}{dt} = \mathbf{r}_s \mathbf{i}_{abcs} + \mathbf{L}_s \frac{d\mathbf{i}_{abcs}}{dt} + \frac{d\psi_m}{dt},$$

where

$$\frac{d\psi_m}{dt} = \psi_m \begin{bmatrix} \cos\theta_r \omega_r \\ \cos\left(\theta_r - \frac{2}{3}\pi\right)\omega_r \\ \cos\left(\theta_r + \frac{2}{3}\pi\right)\omega_r \end{bmatrix}.$$

Cauchy's form differential equations can be found by making use of $\mathbf{L}_s^{-1}$. In particular,

$$\frac{d\mathbf{i}_{abcs}}{dt} = -\mathbf{L}_s^{-1}\mathbf{r}_s\mathbf{i}_{abcs} - \mathbf{L}_s^{-1}\frac{d\psi_m}{dt} + \mathbf{L}_s^{-1}\mathbf{u}_{abcs}.$$

The circuitry-electromagnetic dynamics is given as

$$
\begin{bmatrix} \dfrac{di_{as}}{dt} \\[2mm] \dfrac{di_{bs}}{dt} \\[2mm] \dfrac{di_{cs}}{dt} \end{bmatrix} = \begin{bmatrix} -\dfrac{r_s(2L_{ss}-\bar{L}_m)}{2L_{ss}^2-L_{ss}\bar{L}_m-\bar{L}_m^2} & -\dfrac{r_s\bar{L}_m}{2L_{ss}^2-L_{ss}\bar{L}_m-\bar{L}_m^2} & -\dfrac{r_s\bar{L}_m}{2L_{ss}^2-L_{ss}\bar{L}_m-\bar{L}_m^2} \\[4mm] -\dfrac{r_s\bar{L}_m}{2L_{ss}^2-L_{ss}\bar{L}_m-\bar{L}_m^2} & -\dfrac{r_s(2L_{ss}-\bar{L}_m)}{2L_{ss}^2-L_{ss}\bar{L}_m-\bar{L}_m^2} & -\dfrac{r_s\bar{L}_m}{2L_{ss}^2-L_{ss}\bar{L}_m-\bar{L}_m^2} \\[4mm] -\dfrac{r_s\bar{L}_m}{2L_{ss}^2-L_{ss}\bar{L}_m-\bar{L}_m^2} & -\dfrac{r_s\bar{L}_m}{2L_{ss}^2-L_{ss}\bar{L}_m-\bar{L}_m^2} & -\dfrac{r_s(2L_{ss}-\bar{L}_m)}{2L_{ss}^2-L_{ss}\bar{L}_m-\bar{L}_m^2} \end{bmatrix} \begin{bmatrix} i_{as} \\[2mm] i_{bs} \\[2mm] i_{cs} \end{bmatrix}
$$

$$
+ \begin{bmatrix} -\dfrac{\psi_m(2L_{ss}-\bar{L}_m)}{2L_{ss}^2-L_{ss}\bar{L}_m-\bar{L}_m^2} & -\dfrac{\psi_m\bar{L}_m}{2L_{ss}^2-L_{ss}\bar{L}_m-\bar{L}_m^2} & -\dfrac{\psi_m\bar{L}_m}{2L_{ss}^2-L_{ss}\bar{L}_m-\bar{L}_m^2} \\[4mm] -\dfrac{\psi_m\bar{L}_m}{2L_{ss}^2-L_{ss}\bar{L}_m-\bar{L}_m^2} & -\dfrac{\psi_m(2L_{ss}-\bar{L}_m)}{2L_{ss}^2-L_{ss}\bar{L}_m-\bar{L}_m^2} & -\dfrac{\psi_m\bar{L}_m}{2L_{ss}^2-L_{ss}\bar{L}_m-\bar{L}_m^2} \\[4mm] -\dfrac{\psi_m\bar{L}_m}{2L_{ss}^2-L_{ss}\bar{L}_m-\bar{L}_m^2} & -\dfrac{\psi_m\bar{L}_m}{2L_{ss}^2-L_{ss}\bar{L}_m-\bar{L}_m^2} & -\dfrac{\psi_m(2L_{ss}-\bar{L}_m)}{2L_{ss}^2-L_{ss}\bar{L}_m-\bar{L}_m^2} \end{bmatrix} \begin{bmatrix} \omega_r\cos\theta_r \\[2mm] \omega_r\cos\left(\theta_r-\frac{2}{3}\pi\right) \\[2mm] \omega_r\cos\left(\theta_r+\frac{2}{3}\pi\right) \end{bmatrix}
$$

$$
+ \begin{bmatrix} \dfrac{2L_{ss}-\bar{L}_m}{2L_{ss}^2-L_{ss}\bar{L}_m-\bar{L}_m^2} & \dfrac{\bar{L}_m}{2L_{ss}^2-L_{ss}\bar{L}_m-\bar{L}_m^2} & \dfrac{\bar{L}_m}{2L_{ss}^2-L_{ss}\bar{L}_m-\bar{L}_m^2} \\[4mm] \dfrac{\bar{L}_m}{2L_{ss}^2-L_{ss}\bar{L}_m-\bar{L}_m^2} & \dfrac{2L_{ss}-\bar{L}_m}{2L_{ss}^2-L_{ss}\bar{L}_m-\bar{L}_m^2} & \dfrac{\bar{L}_m}{2L_{ss}^2-L_{ss}\bar{L}_m-\bar{L}_m^2} \\[4mm] \dfrac{\bar{L}_m}{2L_{ss}^2-L_{ss}\bar{L}_m-\bar{L}_m^2} & \dfrac{\bar{L}_m}{2L_{ss}^2-L_{ss}\bar{L}_m-\bar{L}_m^2} & \dfrac{2L_{ss}-\bar{L}_m}{2L_{ss}^2-L_{ss}\bar{L}_m-\bar{L}_m^2} \end{bmatrix} \begin{bmatrix} u_{as} \\[2mm] u_{bs} \\[2mm] u_{cs} \end{bmatrix}.
$$

In expanded form, we have the following nonlinear differential equations, which describe the circuitry-electromagnetic transient behavior

$$
\frac{di_{as}}{dt} = -\frac{r_s(2L_{ss}-\bar{L}_m)}{2L_{ss}^2-L_{ss}\bar{L}_m-\bar{L}_m^2}i_{as} - \frac{r_s\bar{L}_m}{2L_{ss}^2-L_{ss}\bar{L}_m-\bar{L}_m^2}i_{bs} - \frac{r_s\bar{L}_m}{2L_{ss}^2-L_{ss}\bar{L}_m-\bar{L}_m^2}i_{cs}
$$

$$
-\frac{\psi_m(2L_{ss}-\bar{L}_m)}{2L_{ss}^2-L_{ss}\bar{L}_m-\bar{L}_m^2}\omega_r\cos\theta_r - \frac{\psi_m\bar{L}_m}{2L_{ss}^2-L_{ss}\bar{L}_m-\bar{L}_m^2}\omega_r\cos\left(\theta_r-\frac{2}{3}\pi\right)
$$

$$
-\frac{\psi_m\bar{L}_m}{2L_{ss}^2-L_{ss}\bar{L}_m-\bar{L}_m^2}\omega_r\cos\left(\theta_r+\frac{2}{3}\pi\right) + \frac{2L_{ss}-\bar{L}_m}{2L_{ss}^2-L_{ss}\bar{L}_m-\bar{L}_m^2}u_{as}
$$

$$
+\frac{\bar{L}_m}{2L_{ss}^2-L_{ss}\bar{L}_m-\bar{L}_m^2}u_{bs} + \frac{\bar{L}_m}{2L_{ss}^2-L_{ss}\bar{L}_m-\bar{L}_m^2}u_{cs},
$$

$$\frac{di_{bs}}{dt} = -\frac{r_s\overline{L}_m}{2L_{ss}^2 - L_{ss}\overline{L}_m - \overline{L}_m^2}i_{as} - \frac{r_s(2L_{ss} - \overline{L}_m)}{2L_{ss}^2 - L_{ss}\overline{L}_m - \overline{L}_m^2}i_{bs} - \frac{r_s\overline{L}_m}{2L_{ss}^2 - L_{ss}\overline{L}_m - \overline{L}_m^2}i_{cs}$$

$$-\frac{\psi_m\overline{L}_m}{2L_{ss}^2 - L_{ss}\overline{L}_m - \overline{L}_m^2}\omega_r\cos\theta_r - \frac{\psi_m(2L_{ss} - \overline{L}_m)}{2L_{ss}^2 - L_{ss}\overline{L}_m - \overline{L}_m^2}\omega_r\cos\left(\theta_r - \frac{2}{3}\pi\right)$$

$$-\frac{\psi_m\overline{L}_m}{2L_{ss}^2 - L_{ss}\overline{L}_m - \overline{L}_m^2}\omega_r\cos\left(\theta_r + \frac{2}{3}\pi\right) + \frac{\overline{L}_m}{2L_{ss}^2 - L_{ss}\overline{L}_m - \overline{L}_m^2}u_{as}$$

$$+\frac{2L_{ss} - \overline{L}_m}{2L_{ss}^2 - L_{ss}\overline{L}_m - \overline{L}_m^2}u_{bs} + \frac{\overline{L}_m}{2L_{ss}^2 - L_{ss}\overline{L}_m - \overline{L}_m^2}u_{cs},$$

$$\frac{di_{cs}}{dt} = -\frac{r_s\overline{L}_m}{2L_{ss}^2 - L_{ss}\overline{L}_m - \overline{L}_m^2}i_{as} - \frac{r_s\overline{L}_m}{2L_{ss}^2 - L_{ss}\overline{L}_m - \overline{L}_m^2}i_{bs} - \frac{r_s(2L_{ss} - \overline{L}_m)}{2L_{ss}^2 - L_{ss}\overline{L}_m - \overline{L}_m^2}i_{cs}$$

$$-\frac{\psi_m\overline{L}_m}{2L_{ss}^2 - L_{ss}\overline{L}_m - \overline{L}_m^2}\omega_r\cos\theta_r - \frac{\psi_m\overline{L}_m}{2L_{ss}^2 - L_{ss}\overline{L}_m - \overline{L}_m^2}\omega_r\cos\left(\theta_r - \frac{2}{3}\pi\right)$$

$$-\frac{\psi_m(2L_{ss} - \overline{L}_m)}{2L_{ss}^2 - L_{ss}\overline{L}_m - \overline{L}_m^2}\omega_r\cos\left(\theta_r + \frac{2}{3}\pi\right) + \frac{\overline{L}_m}{2L_{ss}^2 - L_{ss}\overline{L}_m - \overline{L}_m^2}u_{as}$$

$$+\frac{\overline{L}_m}{2L_{ss}^2 - L_{ss}\overline{L}_m - \overline{L}_m^2}u_{bs} + \frac{2L_{ss} - \overline{L}_m}{2L_{ss}^2 - L_{ss}\overline{L}_m - \overline{L}_m^2}u_{cs}. \tag{6.23}$$

The transient behavior of the mechanical system must be used. One cannot solve (6.23), where the electrical angular velocity ω_r and angular displacement θ_r are the state variables.

Newton's second law

$$J\frac{d^2\theta_{rm}}{dt^2} = T_e - B_m\omega_{rm} - T_L$$

yields a set of two differential equations

$$\frac{d\omega_{rm}}{dt} = \frac{1}{J}(T_e - B_m\omega_{rm} - T_L) \quad \text{and} \quad \frac{d\theta_{rm}}{dt} = \omega_{rm}.$$

The expression for the electromagnetic torque is found using the coenergy

$$W_c = \frac{1}{2}\begin{bmatrix} i_{as} & i_{bs} & i_{cs} \end{bmatrix}\mathbf{L}_s\begin{bmatrix} i_{as} \\ i_{bs} \\ i_{cs} \end{bmatrix} + \begin{bmatrix} i_{as} & i_{bs} & i_{cs} \end{bmatrix}\begin{bmatrix} \psi_m\sin\theta_r \\ \psi_m\sin\left(\theta_r - \frac{2}{3}\pi\right) \\ \psi_m\sin\left(\theta_r + \frac{2}{3}\pi\right) \end{bmatrix} + W_{PM},$$

where W_{PM} is the energy stored in the permanent magnet.

For round-rotor synchronous machines

$$
\mathbf{L}_s = \begin{bmatrix} L_{ls} + \bar{L}_m & -\frac{1}{2}\bar{L}_m & -\frac{1}{2}\bar{L}_m \\ -\frac{1}{2}\bar{L}_m & L_{ls}' + \bar{L}_m & -\frac{1}{2}\bar{L}_m \\ -\frac{1}{2}\bar{L}_m & -\frac{1}{2}\bar{L}_m & L_{ls} + \bar{L}_m \end{bmatrix}.
$$

The inductance matrix $\mathbf{L}_s$ and W_{PM} are not functions of θ_r. One obtains the following formula for the electromagnetic torque for three-phase P-pole permanent-magnet synchronous motors

$$
T_e = \frac{P}{2}\frac{\partial W_c}{\partial \theta_r} = \frac{P\psi_m}{2}\left(i_{as}\cos\theta_r + i_{bs}\cos\left(\theta_r - \frac{2}{3}\pi\right) + i_{cs}\cos\left(\theta_r + \frac{2}{3}\pi\right)\right). \quad (6.24)
$$

Therefore,

$$
\frac{d\omega_{rm}}{dt} = \frac{P\psi_m}{2J}\left(i_{as}\cos\theta_r + i_{bs}\cos\left(\theta_r - \frac{2}{3}\pi\right) + i_{cs}\cos\left(\theta_r + \frac{2}{3}\pi\right)\right) - \frac{B_m}{J}\omega_{rm} - \frac{1}{J}T_L,
$$

$$
\frac{d\theta_{rm}}{dt} = \omega_{rm}.
$$

Using the electrical angular velocity ω_r and displacement θ_r, related to the mechanical angular velocity and displacement as

$$
\omega_{rm} = \frac{2}{P}\omega_r \quad \text{and} \quad \theta_{rm} = \frac{2}{P}\theta_r,
$$

the following differential equations to describe the *torsional-mechanical* transient dynamics result

$$
\frac{d\omega_r}{dt} = \frac{P^2\psi_m}{4J}\left(i_{as}\cos\theta_r + i_{bs}\cos\left(\theta_r - \frac{2}{3}\pi\right) + i_{cs}\cos\left(\theta_r + \frac{2}{3}\pi\right)\right) - \frac{B_m}{J}\omega_r - \frac{P}{2J}T_L,
$$

$$
\frac{d\theta_r}{dt} = \omega_r. \quad\quad (6.25)
$$

From (6.23) and (6.25), one obtains a nonlinear mathematical model of permanent-magnet synchronous motors in Cauchy's form as given by a system of five highly nonlinear differential equations

$$
\frac{di_{as}}{dt} = -\frac{r_s(2L_{ss}-\bar{L}_m)}{2L_{ss}^2 - L_{ss}\bar{L}_m - \bar{L}_m^2} i_{as} - \frac{r_s\bar{L}_m}{2L_{ss}^2 - L_{ss}\bar{L}_m - \bar{L}_m^2} i_{bs} - \frac{r_s\bar{L}_m}{2L_{ss}^2 - L_{ss}\bar{L}_m - \bar{L}_m^2} i_{cs}
$$

$$
- \frac{\psi_m(2L_{ss}-\bar{L}_m)}{2L_{ss}^2 - L_{ss}\bar{L}_m - \bar{L}_m^2} \omega_r \cos\theta_r - \frac{\psi_m\bar{L}_m}{2L_{ss}^2 - L_{ss}\bar{L}_m - \bar{L}_m^2} \omega_r \cos\left(\theta_r - \frac{2}{3}\pi\right)
$$

$$
- \frac{\psi_m\bar{L}_m}{2L_{ss}^2 - L_{ss}\bar{L}_m - \bar{L}_m^2} \omega_r \cos\left(\theta_r + \frac{2}{3}\pi\right) + \frac{2L_{ss}-\bar{L}_m}{2L_{ss}^2 - L_{ss}\bar{L}_m - \bar{L}_m^2} u_{as}
$$

$$
+ \frac{\bar{L}_m}{2L_{ss}^2 - L_{ss}\bar{L}_m - \bar{L}_m^2} u_{bs} + \frac{\bar{L}_m}{2L_{ss}^2 - L_{ss}\bar{L}_m - \bar{L}_m^2} u_{cs},
$$

$$
\frac{di_{bs}}{dt} = -\frac{r_s\bar{L}_m}{2L_{ss}^2 - L_{ss}\bar{L}_m - \bar{L}_m^2} i_{as} - \frac{r_s(2L_{ss}-\bar{L}_m)}{2L_{ss}^2 - L_{ss}\bar{L}_m - \bar{L}_m^2} i_{bs} - \frac{r_s\bar{L}_m}{2L_{ss}^2 - L_{ss}\bar{L}_m - \bar{L}_m^2} i_{cs}
$$

$$
- \frac{\psi_m\bar{L}_m}{2L_{ss}^2 - L_{ss}\bar{L}_m - \bar{L}_m^2} \omega_r \cos\theta_r - \frac{\psi_m(2L_{ss}-\bar{L}_m)}{2L_{ss}^2 - L_{ss}\bar{L}_m - \bar{L}_m^2} \omega_r \cos\left(\theta_r - \frac{2}{3}\pi\right)
$$

$$
- \frac{\psi_m\bar{L}_m}{2L_{ss}^2 - L_{ss}\bar{L}_m - \bar{L}_m^2} \omega_r \cos\left(\theta_r + \frac{2}{3}\pi\right) + \frac{\bar{L}_m}{2L_{ss}^2 - L_{ss}\bar{L}_m - \bar{L}_m^2} u_{as}
$$

$$
+ \frac{2L_{ss}-\bar{L}_m}{2L_{ss}^2 - L_{ss}\bar{L}_m - \bar{L}_m^2} u_{bs} + \frac{\bar{L}_m}{2L_{ss}^2 - L_{ss}\bar{L}_m - \bar{L}_m^2} u_{cs},
$$

$$
\frac{di_{cs}}{dt} = -\frac{r_s\bar{L}_m}{2L_{ss}^2 - L_{ss}\bar{L}_m - \bar{L}_m^2} i_{as} - \frac{r_s\bar{L}_m}{2L_{ss}^2 - L_{ss}\bar{L}_m - \bar{L}_m^2} i_{bs} - \frac{r_s(2L_{ss}-\bar{L}_m)}{2L_{ss}^2 - L_{ss}\bar{L}_m - \bar{L}_m^2} i_{cs}
$$

$$
- \frac{\psi_m\bar{L}_m}{2L_{ss}^2 - L_{ss}\bar{L}_m - \bar{L}_m^2} \omega_r \cos\theta_r - \frac{\psi_m\bar{L}_m}{2L_{ss}^2 - L_{ss}\bar{L}_m - \bar{L}_m^2} \omega_r \cos\left(\theta_r - \frac{2}{3}\pi\right)
$$

$$
- \frac{\psi_m(2L_{ss}-\bar{L}_m)}{2L_{ss}^2 - L_{ss}\bar{L}_m - \bar{L}_m^2} \omega_r \cos\left(\theta_r + \frac{2}{3}\pi\right) + \frac{\bar{L}_m}{2L_{ss}^2 - L_{ss}\bar{L}_m - \bar{L}_m^2} u_{as}
$$

$$
+ \frac{\bar{L}_m}{2L_{ss}^2 - L_{ss}\bar{L}_m - \bar{L}_m^2} u_{bs} + \frac{2L_{ss}-\bar{L}_m}{2L_{ss}^2 - L_{ss}\bar{L}_m - \bar{L}_m^2} u_{cs},
$$

$$
\frac{d\omega_r}{dt} = \frac{P^2\psi_m}{4J}\left(i_{as}\cos\theta_r + i_{bs}\cos\left(\theta_r - \frac{2}{3}\pi\right) + i_{cs}\cos\left(\theta_r + \frac{2}{3}\pi\right)\right) - \frac{B_m}{J}\omega_r - \frac{P}{2J}T_L,
$$

$$
\frac{d\theta_r}{dt} = \omega_r. \tag{6.26}
$$

The state-space form of (6.26) is

$$
\begin{bmatrix}
\dfrac{di_{as}}{dt} \\[2mm]
\dfrac{di_{bs}}{dt} \\[2mm]
\dfrac{di_{cs}}{dt} \\[2mm]
\dfrac{d\omega_r}{dt} \\[2mm]
\dfrac{d\theta_r}{dt}
\end{bmatrix}
=
\begin{bmatrix}
-\dfrac{r_s(2L_{ss}-\bar{L}_m)}{2L_{ss}^2-L_{ss}\bar{L}_m-\bar{L}_m^2} & -\dfrac{r_s\bar{L}_m}{2L_{ss}^2-L_{ss}\bar{L}_m-\bar{L}_m^2} & -\dfrac{r_s\bar{L}_m}{2L_{ss}^2-L_{ss}\bar{L}_m-\bar{L}_m^2} & 0 & 0 \\[3mm]
-\dfrac{r_s\bar{L}_m}{2L_{ss}^2-L_{ss}\bar{L}_m-\bar{L}_m^2} & -\dfrac{r_s(2L_{ss}-\bar{L}_m)}{2L_{ss}^2-L_{ss}\bar{L}_m-\bar{L}_m^2} & -\dfrac{r_s\bar{L}_m}{2L_{ss}^2-L_{ss}\bar{L}_m-\bar{L}_m^2} & 0 & 0 \\[3mm]
-\dfrac{r_s\bar{L}_m}{2L_{ss}^2-L_{ss}\bar{L}_m-\bar{L}_m^2} & -\dfrac{r_s\bar{L}_m}{2L_{ss}^2-L_{ss}\bar{L}_m-\bar{L}_m^2} & -\dfrac{r_s(2L_{ss}-\bar{L}_m)}{2L_{ss}^2-L_{ss}\bar{L}_m-\bar{L}_m^2} & 0 & 0 \\[3mm]
0 & 0 & 0 & -\dfrac{B_m}{J} & 0 \\[3mm]
0 & 0 & 0 & 1 & 0
\end{bmatrix}
\begin{bmatrix}
i_{as} \\ i_{bs} \\ i_{cs} \\ \omega_r \\ \theta_r
\end{bmatrix}
$$

$$
+
\begin{bmatrix}
-\dfrac{\psi_m(2L_{ss}-\bar{L}_m)}{2L_{ss}^2-L_{ss}\bar{L}_m-\bar{L}_m^2}\omega_r & -\dfrac{\psi_m\bar{L}_m}{2L_{ss}^2-L_{ss}\bar{L}_m-\bar{L}_m^2}\omega_r & -\dfrac{\psi_m\bar{L}_m}{2L_{ss}^2-L_{ss}\bar{L}_m-\bar{L}_m^2}\omega_r \\[3mm]
-\dfrac{\psi_m\bar{L}_m}{2L_{ss}^2-L_{ss}\bar{L}_m-\bar{L}_m^2}\omega_r & -\dfrac{\psi_m(2L_{ss}-\bar{L}_m)}{2L_{ss}^2-L_{ss}\bar{L}_m-\bar{L}_m^2}\omega_r & -\dfrac{\psi_m\bar{L}_m}{2L_{ss}^2-L_{ss}\bar{L}_m-\bar{L}_m^2}\omega_r \\[3mm]
-\dfrac{\psi_m\bar{L}_m}{2L_{ss}^2-L_{ss}\bar{L}_m-\bar{L}_m^2}\omega_r & -\dfrac{\psi_m\bar{L}_m}{2L_{ss}^2-L_{ss}\bar{L}_m-\bar{L}_m^2}\omega_r & -\dfrac{\psi_m(2L_{ss}-\bar{L}_m)}{2L_{ss}^2-L_{ss}\bar{L}_m-\bar{L}_m^2}\omega_r \\[3mm]
\dfrac{P^2\psi_m}{4J}i_{as} & \dfrac{P^2\psi_m}{4J}i_{bs} & \dfrac{P^2\psi_m}{4J}i_{cs} \\[3mm]
0 & 0 & 0
\end{bmatrix}
\begin{bmatrix}
\cos\theta_r \\[2mm]
\cos\left(\theta_r-\dfrac{2}{3}\pi\right) \\[2mm]
\cos\left(\theta_r+\dfrac{2}{3}\pi\right)
\end{bmatrix}
$$

$$
+
\begin{bmatrix}
\dfrac{2L_{ss}-\bar{L}_m}{2L_{ss}^2-L_{ss}\bar{L}_m-\bar{L}_m^2} & \dfrac{\bar{L}_m}{2L_{ss}^2-L_{ss}\bar{L}_m-\bar{L}_m^2} & \dfrac{\bar{L}_m}{2L_{ss}^2-L_{ss}\bar{L}_m-\bar{L}_m^2} \\[3mm]
\dfrac{\bar{L}_m}{2L_{ss}^2-L_{ss}\bar{L}_m-\bar{L}_m^2} & \dfrac{2L_{ss}-\bar{L}_m}{2L_{ss}^2-L_{ss}\bar{L}_m-\bar{L}_m^2} & \dfrac{\bar{L}_m}{2L_{ss}^2-L_{ss}\bar{L}_m-\bar{L}_m^2} \\[3mm]
\dfrac{\bar{L}_m}{2L_{ss}^2-L_{ss}\bar{L}_m-\bar{L}_m^2} & \dfrac{\bar{L}_m}{2L_{ss}^2-L_{ss}\bar{L}_m-\bar{L}_m^2} & \dfrac{2L_{ss}-\bar{L}_m}{2L_{ss}^2-L_{ss}\bar{L}_m-\bar{L}_m^2} \\[3mm]
0 & 0 & 0 \\[3mm]
0 & 0 & 0
\end{bmatrix}
\begin{bmatrix}
u_{as} \\ u_{bs} \\ u_{cs}
\end{bmatrix}
-
\begin{bmatrix}
0 \\ 0 \\ 0 \\ \dfrac{P}{2J} \\ 0
\end{bmatrix}
T_L.
$$

To control the angular velocity, one regulates the currents fed or voltages supplied to the stator *abc* windings. Neglecting the viscous friction coefficient, the analysis of Newton's second law

$$
J\frac{d\omega_{rm}}{dt}=T_e-T_L
$$

indicates that

- The angular velocity ω_{rm} increases (motor accelerates) if $T_e > T_L$,
- The angular velocity ω_{rm} decreases (motor disaccelerates) if $T_e < T_L$,
- The angular velocity ω_{rm} is constant (ω_{rm} = const, e.g., steady-state operation) if $T_e = T_L$.

That is, to regulate motion devices, the electromagnetic torque (6.24) must be changed. A balanced three-phase current set is

$$i_{as} = \sqrt{2}i_M \cos(\omega_r t) = \sqrt{2}i_M \cos(\omega_e t) = \sqrt{2}i_M \cos\theta_r,$$

$$i_{bs} = \sqrt{2}i_M \cos\left(\omega_r t - \frac{2}{3}\pi\right) = \sqrt{2}i_M \cos\left(\omega_e t - \frac{2}{3}\pi\right) = \sqrt{2}i_M \cos\left(\theta_r - \frac{2}{3}\pi\right),$$

$$i_{cs} = \sqrt{2}i_M \cos\left(\omega_r t + \frac{2}{3}\pi\right) = \sqrt{2}i_M \cos\left(\omega_e t + \frac{2}{3}\pi\right) = \sqrt{2}i_M \cos\left(\theta_r + \frac{2}{3}\pi\right).$$

From the trigonometric identity

$$\cos^2\theta_r + \cos^2\left(\theta_r - \frac{2}{3}\pi\right) + \cos^2\left(\theta_r + \frac{2}{3}\pi\right) = \frac{3}{2},$$

one yields

$$T_e = \frac{P\psi_m}{2}\sqrt{2}i_M\left(\cos^2\theta_r + \cos^2\left(\theta_r - \frac{2}{3}\pi\right) + \cos^2\left(\theta_r + \frac{2}{3}\pi\right)\right) = \frac{3P\psi_m}{2\sqrt{2}}i_M.$$

Hence, to ensure the controlled motion and regulate the output mechanical variables (angular velocity and displacement), i_M must be changed. Furthermore, the phase currents i_{as}, i_{bs}, and i_{cs}, which are shifted by $2\pi/3$ are the functions of the electrical angular displacement θ_r, measured by using the Hall-effect sensors.

If the PWM amplifiers are used, one changes the magnitude u_M of the phase voltages u_{as}, u_{bs}, and u_{cs}. The angular displacement θ_r is needed to be measured (or estimated), and the *abc* voltages needed to be supplied are

$$u_{as} = \sqrt{2}u_M \cos(\theta_r + \varphi_u), \quad u_{bs} = \sqrt{2}u_M \cos\left(\theta_r - \frac{2}{3}\pi + \varphi_u\right) \text{ and}$$

$$u_{cs} = \sqrt{2}u_M \cos\left(\theta_r + \frac{2}{3}\pi + \varphi_u\right).$$

Neglecting the circuitry transients (assuming that inductances, delays, transistor transients, and other dynamic behaviors are negligible or small), we have

$$u_{as} = \sqrt{2}u_M \cos\theta_r, \quad u_{bs} = \sqrt{2}u_M \cos\left(\theta_r - \frac{2}{3}\pi\right) \text{ and } u_{cs} = \sqrt{2}u_M \cos\left(\theta_r + \frac{2}{3}\pi\right).$$

Using a set of nonlinear differential equations (6.26), the block diagram is developed and documented in Figure 6.16, where

$$T_s = \frac{r_s(2L_{ss} - \bar{L}_m)}{2L_{ss}^2 - L_{ss}\bar{L}_m - \bar{L}_m^2}.$$

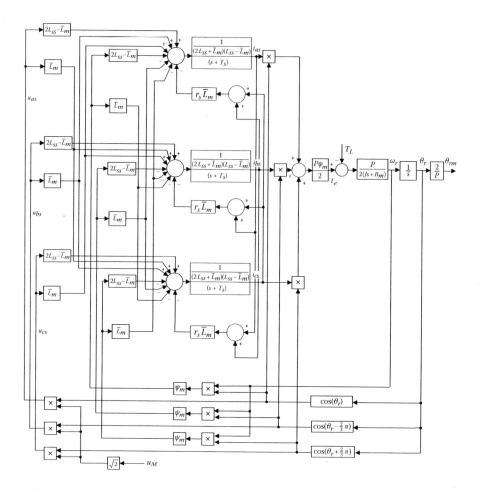

FIGURE 6.16

The s-domain diagram for radial topology three-phase permanent-magnet synchronous motors controlled by applying a three-phase balanced voltage set $u_{as} = \sqrt{2}u_M \cos\theta_r$, $u_{bs} = \sqrt{2}u_M \cos(\theta_r - \frac{2}{3}\pi)$, and $u_{cs} = \sqrt{2}u_M \cos(\theta_r + \frac{2}{3}\pi)$.

Utilizing the basic electromagnetics, we found that to control the angular velocity (in the drive application) of permanent-magnet synchronous motors or the displacement (in servosystem application), one should supply the phase voltages to the stator windings as a function of the angular displacement. The control concepts must be implemented by the hardware. For example, the balance voltage sets can be implemented by the "control logic" which regulates the output stage transistors in PWM amplifiers, while the angular displacement is measured by the Hall-effect sensors. Various PWM amplifiers have been designed and used. The terminals *as, bs* (for two-phase motors) and *as, bs, cs* (for three-phase motors) are connected to the power stage outputs. For three-phase permanent-magnet synchronous motors one varies u_{as}, u_{bs}, and u_{cs} using the measured

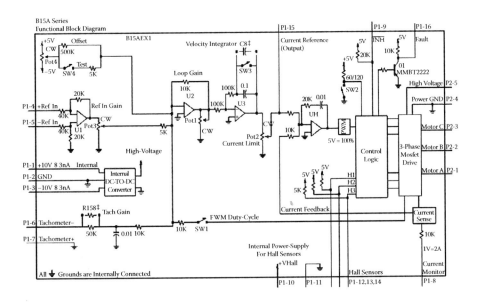

FIGURE 6.17
B15A8 PWM servo amplifier (courtesy of Advanced Motion Controls, http://www.a-m-c.com) [5].

(or observed) rotor angular displacement. Permanent-magnet synchronous electric machines are made in different sizes, and matching high-performance PWM power amplifiers are available. For ~250 W (80 V rated and ~15 A peak) permanent-magnet synchronous motors, the schematics of a B15A8 servo amplifier (20–80 V, ±7.5 A continuous current, ±15 A peak current, 2.5 kHz bandwidth, 129 × 76 × 25 mm dimensions) is documented in Figure 6.17. The motor terminals (phase windings) are connected to P2-1, P2-2, and P2-3. One connects the Hall sensor outputs to P1-12, P1-13, and P1-14. The "control logic" utilizes the measured rotor angular displacement to develop appropriate phase voltages u_{as}, u_{bs} and u_{cs} by driving the transistors. The proportional-integral analog controller is integrated in the amplifier. The reference (command) voltage is supplied to P1-4. The voltage induced by the tachogenerator (proportional to the motor angular velocity) is supplied to P1-6. The reference and measured angular velocities are compared to obtain the tracking error $e(t)$. This $e(t)$ is utilized by the analog proportional-integral controller to develop the control signals which turn transistors *on* and *off*. One can change the proportional and integral feedback gains adjusting the potentiometers (resistors). The amplifier can be used in the servosystem applications, and the angular (or linear) displacement should be supplied to P1-6. Various PWM amplifiers are available from Advanced Motion Controls and other companies.

Small permanent-magnet synchronous motors (from mW to ~1 W) are applied in various applications such as rotating and positioning stages, hard drives, robotics, appliances, and so on. A small ~1 W permanent-magnet synchronous motors, as shown in Figure 6.2c, can be driven by monolithic PWM amplifiers such as MC33035 (40 V and 50 mA) and others [5]. The phase voltages are derived by using the rotor angular displacement decoder. The application-specific ICs drive power MOSFETs as reported in Figure 6.18. The PWM concept is implemented, and the Hall sensor signals are supplied to obtain u_{as}, u_{bs}, and u_{cs}. The representative block diagram, as documented in Figure 6.18a, provides the functional/circuit schematics. The three-phase, six-step full-wave converter topology is implemented. As reported in Figure 6.18b, the closed-loop configuration is realized by integrating the MC33035 and MC33039 utilizing the "Error Amplifier." The proportional controller is implemented. The reader is referred to the Motorola catalogs for detailed specifications, application notes, description, instructions, guidance, and so on.

Example 6.4   Lagrange Equations of Motion and Dynamics of Permanent-Magnet Synchronous Motors

We derived the mathematical model for three-phase permanent-magnet synchronous motors using Kirchhoff's voltage law (to model the

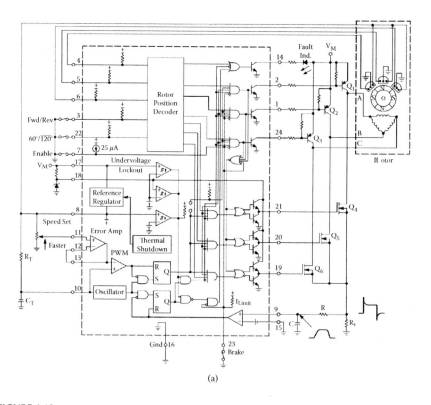

(a)

FIGURE 6.18

Schematics of the MC33035 Brushless DC Motor Controller (permanent-magnet synchronous motor controller) to drive and control permanent-magnet synchronous minimotors (copyright of Motorola, used with permission) [5].

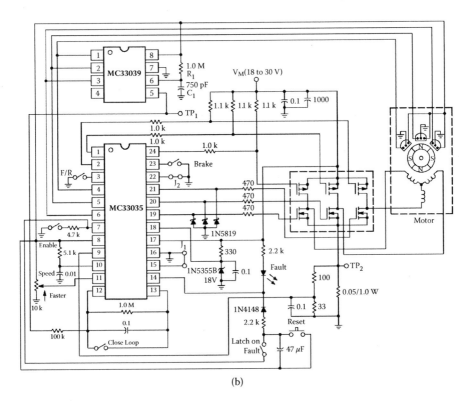

FIGURE 6.18
(Continued).

circuitry-electromagnetic dynamics), Newtonian mechanics (to model the *tor-sional-mechanical* dynamics), and the coenergy concept (to find the electromagnetic torque). Our goal is to develop the equations of motion for three-phase permanent-magnet synchronous motors utilizing Lagrange's concept.

The generalized coordinates are the electric charges in the *abc* stator windings

$$q_1 = \frac{i_{as}}{s}, \quad \dot{q}_1 = i_{as}, \quad q_2 = \frac{i_{bs}}{s}, \quad \dot{q}_2 = i_{bs}, \quad q_3 = \frac{i_{cs}}{s}, \quad \dot{q}_3 = i_{cs},$$

and the angular displacement

$$q_4 = \theta_r, \quad \dot{q}_4 = \omega_r.$$

The generalized forces are the applied voltages to the *abc* windings $Q_1 = u_{as}$, $Q_2 = u_{bs}$, $Q_3 = u_{cs}$, and the load torque $Q_4 = -T_L$.

The resulting Lagrange equations are

$$\frac{d}{dt}\left(\frac{\partial \Gamma}{\partial \dot{q}_1}\right) - \frac{\partial \Gamma}{\partial q_1} + \frac{\partial D}{\partial \dot{q}_1} + \frac{\partial \Pi}{\partial q_1} = Q_1, \quad \frac{d}{dt}\left(\frac{\partial \Gamma}{\partial \dot{q}_2}\right) - \frac{\partial \Gamma}{\partial q_2} + \frac{\partial D}{\partial \dot{q}_2} + \frac{\partial \Pi}{\partial q_2} = Q_2,$$

$$\frac{d}{dt}\left(\frac{\partial \Gamma}{\partial \dot{q}_3}\right) - \frac{\partial \Gamma}{\partial q_3} + \frac{\partial D}{\partial \dot{q}_3} + \frac{\partial \Pi}{\partial q_3} = Q_3, \quad \frac{d}{dt}\left(\frac{\partial \Gamma}{\partial \dot{q}_4}\right) - \frac{\partial \Gamma}{\partial q_4} + \frac{\partial D}{\partial \dot{q}_4} + \frac{\partial \Pi}{\partial q_4} = Q_4.$$

The total kinetic energy includes kinetic energies of electrical and mechanical systems. We have

$$\Gamma = \Gamma_E + \Gamma_M = \frac{1}{2}L_{asas}\dot{q}_1^2 + \frac{1}{2}(L_{asbs} + L_{bsas})\dot{q}_1\dot{q}_2 + \frac{1}{2}(L_{ascs} + L_{csas})\dot{q}_1\dot{q}_3 + \frac{1}{2}L_{bsbs}\dot{q}_2^2$$

$$+ \frac{1}{2}(L_{bscs} + L_{csbs})\dot{q}_2\dot{q}_3 + \frac{1}{2}L_{cscs}\dot{q}_3^2 + \psi_m\dot{q}_1 \sin q_4 + \psi_m\dot{q}_2 \sin\left(q_4 - \frac{2}{3}\pi\right)$$

$$+ \psi_m\dot{q}_3 \sin\left(q_4 + \frac{2}{3}\pi\right) + \frac{1}{2}J\dot{q}_4^2.$$

Therefore,

$$\frac{\partial \Gamma}{\partial q_1} = 0, \quad \frac{\partial \Gamma}{\partial \dot{q}_1} = L_{asas}\dot{q}_1 + \frac{1}{2}(L_{asbs} + L_{bsas})\dot{q}_2 + \frac{1}{2}(L_{ascs} + L_{csas})\dot{q}_3 + \psi_m \sin q_4,$$

$$\frac{\partial \Gamma}{\partial q_2} = 0, \quad \frac{\partial \Gamma}{\partial \dot{q}_2} = \frac{1}{2}(L_{asbs} + L_{bsas})\dot{q}_1 + L_{bsbs}\dot{q}_2 + \frac{1}{2}(L_{bscs} + L_{csbs})\dot{q}_3 + \psi_m \sin\left(q_4 - \frac{2}{3}\pi\right),$$

$$\frac{\partial \Gamma}{\partial q_3} = 0, \quad \frac{\partial \Gamma}{\partial \dot{q}_3} = \frac{1}{2}(L_{ascs} + L_{csas})\dot{q}_1 + \frac{1}{2}(L_{bscs} + L_{csbs})\dot{q}_2 + L_{cscs}\dot{q}_3 + \psi_m \sin\left(q_4 + \frac{2}{3}\pi\right),$$

$$\frac{\partial \Gamma}{\partial q_4} = \psi_m\dot{q}_1 \cos q_4 + \psi_m\dot{q}_2 \cos\left(q_4 - \frac{2}{3}\pi\right) + \psi_m\dot{q}_3 \cos\left(q_4 + \frac{2}{3}\pi\right), \quad \frac{\partial \Gamma}{\partial \dot{q}_4} = J\dot{q}_4.$$

The total potential energy is $\Pi = 0$.

The total dissipated energy is a sum of the heat energy dissipated by the electrical system and the heat energy dissipated by the mechanical system. That is,

$$D = \frac{1}{2}\left(r_s\dot{q}_1^2 + r_s\dot{q}_2^2 + r_s\dot{q}_3^2 + B_m\dot{q}_4^2\right).$$

The differentiation of D with respect to the generalized coordinates gives

$$\frac{\partial D}{\partial \dot{q}_1} = r_s\dot{q}_1, \quad \frac{\partial D}{\partial \dot{q}_2} = r_s\dot{q}_2, \quad \frac{\partial D}{\partial \dot{q}_3} = r_s\dot{q}_3 \quad \text{and} \quad \frac{\partial D}{\partial \dot{q}_4} = B_m\dot{q}_4.$$

Using

$$\dot{q}_1 = i_{as}, \ \dot{q}_2 = i_{bs}, \ \dot{q}_3 = i_{cs}, \ \dot{q}_4 = \omega_r \text{ and } Q_1 = u_{as}, \ Q_2 = u_{bs}, \ Q_3 = u_{cs}, \ Q_4 = -T_L,$$

the Lagrange equations yield four differential equations

$$L_{asas}\frac{di_{as}}{dt} + \frac{1}{2}(L_{asbs} + L_{bsas})\frac{di_{bs}}{dt} + \frac{1}{2}(L_{ascs} + L_{csas})\frac{di_{cs}}{dt} + \psi_m\omega_r\cos\theta_r + r_s i_{as} = u_{as},$$

$$\frac{1}{2}(L_{asbs} + L_{bsas})\frac{di_{as}}{dt} + L_{bsbs}\frac{di_{bs}}{dt} + \frac{1}{2}(L_{bscs} + L_{csbs})\frac{di_{cs}}{dt} + \psi_m\omega_r\cos\left(\theta_r - \frac{2}{3}\pi\right) + r_s i_{bs} = u_{bs},$$

$$\frac{1}{2}(L_{ascs} + L_{csas})\frac{di_{as}}{dt} + \frac{1}{2}(L_{bscs} + L_{csbs})\frac{di_{bs}}{dt} + L_{cscs}\frac{di_{cs}}{dt} + \psi_m\omega_r\cos\left(\theta_r + \frac{2}{3}\pi\right) + r_s i_{cs} = u_{cs},$$

$$J\frac{d^2\theta_r}{dt^2} - \psi_m i_{as}\cos\theta_r - \psi_m i_{bs}\cos\left(\theta_r - \frac{2}{3}\pi\right) - \psi_m i_{cs}\cos\left(\theta_r + \frac{2}{3}\pi\right) + B_m\frac{d\theta_r}{dt} = -T_L.$$

For round-rotor permanent-magnet synchronous motors, one obtains

$$(\bar{L}_{ls} + \bar{L}_m)\frac{di_{as}}{dt} - \frac{1}{2}\bar{L}_m\frac{di_{bs}}{dt} - \frac{1}{2}\bar{L}_m\frac{di_{cs}}{dt} + \psi_m\omega_r\cos\theta_r + r_s i_{as} = u_{as},$$

$$-\frac{1}{2}\bar{L}_m\frac{di_{as}}{dt} + (\bar{L}_{ls} + \bar{L}_m)\frac{di_{bs}}{dt} - \frac{1}{2}\bar{L}_m\frac{di_{cs}}{dt} + \psi_m\omega_r\cos\left(\theta_r - \frac{2}{3}\pi\right) + r_s i_{bs} = u_{bs},$$

$$-\frac{1}{2}\bar{L}_m\frac{di_{as}}{dt} - \frac{1}{2}\bar{L}_m\frac{di_{bs}}{dt} + (\bar{L}_{ls} + \bar{L}_m)\frac{di_{cs}}{dt} + \psi_m\omega_r\cos\left(\theta_r + \frac{2}{3}\pi\right) + r_s i_{cs} = u_{cs},$$

$$J\frac{d\omega_r}{dt} + B_m\omega_r - \psi_m\left[i_{as}\cos\theta_r + i_{bs}\cos\left(\theta_r - \frac{2}{3}\pi\right) + i_{cs}\cos\left(\theta_r + \frac{2}{3}\pi\right)\right] = -T_L,$$

$$\frac{d\theta_r}{dt} = \omega_r.$$

From

$$\frac{\partial\Gamma}{\partial q_4} = \psi_m\dot{q}_1\cos q_4 + \psi_m\dot{q}_2\cos\left(q_4 - \frac{2}{3}\pi\right) + \psi_m\dot{q}_3\cos\left(q_4 + \frac{2}{3}\pi\right),$$

which is a term of the fourth Lagrange equation

$$\frac{d}{dt}\left(\frac{\partial\Gamma}{\partial\dot{q}_4}\right) - \frac{\partial\Gamma}{\partial q_4} + \frac{\partial D}{\partial\dot{q}_4} + \frac{\partial\Pi}{\partial q_4} = Q_4,$$

one concludes that the electromagnetic torque is given as

$$T_e = \psi_m \left[i_{as} \cos\theta_r + i_{bs} \cos\left(\theta_r - \frac{2}{3}\pi \right) + i_{cs} \cos\left(\theta_r + \frac{2}{3}\pi \right) \right].$$

Differential equations in Cauchy's form result, as given by (6.26) for *P*-pole permanent-magnet synchronous motors. It was demonstrated that by applying Lagrange's concept, a complete mathematical model for permanent-magnet synchronous motors is straightforwardly developed. ☐

Example 6.5   Simulation and Analysis of a Permanent-Magnet Synchronous Motor in SIMULINK

The radial topology permanent-magnet synchronous motors are described by five nonlinear differential equations (6.26). The following phase voltages, as functions of θ_r, are applied to guarantee the balanced operation

$$u_{as} = \sqrt{2}u_M \cos\theta_r, \quad u_{bs} = \sqrt{2}u_M \cos\left(\theta_r - \frac{2}{3}\pi \right) \text{ and } u_{cs} = \sqrt{2}u_M \cos\left(\theta_r + \frac{2}{3}\pi \right).$$

The SIMULINK diagram to simulate permanent-magnet synchronous motors is documented in Figure 6.19. To ensure the greatest degree of flexibility and effectiveness, the motor parameters may be embedded by utilizing the corresponding symbols rather than the numerical values.

The numerical values of the motor parameters must be uploaded. For a 500 W four-pole permanent-magnet synchronous motor, the parameters are: $u_M = 50$, $r_s = 1$ ohm, $L_{ss} = 0.005$ H, $L_{ls} = 0.0005$ H, $\bar{L}_m = 0.0045$ H, $\psi_m = 0.15$ V-sec/rad (N-m/A), $B_m = 0.0005$ N-m-sec/rad, and $J = 0.0015$ kg-m^2.

The motor parameters can be measured and derived by performing the experimental results. One can measure the stator resistance r_s. The constant ψ_m is calculated using the induced *emf* (terminal phase voltage generated operating machine as a generator) at the steady-state ω_r. The friction coefficient B_m is found by measuring the magnitude of the phase currents at no load (one applies the expression $T_e = B_m\omega_r$ at no load when $T_L = 0$). The moment of inertia J can be estimated as $m_{rotor}R_{rotor}^2/2$, or derived by using the disacceleration or acceleration regime. In the disacceleration operation if $u_M = 0$ and the load is known, we have

$$\frac{d\omega_{rm}}{dt} = \frac{1}{J}(T_e - B_m\omega_{rm} - T_L) = \frac{1}{J}(-B_m\omega_{rm} - T_L),$$

and, for $T_L = 0$ one yields

$$\frac{d\omega_{rm}}{dt} = -\frac{B_m}{J}\omega_{rm},$$

which gives the value of *J*.

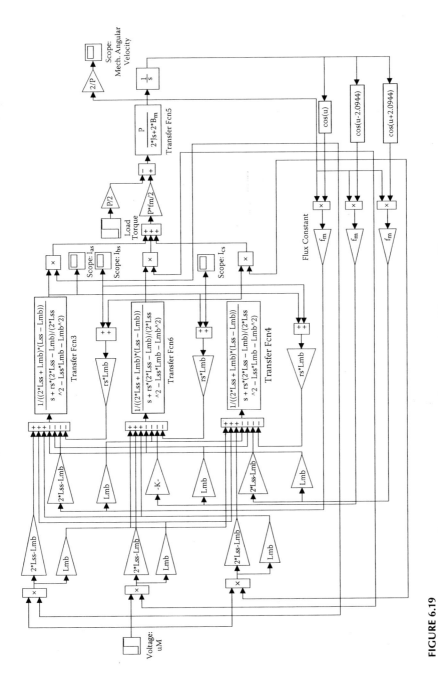

FIGURE 6.19
Simulink diagram to simulate permanent-magnet synchronous motors (ch6_02.mdl).

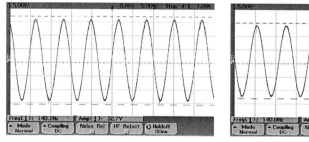
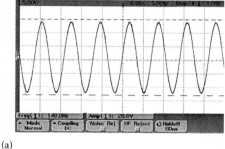

(a)

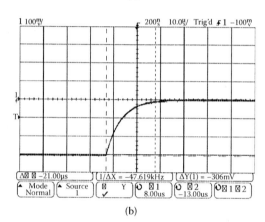

(b)

FIGURE 6.20

(a) Induced $emf_{as\omega} = -\psi_m\cos\theta_r\omega_r$ (terminal phase voltage) if $\omega_r = $ const; (b) current in the phase winding for the step voltage applied (motor at the stall).

The induced *emf* waveforms at the steady-state ω_r are illustrated in Figure 6.20a for the unloaded generator and the rated load. From $\psi_{asm} = \psi_m\sin\theta_r$, one finds that in the steady-state $emf_{as\omega} = -\psi_m\cos\theta_r\omega_r$. The measured magnitude $\psi_m\omega_r$ of the induced *emf* yields the unknown ψ_m. One can obtain the self- and mutual inductances. In particular, for the RL series circuit, which represents the winding, the current $i(t)$ transient depends on the values of R and L. By measuring the current (as the voltage across the resistor r_R) for a non-rotating motor, one obtains $i(t)$ supplying the specified voltage waveform. For the step voltage, one has

$$i(t) = \frac{u}{R}(1 - e^{-(R/L)t}).$$

The time constant is $\tau = L/R$, $L = 2L_{ss}$, and $R = 2r_s + r_R$ because two phases are in series for the wye-connected motor. For the first-order RL circuit, the time delay corresponds to $0.632i_{\text{steady-state}}$. Hence,

$$L_{ss} = \tau(2r_s + r_R)/2.$$

The oscilloscope data for the voltage across the resistor r_R, which corresponds to the current evolution, is reported in Figure 6.20b. For $r_R = 1.97 \times 10^3$ ohm, one obtains $\tau = 5.2 \times 10^{-6}$ sec. Hence, $L_{ss} = 0.0051$ H.

The motor dynamics is studied as the motor accelerates with the rated voltage applied (the magnitude is $\sqrt{2}\,50$ V). At $t = 0$ sec, the load torque is $T_{L0} = 0.1$ N-m. At $t = 0.5$ sec, the load becomes $T_L = 0.5$ N-m. The motor parameters are uploaded using the following statement:

```
% Parameters of the permanent-magnet synchronous motor
P=4; uM=50; rs=1; Lss=0.005; Lls=0.0005; fm=0.15; Bm=0.0005; J=0.0015;
Lmb=Lss-Lls;
```

The motor accelerates from stall, $T_{L0} = 0.1$ N-m for $t \in [0\ 0.5)$ sec, and $T_L = 0.5$ N-m is applied at $t = 0.5$ sec. Figure 6.21 illustrates the evolution of the

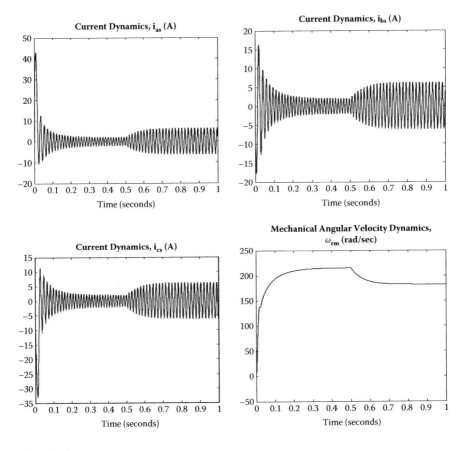

FIGURE 6.21
Transient dynamics of the permanent-magnet synchronous motor variables.

phase currents and mechanical angular velocity. The motor reaches the steady-state mechanical angular velocity (215.3 rad/sec with the load $T_{L0}=0.1$ N-m) at 0.3 sec. The angular velocity reduces and phase currents magnitude increases as the higher load is applied. The current and angular velocity dynamics, reported in Figure 6.21, allow one to assess the current magnitude, efficiency, acceleration rate, starting capabilities, and so on.

The state variables (phase currents, angular velocity, and angular displacement) can be plotted, and the needed calculations (torque assessment, efficiency estimates, etc.), can be performed. The plotting statements are

```
% Plots of the transient dynamics for the permanent-magnet synchronous motor
plot(Ias(:,1),Ias(:,2)); xlabel('Time [seconds]','FontSize',14);
title('Current Dynamics, i _ a _ s [A]','FontSize',14); pause;
plot(Ibs(:,1),Ibs(:,2)); xlabel('Time [seconds]','FontSize',14);
title('Current Dynamics, i _ b _ s [A]','FontSize',14); pause;
plot(Ics(:,1),Ics(:,2)); xlabel('Time [seconds]','FontSize',14);
title('Current Dynamics, i _ c _ s [A]','FontSize',14); pause;
plot(wrm(:,1),wrm(:,2)); xlabel('Time [seconds]','FontSize',14);
title('Mechanical Angular Velocity Dynamics, \omega _ r _ m [rad/sec]','FontSize',14);
```
□

We will develop the equations of motion for three-phase permanent-magnet synchronous generators as shown in Figure 6.15. We apply Kirchhoff's second law (6.20) with the flux linkages as given by (6.21). Newton's second law of motion

$$J\frac{d^2\theta_{rm}}{dt^2} = -T_e - B_m\omega_{rm} + T_{pm}$$

yields

$$\frac{d\omega_{rm}}{dt} = \frac{1}{J}(-T_e - B_m\omega_{rm} + T_{pm}), \quad \frac{d\theta_{rm}}{dt} = \omega_{rm}.$$

Here, T_{pm} denotes the toque of the prime mover, which rotates the generator, whereas the electromagnetic torque T_e can be viewed as the load torque.

The induced *emfs*, as terminal voltages (u_{as}, u_{bs} and u_{cs}), can be viewed as the generator outputs. For the symmetric resistive load, the stator phase resistances r_{as}, r_{bs} and r_{cs} are in series with the three-phase load resistors R_L. We obtain the following set of differential equations for the permanent-magnet synchronous generators

$$\frac{di_{as}}{dt} = -\frac{(r_s + R_L)(2L_{ss} - \bar{L}_m)}{2L_{ss}^2 - L_{ss}\bar{L}_m - \bar{L}_m^2} i_{as} - \frac{(r_s + R_L)\bar{L}_m}{2L_{ss}^2 - L_{ss}\bar{L}_m - \bar{L}_m^2} i_{bs} - \frac{(r_s + R_L)\bar{L}_m}{2L_{ss}^2 - L_{ss}\bar{L}_m - \bar{L}_m^2} i_{cs}$$

$$+ \frac{\psi_m(2L_{ss} - \bar{L}_m)}{2L_{ss}^2 - L_{ss}\bar{L}_m - \bar{L}_m^2} \omega_r \cos\theta_r + \frac{\psi_m \bar{L}_m}{2L_{ss}^2 - L_{ss}\bar{L}_m - \bar{L}_m^2} \omega_r \cos\left(\theta_r - \frac{2}{3}\pi\right)$$

$$+ \frac{\psi_m \bar{L}_m}{2L_{ss}^2 - L_{ss}\bar{L}_m - \bar{L}_m^2} \omega_r \cos\left(\theta_r + \frac{2}{3}\pi\right),$$

$$\frac{di_{bs}}{dt} = -\frac{(r_s + R_L)\bar{L}_m}{2L_{ss}^2 - L_{ss}\bar{L}_m - \bar{L}_m^2} i_{as} - \frac{(r_s + R_L)(2L_{ss} - \bar{L}_m)}{2L_{ss}^2 - L_{ss}\bar{L}_m - \bar{L}_m^2} i_{bs} - \frac{(r_s + R_L)\bar{L}_m}{2L_{ss}^2 - L_{ss}\bar{L}_m - \bar{L}_m^2} i_{cs}$$

$$+ \frac{\psi_m \bar{L}_m}{2L_{ss}^2 - L_{ss}\bar{L}_m - \bar{L}_m^2} \omega_r \cos\theta_r + \frac{\psi_m(2L_{ss} - \bar{L}_m)}{2L_{ss}^2 - L_{ss}\bar{L}_m - \bar{L}_m^2} \omega_r \cos\left(\theta_r - \frac{2}{3}\pi\right)$$

$$+ \frac{\psi_m \bar{L}_m}{2L_{ss}^2 - L_{ss}\bar{L}_m - \bar{L}_m^2} \omega_r \cos\left(\theta_r + \frac{2}{3}\pi\right),$$

$$\frac{di_{cs}}{dt} = -\frac{(r_s + R_L)\bar{L}_m}{2L_{ss}^2 - L_{ss}\bar{L}_m - \bar{L}_m^2} i_{as} - \frac{(r_s + R_L)\bar{L}_m}{2L_{ss}^2 - L_{ss}\bar{L}_m - \bar{L}_m^2} i_{bs} - \frac{(r_s + R_L)(2L_{ss} - \bar{L}_m)}{2L_{ss}^2 - L_{ss}\bar{L}_m - \bar{L}_m^2} i_{cs}$$

$$+ \frac{\psi_m \bar{L}_m}{2L_{ss}^2 - L_{ss}\bar{L}_m - \bar{L}_m^2} \omega_r \cos\theta_r + \frac{\psi_m \bar{L}_m}{2L_{ss}^2 - L_{ss}\bar{L}_m - \bar{L}_m^2} \omega_r \cos\left(\theta_r - \frac{2}{3}\pi\right)$$

$$+ \frac{\psi_m(2L_{ss} - \bar{L}_m)}{2L_{ss}^2 - L_{ss}\bar{L}_m - \bar{L}_m^2} \omega_r \cos\left(\theta_r + \frac{2}{3}\pi\right),$$

$$\frac{d\omega_r}{dt} = -\frac{P^2 \psi_m}{4J}\left(i_{as}\cos\theta_r + i_{bs}\cos\left(\theta_r - \frac{2}{3}\pi\right) + i_{cs}\cos\left(\theta_r + \frac{2}{3}\pi\right)\right) - \frac{B_m}{J}\omega_r + \frac{P}{2J}T_{pm},$$

$$\frac{d\theta_r}{dt} = \omega_r.$$

Example 6.6

The vector and matrix notations have been used. In the following section, vectors and matrices are used to ensure the compactness and simplicity of mathematical deviations. The derived differential equations for radial topology permanent-magnet synchronous generators can be written in the state-space form as

$$
\begin{bmatrix} \dfrac{di_{as}}{dt} \\[2mm] \dfrac{di_{bs}}{dt} \\[2mm] \dfrac{di_{cs}}{dt} \\[2mm] \dfrac{d\omega_r}{dt} \\[2mm] \dfrac{d\theta_r}{dt} \end{bmatrix} =
\begin{bmatrix}
-\dfrac{(r_s+R_L)(2L_{ss}-\bar{L}_m)}{2L_{ss}^2-L_{ss}\bar{L}_m-\bar{L}_m^2} & -\dfrac{(r_s+R_L)\bar{L}_m}{2L_{ss}^2-L_{ss}\bar{L}_m-\bar{L}_m^2} & -\dfrac{(r_s+R_L)\bar{L}_m}{2L_{ss}^2-L_{ss}\bar{L}_m-\bar{L}_m^2} & 0 & 0 \\[3mm]
-\dfrac{(r_s+R_L)\bar{L}_m}{2L_{ss}^2-L_{ss}\bar{L}_m-\bar{L}_m^2} & -\dfrac{(r_s+R_L)(2L_{ss}-\bar{L}_m)}{2L_{ss}^2-L_{ss}\bar{L}_m-\bar{L}_m^2} & -\dfrac{(r_s+R_L)\bar{L}_m}{2L_{ss}^2-L_{ss}\bar{L}_m-\bar{L}_m^2} & 0 & 0 \\[3mm]
-\dfrac{(r_s+R_L)\bar{L}_m}{2L_{ss}^2-L_{ss}\bar{L}_m-\bar{L}_m^2} & -\dfrac{(r_s+R_L)\bar{L}_m}{2L_{ss}^2-L_{ss}\bar{L}_m-\bar{L}_m^2} & -\dfrac{(r_s+R_L)(2L_{ss}-\bar{L}_m)}{2L_{ss}^2-L_{ss}\bar{L}_m-\bar{L}_m^2} & 0 & 0 \\[3mm]
0 & 0 & 0 & -\dfrac{B_m}{J} & 0 \\[3mm]
0 & 0 & 0 & 1 & 0
\end{bmatrix}
\begin{bmatrix} i_{as} \\ i_{bs} \\ i_{cs} \\ \omega_r \\ \theta_r \end{bmatrix}
$$

$$
+\begin{bmatrix}
\dfrac{\psi_m(2L_{ss}-\bar{L}_m)}{2L_{ss}^2-L_{ss}\bar{L}_m-\bar{L}_m^2}\omega_r & \dfrac{\psi_m\bar{L}_m}{2L_{ss}^2-L_{ss}\bar{L}_m-\bar{L}_m^2}\omega_r & \dfrac{\psi_m\bar{L}_m}{2L_{ss}^2-L_{ss}\bar{L}_m-\bar{L}_m^2}\omega_r \\[3mm]
\dfrac{\psi_m\bar{L}_m}{2L_{ss}^2-L_{ss}\bar{L}_m-\bar{L}_m^2}\omega_r & \dfrac{\psi_m(2L_{ss}-\bar{L}_m)}{2L_{ss}^2-L_{ss}\bar{L}_m-\bar{L}_m^2}\omega_r & \dfrac{\psi_m\bar{L}_m}{2L_{ss}^2-L_{ss}\bar{L}_m-\bar{L}_m^2}\omega_r \\[3mm]
\dfrac{\psi_m\bar{L}_m}{2L_{ss}^2-L_{ss}\bar{L}_m-\bar{L}_m^2}\omega_r & \dfrac{\psi_m\bar{L}_m}{2L_{ss}^2-L_{ss}\bar{L}_m-\bar{L}_m^2}\omega_r & \dfrac{\psi_m(2L_{ss}-\bar{L}_m)}{2L_{ss}^2-L_{ss}\bar{L}_m-\bar{L}_m^2}\omega_r \\[3mm]
-\dfrac{P^2\psi_m}{4J}i_{as} & -\dfrac{P^2\psi_m}{4J}i_{bs} & -\dfrac{P^2\psi_m}{4J}i_{cs} \\[3mm]
0 & 0 & 0
\end{bmatrix}
\begin{bmatrix} \cos\theta_r \\ \cos\left(\theta_r-\frac{2}{3}\pi\right) \\ \cos\left(\theta_r+\frac{2}{3}\pi\right) \end{bmatrix}
+\begin{bmatrix} 0 \\ 0 \\ 0 \\ \dfrac{P}{2J} \\ 0 \end{bmatrix} T_{pm}.
$$

One may find that the state-space notations may ensure enhanced descriptive features. □

6.3.3 Mathematical Models of Permanent-Magnet Synchronous Machines in the Arbitrary, Rotor, and Synchronous Reference Frames

In the arbitrary reference frame (not specifying θ and ω), we fix the reference frame with the rotor. As reported in Table 5.1., the *direct* Park transformations are

$$
\mathbf{u}_{qd0s}=\mathbf{K}_s\mathbf{u}_{abcs},\quad \mathbf{i}_{qd0s}=\mathbf{K}_s\mathbf{i}_{abcs},\quad \mathbf{\psi}_{qd0s}=\mathbf{K}_s\mathbf{\psi}_{abcs},
$$

$$
\mathbf{K}_s=\frac{2}{3}\begin{bmatrix}
\cos\theta & \cos\left(\theta-\frac{2}{3}\pi\right) & \cos\left(\theta+\frac{2}{3}\pi\right) \\[2mm]
\sin\theta & \sin\left(\theta-\frac{2}{3}\pi\right) & \sin\left(\theta+\frac{2}{3}\pi\right) \\[2mm]
\frac{1}{2} & \frac{1}{2} & \frac{1}{2}
\end{bmatrix}.
$$

The differential equation (6.20)

$$
\mathbf{u}_{abcs}=\mathbf{r}_s\mathbf{i}_{abcs}+\frac{d\mathbf{\psi}_{abcs}}{dt}
$$

is rewritten in the $qd0$ quantities as

$$\mathbf{K}_s^{-1}\mathbf{u}_{qd0s} = \mathbf{r}_s\mathbf{K}_s^{-1}\mathbf{i}_{qd0s} + \frac{d\left(\mathbf{K}_s^{-1}\boldsymbol{\psi}_{qd0s}\right)}{dt}, \quad \mathbf{K}_s^{-1} = \begin{bmatrix} \cos\theta & \sin\theta & 1 \\ \cos\left(\theta - \frac{2}{3}\pi\right) & \sin\left(\theta - \frac{2}{3}\pi\right) & 1 \\ \cos\left(\theta + \frac{2}{3}\pi\right) & \sin\left(\theta + \frac{2}{3}\pi\right) & 1 \end{bmatrix}.$$

Multiplying left and right sides by $\mathbf{K}_s$, one obtains

$$\mathbf{K}_s\mathbf{K}_s^{-1}\mathbf{u}_{qd0s} = \mathbf{K}_s\mathbf{r}_s\mathbf{K}_s^{-1}\mathbf{i}_{qd0s} + \mathbf{K}_s\frac{d\mathbf{K}_s^{-1}}{dt}\boldsymbol{\psi}_{qd0s} + \mathbf{K}_s\mathbf{K}_s^{-1}\frac{d\boldsymbol{\psi}_{qd0s}}{dt}.$$

The matrix $\mathbf{r}_s$ is diagonal. Thus $\mathbf{K}_s\mathbf{r}_s\mathbf{K}_s^{-1} = \mathbf{r}_s$.
From

$$\frac{d\mathbf{K}_s^{-1}}{dt} = \omega \begin{bmatrix} -\sin\theta & \cos\theta & 0 \\ -\sin\left(\theta - \frac{2}{3}\pi\right) & \cos\left(\theta - \frac{2}{3}\pi\right) & 0 \\ -\sin\left(\theta + \frac{2}{3}\pi\right) & \cos\left(\theta + \frac{2}{3}\pi\right) & 0 \end{bmatrix},$$

we have

$$\mathbf{K}_s\frac{d\mathbf{K}_s^{-1}}{dt} = \omega \begin{bmatrix} 0 & 1 & 0 \\ -1 & 0 & 0 \\ 0 & 0 & 0 \end{bmatrix}.$$

Hence, using the *direct* Park transformations, the resulting circuitry-electromagnetic equation of motion is

$$\mathbf{u}_{qd0s} = \mathbf{r}_s\mathbf{i}_{qd0s} + \omega \begin{bmatrix} \psi_{ds} \\ -\psi_{qs} \\ 0 \end{bmatrix} + \frac{d\boldsymbol{\psi}_{qd0s}}{dt}. \tag{6.27}$$

The *quadrature-*, *direct-*, and *zero*-axis components of stator flux linkages are $\boldsymbol{\psi}_{qd0s} = \mathbf{K}_s\boldsymbol{\psi}_{abcs}$. Using

$$\boldsymbol{\psi}_{abcs} = \mathbf{L}_s\mathbf{i}_{abcs} + \boldsymbol{\psi}_m = \begin{bmatrix} L_{ls} + \bar{L}_m & -\frac{1}{2}\bar{L}_m & -\frac{1}{2}\bar{L}_m \\ -\frac{1}{2}\bar{L}_m & L_{ls} + \bar{L}_m & -\frac{1}{2}\bar{L}_m \\ -\frac{1}{2}\bar{L}_m & -\frac{1}{2}\bar{L}_m & L_{ls} + \bar{L}_m \end{bmatrix} \begin{bmatrix} i_{as} \\ i_{bs} \\ i_{cs} \end{bmatrix} + \psi_m \begin{bmatrix} \sin\theta_r \\ \sin\left(\theta_r - \frac{2}{3}\pi\right) \\ \sin\left(\theta_r + \frac{2}{3}\pi\right) \end{bmatrix},$$

one obtains

$$\psi_{qd0s} = \mathbf{K}_s \mathbf{L}_s \mathbf{K}_s^{-1} \mathbf{i}_{qd0s} + \mathbf{K}_s \psi_m, \tag{6.28}$$

where

$$\mathbf{K}_s \mathbf{L}_s \mathbf{K}_s^{-1} = \begin{bmatrix} L_{ls} + \frac{3}{2}\bar{L}_m & 0 & 0 \\ 0 & L_{ls} + \frac{3}{2}\bar{L}_m & 0 \\ 0 & 0 & L_{ls} \end{bmatrix}$$

and

$$\mathbf{K}_s \psi_m = \frac{2}{3} \begin{bmatrix} \cos\theta & \cos\left(\theta - \frac{2}{3}\pi\right) & \cos\left(\theta + \frac{2}{3}\pi\right) \\ \sin\theta & \sin\left(\theta - \frac{2}{3}\pi\right) & \sin\left(\theta + \frac{2}{3}\pi\right) \\ \frac{1}{2} & \frac{1}{2} & \frac{1}{2} \end{bmatrix} \psi_m \begin{bmatrix} \sin\theta_r \\ \sin\left(\theta_r - \frac{2}{3}\pi\right) \\ \sin\left(\theta_r + \frac{2}{3}\pi\right) \end{bmatrix}$$

$$= \psi_m \begin{bmatrix} -\sin(\theta - \theta_r) \\ \cos(\theta - \theta_r) \\ 0 \end{bmatrix}.$$

From (6.28) we obtain the following expression

$$\psi_{qd0s} = \begin{bmatrix} L_{ls} + \frac{3}{2}\bar{L}_m & 0 & 0 \\ 0 & L_{ls} + \frac{3}{2}\bar{L}_m & 0 \\ 0 & 0 & L_{ls} \end{bmatrix} \mathbf{i}_{qd0s} + \psi_m \begin{bmatrix} -\sin(\theta - \theta_r) \\ \cos(\theta - \theta_r) \\ 0 \end{bmatrix}$$

which is substituted in (6.27) yielding

$$\mathbf{u}_{qd0s} = \mathbf{r}_s \mathbf{i}_{qd0s} + \omega \begin{bmatrix} \psi_{ds} \\ -\psi_{qs} \\ 0 \end{bmatrix} + \begin{bmatrix} L_{ls} + \frac{3}{2}\bar{L}_m & 0 & 0 \\ 0 & L_{ls} + \frac{3}{2}\bar{L}_m & 0 \\ 0 & 0 & L_{ls} \end{bmatrix} \frac{d\mathbf{i}_{qd0s}}{dt} + \psi_m \frac{d\begin{bmatrix} -\sin(\theta - \theta_r) \\ \cos(\theta - \theta_r) \\ 0 \end{bmatrix}}{dt}.$$

The differential equations that model the circuitry-electromagnetic dynamics in the *arbitrary* reference frame are given as

$$\mathbf{u}_{qd0s} = \mathbf{r}_s \mathbf{i}_{qd0s} + \omega \begin{bmatrix} \psi_{ds} \\ -\psi_{qs} \\ 0 \end{bmatrix} + \begin{bmatrix} L_{ls} + \frac{3}{2}\bar{L}_m & 0 & 0 \\ 0 & L_{ls} + \frac{3}{2}\bar{L}_m & 0 \\ 0 & 0 & L_{ls} \end{bmatrix} \frac{d\mathbf{i}_{qd0s}}{dt} + \psi_m \frac{d\begin{bmatrix} -\sin(\theta - \theta_r) \\ \cos(\theta - \theta_r) \\ 0 \end{bmatrix}}{dt}.$$

$$\tag{6.29}$$

In the rotor reference frame, the angular velocity of the reference frame is $\omega = \omega_r = \omega_e$, and $\theta = \theta_r$. The Park transformation matrix is

$$\mathbf{K}_s^r = \frac{2}{3}\begin{bmatrix} \cos\theta_r & \cos\left(\theta_r - \frac{2}{3}\pi\right) & \cos\left(\theta_r + \frac{2}{3}\pi\right) \\ \sin\theta_r & \sin\left(\theta_r - \frac{2}{3}\pi\right) & \sin\left(\theta_r + \frac{2}{3}\pi\right) \\ \frac{1}{2} & \frac{1}{2} & \frac{1}{2} \end{bmatrix}.$$

Correspondingly, one finds

$$\mathbf{K}_s^r \boldsymbol{\psi}_m = \frac{2}{3}\begin{bmatrix} \cos\theta_r & \cos\left(\theta_r - \frac{2}{3}\pi\right) & \cos\left(\theta_r + \frac{2}{3}\pi\right) \\ \sin\theta_r & \sin\left(\theta_r - \frac{2}{3}\pi\right) & \sin\left(\theta_r + \frac{2}{3}\pi\right) \\ \frac{1}{2} & \frac{1}{2} & \frac{1}{2} \end{bmatrix}\psi_m \begin{bmatrix} \sin\theta_r \\ \sin\left(\theta_r - \frac{2}{3}\pi\right) \\ \sin\left(\theta_r + \frac{2}{3}\pi\right) \end{bmatrix} = \begin{bmatrix} 0 \\ \psi_m \\ 0 \end{bmatrix}.$$

From (6.28), we have

$$\boldsymbol{\psi}_{qd0s}^r = \begin{bmatrix} L_{ls} + \frac{3}{2}\overline{L}_m & 0 & 0 \\ 0 & L_{ls} + \frac{3}{2}\overline{L}_m & 0 \\ 0 & 0 & L_{ls} \end{bmatrix}\mathbf{i}_{qd0s}^r + \begin{bmatrix} 0 \\ \psi_m \\ 0 \end{bmatrix},$$

or

$$\psi_{qs}^r = \left(L_{ls} + \frac{3}{2}\overline{L}_m\right)i_{qs}^r, \quad \psi_{ds}^r = \left(L_{ls} + \frac{3}{2}\overline{L}_m\right)i_{ds}^r + \psi_m \quad \text{and} \quad \psi_{0s}^r = L_{ls}i_{0s}^r.$$

Hence, in the rotor reference frame, using (6.29), one finds

$$\frac{di_{qs}^r}{dt} = -\frac{r_s}{L_{ls} + \frac{3}{2}\overline{L}_m}i_{qs}^r - \frac{\psi_m}{L_{ls} + \frac{3}{2}\overline{L}_m}\omega_r - i_{ds}^r\omega_r + \frac{1}{L_{ls} + \frac{3}{2}\overline{L}_m}u_{qs}^r,$$

$$\frac{di_{ds}^r}{dt} = -\frac{r_s}{L_{ls} + \frac{3}{2}\overline{L}_m}i_{ds}^r + i_{qs}^r\omega_r + \frac{1}{L_{ls} + \frac{3}{2}\overline{L}_m}u_{ds}^r,$$

$$\frac{di_{0s}^r}{dt} = -\frac{r_s}{L_{ls}}i_{0s}^r + \frac{1}{L_{ls}}u_{0s}^r. \tag{6.30}$$

The electromagnetic torque, expressed in the *machine* variables by (6.24) as

$$T_e = \frac{P\psi_m}{2}\left[i_{as}\cos\theta_r + i_{bs}\cos\left(\theta_r - \frac{2}{3}\pi\right) + i_{cs}\cos\left(\theta_r + \frac{2}{3}\pi\right)\right],$$

should be given using the qd0 current components. Using the *inverse* Park transformation

$$
\begin{bmatrix} i_{as} \\ i_{bs} \\ i_{cs} \end{bmatrix} = \begin{bmatrix} \cos\theta_r & \sin\theta_r & 1 \\ \cos\left(\theta_r - \frac{2}{3}\pi\right) & \sin\left(\theta_r - \frac{2}{3}\pi\right) & 1 \\ \cos\left(\theta_r + \frac{2}{3}\pi\right) & \sin\left(\theta_r + \frac{2}{3}\pi\right) & 1 \end{bmatrix} \begin{bmatrix} i_{qs}^r \\ i_{ds}^r \\ i_{0s}^r \end{bmatrix},
$$

and substituting

$$
i_{as} = \cos\theta_r i_{qs}^r + \sin\theta_r i_{ds}^r + i_{0s}^r, \ i_{bs} = \cos\left(\theta_r - \frac{2}{3}\pi\right) i_{qs}^r + \sin\left(\theta_r - \frac{2}{3}\pi\right) i_{ds}^r + i_{0s}^r \text{ and}
$$

$$
i_{cs} = \cos\theta\left(\theta_r + \frac{2}{3}\pi\right) i_{qs}^r + \sin\left(\theta_r + \frac{2}{3}\pi\right) i_{ds}^r + i_{0s}^r,
$$

in (6.24), one finds

$$
T_e = \frac{3P\psi_m}{4} i_{qs}^r. \tag{6.31}
$$

For *P*-pole permanent-magnet synchronous motors, from (6.25) and (6.31), the *torsional-mechanical* dynamics is

$$
\frac{d\omega_r}{dt} = \frac{3P^2\psi_m}{8J} i_{qs}^r - \frac{B_m}{J}\omega_r - \frac{P}{2J}T_L,
$$

$$
\frac{d\theta_r}{dt} = \omega_r. \tag{6.32}
$$

The differential equations (6.30) and (6.32) result in the following description of three-phase permanent-magnet synchronous motors in the rotor reference frame

$$
\frac{di_{qs}^r}{dt} = -\frac{r_s}{L_{ls} + \frac{3}{2}\bar{L}_m} i_{qs}^r - \frac{\psi_m}{L_{ls} + \frac{3}{2}\bar{L}_m}\omega_r - i_{ds}^r\omega_r + \frac{1}{L_{ls} + \frac{3}{2}\bar{L}_m} u_{qs}^r,
$$

$$
\frac{di_{ds}^r}{dt} = -\frac{r_s}{L_{ls} + \frac{3}{2}\bar{L}_m} i_{ds}^r + i_{qs}^r\omega_r + \frac{1}{L_{ls} + \frac{3}{2}\bar{L}_m} u_{ds}^r,
$$

$$
\frac{di_{0s}^r}{dt} = -\frac{r_s}{L_{ls}} i_{0s}^r + \frac{1}{L_{ls}} u_{0s}^r,
$$

$$
\frac{d\omega_r}{dt} = \frac{3P^2\psi_m}{8J} i_{qs}^r - \frac{B_m}{J}\omega_r - \frac{P}{2J}T_L,
$$

$$
\frac{d\theta_r}{dt} = \omega_r. \tag{6.33}
$$

A balanced three-phase current set is

$$i_{as} = \sqrt{2}i_M \cos\theta_r, \quad i_{bs} = \sqrt{2}i_M \cos\left(\theta_r - \frac{2}{3}\pi\right) \quad \text{and} \quad i_{cs} = \sqrt{2}i_M \cos\left(\theta_r + \frac{2}{3}\pi\right).$$

Using the *direct* Park transformation

$$\begin{bmatrix} i_{qs}^r \\ i_{ds}^r \\ i_{0s}^r \end{bmatrix} = \frac{2}{3} \begin{bmatrix} \cos\theta_r & \cos\left(\theta_r - \frac{2}{3}\pi\right) & \cos\left(\theta_r + \frac{2}{3}\pi\right) \\ \sin\theta_r & \sin\left(\theta_r - \frac{2}{3}\pi\right) & \sin\left(\theta_r + \frac{2}{3}\pi\right) \\ \frac{1}{2} & \frac{1}{2} & \frac{1}{2} \end{bmatrix} \begin{bmatrix} i_{as} \\ i_{bs} \\ i_{cs} \end{bmatrix},$$

one obtains the *quadrature, direct,* and *zero* axis current components to ensure the balanced operation. From

$$\begin{bmatrix} i_{qs}^r \\ i_{ds}^r \\ i_{0s}^r \end{bmatrix} = \frac{2}{3} \begin{bmatrix} \cos\theta_r & \cos\left(\theta_r - \frac{2}{3}\pi\right) & \cos\left(\theta_r + \frac{2}{3}\pi\right) \\ \sin\theta_r & \sin\left(\theta_r - \frac{2}{3}\pi\right) & \sin\left(\theta_r + \frac{2}{3}\pi\right) \\ \frac{1}{2} & \frac{1}{2} & \frac{1}{2} \end{bmatrix} \begin{bmatrix} \sqrt{2}i_M \cos\theta_r \\ \sqrt{2}i_M \cos\left(\theta_r - \frac{2}{3}\pi\right) \\ \sqrt{2}i_M \cos\left(\theta_r + \frac{2}{3}\pi\right) \end{bmatrix},$$

we have

$$i_{qs}^r = \sqrt{2}i_M, \quad i_{ds}^r = 0 \quad \text{and} \quad i_{0s}^r = 0.$$

The *abc* voltages may be supplied with advanced phase shifting as

$$u_{as} = \sqrt{2}u_M \cos(\theta_r + \varphi_u), \quad u_{bs} = \sqrt{2}u_M \cos\left(\theta_r - \frac{2}{3}\pi + \varphi_u\right) \quad \text{and}$$

$$u_{cs} = \sqrt{2}u_M \cos\left(\theta_r + \frac{2}{3}\pi + \varphi_u\right)$$

The *direct* Park transformation and voltage balanced set yield

$$\begin{bmatrix} u_{qs}^r \\ u_{ds}^r \\ u_{0s}^r \end{bmatrix} = \frac{2}{3} \begin{bmatrix} \cos\theta_r & \cos\left(\theta_r - \frac{2}{3}\pi\right) & \cos\left(\theta_r + \frac{2}{3}\pi\right) \\ \sin\theta_r & \sin\left(\theta_r - \frac{2}{3}\pi\right) & \sin\left(\theta_r + \frac{2}{3}\pi\right) \\ \frac{1}{2} & \frac{1}{2} & \frac{1}{2} \end{bmatrix} \begin{bmatrix} \sqrt{2}u_M \cos(\theta_r + \varphi_u) \\ \sqrt{2}u_M \cos\left(\theta_r - \frac{2}{3}\pi + \varphi_u\right) \\ \sqrt{2}u_M \cos\left(\theta_r + \frac{2}{3}\pi + \varphi_u\right) \end{bmatrix}.$$

The matrix multiplication and trigonometric identities result in

$$u_{qs}^r = \sqrt{2}u_M \cos\varphi_u, \quad u_{ds}^r = -\sqrt{2}u_M \sin\varphi_u \quad \text{and} \quad u_{0s}^r = 0.$$

Because of small inductances, $\varphi_{u} \approx 0$. Hence,

$$u_{qs}^{r} = \sqrt{2}u_{M}, \quad u_{ds}^{r} = 0 \quad \text{and} \quad u_{0s}^{r} = 0.$$

In the synchronous reference frame, one specifies the angular velocity of the reference frame to be $\omega = \omega_{e}$. Hence, $\theta = \theta_{e}$. The Park transformation matrix is

$$\mathbf{K}_{s}^{e} = \frac{2}{3} \begin{bmatrix} \cos\theta_{e} & \cos\left(\theta_{e} - \frac{2}{3}\pi\right) & \cos\left(\theta_{e} + \frac{2}{3}\pi\right) \\ \sin\theta_{e} & \sin\left(\theta_{e} - \frac{2}{3}\pi\right) & \sin\left(\theta_{e} + \frac{2}{3}\pi\right) \\ \frac{1}{2} & \frac{1}{2} & \frac{1}{2} \end{bmatrix}.$$

Substituting $\omega_{r} = \omega_{e}$ in (6.33) we have the following system of differential equations, which describe the permanent-magnet motor dynamics in the synchronous reference frame

$$\frac{di_{qs}^{e}}{dt} = -\frac{r_{s}}{L_{ls} + \frac{3}{2}\bar{L}_{m}} i_{qs}^{e} - \frac{\psi_{m}}{L_{ls} + \frac{3}{2}\bar{L}_{m}} \omega_{r} - i_{ds}^{e}\omega_{r} + \frac{1}{L_{ls} + \frac{3}{2}\bar{L}_{m}} u_{qs}^{e},$$

$$\frac{di_{ds}^{e}}{dt} = -\frac{r_{s}}{L_{ls} + \frac{3}{2}\bar{L}_{m}} i_{ds}^{e} + i_{qs}^{e}\omega_{r} + \frac{1}{L_{ls} + \frac{3}{2}\bar{L}_{m}} u_{ds}^{e},$$

$$\frac{di_{0s}^{e}}{dt} = -\frac{r_{s}}{L_{ls}} i_{0s}^{e} + \frac{1}{L_{ls}} u_{0s}^{e},$$

$$\frac{d\omega_{r}}{dt} = \frac{3P^{2}\psi_{m}}{8J} i_{qs}^{e} - \frac{B_{m}}{J}\omega_{r} - \frac{P}{2J} T_{L},$$

$$\frac{d\theta_{r}}{dt} = \omega_{r}.$$

6.3.4 Advanced Topics in Analysis of Permanent-Magnet Synchronous Machines

We documented analysis tasks (modeling, parameter deviations, simulation, etc.) by assuming an optimal structural design of permanent-magnet synchronous machines, linearity of magnetic system, and so on. The designer can achieve near-optimal design in the specified operating envelope. However, the undesired effects (magnetic field nonuniformity, nonlinear *B-H* characteristic, nonlinear magnetic system, saturation, eccentricity, etc.) may significantly affect and degrade electromechanical motion devices performance and capabilities. In this section, we document how to perform the advanced studies examining some practical cases. We study the effect of near-optimal distribution of flux linkages. Other problems faced can be approached and solved by applying the basic electromagnetics and mechanics reported.

Using the Kirchhoff second law (6.20), the circuitry-electromagnetic dynamics was studied assuming an optimal (ideal sinusoidal) distribution of the flux linkages established by permanent magnets as viewed from the windings. In particular, we let

$$\psi_{asm} = \psi_m \sin \theta_r, \quad \psi_{bsm} = \psi_m \sin \left(\theta_r - \frac{2}{3}\pi \right) \text{ and}$$

$$\psi_{csm} = \psi_m \sin \left(\theta_r + \frac{2}{3}\pi \right).$$

The distribution of flux linkages in the specified operating envelope (loads, angular velocity, etc.) can be obtained experimentally by examining the induced *emf* by rotating and loading permanent-magnet synchronous machines used as generators. These results were reported in Example 6.5. The electric machine characteristics and parameters found for any permanent-magnet synchronous generator are valid if this motion device is utilized as a motor, and vice versa. In general, the flux linkages obey

$$\psi_{asm} = \psi_m \sum_{n=1}^{\infty} \left(a_{asn} \sin^{2n-1} \theta_r + b_{asn} \cos^{2n-1} \theta_r \right),$$

$$\psi_{bsm} = \psi_m \sum_{n=1}^{\infty} \left(a_{bsn} \sin^{2n-1} \left(\theta_r - \frac{2}{3}\pi \right) + b_{bsn} \cos^{2n-1} \left(\theta_r - \frac{2}{3}\pi \right) \right),$$

$$\psi_{csm} = \psi_m \sum_{n=1}^{\infty} \left(a_{csn} \sin^{2n-1} \left(\theta_r + \frac{2}{3}\pi \right) + b_{csn} \cos^{2n-1} \left(\theta_r + \frac{2}{3}\pi \right) \right), \quad (6.34)$$

where a_n and b_n are the coefficients that depend on the operating envelope, structural design, materials, fabrication technology, and many other factors, for example, $a_n(\mathbf{E,D,B,H,i}_{abcs}, \omega_r, \mathbf{T}_L, \varepsilon, \mu, \Sigma)$ and $b_n(\mathbf{E,D,B,H,i}_{abcs}, \omega_r, \mathbf{T}_L, \varepsilon, \mu, \Sigma)$; Σ denotes the machine structural design, topology, sizing, materials, and other factors.

In practice, (6.34) usually results in

$$\psi_{asm} = \psi_m \sum_{n=1}^{\infty} a_n \sin^{2n-1} \theta_r, \quad \psi_{bsm} = \psi_m \sum_{n=1}^{\infty} a_n \sin^{2n-1} \left(\theta_r - \frac{2}{3}\pi \right) \text{ and}$$

$$\psi_{csm} = \psi_m \sum_{n=1}^{\infty} a_n \sin^{2n-1} \left(\theta_r + \frac{2}{3}\pi \right). \quad (6.35)$$

From (6.35), one yields the expressions for the flux linkages as

$$
\begin{bmatrix} \psi_{as} \\ \psi_{bs} \\ \psi_{cs} \end{bmatrix} = \begin{bmatrix} L_{ls} + \bar{L}_m & -\tfrac{1}{2}\bar{L}_m & -\tfrac{1}{2}\bar{L}_m \\ -\tfrac{1}{2}\bar{L}_m & L_{ls} + \bar{L}_m & -\tfrac{1}{2}\bar{L}_m \\ -\tfrac{1}{2}\bar{L}_m & -\tfrac{1}{2}\bar{L}_m & L_{ls} + \bar{L}_m \end{bmatrix} \begin{bmatrix} i_{as} \\ i_{bs} \\ i_{cs} \end{bmatrix} + \psi_m \begin{bmatrix} \displaystyle\sum_{n=1}^{\infty} a_n \sin^{2n-1}\theta_r \\ \displaystyle\sum_{n=1}^{\infty} a_n \sin^{2n-1}\left(\theta_r - \tfrac{2}{3}\pi\right) \\ \displaystyle\sum_{n=1}^{\infty} a_n \sin^{2n-1}\left(\theta_r + \tfrac{2}{3}\pi\right) \end{bmatrix}.
$$

$$(6.36)$$

Using (6.36), the total derivatives

$$d\psi_{as}/dt, \quad d\psi_{bs}/dt \quad \text{and} \quad d\psi_{cs}/dt$$

can be found and used in (6.20)

$$u_{as} = r_s i_{as} + \frac{d\psi_{as}}{dt}, \quad u_{bs} = r_s i_{bs} + \frac{d\psi_{bs}}{dt}, \quad u_{cs} = r_s i_{cs} + \frac{d\psi_{cs}}{dt},$$

yielding the circuitry-electromagnetic equations of motion in the *machine* variables.

One needs to derive the expression for the electromagnetic torque. From

$$T_e = \frac{P}{2}\frac{\partial W_c}{\partial \theta_r},$$

we obtain

$$
\begin{aligned}
T_e = \frac{P\psi_m}{2}\Bigg[& i_{as}\sum_{n=1}^{\infty}(2n-1)a_n\cos\theta_r\sin^{2n-2}\theta_r \\
& + i_{bs}\sum_{n=1}^{\infty}(2n-1)a_n\cos\left(\theta_r - \frac{2}{3}\pi\right)\sin^{2n-2}\left(\theta_r - \frac{2}{3}\pi\right) \\
& + i_{cs}\sum_{n=1}^{\infty}(2n-1)a_n\cos\left(\theta_r + \frac{2}{3}\pi\right)\sin^{2n-2}\left(\theta_r + \frac{2}{3}\pi\right)\Bigg].
\end{aligned}
$$

$$(6.37)$$

The Newtonian mechanics and (6.37) results in the *torsional-mechanical dynamics* as

$$\frac{d\omega_r}{dt} = \frac{P^2\psi_m}{4J}\left[i_{as}\sum_{n=1}^{\infty}(2n-1)a_n\cos\theta_r\sin^{2n-2}\theta_r \right.$$

$$+ i_{bs}\sum_{n=1}^{\infty}(2n-1)a_n\cos\left(\theta_r - \frac{2}{3}\pi\right)\sin^{2n-2}\left(\theta_r - \frac{2}{3}\pi\right)$$

$$\left. + i_{cs}\sum_{n=1}^{\infty}(2n-1)a_n\cos\left(\theta_r + \frac{2}{3}\pi\right)\sin^{2n-2}\left(\theta_r + \frac{2}{3}\pi\right) \right]$$

$$- \frac{B_m}{J}\omega_r - \frac{P}{2J}T_L,$$

$$\frac{d\theta_r}{dt} = \omega_r. \tag{6.38}$$

The current and voltage sets to be applied are derived using the expression for T_e. The trigonometric identities are applied. From (6.37), one finds the following phase currents and voltages which theoretically guarantee balanced operations

$$i_{as} = \sqrt{2}i_M\cos\theta_r\left(\sum_{n=1}^{\infty}(2n-1)a_n\sin^{2n-2}\theta_r\right)^{-1},$$

$$i_{bs} = \sqrt{2}i_M\cos\left(\theta_r - \frac{2}{3}\pi\right)\left(\sum_{n=1}^{\infty}(2n-1)a_n\sin^{2n-2}\left(\theta_r - \frac{2}{3}\pi\right)\right)^{-1},$$

$$i_{cs} = \sqrt{2}i_M\cos\left(\theta_r + \frac{2}{3}\pi\right)\left(\sum_{n=1}^{\infty}(2n-1)a_n\sin^{2n-2}\left(\theta_r + \frac{2}{3}\pi\right)\right)^{-1}, \tag{6.39}$$

and

$$u_{as} = \sqrt{2}u_M\cos\theta_r\left(\sum_{n=1}^{\infty}(2n-1)a_n\sin^{2n-2}\theta_r\right)^{-1},$$

$$u_{bs} = \sqrt{2}u_M\cos\left(\theta_r - \frac{2}{3}\pi\right)\left(\sum_{n=1}^{\infty}(2n-1)a_n\sin^{2n-2}\left(\theta_r - \frac{2}{3}\pi\right)\right)^{-1},$$

$$u_{cs} = \sqrt{2}u_M\cos\left(\theta_r + \frac{2}{3}\pi\right)\left(\sum_{n=1}^{\infty}(2n-1)a_n\sin^{2n-2}\left(\theta_r + \frac{2}{3}\pi\right)\right)^{-1}. \tag{6.40}$$

The phase currents and voltages are constrained. Therefore, the saturation effect (bounds) should be integrated using the conditional statements. The singularity problem can be resolved by normalizing sets (6.39) and (6.40).

It should be emphasized that the current and voltage sets (6.39) and (6.40) are the mathematical expressions to implement, if possible, by the hardware (power electronics, DSP, etc.) and software. Utilizing specific output stages (usually 6- or 12-step) and converter topologies (hard- or soft-switching, etc.), although one strives to ensure the appropriate solutions, hardware largely defines the voltage waveforms. Advanced DSPs are required to integrate the largely empirical $a_n(\mathbf{E,D,B,H},i_{abcs},\omega_r,\mathbf{T}_L,\varepsilon,\mu,\Sigma)$ as conditional logics and look-up table to realize (6.39) or (6.40). Within the existing converter topologies one cannot ensure sinusoidal-like voltage waveforms. Furthermore, the PWM concept implies the voltage *averaging*, and the rated solid-state device voltage, current, switching frequency, and other characteristics affect voltage waveforms. The hardware-centered (implementable) phase voltages must be used in the performance analysis and capabilities assessments. Even the ideal sinusoidal voltage set

$$u_{as} = \sqrt{2}u_M \cos\theta_r, \; u_{bs} = \sqrt{2}u_M \cos\left(\theta_r + \frac{2}{3}\pi\right) \quad \text{and}$$

$$u_{cs} = \sqrt{2}u_M \cos\left(\theta_r - \frac{2}{3}\pi\right)$$

is not implementable. However, we advanced the analysis tasks providing the justification and foundation for advanced studies. One may integrate the derived basic electromagnetics, control concepts, and hardware solutions ensuring near-optimal or *achievable* performance.

Example 6.7

For two-phase permanent-magnet synchronous machines, it is desired to ensure the machine design which leads to

$$\psi_{asm} = \psi_m \sin\theta_r \quad \text{and} \quad \psi_{bsm} = \psi_m \cos\theta_r.$$

In fact, the derived electromagnetic torque

$$T_e = \frac{P\psi_m}{2}(\cos\theta_r i_{as} - \sin\theta_r i_{bs})$$

yields the balanced current set as $i_{as} = i_M\cos\theta_r$ and $i_{bs} = -i_M\sin\theta_r$.

Assume the *ab* flux linkages, established by the permanent magnets as viewed from the windings, are

$$\psi_{asm} = \psi_m \sum_{n=1}^{\infty} a_n \sin^{2n-1}\theta_r \quad \text{and} \quad \psi_{bsm} = \psi_m \sum_{n=1}^{\infty} a_n \cos^{2n-1}\theta_r.$$

The electromagnetic torque is given as

$$T_e = \frac{P\psi_m}{2}\left[i_{as}\sum_{n=1}^{\infty}(2n-1)a_n\cos\theta_r\sin^{2n-2}\theta_r - i_{bs}\sum_{n=1}^{\infty}(2n-1)a_n\sin\theta_r\cos^{2n-2}\theta_r\right].$$

Let $a_1 \neq 1$, $a_2 \neq 0$, and $\forall a_n = 0$, $n > 2$. Hence,

$$T_e = \frac{P\psi_m}{2}\left[i_{as}\cos\theta_r\left(a_1 + 3a_2\sin^2\theta_r\right) - i_{bs}\sin\theta_r\left(a_1 + 3a_2\cos^2\theta_r\right)\right].$$

The phase voltages u_{as} and u_{bs}, as functions of θ_r, which ensure the near-balanced operating conditions, are

$$u_{as} = \frac{u_M\cos\theta_r}{(a_1 + 3a_2\sin^2\theta_r + \varepsilon)} \quad\text{if}\quad \left|\frac{u_M\cos\theta_r}{(a_1 + 3a_2\sin^2\theta_r + \varepsilon)}\right| \leq u_{max}, \quad u_{as} = u_{max}, \text{ or}$$

$$u_{as} = -u_{max} \text{ otherwise,}$$

$$u_{bs} = \frac{u_M\sin\theta_r}{(a_1 + 3a_2\cos^2\theta_r + \varepsilon)} \quad\text{if}\quad \left|\frac{-u_M\sin\theta_r}{(a_1 + 3a_2\cos^2\theta_r + \varepsilon)}\right| \leq u_{max}, \quad u_{bs} = -u_{max}, \text{ or}$$

$$u_{bs} = u_{max} \text{ otherwise.}$$

If $a_1 \gg a_2$, one may utilize $u_{as} = u_M\cos\theta_r$ and $u_{bs} = -u_M\sin\theta_r$. $\quad\square$

Example 6.8   Simulation and Analysis of a Permanent-Magnet Synchronous Motor

We simulate and study the performance of a radial topology three-phase synchronous motor. The motor parameters are: $P = 4$, $u_M = 50$, $r_s = 1$ ohm, $L_{ss} = 0.002$ H, $L_{ls} = 0.0002$ H, $\bar{L}_m = 0.0018$ H, $\psi_m = 0.1$ V-sec/rad (N-m/A), $B_m = 0.00008$ N-m-sec/rad, and $J = 0.00004$ kg-m². For no load and light load conditions (T_L is ~ 0.1 N-m), the constants are $a_1 = 1$ and all other a_n are zeros ($\forall a_n = 0$, $n > 1$). For the loaded motor, we have $a_1 = 1$, $a_2 = 0.05$, $a_3 = 0.02$, and $\forall a_n = 0$, $n > 3$. The parameters are uploaded as

```
% Optimal distribution: Light loads
psim=0.1; a1=1; a2=0; a3=0;
% Near-optimal distribution: Heavy loads
% psim=0.1; a1=1; a2=0.05; a3=0.02;
% Motor parameters
P=4; uM=50; rs=1; Lss=0.002; Lls=0.0002; Bm=0.00008; J=0.00004;
Lmb=Lss-Lls;
```

The studied permanent-magnet synchronous motor is described by five nonlinear differential equations; see (6.20) and (6.38). For $a_1 \neq 0$, $a_2 \neq 0$, $a_3 \neq 0$, and $\forall a_n = 0$, $n > 3$, we have

$$
\psi_{as} = L_{ss}i_{as} - \frac{1}{2}\bar{L}_m i_{bs} - \frac{1}{2}\bar{L}_m i_{cs} + \psi_m\left(a_1\sin\theta_r + a_2\sin^3\theta_r + a_3\sin^5\theta_r\right),
$$

$$
\psi_{bs} = -\frac{1}{2}\bar{L}_m i_{as} + L_{ss}i_{bs} - \frac{1}{2}\bar{L}_m i_{cs} + \psi_m\left(a_1\sin\left(\theta_r + \frac{2}{3}\pi\right) + a_2\sin^3\left(\theta_r + \frac{2}{3}\pi\right)\right.
$$
$$
\left. + a_3\sin^5\left(\theta_r + \frac{2}{3}\pi\right)\right),
$$

$$
\psi_{cs} = -\frac{1}{2}\bar{L}_m i_{as} - \frac{1}{2}\bar{L}_m i_{bs} + L_{ss}i_{cs} + \psi_m\left(a_1\sin\left(\theta_r - \frac{2}{3}\pi\right) + a_2\sin^3\left(\theta_r - \frac{2}{3}\pi\right)\right.
$$
$$
\left. + a_3\sin^5\left(\theta_r - \frac{2}{3}\pi\right)\right). \tag{6.41}
$$

Using the Kirchhoff second law (6.20) and (6.41), we obtain

$$
u_{as} = r_s i_{as} + L_{ss}\frac{di_{as}}{dt} - \frac{1}{2}\bar{L}_m\frac{di_{bs}}{dt} - \frac{1}{2}\bar{L}_m\frac{di_{cs}}{dt}
$$
$$
+ \psi_m\cos\theta_r\left(a_1 + 3a_2\sin^2\theta_r + 5a_3\sin^4\theta_r\right)\omega_r,
$$

$$
u_{bs} = r_s i_{bs} - \frac{1}{2}\bar{L}_m\frac{di_{as}}{dt} + L_{ss}\frac{di_{bs}}{dt} - \frac{1}{2}\bar{L}_m\frac{di_{cs}}{dt}
$$
$$
+ \psi_m\cos\left(\theta_r + \frac{2}{3}\pi\right)\left(a_1 + 3a_2\sin^2\left(\theta_r + \frac{2}{3}\pi\right) + 5a_3\sin^4\left(\theta_r + \frac{2}{3}\pi\right)\right)\omega_r,
$$

$$
u_{cs} = r_s i_{cs} - \frac{1}{2}\bar{L}_m\frac{di_{as}}{dt} - \frac{1}{2}\bar{L}_m\frac{di_{bs}}{dt} + L_{ss}\frac{di_{cs}}{dt}
$$
$$
+ \psi_m\cos\left(\theta_r - \frac{2}{3}\pi\right)\left(a_1 + 3a_2\sin^2\left(\theta_r - \frac{2}{3}\pi\right) + 5a_3\sin^4\left(\theta_r - \frac{2}{3}\pi\right)\right)\omega_r.
$$

Therefore,

$$
\frac{di_{as}}{dt} = \frac{1}{L_{ss}}\left[-r_s i_{as} + \frac{1}{2}\bar{L}_m\frac{di_{bs}}{dt} + \frac{1}{2}\bar{L}_m\frac{di_{cs}}{dt}\right.
$$
$$
\left. - \psi_m\cos\theta_r\left(a_1 + 3a_2\sin^2\theta_r + 5a_3\sin^4\theta_r\right)\omega_r + u_{as}\right],
$$

$$\frac{di_{bs}}{dt} = \frac{1}{L_{ss}}\left[-r_s i_{bs} + \frac{1}{2}\bar{L}_m \frac{di_{as}}{dt} + \frac{1}{2}\bar{L}_m \frac{di_{cs}}{dt} \right.$$

$$-\psi_m \cos\left(\theta_r + \frac{2}{3}\pi\right)\left(a_1 + 3a_2 \sin^2\left(\theta_r + \frac{2}{3}\pi\right)\right.$$

$$\left. + 5a_3 \sin^4\left(\theta_r + \frac{2}{3}\pi\right)\right)\omega_r + u_{bs} \right].$$

$$\frac{di_{bs}}{dt} = \frac{1}{L_{ss}}\left[-r_s i_{cs} + \frac{1}{2}\bar{L}_m \frac{di_{as}}{dt} + \frac{1}{2}\bar{L}_m \frac{di_{bs}}{dt} \right.$$

$$-\psi_m \cos\left(\theta_r - \frac{2}{3}\pi\right)\left(a_1 + 3a_2 \sin^2\left(\theta_r - \frac{2}{3}\pi\right)\right.$$

$$\left. + 5a_3 \sin^4\left(\theta_r - \frac{2}{3}\pi\right)\right)\omega_r + u_{cs} \right]. \tag{6.42}$$

From (6.37) and (6.41), the expression for the electromagnetic torque is

$$T_e = \frac{P\psi_m}{2}\left[i_{as} \cos\theta_r \left(a_1 + 3a_2 \sin^2\theta_r + 5a_3 \sin^4\theta_r\right)\right.$$

$$+ i_{bs} \cos\left(\theta_r + \frac{2}{3}\pi\right)\left(a_1 + 3a_2 \sin^2\left(\theta_r + \frac{2}{3}\pi\right) + 5a_3 \sin^4\left(\theta_r + \frac{2}{3}\pi\right)\right)$$

$$\left. + i_{cs} \cos\left(\theta_r - \frac{2}{3}\pi\right)\left(a_1 + 3a_2 \sin^2\left(\theta_r - \frac{2}{3}\pi\right) + 5a_3 \sin^4\left(\theta_r - \frac{2}{3}\pi\right)\right) \right].$$

The *torsional-mechanical* equations of motion (6.38) are

$$\frac{d\omega_r}{dt} = \frac{P^2\psi_m}{4J}\left[i_{as} \cos\theta_r \left(a_1 + 3a_2 \sin^2\theta_r + 5a_3 \sin^4\theta_r\right)\right.$$

$$+ i_{bs} \cos\left(\theta_r + \frac{2}{3}\pi\right)\left(a_1 + 3a_2 \sin^2\left(\theta_r + \frac{2}{3}\pi\right) + 5a_3 \sin^4\left(\theta_r + \frac{2}{3}\pi\right)\right)$$

$$\left. + i_{cs} \cos\left(\theta_r - \frac{2}{3}\pi\right)\left(a_1 + 3a_2 \sin^2\left(\theta_r - \frac{2}{3}\pi\right) + 5a_3 \sin^4\left(\theta_r - \frac{2}{3}\pi\right)\right) \right]$$

$$-\frac{B_m}{J}\omega_r - \frac{P}{2J}T_L,$$

$$\frac{d\theta_r}{dt} = \omega_r. \tag{6.43}$$

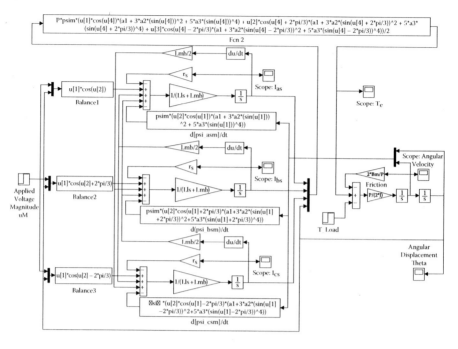

FIGURE 6.22
Simulink block diagram to simulate permanent-magnet synchronous motors when $a_1 \neq 0$, $a_2 \neq 0$, $a_3 \neq 0$, and $\forall a_n = 0$, $n > 3$ (ch6 _ 03.mdl).

Using (6.42) and (6.43), the Simulink block diagram to simulate permanent-magnet synchronous motors ($a_1 \neq 0$, $a_2 \neq 0$, $a_3 \neq 0$, and $\forall a_n = 0$, $n > 3$) is developed as reported in Figure 6.22. The following phase voltages, as functions of θ_r, are supplied

$$u_{as} = \sqrt{2}u_M \cos\theta_r, \quad u_{bs} = \sqrt{2}u_M \cos\left(\theta_r + \frac{2}{3}\pi\right) \quad \text{and}$$

$$u_{as} = \sqrt{2}u_M \cos\left(\theta_r - \frac{2}{3}\pi\right).$$

This voltage set can be used because $a_1 \gg (2n-1)a_n$, $\forall n > 1$.

The motor dynamics is studied as the motor accelerates from stall with the rated voltage applied ($u_M = 50$ V) at no load. The load torque $T_L = 0.1$ N-m is applied at $t = 0.025$ sec. Figure 6.23 illustrates the evolution of the phase currents and electrical angular velocity. The motor reaches the steady-state ω_r (500 rad/sec with no load, and 490 rad/sec if $T_L = 0.1$ N-m, respectively) within 0.02 sec. The angular velocity reduces and phase currents magnitude increases as T_L is applied. The current dynamics, reported in Figure 6.23,

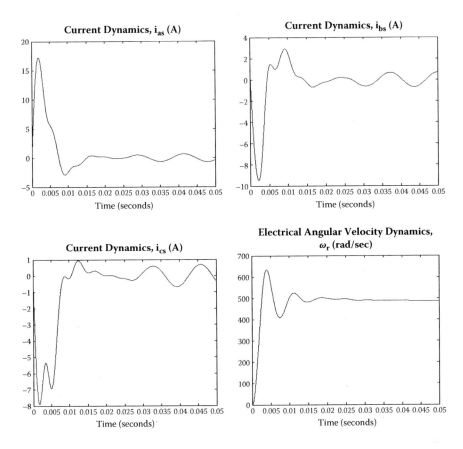

FIGURE 6.23
Transient dynamics of a permanent-magnet synchronous motor variables for the light load
($T_L = 0.1$ N-m at $t = 0.025$ sec).

allows one to assess the motor performance and capabilities when $a_1 = 1$ and
$\forall a_n = 0, n > 1$.

The phase currents and angular velocity are plotted using the following
statements

```
% Plots of the transient dynamics for a permanent-magnet synchronous motor
plot(Ias(:,1),Ias(:,2)); xlabel('Time [seconds]','FontSize',14);
title('Current Dynamics, i _ a _ s [A]','FontSize',14); pause;
plot(Ibs(:,1),Ibs(:,2)); xlabel('Time [seconds]','FontSize',14);
title('Current Dynamics, i _ b _ s [A]','FontSize',14); pause;
plot(Ics(:,1),Ics(:,2)); xlabel('Time [seconds]','FontSize',14);
title('Current Dynamics, i _ c _ s [A]','FontSize',14); pause;
plot(wrm(:,1),wrm(:,2)); axis([0,0.05,0,700]); xlabel('Time [seconds]','FontSize',14);
title('Electrical Angular Velocity Dynamics, \omega _ r [rad/sec]','FontSize',14);
```

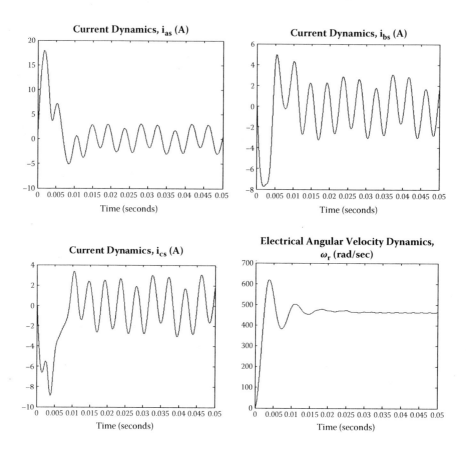

FIGURE 6.24
Transient dynamics of a permanent-magnet synchronous motor variables for the rated load
($T_{L0} = 0.2$ N-m and $T_L = 0.5$ N-m at $t = 0.025$ sec).

For the loaded motor, a_n coefficients (found by examining the induced *emf*
when the motor is operated as a generator) are $a_1 = 1$, $a_2 = 0.05$, $a_3 = 0.02$,
and $\forall a_n = 0$, $n > 3$. The motor accelerates with the rated voltage applied, and
$u_M = 50$ V. At $t = 0$ sec, the load torque is $T_{L0} = 0.2$ N-m. At $t = 0.025$ sec, the
load increases to $T_L = 0.5$ N-m. The evolution of the phase currents and the
electrical angular velocity are documented in Figure 6.24. The motor reaches
the steady-state operation within 0.02 sec. One observes the phase current
chattering and the electromagnetic torque ripple. This leads to the reduction of efficiency, losses, vibration, noise, heating, and so on. There are many
other secondary effects which degrade the motor performance, for example,
for the overheated motor, ψ_m reduces leading to the reduction of T_e as well as
torque density. Using the conditional statements and look-up table, the near-
balanced voltage sets can be implemented in a full operating envelope. The

closed-loop electromechanical systems must be designed to ensure optimal performance guaranteeing *achievable* capabilities. □

Example 6.9 Modeling, Analysis, and Simulation of a Permanent-Magnet Synchronous Generator

Let the experimental results indicate that in the full operating envelope $a_1 \neq 0$, $a_2 \neq 0$, $a_3 \neq 0$, and $\forall a_n = 0$, $n > 3$. The equations of motion for permanent-magnet synchronous generators is developed by using the circuitry-electromagnetic equations (6.20) with (6.36), expression for the electromagnetic torque (6.37), as well as the *torsional-mechanical* dynamics

$$
\frac{d\omega_r}{dt} = -T_e - \frac{B_m}{J}\omega_r + \frac{P}{2J}T_{pm} = -\frac{P^2\psi_m}{4J}\left[i_{as} \sum_{n=1}^{\infty}(2n-1)a_n \cos\theta_r \sin^{2n-2}\theta_r \right.
$$

$$
+ i_{bs} \sum_{n=1}^{\infty}(2n-1)a_n \cos\left(\theta_r - \frac{2}{3}\pi\right)\sin^{2n-2}\left(\theta_r - \frac{2}{3}\pi\right)
$$

$$
\left. + i_{cs} \sum_{n=1}^{\infty}(2n-1)a_n \cos\left(\theta_r + \frac{2}{3}\pi\right)\sin^{2n-2}\left(\theta_r + \frac{2}{3}\pi\right) \right] - \frac{B_m}{J}\omega_r + \frac{P}{2J}T_{pm},
$$

$$
\frac{d\theta_r}{dt} = \omega_r.
$$

The torque developed by the prime mover T_{pm} rotates the generator, and the terminal phase voltages are the induced *emfs*. We simulate and analyze a radial topology three-phase permanent-magnet synchronous generator driven by a prime mover, which develops the constant torque $T_{pm} = 0.05$ N-m. The generator parameters are the same as used in Example 6.8. Using the expressions for the flux linkages (6.41), one finds the circuit-electromagnetic dynamics as

$$
\frac{di_{as}}{dt} = \frac{1}{L_{ss}}\left[-(r_s + R_L)i_{as} + \frac{1}{2}\bar{L}_m\frac{di_{bs}}{dt} + \frac{1}{2}\bar{L}_m\frac{di_{cs}}{dt} \right.
$$

$$
\left. - \psi_m \cos\theta_r (a_1 + 3a_2 \sin^2\theta_r + 5a_3 \sin^4\theta_r)\omega_r \right],
$$

$$
\frac{di_{bs}}{dt} = \frac{1}{L_{ss}}\left[-(r_s + R_L)i_{bs} + \frac{1}{2}\bar{L}_m\frac{di_{as}}{dt} + \frac{1}{2}\bar{L}_m\frac{di_{cs}}{dt} \right.
$$

$$
\left. - \psi_m \cos\left(\theta_r + \frac{2}{3}\pi\right)\left(a_1 + 3a_2 \sin^2\left(\theta_r + \frac{2}{3}\pi\right) + 5a_3 \sin^4\left(\theta_r + \frac{2}{3}\pi\right)\right)\omega_r \right],
$$

$$\frac{di_{cs}}{dt} = \frac{1}{L_{ss}}\left[-(r_s + R_L)i_{cs} + \frac{1}{2}\bar{L}_m\frac{di_{as}}{dt} + \frac{1}{2}\bar{L}_m\frac{di_{bs}}{dt}\right.$$

$$\left. -\psi_m\cos\left(\theta_r - \frac{2}{3}\pi\right)\left(a_1 + 3a_2\sin^2\left(\theta_r - \frac{2}{3}\pi\right) + 5a_3\sin^4\left(\theta_r - \frac{2}{3}\pi\right)\right)\omega_r\right].$$

The *torsional-mechanical* equations are

$$\frac{d\omega_r}{dt} = -\frac{P^2\psi_m}{4J}\left[i_{as}\cos\theta_r(a_1 + 3a_2\sin^2\theta_r + 5a_3\sin^4\theta_r)\right.$$

$$+ i_{bs}\cos\left(\theta_r + \frac{2}{3}\pi\right)\left(a_1 + 3a_2\sin^2\left(\theta_r + \frac{2}{3}\pi\right) + 5a_3\sin^4\left(\theta_r + \frac{2}{3}\pi\right)\right)$$

$$\left. + i_{cs}\cos\left(\theta_r - \frac{2}{3}\pi\right)\left(a_1 + 3a_2\sin^2\left(\theta_r - \frac{2}{3}\pi\right) + 5a_3\sin^4\left(\theta_r - \frac{2}{3}\pi\right)\right)\right]$$

$$- \frac{B_m}{J}\omega_r + \frac{P}{2J}T_L,$$

$$\frac{d\theta_r}{dt} = \omega_r.$$

For no load and light load conditions, when $R_L\in[100 \; \infty]$ ohm, one has $a_1 = 1$ and $\forall a_n = 0$, $n > 1$. For the heavy loaded generator, as $R_L\in[10 \quad 75)$ ohm, $a_1 = 1$, $a_2 = 0.05$, $a_3 = 0.02$, and $\forall a_n = 0$, $n > 3$. To simulate permanent-magnet synchronous generators ($a_1 \neq 0$, $a_2 \neq 0$, $a_3 \neq 0$, and $\forall a_n = 0$, $n > 3$), described by five nonlinear differential equations, the corresponding Simulink diagram is illustrated in Figure 6.25.

The generator dynamics and voltage generation are studied as the generator accelerates from stall with $T_{pm} = 0.05$ N-m. The symmetric three-phase load resistors $R_L = 150$ ohm and $R_L = 100$ ohm are inserted at $t = 0$ sec and $t = 0.5$ sec, respectively. Figure 6.26 illustrates the evolution of the induced *emfs* (terminal phase voltages) and the electrical angular velocity. The generator reaches the steady-state ω_r when $T_{pm} = T_e + T_{friction}$. The generator is loaded by the symmetric wye-connected three-phase resistors R_L, which are in series with r_s of the *abc* phases. As the load increases (R_L are reduced), the angular velocity and induced *emfs* decrease. The generator dynamics, reported in Figure 6.26, also allows one to assess the lightly loaded generator performance and capabilities when $a_1 = 1$ and $\forall a_n = 0$, $n > 1$.

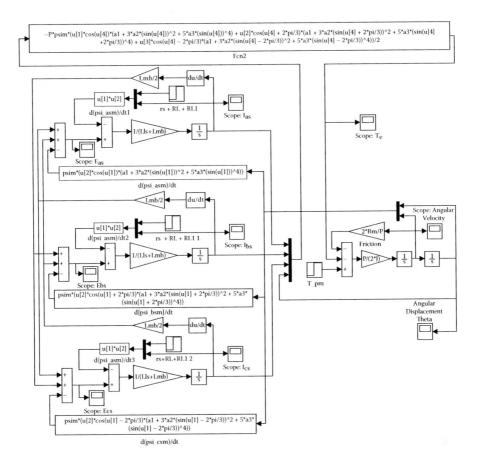

FIGURE 6.25
Simulink diagram to simulate permanent-magnet synchronous generators if $a_1 \neq 0$, $a_2 \neq 0$, $a_3 \neq 0$, and $\forall a_n = 0$, $n > 3$ (ch6 _ 04.mdl).

The induced *emfs* and angular velocity are plotted using the following statements

```
% Plots of the transient dynamics for a permanent-magnet synchronous motor
plot(Eas(:,1),Eas(:,2)); xlabel('Time [seconds]','FontSize',14);
title('Induced Voltage,emf _ a _ s [V]','FontSize',14); pause;
plot(Ebs(:,1),Ebs(:,2)); xlabel('Time [seconds]','FontSize',14);
title('Induced Voltage,emf _ b _ s [V]','FontSize',14); pause;
plot(Ecs(:,1),Ecs(:,2)); xlabel('Time [seconds]','FontSize',14);
title('Induced Voltage,emf _ c _ s [V]','FontSize',14); pause;
plot(wrm(:,1),wrm(:,2)); xlabel('Time [seconds]','FontSize',14);
title('Electrical Angular Velocity, \omega _ r [rad/sec]','FontSize',14);
```

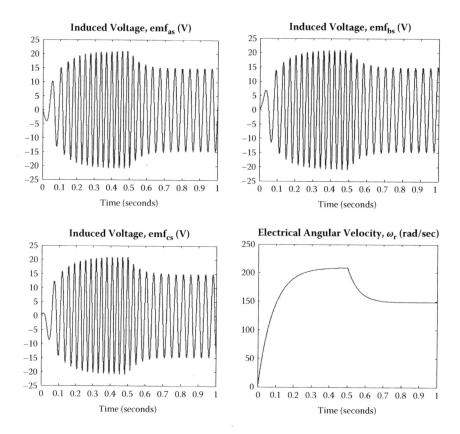

FIGURE 6.26
Transient dynamics of a permanent-magnet synchronous generator ($T_{pm} = 0.05$ N-m, $R_L = 150$ ohm at $t = 0$ sec, and $R_L = 100$ ohm at $t = 0.5$ sec).

For the loaded generator, when $R_L \in [10 \quad 75)$ ohm (peak and rated load, respectively), we have $a_1 = 1$, $a_2 = 0.05$, $a_3 = 0.02$, and $\forall a_n = 0$, $n > 3$. The generator performance is studied if $R_L = 25$ ohm at $t = 0$ sec, and $R_L = 75$ ohm at $t = 0.5$ sec. The evolution of the induced *emfs* and the electrical angular velocity are documented in Figure 6.27. As R_L increases at $t = 0.5$ sec, the load reduces, and the angular velocity increases. This results in the increase of terminal phase voltages. The induced *emfs* can be significantly distorted if the coefficients a_n, $n > 1$ are relatively high as compared to a_1. This results in the efficiency reduction, losses, and other undesirable phenomena. Therefore, permanent-magnet synchronous machines are designed with the attempt to guarantee near-optimal design ensuring $a_1 \gg (2n - 1)a_n$, $\forall n > 1$ in the full operating envelope. This objective, in general, can be achieved through a coherent structural machine design, advanced technologies, enhanced processes, and appropriate materials.

$\square$

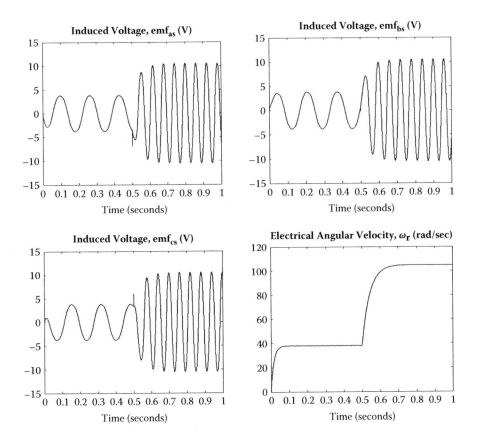

FIGURE 6.27

Transient dynamics of a permanent-magnet synchronous generator ($T_{pm} = 0.05$ N-m, $R_L = 25$ ohm at $t = 0$ sec, and $R_L = 75$ ohm at $t = 0.5$ sec).

6.4 Axial Topology Permanent-Magnet Synchronous Machines

In some automotive, avionics, biotechnology, energy systems, marine, medical, robotics, and other applications, axial topology permanent-magnet synchronous machines could be a preferable solution. The axial topology permanent-magnet dc electromechanical motion devices were examined in Section 4.2. In synchronous machines, the stationary magnetic field is established by the permanent magnets on the rotor, and the ac voltages are applied to the stator windings as functions of θ_r. The image of a three-phase axial topology permanent-magnet synchronous machine is reported in Figure 6.28.

FIGURE 6.28
Stator (with planar windings) and rotor (with magnet segments array) of a three-phase axial topology permanent-magnet synchronous machine.

It is believed that the *Escherichia coli* biomotor possesses an axial topology. The *E. coli* filament is driven by a 45 nm rotor of the biomotor embedded in the cell wall. The cytoplasmic membrane forms a "stator." This biomotor structure integrates more than 20 proteins, and the electrochemomechanical motion device may operate because of the axial *protonomotive* force resulting from the proton flux (see Figure 6.29). There is a limited knowledge even for

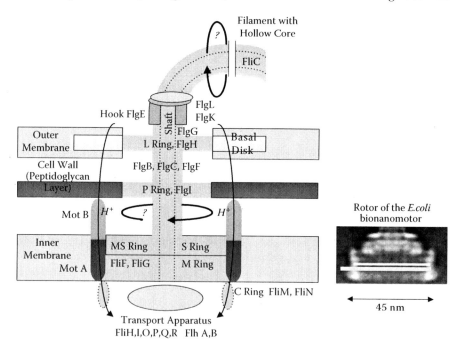

FIGURE 6.29
E. coli biomotor (motor-bearing-coupling-flagella complex formed by different proteins) and rotor image.

the most widely examined *E. coli* biomotor to make conclusive conclusions. With some degree of confidence, one may guess that the biomotor has an axial topology (motor utilizes the proton or sodium gradient, maintained across the cell's inner membrane as the energy source, and the torque is developed due to the axial flux). The efficiency and capabilities of the *E. coli* bacteria propulsion system with controlled biomotor are remarkable. It is of a great interest to comprehend and prototype the torque generation, energy conversion, bearing, sensing-feedback-control, and other mechanisms. Complex electro-chemo-mechanical energy generation, harvesting, and conversion, as well other transitions and phenomena inherently exhibited by biosystems are far-reaching research. However, as reported in Figures 6.28 and 6.29, the topology of axial topology permanent-magnet synchronous machine typifies, to some degree, axial biomotors.

Axial topology permanent-magnet synchronous machines can be fabricated in different sizes. In contrast to the conventional technology as applied to a device as illustrated in Figure 6.28, micro- and miniscale axial topology permanent-magnet synchronous machines have been fabricated using surface and bulk micromachining and CMOS technology. The advantages of axial topology electromechanical motion devices are excellent performance and capabilities, as well as affordability as a result of fabrication, assembly, and packaging simplicity. This simplicity results because: (1) magnets are flat (planar) and made as planar segmented arrays; (2) there are no strict three-dimensional shape and magnetization requirements imposed on magnets; (3) rotor back ferromagnetic material is not required; (4) it is easy to make planar windings on the flat (planar) stator. The axial topology permanent-magnet synchronous machines are reported in Figure 6.30. Two- and three-phase axial permanent-magnet synchronous machines are illustrated in Figures 6.28 and 6.30. For three-phase synchronous machines, one supplies phase voltages u_{as}, u_{bs} and u_{cs} as functions of θ_r. For two-phase machines, phase voltages u_{as} and u_{bs} are applied.

We cover the basic physics, modeling, and analysis of axial topology permanent-magnet synchronous machines. The fundamentals and basic electromagnetics are with the correspondence as covered for axial topology dc electric machines (see Section 4.2). One recalls a current loop in the magnetic field, which is produced by the permanent magnets. Assuming that the magnetic flux is constant through the magnetic plane (current loop), the torque on a planar current loop of any size and shape in the uniform magnetic field is

$$\vec{T} = i\vec{s} \times \vec{B} = \vec{m} \times \vec{B}.$$

In permanent-magnet dc motion devices, brushes are needed to supply the voltage to the armature windings on the rotor, and a commutator is utilized. In contrast, in permanent-magnet synchronous motors, phase windings are on the stator (stationary part). Correspondingly, brushes are not needed. However, to develop the electromagnetic torque, the phase voltages

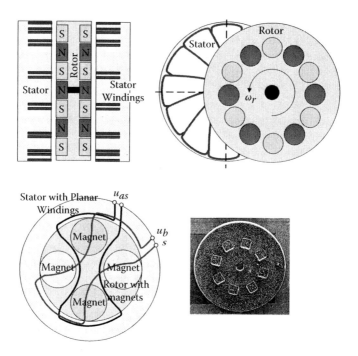

FIGURE 6.30
Axial topology permanent-magnet synchronous machines and fabricated ministructure (rotor or stator) with planar segmented magnets.

are supplied as functions of the angular displacement θ_r. Hence, there is a need for the angular displacement sensors and advanced power amplifiers.

Consider axial topology permanent-magnet synchronous machines. The *effective* phase flux density varies as a function of θ_r because of the angular displacement of the rotor with magnets relative to the stator with windings. For three-phase axial topology permanent-magnet synchronous machines, depending on the magnet magnetization, geometry and shape, one applies distinct expressions for the *effective* phase flux densities $B_{as}(\theta_r)$, $B_{bs}(\theta_r)$, and $B_{cs}(\theta_r)$, which are periodic functions of θ_{rm}. For ac machines, we use the electric angular displacement θ_r which is related to the mechanical angular displacement θ_{rm} as $\theta_{rm} = 2\theta_r/N_m$, where N_m is the number of magnets (segments). If an optimal structural design is accomplished and magnets are properly magnetized, one may have

$$B_{as}(\theta_r) = B_{\max} \sin(\theta_r), \quad B_{bs}(\theta_r) = B_{\max} \sin\left(\theta_r - \frac{2}{3}\pi\right) \text{ and}$$

$$B_{cs}(\theta_r) = B_{\max} \sin\left(\theta_r + \frac{2}{3}\pi\right).$$

where B_{max} is the maximum effective flux density produced by the magnets as viewed from the winding (B_{max} depends on the magnets used, magnet-winding separation, temperature, etc.).

For the specific magnet (magnet array) topologies, the particular *effective* phase flux densities result. For example, one may find

$$B_{as}(\theta_r) = B_{max} \left| \sin \theta_r \right|, \; B_{bs}(\theta_r) = B_{max} \left| \sin\left(\theta_r - \frac{2}{3}\pi\right) \right| \text{ and}$$

$$B_{cs}(\theta_r) = B_{max} \left| \sin\left(\theta_r + \frac{2}{3}\pi\right) \right|.$$

Using experimentally and analytically derived $a_n(\mathbf{E,D,B,H,i}_{abcs}, \omega_r, \mathbf{T}_L, \varepsilon, \mu, \Sigma)$, we have

$$B_{as} = B_{max} \sum_{n=1}^{\infty} a_n \sin^{2n-1}\theta_r, \quad B_{bs} = B_{max} \sum_{n=1}^{\infty} a_n \sin^{2n-1}\left(\theta_r - \frac{2}{3}\pi\right) \text{ and}$$

$$B_{cs} = B_{max} \sum_{n=1}^{\infty} a_n \sin^{2n-1}\left(\theta_r + \frac{2}{3}\pi\right). \tag{6.44}$$

From (6.44), using the number of turns and the *effective* area, one obtains the expression for the flux linkages

$$\begin{bmatrix} \psi_{as} \\ \psi_{bs} \\ \psi_{cs} \end{bmatrix} = \begin{bmatrix} L_{ss} & 0 & 0 \\ 0 & L_{ss} & 0 \\ 0 & 0 & L_{ss} \end{bmatrix} \begin{bmatrix} i_{as} \\ i_{bs} \\ i_{cs} \end{bmatrix} + \psi_m \begin{bmatrix} \sum_{n=1}^{\infty} a_n \sin^{2n-1}\theta_r \\ \sum_{n=1}^{\infty} a_n \sin^{2n-1}\left(\theta_r - \frac{2}{3}\pi\right) \\ \sum_{n=1}^{\infty} a_n \sin^{2n-1}\left(\theta_r + \frac{2}{3}\pi\right) \end{bmatrix}, \tag{6.45}$$

as was reported in Section 6.3.4. In (6.45) the mutual inductances between the planar windings are zero (or negligibly small), whereas there are mutual inductances $-\bar{L}_m/2$ in (6.36) due to the ferromagnetic core of radial topology electric machines. Substituting (6.45) in Kirchhoff's second law (6.20), the circuitry-electromagnetic equations of motion in the *machine* variables result.

The expression

$$T_e = \frac{N_m}{2} \frac{\partial W_c}{\partial \theta_r}$$

yields

$$T_e = \frac{N_m \psi_m}{2} \left[i_{as} \sum_{n=1}^{\infty} (2n-1)a_n \cos\theta_r \sin^{2n-2}\theta_r \right.$$

$$+ i_{bs} \sum_{n=1}^{\infty} (2n-1)a_n \cos\left(\theta_r - \frac{2}{3}\pi\right) \sin^{2n-2}\left(\theta_r - \frac{2}{3}\pi\right)$$

$$\left. + i_{cs} \sum_{n=1}^{\infty} (2n-1)a_n \cos\left(\theta_r + \frac{2}{3}\pi\right) \sin^{2n-2}\left(\theta_r + \frac{2}{3}\pi\right) \right]. \qquad (6.46)$$

Hence, the *torsional-mechanical* dynamics is

$$\frac{d\omega_r}{dt} = \frac{N_m^2 \psi_m}{4J} \left[i_{as} \sum_{n=1}^{\infty} (2n-1)a_n \cos\theta_r \sin^{2n-2}\theta_r \right.$$

$$+ i_{bs} \sum_{n=1}^{\infty} (2n-1)a_n \cos\left(\theta_r - \frac{2}{3}\pi\right) \sin^{2n-2}\left(\theta_r - \frac{2}{3}\pi\right)$$

$$\left. + i_{cs} \sum_{n=1}^{\infty} (2n-1)a_n \cos\left(\theta_r + \frac{2}{3}\pi\right) \sin^{2n-2}\left(\theta_r + \frac{2}{3}\pi\right) \right] - \frac{B_m}{J}\omega_r - \frac{N_m}{2J}T_L,$$

$$\frac{d\theta_r}{dt} = \omega_r. \qquad (6.47)$$

The near-balanced current and voltage sets are given by (6.39) and (6.40).

Example 6.10   Axial Topology Single-Phase Permanent-Magnet Synchronous Machine

Consider a single-phase permanent-magnet synchronous motor with the segmented array of the permanent magnets as illustrated in Figure 6.31.

Assume for a single-phase machine that the *effective* flux density is

$$B_{as}(\theta_{rm}) = B_{\max} \sin\left(\frac{1}{2}N_m\theta_{rm}\right) \quad \text{or} \quad B_{as}(\theta_r) = B_{\max} \sin\theta_r.$$

The electromagnetic torque, developed by single-phase axial topology permanent-magnet synchronous motors is found by using

$$\vec{T} = i\vec{s} \times \vec{B} = \vec{m} \times \vec{B} \quad \text{or} \quad \vec{T} = \vec{R} \times \vec{F}.$$

Alternatively, the expression for coenergy

$$W_c(i_{as}, \theta_r) = NA_{eq}B_{as}(\theta_r)i_{as}$$

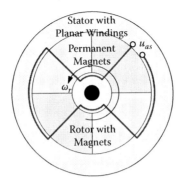

FIGURE 6.31
Axial topology single-phase permanent-magnet synchronous machine.

can be used to derive T_e. Here, N is the number of turns; A_{eq} is the *effective* area, which takes into account magnetic field nonuniformity, magnet non-uniformity, etc. We find

$$T_e = \frac{\partial W_c(i_{as}, \theta_r)}{\partial \theta_r} = \frac{N_m \Psi_m}{2} i_{as} \cos \theta_r.$$

The electromagnetic torque is not developed by the synchronous motors if one feeds the dc current or voltage to the winding. The average value of T_e is not equal to zero if the current is a function of the rotor displacement θ_r. As an illustration, we fed the phase current $i_{as} = i_M \cos \theta_r$ to the motor winding. The electromagnetic torque is

$$T_e = \frac{N_m \Psi_m}{2} i_M \cos^2 \theta_r$$

and $T_{e\,av} \neq 0$. However, there will be the torque ripple.

The equations of motion are found by using the following Kirchhoff's and Newton's second laws

$$u_{as} = r_s i_{as} + \frac{d\psi_{as}}{dt} \quad \text{and} \quad J \frac{d^2 \theta_{rm}}{dt^2} = T_e - B_m \omega_{rm} - T_L.$$

From the flux linkage equation $\psi_{as} = L_{ss} i_{as} + \psi_m \sin \theta_r$, we have

$$\frac{di_{as}}{dt} = \frac{1}{L_{ss}} (-r_s i_{as} - \psi_m \cos \theta_r \omega_r + u_{as})$$

$$\frac{d\omega_r}{dt} = \frac{N_m^2 \Psi_m}{4J} i_{as} \cos \theta_r - \frac{B_m}{J} \omega_r - \frac{N_m}{2J} T_L,$$

$$\frac{d\theta_r}{dt} = \omega_r.$$

□

Example 6.11 Axial Topology Two-Phase Permanent-Magnet Synchronous Motors

Consider two-phase permanent-magnet synchronous machines, as depicted in Figure 6.30. The *effective* phase flux densities are

$$B_{as} = B_{max}\sin\theta_r \quad \text{and} \quad B_{bs} = B_{max}\cos\theta_r.$$

The electromagnetic torque is found to be

$$T_e = \frac{N_m \Psi_m}{2}(i_{as}\cos\theta_r - i_{bs}\sin\theta_r).$$

The balanced current set is

$$i_{as} = i_M\cos\theta_r, \quad i_{bs} = -i_M\sin\theta_r.$$

Hence, the T_e is maximized, and from

$$T_e = \frac{N_m \Psi_m}{2}i_M$$

one concludes that there is no torque ripple.

Therefore, for two-phase permanent-magnet synchronous machines it is desired to ensure the structural design, which leads to $B_{as} = B_{max}\sin\theta_r$ and $B_{bs} = B_{max}\cos\theta_r$. The phase currents or voltages are the functions of the electrical angular displacement θ_r measured by the Hall-effect sensors or calculated by the observers in the sensorless control.

As an illustrative example, consider the case when $a_3 = 1$ and all other $\forall a_n = 0$. This flux distribution may correspond to the rotor with planar segmented magnets documented in the image in Figure 6.30. One obtains

$$\Psi_{asm} = \Psi_m\sin^5\theta_r \quad \text{and} \quad \Psi_{bsm} = \Psi_m\cos^5\theta_r.$$

The electromagnetic torque is

$$T_e = \frac{5N_m \Psi_m}{2}\left(i_{as}\cos\theta_r\sin^4\theta_r - i_{bs}\sin\theta_r\cos^4\theta_r\right).$$

Assume the motor parameters are

$$N = 20, \quad A_{eq} = 0.001, \quad B_{max} = 1, \quad \text{and} \quad N_m = 8.$$

Let the phase currents be

$$i_{as} = i_M\cos\theta_r \quad \text{and} \quad i_{bs} = -i_M\sin\theta_r, \quad i_M = 2 \text{ A}.$$

The expression for the electromagnetic torque, deviations, and plotting are performed using the Symbolic Math Toolbox. We apply the equation for the coenergy to obtain the electromagnetic torque performing the differentiation

$$T_e = \frac{\partial W_c(i_{as}, \theta_r)}{\partial \theta_r}.$$

The following MATLAB file with comments is used

```
% To use a symbolic variable, create an object of type SYM
x=sym('x');
N=20; Aeq=0.001; Bmax=1; psim=N*Aeq*Bmax; Nm=8; iM=2;
y1=N*Aeq*Bmax*sin(x)^5; y2=N*Aeq*Bmax*cos(x)^5;
% Differentiate y1 and y2 using the DIFF command
d1=diff(y1); d2=diff(y2);
% Phase currents
ias=iM*cos(x); ibs=-iM*sin(x);
% Derive and plot the electromagnetic torque
Te=Nm*(d1*ias+d2*ibs)/2, Te=simplify(Te), ezplot(Te)
```

The results of the calculations are given in the Command Window as reported below

```
Te = 4/5*sin(x)^4*cos(x)^2+4/5*cos(x)^4*sin(x)^2
Te = -4/5*(-1+cos(x)^2)*cos(x)^2
```

One concludes that the electromagnetic torque is

$$T_e = -\frac{4}{5}\left(-1 + \cos^2\theta_r\right)\cos^2\theta_r \text{ N-m.}$$

The plot for the electromagnetic torque is reported in Figure 6.32a. The electromagnetic torque varies as a sinusoidal-like function of the rotor angular displacement. The torque ripple is an undesirable phenomena due to losses, noise, vibration, and so on. To minimize the torque ripple, one can perform structural redesign attempting to ensure sinusoidal flux distribution. Alternatively, we may derive and feed (if implementable) the proper phase currents.
    Having obtained

$$T_e = \frac{5N_m\psi_m}{2}\left(i_{as}\cos\theta_r\sin^4\theta_r - i_{bs}\sin\theta_r\cos^4\theta_r\right),$$

the current set

$$i_{as} = \frac{i_M\cos\theta_r}{\sin^4\theta_r} \quad \text{and} \quad i_{bs} = \frac{-i_M\sin\theta_r}{\cos^4\theta_r},$$

leads to the maximization of the electromagnetic torque and elimination of the torque ripple. Using

```
% Phase currents
ias=iM*cos(x)/sin(x)^4; ibs=-iM*sin(x)/cos(x)^4;
```

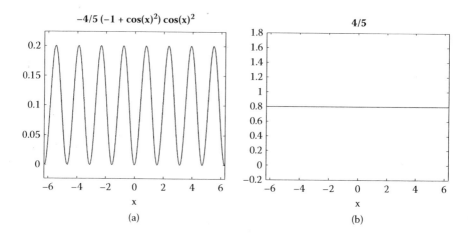

FIGURE 6.32
Electromagnetic torque: (a) $T_e = -\frac{4}{5}(-1+\cos^2\theta_r)\cos^2\theta_r$ N-m; (b) $T_e = 0.8$ N-m.

```
% Derive and plot the electromagnetic torque
Te=Nm*(d1*ias+d2*ibs)/2, Te=simplify(Te), ezplot(Te)
```

we obtain the following results

```
Te = 4/5*cos(x)^2+4/5*sin(x)^2
Te = 4/5
```

That is, $T_e = 0.8$ N-m. This T_e is documented in Figure 6.32b. However, because of the singularity and saturation, it is impossible to implement the current set $i_{as} = i_M\cos\theta_r/\sin^4\theta_r$ and $i_{bs} = -i_M\sin\theta_r/\cos^4\theta_r$. The following near-balanced current set can be applied

$$i_{as} = \frac{i_M\cos\theta_r}{(\sin^4\theta_r+\varepsilon)} \quad \text{if} \left|\frac{i_M\cos\theta_r}{(\sin^4\theta_r+\varepsilon)}\right| \le i_{max}, \quad i_{as} = i_{max} \quad \text{or}$$

$$i_{as} = -i_{max} \quad \text{otherwise,}$$

$$i_{bs} = \frac{i_M\sin\theta_r}{(\cos^4\theta_r+\varepsilon)} \quad \text{if} \left|\frac{i_M\sin\theta_r}{(\cos^4\theta_r+\varepsilon)}\right| \le i_{max}, i_{bs} = -i_{max} \quad \text{or}$$

$$i_{bs} = i_{max} \quad \text{otherwise.}$$

$\square$

Example 6.12   Axial Topology Three-Phase Permanent-Magnet Synchronous Motors

Our goal is to model, simulate, and analyze three-phase synchronous motors. In the full operating envelope, assume $a_1 \neq 0$, $a_2 \neq 0$, $a_3 \neq 0$, whereas $\forall a_n = 0$,

$n > 3$. Therefore, from (6.44) and (6.45) one finds

$$\psi_{as} = L_{ss}i_{as} + L_{asbs}i_{bs} + L_{ascs}i_{cs} + \psi_m\left(a_1\sin\theta_r + a_2\sin^3\theta_r + a_3\sin^5\theta_r\right),$$

$$\psi_{bs} = L_{bsas}i_{as} + L_{ss}i_{bs} + L_{bscs}i_{cs} + \psi_m\left(a_1\sin\left(\theta_r + \frac{2}{3}\pi\right) + a_2\sin^3\left(\theta_r + \frac{2}{3}\pi\right)\right.$$

$$\left. + a_3\sin^5\left(\theta_r + \frac{2}{3}\pi\right)\right),$$

$$\psi_{cs} = L_{csas}i_{as} + L_{csbs}i_{bs} + L_{ss}i_{cs} + \psi_m\left(a_1\sin\left(\theta_r - \frac{2}{3}\pi\right) + a_2\sin^3\left(\theta_r - \frac{2}{3}\pi\right)\right.$$

$$\left. + a_3\sin^5\left(\theta_r - \frac{2}{3}\pi\right)\right).$$

The mutual inductances between the phase windings are zero. Using the total derivatives for the flux linkages in the Kirchhoff second law (6.20), we have

$$\frac{di_{as}}{dt} = \frac{1}{L_{ss}}\left[-r_s i_{as} - \psi_m\cos\theta_r\left(a_1 + 3a_2\sin^2\theta_r + 5a_3\sin^4\theta_r\right)\omega_r + u_{as}\right],$$

$$\frac{di_{bs}}{dt} = \frac{1}{L_{ss}}\left[-r_s i_{bs} - \psi_m\cos\left(\theta_r + \frac{2}{3}\pi\right)\left(a_1 + 3a_2\sin^2\left(\theta_r + \frac{2}{3}\pi\right)\right.\right.$$

$$\left.\left. + 5a_3\sin^4\left(\theta_r + \frac{2}{3}\pi\right)\right)\omega_r + u_{bs}\right],$$

$$\frac{di_{cs}}{dt} = \frac{1}{L_{ss}}\left[-r_s i_{cs} - \psi_m\cos\left(\theta_r - \frac{2}{3}\pi\right)\left(a_1 + 3a_2\sin^2\left(\theta_r - \frac{2}{3}\pi\right)\right.\right.$$

$$\left.\left. + 5a_3\sin^4\left(\theta_r - \frac{2}{3}\pi\right)\right)\omega_r + u_{cs}\right]. \tag{6.48}$$

From the expression for the electromagnetic torque (6.46), one yields

$$T_e = \frac{N_m\psi_m}{2}\left[i_{as}\cos\theta_r\left(a_1 + 3a_2\sin^2\theta_r + 5a_3\sin^4\theta_r\right)\right.$$

$$+ i_{bs}\cos\left(\theta_r + \frac{2}{3}\pi\right)\left(a_1 + 3a_2\sin^2\left(\theta_r + \frac{2}{3}\pi\right) + 5a_3\sin^4\left(\theta_r + \frac{2}{3}\pi\right)\right)$$

$$\left. + i_{cs}\cos\left(\theta_r - \frac{2}{3}\pi\right)\left(a_1 + 3a_2\sin^2\left(\theta_r - \frac{2}{3}\pi\right) + 5a_3\sin^4\left(\theta_r - \frac{2}{3}\pi\right)\right)\right].$$

The *torsional-mechanical* equations of motion (6.47) are

$$\frac{d\omega_r}{dt} = \frac{N_m^2 \psi_m}{4J} \left[i_{as} \cos\theta_r \left(a_1 + 3a_2 \sin^2\theta_r + 5a_3 \sin^4\theta_r \right) \right.$$

$$+ i_{bs} \cos\left(\theta_r + \frac{2}{3}\pi\right) \left(a_1 + 3a_2 \sin^2\left(\theta_r + \frac{2}{3}\pi\right) + 5a_3 \sin^4\left(\theta_r + \frac{2}{3}\pi\right) \right)$$

$$\left. + i_{cs} \cos\left(\theta_r - \frac{2}{3}\pi\right) \left(a_1 + 3a_2 \sin^2\left(\theta_r - \frac{2}{3}\pi\right) + 5a_3 \sin^4\left(\theta_r - \frac{2}{3}\pi\right) \right) \right]$$

$$- \frac{B_m}{J}\omega_r - \frac{P}{2J}T_L,$$

$$\frac{d\theta_r}{dt} = \omega_r. \tag{6.49}$$

Using (6.48) and (6.49), the Simulink block diagram to simulate axial topology permanent-magnet synchronous motors ($a_1 \neq 0$, $a_2 \neq 0$, $a_3 \neq 0$, $\forall a_n = 0$,

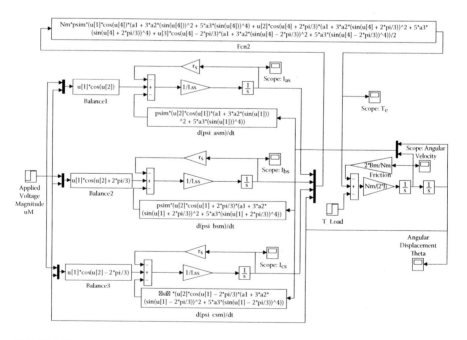

FIGURE 6.33
Simulink diagram for axial topology permanent-magnet synchronous motors ($a_1 \neq 0$, $a_2 \neq 0$, $a_3 \neq 0$, whereas $\forall a_n = 0$, $n > 3$) (ch6_05.mdl).

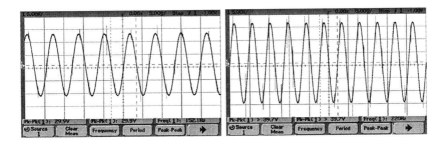

FIGURE 6.34
Induced $emf_{as\omega} = \psi_m \cos\theta_r \omega_r$ if the generator is rotated at 956 and 1382 rad/sec.

$n > 3$) is built as depicted in Figure 6.33. The phase voltages supplied are

$$u_{as} = \sqrt{2}u_M \cos\theta_r, \quad u_{bs} = \sqrt{2}u_M \cos\left(\theta_r + \frac{2}{3}\pi\right) \quad \text{and}$$

$$u_{cs} = \sqrt{2}u_M \cos\left(\theta_r - \frac{2}{3}\pi\right)$$

The 50 V motor parameters are experimentally found. We have $N_m = 8$, $r_s = 13.5$ ohm, $L_{ss} = 0.035$ H, $\psi_m = 0.03$ V-sec/rad (N-m/A), $B_m = 0.0000005$ N-m-sec/rad, and $J = 0.00001$ kg-m^2. The value of ψ_m is found by measuring the induced *emf* when the synchronous machine was rotated by the prime mover. The experimental results are illustrated in Figure 6.34. At no load, for two different steady-state ω_r (956 and 1382 rad/sec) the terminal phase voltage is 29.9 and 39.7 V, respectively. Even for the same load, ψ_m varies due to different operating envelope, and $\psi_m \in [0.029 \ 0.031]$ V-sec/rad assuming that $a_1 \neq 0$, $\forall a_n = 0$, $n > 1$.

For no load and light load conditions (T_L is ~ 0.01 N-m), the constants are $a_1 = 1$ and all other a_n are zeros ($\forall a_n = 0$, $n > 1$). For the loaded motor, we have $a_1 = 0.85$, $a_2 = 0.06$, $a_3 = 0.04$, and $\forall a_n = 0$, $n > 3$. The parameters are uploaded as

```
% Optimal distribution: Light loads
psim=0.03; a1=1; a2=0; a3=0;
% Near-optimal distribution: Heavy loads
psim=0.03; a1=0.85; a2=0.06; a3=0.04;
% Motor parameters
Nm=8; uM=50; rs=13.5; Lss=0.035; Bm=0.0000005; J=0.00001;
```

The motor dynamics is studied as the motor accelerates from stall with the rated voltage applied ($u_M = 50$ V) if $T_{L0} = 0.005$ N-m. The load torque $T_L = 0.01$ N-m is applied at $t = 0.5$ sec. Figure 6.35 illustrates the evolution of $i_{as}(t)$, $i_{bs}(t)$, $i_{cs}(t)$, and $\omega_r(t)$.

At $T_L = 0.005$ N-m, the steady-state ω_r is 1470 rad/sec, and the angular velocity decreases as the load applied at $t = 0.5$ sec.

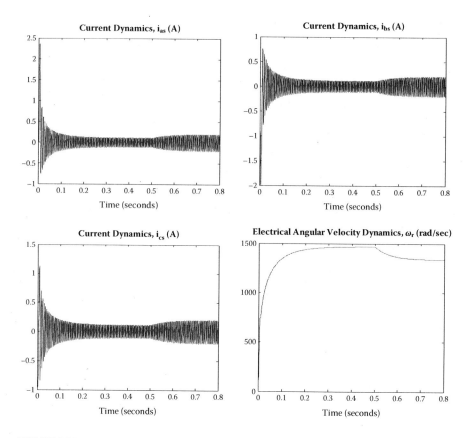

FIGURE 6.35
Transient dynamics of an axial topology permanent-magnet synchronous motor if $a_1 = 1$ and $\forall a_n = 0, n > 1$.

The phase currents and angular velocity are plotted using the following statements

```
% Plots of the transient dynamics for a permanent-magnet synchronous motor
plot(Ias(:,1),Ias(:,2)); xlabel('Time [seconds]','FontSize',14);
title('Current Dynamics, i_a_s [A]','FontSize',14); pause;
plot(Ibs(:,1),Ibs(:,2)); xlabel('Time [seconds]','FontSize',14);
title('Current Dynamics, i_b_s [A]','FontSize',14); pause;
plot(Ics(:,1),Ics(:,2)); xlabel('Time [seconds]','FontSize',14);
title('Current Dynamics, i_c_s [A]','FontSize',14); pause;
plot(wrm(:,1),wrm(:,2)); axis([0,0.8,0,1500]); xlabel('Time [seconds]','FontSize',14);
title('Electrical Angular Velocity Dynamics, \omega_r [rad/sec]','FontSize',14);
```

For the loaded motor, we have $a_1 = 0.85$, $a_2 = 0.06$, $a_3 = 0.04$, and $\forall a_n = 0, n > 3$. The motor accelerates with the rated voltage applied. At $t = 0$ sec, the load

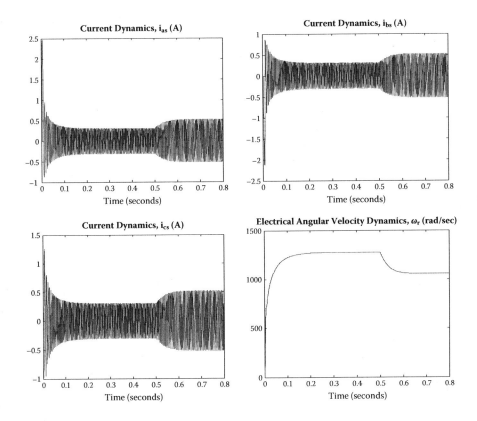

FIGURE 6.36
Transient dynamics of an axial topology permanent-magnet synchronous motor ($a_1 = 0.85$, $a_2 = 0.06$, $a_3 = 0.04$, and $\forall a_n = 0$, $n > 3$), $T_{L0} = 0.015$ N-m and $T_L = 0.03$ N-m at $t = 0.5$ sec.

torque is $T_{L0} = 0.015$ N-m, and $T_L = 0.03$ N-m is applied at $t = 0.5$ sec. The evolutions of the motor variables are reported in Figure 6.36. One observes the phase current chattering and the electromagnetic torque ripple. These undesirable phenomena can be minimized refining the phase voltages supplied.

□

6.5 Conventional Three-Phase Synchronous Machines

Consider three-phase synchronous machines which are widely used in high-power drives and power generation systems. There are significant challenges to design and build ~100 kW and higher permanent-magnet

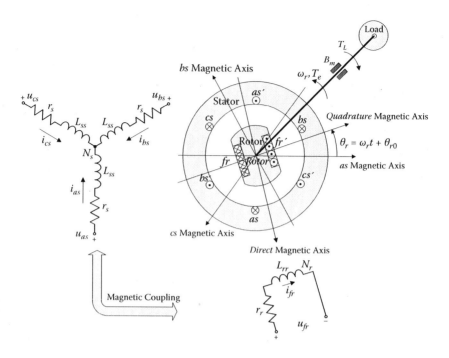

FIGURE 6.37
Three-phase wye-connected synchronous motor.

synchronous machines, whereas mega-watt and much higher power requirements are common. In those applications, particularly in power systems, conventional three-phase machines are utilized. Symmetric synchronous motors and generators with sinusoidally distributed windings are illustrated in Figures 6.37 and 6.38. The stator and rotor windings are denoted as *as*, *bs*, *cs*, and *fr*. The dc voltage is supplied to the rotor winding. Therefore, brushes are needed. The angular velocity of conventional synchronous machines usually is relatively low, and challenges do not arise to supply the dc voltage to the *fr* winding. Alternative solutions exist to induce the voltage supplied to the rotor *fr* winding using *exciters* (additional electric machines), permanent magnets, and so on. Furthermore, high-power machines have several field and compensating windings. All these possible solutions are typified within the considered case because alternative structural designs do not change the basic electromagnetics and operating features.

The application of Kirchhoff's voltage law for stator and rotor circuitry results in four differential equations for the stator (subscripts *as*, *bs*, and *cs*) and rotor (subscript *fr*) windings. We have

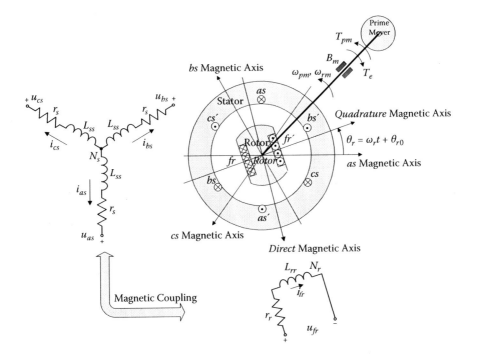

FIGURE 6.38
Three-phase wye-connected synchronous generator.

$$u_{as} = r_s i_{as} + \frac{d\psi_{as}}{dt}, \quad u_{bs} = r_s i_{bs} + \frac{d\psi_{bs}}{dt}, \quad u_{cs} = r_s i_{cs} + \frac{d\psi_{cs}}{dt},$$

$$u_{fr} = r_r i_{fr} + \frac{d\psi_{fr}}{dt}, \tag{6.50}$$

where the flux linkages are given as

$$\psi_{as} = L_{asas} i_{as} + L_{asbs} i_{bs} + L_{ascs} i_{cs} + L_{asfr} i_{fr}, \quad \psi_{bs} = L_{bsas} i_{as} + L_{bsbs} i_{bs} + L_{bscs} i_{cs} + L_{bsfr} i_{fr},$$

$$\psi_{cs} = L_{csas} i_{as} + L_{csbs} i_{bs} + L_{cscs} i_{cs} + L_{csfr} i_{fr}, \quad \psi_{fr} = L_{fras} i_{as} + L_{frbs} i_{bs} + L_{frcs} i_{cs} + L_{frfr} i_{fr}.$$

From (6.50), in matrix form, one obtains

$$\mathbf{u}_{abcs} = \mathbf{r}_s \mathbf{i}_{abcs} + \frac{d\psi_{abcs}}{dt}, \quad u_{fr} = r_r i_{fr} + \frac{d\psi_{fr}}{dt}.$$

For synchronous reluctance machines, the self- and mutual inductances between the stator windings were derived in Section 6.2.2. Studying the stator and rotor magnetic coupling, under the assumptions emphasized earlier, we obtain the following expressions for the stator and rotor flux linkages

$$\psi_{as} = (L_{ls} + \bar{L}_m - L_{\Delta m}\cos 2\theta_r)i_{as} + \left(-\frac{1}{2}\bar{L}_m - L_{\Delta m}\cos 2\left(\theta_r - \frac{1}{3}\pi\right)\right)i_{bs}$$

$$+ \left(-\frac{1}{2}\bar{L}_m - L_{\Delta m}\cos 2\left(\theta_r + \frac{1}{3}\pi\right)\right)i_{cs} + L_{md}\sin\theta_r i_{fr},$$

$$\psi_{bs} = \left(-\frac{1}{2}\bar{L}_m - L_{\Delta m}\cos 2\left(\theta_r - \frac{1}{3}\pi\right)\right)i_{as} + \left(L_{ls} + \bar{L}_m - L_{\Delta m}\cos 2\left(\theta_r - \frac{2}{3}\pi\right)\right)i_{bs}$$

$$+ \left(-\frac{1}{2}\bar{L}_m - L_{\Delta m}\cos 2\theta_r\right)i_{cs} + L_{md}\sin\left(\theta_r - \frac{2}{3}\pi\right)i_{fr},$$

$$\psi_{cs} = \left(-\frac{1}{2}\bar{L}_m - L_{\Delta m}\cos 2\left(\theta_r + \frac{1}{3}\pi\right)\right)i_{as} + \left(-\frac{1}{2}\bar{L}_m - L_{\Delta m}\cos 2\theta_r\right)i_{bs}$$

$$+ \left(L_{ls} + \bar{L}_m - L_{\Delta m}\cos 2\left(\theta_r + \frac{2}{3}\pi\right)\right)i_{cs} + L_{md}\sin\left(\theta_r + \frac{2}{3}\pi\right)i_{fr},$$

$$\psi_{fr} = L_{md}\sin\theta_r i_{as} + L_{md}\sin\left(\theta_r - \frac{2}{3}\pi\right)i_{bs} + L_{md}\sin\left(\theta_r + \frac{2}{3}\pi\right)i_{cs} + (L_{lf} + L_{mf})i_{fr},$$

$$(6.51)$$

where

$$L_{md} = \frac{N_r N_s}{\Re_{md}} \quad \text{and} \quad L_{mf} = \frac{N_r^2}{\Re_{md}}.$$

The self- and mutual inductance mapping $\mathbf{L}_{abcs/fr}(\theta_r)$ is

$$\mathbf{L}_{\frac{abcs}{fr}}(\theta_r) = \begin{bmatrix} L_{ls} + \bar{L}_m - L_{\Delta m}\cos 2\theta_r & -\frac{1}{2}\bar{L}_m - L_{\Delta m}\cos 2(\theta_r - \frac{1}{3}\pi) & -\frac{1}{2}\bar{L}_m - L_{\Delta m}\cos 2(\theta_r + \frac{1}{3}\pi) & L_{md}\sin\theta_r \\ -\frac{1}{2}\bar{L}_m - L_{\Delta m}\cos 2(\theta_r - \frac{1}{3}\pi) & L_{ls} + \bar{L}_m - L_{\Delta m}\cos 2(\theta_r - \frac{2}{3}\pi) & -\frac{1}{2}\bar{L}_m - L_{\Delta m}\cos 2\theta_r & L_{md}\sin(\theta_r - \frac{2}{3}\pi) \\ -\frac{1}{2}\bar{L}_m - L_{\Delta m}\cos 2(\theta_r + \frac{1}{3}\pi) & -\frac{1}{2}\bar{L}_m - L_{\Delta m}\cos 2\theta_r & L_{ls} + \bar{L}_m - L_{\Delta m}\cos 2(\theta_r + \frac{2}{3}\pi) & L_{md}\sin(\theta_r + \frac{2}{3}\pi) \\ L_{md}\sin\theta_r & L_{md}\sin(\theta_r - \frac{2}{3}\pi) & L_{md}\sin(\theta_r + \frac{2}{3}\pi) & L_{lf} + L_{mf} \end{bmatrix}.$$

One obtains

$$\mathbf{\psi}_{\underset{fr}{abcs}} = \mathbf{L}_{\underset{fr}{abcs}} \mathbf{i}_{\underset{fr}{abcs}}$$

$$= \begin{bmatrix} L_{ls}+\bar{L}_m-L_{\Delta m}\cos 2\theta_r & -\frac{1}{2}\bar{L}_m-L_{\Delta m}\cos 2(\theta_r-\frac{1}{3}\pi) & -\frac{1}{2}\bar{L}_m-L_{\Delta m}\cos 2(\theta_r+\frac{1}{3}\pi) & L_{md}\sin\theta_r \\ -\frac{1}{2}\bar{L}_m-L_{\Delta m}\cos 2(\theta_r-\frac{1}{3}\pi) & L_{ls}+\bar{L}_m-L_{\Delta m}\cos 2(\theta_r-\frac{2}{3}\pi) & -\frac{1}{2}\bar{L}_m-L_{\Delta m}\cos 2\theta_r & L_{md}\sin(\theta_r-\frac{2}{3}\pi) \\ -\frac{1}{2}\bar{L}_m-L_{\Delta m}\cos 2(\theta_r+\frac{1}{3}\pi) & -\frac{1}{2}\bar{L}_m-L_{\Delta m}\cos 2\theta_r & L_{ls}+\bar{L}_m-L_{\Delta m}\cos 2(\theta_r+\frac{2}{3}\pi) & L_{md}\sin(\theta_r+\frac{2}{3}\pi) \\ L_{md}\sin\theta_r & L_{md}\sin(\theta_r-\frac{2}{3}\pi) & L_{md}\sin(\theta_r+\frac{2}{3}\pi) & L_{lf}+L_{mf} \end{bmatrix} \begin{bmatrix} i_{as} \\ i_{bs} \\ i_{cs} \\ i_{fr} \end{bmatrix}.$$

From (6.50) and (6.51), we have a set of four differential equations

$$u_{as} = r_s i_{as} + \frac{d\left((L_{ls}+\bar{L}_m-L_{\Delta m}\cos 2\theta_r)i_{as}+\left(-\frac{1}{2}\bar{L}_m-L_{\Delta m}\cos 2\left(\theta_r-\frac{1}{3}\pi\right)\right)i_{bs}+\left(-\frac{1}{2}\bar{L}_m-L_{\Delta m}\cos 2\left(\theta_r+\frac{1}{3}\pi\right)\right)i_{cs}+L_{md}\sin\theta_r i_{fr}\right)}{dt},$$

$$u_{bs} = r_s i_{bs} + \frac{d\left(\left(-\frac{1}{2}\bar{L}_m-L_{\Delta m}\cos 2\left(\theta_r-\frac{1}{3}\pi\right)\right)i_{as}+\left(L_{ls}+\bar{L}_m-L_{\Delta m}\cos 2\left(\theta_r-\frac{2}{3}\pi\right)\right)i_{bs}+\left(-\frac{1}{2}\bar{L}_m-L_{\Delta m}\cos 2\theta_r\right)i_{cs}+L_{md}\sin\left(\theta_r-\frac{2}{3}\pi\right)i_{fr}\right)}{dt},$$

$$u_{cs} = r_s i_{cs} + \frac{d\left(\left(-\frac{1}{2}\bar{L}_m-L_{\Delta m}\cos 2\left(\theta_r+\frac{1}{3}\pi\right)\right)i_{as}+\left(-\frac{1}{2}\bar{L}_m-L_{\Delta m}\cos 2\theta_r\right)i_{bs}+\left(L_{ls}+\bar{L}_m-L_{\Delta m}\cos 2\left(\theta_r+\frac{2}{3}\pi\right)\right)i_{cs}+L_{md}\sin\left(\theta_r+\frac{2}{3}\pi\right)i_{fr}\right)}{dt},$$

$$u_{fr} = r_r i_{fr} + \frac{d\left(L_{md}\sin\theta_r i_{as}+L_{md}\sin\left(\theta_r-\frac{2}{3}\pi\right)i_{bs}+L_{md}\sin\left(\theta_r+\frac{2}{3}\pi\right)i_{cs}+(L_{lf}+L_{mf})i_{fr}\right)}{dt}.$$

For round-rotor motors $L_{\Delta m}=0$, yielding

$$\mathbf{\psi}_{\underset{fr}{abcs}} = \mathbf{L}_{\underset{fr}{abcs}} \mathbf{i}_{\underset{fr}{abcs}} = \begin{bmatrix} L_{ls}+\bar{L}_m & -\frac{1}{2}\bar{L}_m & -\frac{1}{2}\bar{L}_m & L_{md}\sin\theta_r \\ -\frac{1}{2}\bar{L}_m & L_{ls}+\bar{L}_m & -\frac{1}{2}\bar{L}_m & L_{md}\sin(\theta_r-\frac{2}{3}\pi) \\ -\frac{1}{2}\bar{L}_m & -\frac{1}{2}\bar{L}_m & L_{ls}+\bar{L}_m & L_{md}\sin(\theta_r+\frac{2}{3}\pi) \\ L_{md}\sin\theta_r & L_{md}\sin(\theta_r-\frac{2}{3}\pi) & L_{md}\sin(\theta_r+\frac{2}{3}\pi) & L_{lf}+L_{mf} \end{bmatrix} \begin{bmatrix} i_{as} \\ i_{bs} \\ i_{cs} \\ i_{fr} \end{bmatrix}.$$

$$(6.52)$$

Using (6.50) and (6.52), Cauchy's form of differential equations to describe the circuitry-electromagnetic dynamics of the conventional three-phase synchronous motors is derived. We apply the Symbolic Math Toolbox [7, 8]. The following file is developed to accomplish our goal.

```
L=sym('[Lls+Lmb,-Lmb/2,-Lmb/2,Lmd*S1; -Lmb/2,Lls+Lmb,-Lmb/2,Lmd*S2; -Lmb/2,-Lmb/2,
    Lls+Lmb,Lmd*S3; Lmd*S1,Lmd*S2,Lmd*S3,Llf+Lmf]');
R=sym('[-rs,0, 0, 0;0,-rs,0,0;0,0,-rs,0;0,0,0,-rr]');
I=sym('[ias; ibs; ics; ifr]');
V=sym('[uas;ubs;ucs;ufr]');
K=sym('[Lmd*C1*wr*ifr;Lmd*C2*wr*ifr;Lmd*C3*wr*ifr;Lmd*wr*(ias*C1+ibs*C2+ics*C3)]');
L1=inv(L); L2=simplify(L1);
FS1=L2*R*I; FS2=simplify(FS1);
FS3=L2*V; FS4=simplify(FS3);
FS5=L2*K; FS6=simplify(FS5);
FS7=FS2+FS4-FS6; FS=simplify(FS7)
pretty(FS)
```

The notations used are:

$$S1 = \sin\theta_r, \quad S2 = \sin\left(\theta_r - \frac{2}{3}\pi\right), \quad S3 = \sin\left(\theta_r + \frac{2}{3}\pi\right)$$

$$C1 = \cos\theta_r, \quad C2 = \cos\left(\theta_r - \frac{2}{3}\pi\right), \quad C3 = \cos\left(\theta_r + \frac{2}{3}\pi\right),$$

$$\text{Lls} = L_{ls}, \quad \text{Lmb} = \bar{L}_m, \quad \text{Lmd} = L_{md}, \quad \text{Lmf} = L_{mf}, \quad \text{and} \quad \text{Llf} = L_{lf}.$$

The results are displayed in the Command Window in the symbolic form. Using the trigonometric identities and grouping like terms, the following differential equations are found

$$\frac{di_{as}}{dt} = \frac{1}{(2L_{ls} + 3\bar{L}_m)\left(3L_{md}^2 L_{ls} - \left(2L_{ls}^2 + 3L_{ls}\bar{L}_m\right)L_{ff}\right)}\Bigg[\left(3L_{md}^2\bar{L}_m + 4L_{ls}L_{md}^2\right.$$

$$-\left(3\bar{L}_m^2 + 4L_{ls}^2 + 8L_{ls}\bar{L}_m\right)L_{ff} + 2L_{md}^2 L_{ls}\cos 2\theta_r\Big)(-r_s i_{as} + u_{as}) + \Big(3L_{md}^2\bar{L}_m$$

$$+ L_{ls}L_{md}^2 - \left(3\bar{L}_m^2 + 2L_{ls}\bar{L}_m\right)L_{ff} + 2L_{md}^2 L_{ls}\cos 2\left(\theta_r - \frac{1}{3}\pi\right)\Big)(-r_s i_{bs} + u_{bs})$$

$$+ \Big(3L_{md}^2\bar{L}_m + L_{ls}L_{md}^2 - \left(3\bar{L}_m^2 + 2L_{ls}\bar{L}_m\right)L_{ff}$$

$$+ 2L_{md}^2 L_{ls}\cos 2\left(\theta_r + \frac{1}{3}\pi\right)\Big)(-r_s i_{cs} + u_{cs})$$

$$+ \left(6L_{ls}L_{md}\bar{L}_m + 4L_{ls}^2 L_{md}\right)\sin\theta_r(-r_r i_{fr} + u_{fr})$$

$$- \left(6L_{ls}L_{md}^2\bar{L}_m + 4L_{ls}^2 L_{md}^2\right)\Big(i_{as}\cos\theta_r + i_{bs}\cos\left(\theta_r - \frac{2}{3}\pi\right)$$

$$+ i_{cs}\cos\left(\theta_r + \frac{2}{3}\pi\right)\Big)\omega_r \sin\theta_r$$

$$+ \left(\left(6L_{md}L_{ls}\bar{L}_m + 4L_{md}L_{ls}^2\right)L_{ff} - 6L_{md}^3 L_{ls}\right)i_{fr}\omega_r \cos\theta_r\Bigg],$$

$$\frac{di_{bs}}{dt} = \frac{1}{(2L_{ls} + 3\bar{L}_m)\left(3L_{md}^2 L_{ls} - \left(2L_{ls}^2 + 3L_{ls}\bar{L}_m\right)L_{ff}\right)}\Bigg[\left(3L_{md}^2\bar{L}_m + L_{ls}L_{md}^2\right.$$

$$-\left(3\bar{L}_m^2 + 2L_{ls}\bar{L}_m\right)L_{ff} + 2L_{md}^2 L_{ls}\cos 2\left(\theta_r - \frac{1}{3}\pi\right)\Big)(-r_s i_{as} + u_{as})$$

$$+\left(3L_{md}^2\bar{L}_m + 4L_{ls}L_{md}^2 - \left(3\bar{L}_m^2 + 4L_{ls}^2 + 8L_{ls}\bar{L}_m\right)L_{ff}\right.$$

$$\left.+ 2L_{md}^2L_{ls}\cos 2\left(\theta_r - \frac{2}{3}\pi\right)\right)(-r_s i_{bs} + u_{bs})$$

$$+\left(3L_{md}^2\bar{L}_m + L_{ls}L_{md}^2 - \left(3\bar{L}_m^2 + 2L_{ls}\bar{L}_m\right)L_{ff} + 2L_{md}^2L_{ls}\cos 2\theta_r\right)(-r_s i_{cs} + u_{cs})$$

$$+\left(6L_{ls}L_{md}\bar{L}_m + 4L_{ls}^2L_{md}\right)\sin\left(\theta_r - \frac{2}{3}\pi\right)(-r_r i_{fr} + u_{fr})$$

$$-\left(6L_{ls}L_{md}^2\bar{L}_m + 4L_{ls}^2L_{md}^2\right)\left(i_{as}\cos\theta_r + i_{bs}\cos\left(\theta_r - \frac{2}{3}\pi\right)\right.$$

$$\left.+ i_{cs}\cos\left(\theta_r + \frac{2}{3}\pi\right)\right)\omega_r\sin\left(\theta_r - \frac{2}{3}\pi\right)$$

$$+\left(\left(6L_{md}L_{ls}\bar{L}_m + 4L_{md}L_{ls}^2\right)L_{ff} - 6L_{md}^3 L_{ls}\right)i_{fr}\omega_r\cos\left(\theta_r - \frac{2}{3}\pi\right)\Bigg],$$

$$\frac{di_{cs}}{dt} = \frac{1}{\left(2L_{ls} + 3\bar{L}_m\right)\left(3L_{md}^2L_{ls} - \left(2L_{ls}^2 + 3L_{ls}\bar{L}_m\right)L_{ff}\right)}\left[\left(3L_{md}^2\bar{L}_m + L_{ls}L_{md}^2\right.\right.$$

$$\left.- \left(3\bar{L}_m^2 + 2L_{ls}\bar{L}_m\right)L_{ff} + 2L_{md}^2L_{ls}\cos 2\left(\theta_r + \frac{1}{3}\pi\right)\right)(-r_s i_{as} + u_{as}) + \left(3L_{md}^2\bar{L}_m\right.$$

$$\left.+ L_{ls}L_{md}^2 - \left(3\bar{L}_m^2 + 2L_{ls}\bar{L}_m\right)L_{ff} + 2L_{md}^2L_{ls}\cos 2\theta_r\right)(-r_s i_{bs} + u_{bs}) + \left(3L_{md}^2\bar{L}_m\right.$$

$$\left.+ 4L_{ls}L_{md}^2 - \left(3\bar{L}_m^2 + 4L_{ls}^2 + 8L_{ls}\bar{L}_m\right)L_{ff} + 2L_{md}^2L_{ls}\cos 2\left(\theta_r + \frac{2}{3}\pi\right)\right)(-r_s i_{cs} + u_{cs})$$

$$+\left(6L_{ls}L_{md}\bar{L}_m + 4L_{ls}^2L_{md}\right)\sin\left(\theta_r + \frac{2}{3}\pi\right)(-r_r i_{fr} + u_{fr})$$

$$-\left(6L_{ls}L_{md}^2\bar{L}_m + 4L_{ls}^2L_{md}^2\right)\left(i_{as}\cos\theta_r + i_{bs}\cos\left(\theta_r - \frac{2}{3}\pi\right)\right.$$

$$\left.+ i_{cs}\cos\left(\theta_r + \frac{2}{3}\pi\right)\right)\omega_r\sin\left(\theta_r + \frac{2}{3}\pi\right)$$

$$+\left(\left(6L_{md}L_{ls}\bar{L}_m + 4L_{md}L_{ls}^2\right)L_{ff} - 6L_{md}^3 L_{ls}\right)i_{fr}\omega_r\cos\left(\theta_r + \frac{2}{3}\pi\right)\Bigg],$$

$$\frac{di_{fr}}{dt} = \frac{1}{3L_{md}^2 L_{ls} - \left(2L_{ls}^2 + 3L_{ls}\bar{L}_m\right)L_{ff} + 3L_{md}^2 L_{ls}} \left[2L_{md}L_{ls}\left(\sin\theta_r(-r_s i_{as} + u_{as})\right.\right.$$

$$+ \sin\left(\theta_r - \frac{2}{3}\pi\right)(-r_s i_{bs} + u_{bs}) + \sin\left(\theta_r + \frac{2}{3}\pi\right)(-r_s i_{cs} + u_{cs})\right)$$

$$- \left(2L_{ls}^2 + 3\bar{L}_m L_{ls}\right)(-r_s i_{fr} + u_{fr}) + \left(3L_{ls}L_{md}\bar{L}_m + 2L_{ls}^2 L_{md}\right)\left(i_{as}\omega_r \cos\theta_r\right.$$

$$+ i_{bs}\omega_r \cos\left(\theta_r - \frac{2}{3}\pi\right) + i_{cs}\omega_r \cos\left(\theta_r + \frac{2}{3}\pi\right)\right)\Bigg]. \tag{6.53}$$

Here, $L_{ff} = L_{lf} + L_{mf}$.
In matrix form, differential equations (6.53) are given as

$$\begin{bmatrix} \dfrac{di_{as}}{dt} \\[6pt] \dfrac{di_{bs}}{dt} \\[6pt] \dfrac{di_{cs}}{dt} \\[6pt] \dfrac{di_{fr}}{dt} \end{bmatrix} = \begin{bmatrix} -\dfrac{r_s L_{Ds}}{L_{\Sigma s}} & -\dfrac{r_s L_{Ms}}{L_{\Sigma s}} & -\dfrac{r_s L_{Ms}}{L_{\Sigma s}} & 0 \\[8pt] -\dfrac{r_s L_{Ms}}{L_{\Sigma s}} & -\dfrac{r_s L_{Ds}}{L_{\Sigma s}} & -\dfrac{r_s L_{Ms}}{L_{\Sigma s}} & 0 \\[8pt] -\dfrac{r_s L_{Ms}}{L_{\Sigma s}} & -\dfrac{r_s L_{Ms}}{L_{\Sigma s}} & -\dfrac{r_s L_{Ds}}{L_{\Sigma s}} & 0 \\[8pt] 0 & 0 & 0 & \dfrac{r_r(2L_{ls}^2 + 3\bar{L}_m L_{ls})}{L_{\Sigma f}} \end{bmatrix} \begin{bmatrix} i_{as} \\[6pt] i_{bs} \\[6pt] i_{cs} \\[6pt] i_{fr} \end{bmatrix}$$

$$+ \begin{bmatrix} -\dfrac{2r_s l_{md}^2 L_{ls}}{L_{\Sigma s}}\cos 2\theta_r & -\dfrac{2r_s l_{md}^2 L_{ls}}{L_{\Sigma s}}\cos 2\left(\theta_r - \frac{1}{3}\pi\right) & -\dfrac{2r_s l_{md}^2 L_{ls}}{L_{\Sigma s}}\cos 2\left(\theta_r + \frac{1}{3}\pi\right) & -\dfrac{r_r L_{Mf}}{L_{\Sigma s}}\sin\theta_r \\[10pt] -\dfrac{2r_s l_{md}^2 L_{ls}}{L_{\Sigma s}}\cos 2\left(\theta_r - \frac{1}{3}\pi\right) & -\dfrac{2r_s l_{md}^2 L_{ls}}{L_{\Sigma s}}\cos 2\left(\theta_r - \frac{2}{3}\pi\right) & -\dfrac{2r_s l_{md}^2 L_{ls}}{L_{\Sigma s}}\cos 2\theta_r & -\dfrac{r_r L_{Mf}}{L_{\Sigma s}}\sin\left(\theta_r - \frac{2}{3}\pi\right) \\[10pt] -\dfrac{2r_s l_{md}^2 L_{ls}}{L_{\Sigma s}}\cos 2\left(\theta_r + \frac{1}{3}\pi\right) & -\dfrac{2r_s l_{md}^2 L_{ls}}{L_{\Sigma s}}\cos 2\theta_r & -\dfrac{2r_s l_{md}^2 L_{ls}}{L_{\Sigma s}}\cos 2\left(\theta_r + \frac{2}{3}\pi\right) & -\dfrac{r_r L_{Mf}}{L_{\Sigma s}}\sin\left(\theta_r + \frac{2}{3}\pi\right) \\[10pt] -\dfrac{2r_r L_{md}L_{ls}}{L_{\Sigma f}}\sin\theta_r & -\dfrac{2r_r L_{md}L_{ls}}{L_{\Sigma f}}\sin\left(\theta_r - \frac{2}{3}\pi\right) & -\dfrac{2r_r L_{md}L_{ls}}{L_{\Sigma f}}\sin\left(\theta_r + \frac{2}{3}\pi\right) & 0 \end{bmatrix} \begin{bmatrix} i_{as} \\[6pt] i_{bs} \\[6pt] i_{cs} \\[6pt] i_{fr} \end{bmatrix}$$

$$+ \begin{bmatrix} -\dfrac{6L_{ls}l_{md}^2\bar{L}_m + 4L_{ls}^2 l_{md}^2}{L_{\Sigma s}}\sin\theta_r & -\dfrac{6L_{ls}l_{md}^2\bar{L}_m + 4L_{ls}^2 l_{md}^2}{L_{\Sigma s}}\sin\theta_r & -\dfrac{6L_{ls}l_{md}^2\bar{L}_m + 4L_{ls}^2 l_{md}^2}{L_{\Sigma s}}\sin\theta_r \\[10pt] -\dfrac{6L_{ls}l_{md}^2\bar{L}_m + 4L_{ls}^2 l_{md}^2}{L_{\Sigma s}}\sin(\theta_r - \frac{2}{3}\pi) & -\dfrac{6L_{ls}l_{md}^2\bar{L}_m + 4L_{ls}^2 l_{md}^2}{L_{\Sigma s}}\sin(\theta_r - \frac{2}{3}\pi) & -\dfrac{6L_{ls}l_{md}^2\bar{L}_m + 4L_{ls}^2 l_{md}^2}{L_{\Sigma s}}\sin(\theta_r - \frac{2}{3}\pi) \\[10pt] -\dfrac{6L_{ls}l_{md}^2\bar{L}_m + 4L_{ls}^2 l_{md}^2}{L_{\Sigma s}}\sin(\theta_r + \frac{2}{3}\pi) & -\dfrac{6L_{ls}l_{md}^2\bar{L}_m + 4L_{ls}^2 l_{md}^2}{L_{\Sigma s}}\sin(\theta_r + \frac{2}{3}\pi) & -\dfrac{6L_{ls}l_{md}^2\bar{L}_m + 4L_{ls}^2 l_{md}^2}{L_{\Sigma s}}\sin(\theta_r + \frac{2}{3}\pi) \\[10pt] \dfrac{3L_{ls}L_{md}\bar{L}_m + 2L_{ls}^2 L_{md}}{L_{\Sigma f}} & \dfrac{3L_{ls}L_{md}\bar{L}_m + 2L_{ls}^2 L_{md}}{L_{\Sigma f}} & \dfrac{3L_{ls}L_{md}\bar{L}_m + 2L_{ls}^2 L_{md}}{L_{\Sigma f}} \end{bmatrix}$$

$$\times \begin{bmatrix} i_{as}\omega_r \cos\theta_r \\[6pt] i_{bs}\omega_r \cos(\theta_r - \frac{2}{3}\pi) \\[6pt] i_{cs}\omega_r \cos(\theta_r + \frac{2}{3}\pi) \end{bmatrix} + \begin{bmatrix} \dfrac{\left(6L_{md}L_{ls}\bar{L}_m + 4L_{md}^2 l_{ls}^2\right)L_{ff} - 6L_{md}^3 L_{ls}}{L_{\Sigma s}}i_{fr}\omega_r \cos\theta_r \\[10pt] \dfrac{\left(6L_{md}L_{ls}\bar{L}_m + 4L_{md}^2 l_{ls}^2\right)L_{ff} - 6L_{md}^3 L_{ls}}{L_{\Sigma s}}i_{fr}\omega_r \cos(\theta_r - \frac{2}{3}\pi) \\[10pt] \dfrac{\left(6L_{md}L_{ls}\bar{L}_m + 4L_{md}^2 l_{ls}^2\right)L_{ff} - 6L_{md}^3 L_{ls}}{L_{\Sigma s}}i_{fr}\omega_r \cos(\theta_r + \frac{2}{3}\pi) \\[10pt] 0 \end{bmatrix}$$

$$
+ \begin{bmatrix}
\dfrac{L_{Ds} + 2I_{md}^2 L_{ls} \cos 2\theta_r}{L_{\Sigma s}} & \dfrac{L_{Ms} + 2I_{md}^2 L_{ls} \cos 2(\theta_r - \frac{1}{3}\pi)}{L_{\Sigma s}} & \dfrac{L_{Ms} + 2I_{md}^2 L_{ls} \cos 2(\theta_r + \frac{1}{3}\pi)}{L_{\Sigma s}} & \dfrac{L_{Mf}}{L_{\Sigma s}} \sin\theta_r \\[4mm]
\dfrac{L_{Ms} + 2I_{md}^2 L_{ls} \cos 2(\theta_r - \frac{1}{3}\pi)}{L_{\Sigma s}} & \dfrac{L_{Ds} + 2I_{md}^2 L_{ls} \cos 2(\theta_r - \frac{2}{3}\pi)}{L_{\Sigma s}} & \dfrac{L_{Ms} + 2I_{md}^2 L_{ls} \cos 2\theta_r}{L_{\Sigma s}} & \dfrac{L_{Mf}}{L_{\Sigma s}} \sin(\theta_r - \frac{2}{3}\pi) \\[4mm]
\dfrac{L_{Ms} + 2I_{md}^2 L_{ls} \cos 2(\theta_r + \frac{1}{3}\pi)}{L_{\Sigma s}} & \dfrac{L_{Ms} + 2I_{md}^2 L_{ls} \cos 2\theta_r}{L_{\Sigma s}} & \dfrac{L_{Ds} + 2I_{md}^2 L_{ls} \cos 2(\theta_r + \frac{2}{3}\pi)}{L_{\Sigma s}} & \dfrac{L_{Mf}}{L_{\Sigma s}} \sin(\theta_r + \frac{2}{3}\pi) \\[4mm]
\dfrac{2L_{md}L_{ls}\sin\theta_r}{L_{\Sigma f}} & \dfrac{2L_{md}L_{ls}\sin(\theta_r - \frac{2}{3}\pi)}{L_{\Sigma f}} & \dfrac{2L_{md}L_{ls}\sin(\theta_r + \frac{2}{3}\pi)}{L_{\Sigma f}} & -\dfrac{2I_{ls}^2 + 3\overline{L}_m L_{ls}}{L_{\Sigma f}}
\end{bmatrix}
\begin{bmatrix} u_{as} \\ u_{bs} \\ u_{cs} \\ u_{fr} \end{bmatrix},
$$

where the following notations are used

$$
L_{Ds} = 3L_{md}^2 \overline{L}_m + 4L_{ls}L_{md}^2 - (3\overline{L}_m^2 + 4L_{ls}^2 + 8L_{ls}\overline{L}_m)(L_{lf} + L_{mf}),
$$

$$
L_{Ms} = 3L_{md}^2 \overline{L}_m + L_{ls}L_{md}^2 - (3\overline{L}_m^2 + 2L_{ls}\overline{L}_m)(L_{lf} + L_{mf}),
$$

$$
L_{\Sigma s} = (2L_{ls} + 3\overline{L}_m)\left[3L_{md}^2 L_{ls} - \left(2L_{ls}^2 + 3L_{ls}\overline{L}_m\right)(L_{lf} + L_{mf}) \right],
$$

$$
L_{\Sigma f} = 3L_{md}^2 L_{ls} - \left(2L_{ls}^2 + 3L_{ls}\overline{L}_m\right)(L_{lf} + L_{mf}),
$$

$$
L_{Mf} = 6L_{ls}L_{md}\overline{L}_m + 4L_{ls}^2 L_{md}.
$$

The *torsional-mechanical* equations of motion are

$$
\frac{d\omega_r}{dt} = \frac{P}{2J}T_e - \frac{B_m}{J}\omega_r - \frac{P}{2J}T_L,
$$

$$
\frac{d\theta_r}{dt} = \omega_r. \tag{6.54}
$$

The electromagnetic torque developed by *P*-pole three-phase synchronous motors is found using

$$
T_e = \frac{P}{2}\frac{\partial W_c(i_{as}, i_{bs}, i_{cs}, i_{fr}, \theta_r)}{\partial \theta_r} = \frac{P}{2}\frac{\partial\left(\frac{1}{2}\begin{bmatrix} i_{as} & i_{bs} & i_{cs} & i_{fr} \end{bmatrix} \mathbf{L}_{abcs/fr} \begin{bmatrix} i_{as} \\ i_{bs} \\ i_{cs} \\ i_{fr} \end{bmatrix}\right)}{\partial \theta_r}.
$$

which yields

$$T_e = \frac{P}{2}\left[L_{\Delta m}\left(i_{as}^2 \sin 2\theta_r + 2i_{as}i_{bs}\sin 2\left(\theta_r - \frac{1}{3}\pi\right) + 2i_{as}i_{cs}\sin 2\left(\theta_r + \frac{1}{3}\pi\right)\right.\right.$$

$$+ i_{bs}^2 \sin 2\left(\theta_r - \frac{2}{3}\pi\right) + 2i_{bs}i_{cs}\sin 2\theta_r + i_{cs}^2 \sin 2\left(\theta_r + \frac{2}{3}\pi\right)\bigg)$$

$$+ L_{md}i_{fr}\left(i_{as}\cos\theta_r + i_{bs}\cos\left(\theta_r - \frac{2}{3}\pi\right) + i_{cs}\cos\left(\theta_r + \frac{2}{3}\pi\right)\right)\bigg].$$

Using the trigonometric identities, an alternative equation for T_e is

$$T_e = \frac{P}{2}\left[\frac{L_{md} - L_{mq}}{3}\left(\left(i_{as}^2 - \frac{1}{2}i_{bs}^2 - \frac{1}{2}i_{cs}^2 - i_{as}i_{bs} - i_{as}i_{cs} + 2i_{bs}i_{cs}\right)\sin 2\theta_r \right.\right.$$

$$+ \frac{\sqrt{3}}{2}\left(i_{bs}^2 - i_{cs}^2 - 2i_{as}i_{bs} + 2i_{as}i_{cs}\right)\cos 2\theta_r \bigg) + L_{md}i_{fr}\left(\left(i_{as} - \frac{1}{2}i_{bs} - \frac{1}{2}i_{cs}\right)\cos\theta_r \right.$$

$$+ \frac{\sqrt{3}}{2}(i_{bs} - i_{cs})\sin\theta_r \bigg)\bigg].$$

For round-rotor synchronous machines $L_{\Delta m} = 0$, and

$$T_e = \frac{PL_{md}}{2}i_{fr}\left[i_{as}\cos\theta_r + i_{bs}\cos\left(\theta_r - \frac{2}{3}\pi\right) + i_{cs}\cos\left(\theta_r + \frac{2}{3}\pi\right)\right]. \qquad (6.55)$$

Applying the balanced three-phase sinusoidal current set

$$i_{as} = \sqrt{2}i_M \cos\theta_r, \quad i_{bs} = \sqrt{2}i_M \cos\left(\theta_r - \frac{2}{3}\pi\right) \quad \text{and} \quad i_{cs} = \sqrt{2}i_M \cos\left(\theta_r + \frac{2}{3}\pi\right),$$

we have

$$T_e = \frac{PL_{md}}{2}i_{fr}\sqrt{2}i_M\left(\cos^2\theta_r + \cos^2\left(\theta_r - \frac{2}{3}\pi\right) + \cos^2\left(\theta_r + \frac{2}{3}\pi\right)\right) = \frac{3PL_{md}}{2\sqrt{2}}i_{fr}i_M.$$

From (6.54) and (6.55), one yields

$$\frac{d\omega_r}{dt} = \frac{P^2 L_{md}}{4J}i_{fr}\left[i_{as}\cos\theta_r + i_{bs}\cos\left(\theta_r - \frac{2}{3}\pi\right) + i_{cs}\cos\left(\theta_r + \frac{2}{3}\pi\right)\right] - \frac{B_m}{J}\omega_r - \frac{P}{2J}T_L,$$

$$\frac{d\theta_r}{dt} = \omega_r. \qquad (6.56)$$

Combining (6.53) and (6.56), the nonlinear differential equations are found to model the transient dynamics of three-phase synchronous motors.

Example 6.13

Study the dynamics of a two-pole synchronous motor with the following parameters: $r_s = 0.25$ ohm, $r_r = 0.5$ ohm, $L_{mqs} = 0.00095$ H, $L_{mds} = 0.00095$ H, $L_{ls} = 0.0001$ H, $L_{mf} = 0.002$ H, $L_{lf} = 0.0002$ H, $L_{md} = 0.004$ H, $J = 0.003$, kg-m^2 and $B_m = 0.0007$ N-m-sec/rad. The phase voltages applied to the stator windings are

$$u_{as}(t) = \sqrt{2}\,150\cos(377t), \quad u_{bs}(t) = \sqrt{2}\,150\cos\left(377t - \frac{2}{3}\pi\right) \quad \text{and}$$

$$u_{cs}(t) = \sqrt{2}\,150\cos\left(377t + \frac{2}{3}\pi\right).$$

The dc field voltage is $u_{fr} = 5$ V. The load conditions are specified to be

$$T_L = \begin{cases} 1 \text{ N-m}, \ \forall t \in [0 \quad 0.55) \text{ sec} \\ 2 \text{ N-m}, \ \forall t \in [0.55 \quad 0.7] \text{ sec} \end{cases}.$$

The derived differential equations in Cauchy's form, as given by (6.53) and (6.56), are used to simulate and analyze conventional synchronous motors. Two MATLAB files are developed; the file ch6 _ 06.m is

```
tspan = [0 0.7]; y0=[0 0 0 0 0 0]';
options = odeset('RelTol',5e-3,'AbsTol',[1e-4 1e-4 1e-4 1e-4 1e-4 1e-4]);
[t,y] = ode45('ch6 _ 07',tspan,y0,options);
plot(t,y(:,1)); axis([0,0.7,-500,500]);
xlabel('Time [seconds]','FontSize',14);
title('Current Dynamics, i _ a _ s [A]','FontSize',14); pause;
plot(t,y(:,2)); axis([0,0.7,-500,500]);
xlabel('Time [seconds]','FontSize',14); title('Current Dynamics, i_b_s [A]'; pause;
plot(t,y(:,3)); axis([0,0.7,-500,500]);
xlabel('Time [seconds]','FontSize',14);
title('Current Dynamics, i _ c _ s [A]','FontSize',14); pause;
plot(t,y(:,4)); axis([0,0.7,-150,150]);
xlabel('Time [seconds]','FontSize',14);
title('Current Dynamics, i _ f _ r [A]','FontSize',14); pause;
plot(t,y(:,5)); axis([0,0.7,0,400]);
xlabel('Time [seconds]','FontSize',14);
title('Electrical Angular Velocity Dynamics, \omega _ r [rad/sec]','FontSize',14);
```

whereas the file ch6 _ 07.m is

```
function yprime=difer(t,y);
% Stator leakage and magnetizing inductances
Lls=0.0001; Lmqs=0.00095; Lmds=0.00095;
% Rotor leakage and magnetizing inductances
Llf=0.0002; Lmf=0.002;
% Mutual inductance
Lmd=0.0004;
```

```
% Average value of the magnetizing inductance
Lmb=(Lmqs+Lmds)/3;
% Resistances of the stator and rotor windings
rs=0.25; rr=0.5;
% Equivalent moment of inertia and viscous friction coefficient
J=0.003; Bm=0.0007;
% Number of poles
P=2;
% Magnitude of the supplied phase voltages and angular frequency
um=sqrt(2)*150; w=377;
% Voltage applied to the rotor field winding
ufr=5;
% Load torque, applied at time tTl sec
if t<=0.55
Tl=1;
else
Tl=2;
end
% Applied phase voltages to the abc windings
uas=um*cos(w*t);
ubs=um*cos(w*t-2*pi/3);
ucs=um*cos(w*t+2*pi/3);
% Numerical expressions used
Ld=2*Lls*Lmd^2; Ldr=2*Lls*Lmd;
Lss=(-3*Lmb^2-4*Lls^2-8*Lls*Lmb)*(Lmf+Llf)+3*Lmd^2*Lmb+4*Lls*Lmd^2;
Ldens=(2*Lls+3*Lmb)*((-2*Lls^2-3*Lmb*Lls)*(Llf+Lmf)+3*Lmd^2*Lls);
Ldenr=(-2*Lls^2-3*Lmb*Lls)*(Llf+Lmf)+3*Lmd^2*Lls;
Lms=(-3*Lmb^2-2*Lls*Lmb)*(Lmf+Llf)+3*Lmd^2*Lmb+Lls*Lmd^2;
Lrr=6*Lls*Lmd*Lmb+4*Lls^2*Lmd;
Llr=-(2*Lls^2+3*Lmb*Lls);
% Variables used
S1=sin(y(6,:)); S2=sin(y(6,:)-2*pi/3); S3=sin(y(6,:)+2*pi/3);
IUas=-rs*y(1,:)+uas; IUbs=-rs*y(2,:)+ubs; IUcs=-rs*y(3,:)+ucs;
IUfr=-rr*y(4,:)+ufr;
C1=cos(y(6,:)); C2=cos(y(6,:)-2*pi/3); C3=cos(y(6,:)+2*pi/3);
C12=cos(2*y(6,:)); C22=cos(2*y(6,:)-2*pi/3); C32=cos(2*y(6,:)+2*pi/3);
Nsts=(-6*Lmd^2*Lls*Lmb-4*Lmd^2*Lls^2)*y(5,:).*(C1.*y(1,:)+C2.*y(2,:)+C3.*y(3,:));
Nstr=(3*Lmd*Lmb*Lls+2*Lmd*Lls^2)*y(5,:).*(C1.*y(1,:)+C2.*y(2,:)+C3.*y(3,:));
Nct=((6*Lmd*Lls*Lmb+4*Lmd*Lls^2)*(Llf+Lmf)-6*Lmd^3*Lls)*y(5,:).*y(4,:);
Te=P*(Lmd*y(4,:).*(y(1,:).*C1+y(2,:).*C2+y(3,:).*C3))/2;
% Differential equations
yprime=[((Lss+Ld*C12)*IUas+(Lms+Ld*C22)*IUbs+...
(Lms+Ld*C32)*IUcs+Lrr*S1*IUfr+Nsts*S1+Nct*C1)/Ldens;...
((Lms+Ld*C22)*IUas+(Lss+Ld*C32)*IUbs+(Lms+Ld*C12)*IUcs+...
  Lrr*S2*IUfr+Nsts*S2+Nct*C2)/Ldens;...
((Lms+Ld*C32)*IUas+(Lms+Ld*C12)*IUbs+(Lss+Ld*C22)*IUcs+...
  Lrr*S3*IUfr+Nsts*S3+Nct*C3)/Ldens;...
(Ldr*S1*IUas+Ldr*S2*IUbs+Ldr*S3*IUcs+Llr*IUfr+Nstr)/Ldenr;...
(P*Te-2*Bm*y(5,:)-P*Tl)/(2*J);...
y(5,:)];
```

The transient dynamics, found by numerically solving the differential equations (6.53) and (6.56), is illustrated in Figure 6.39. A two-pole motor

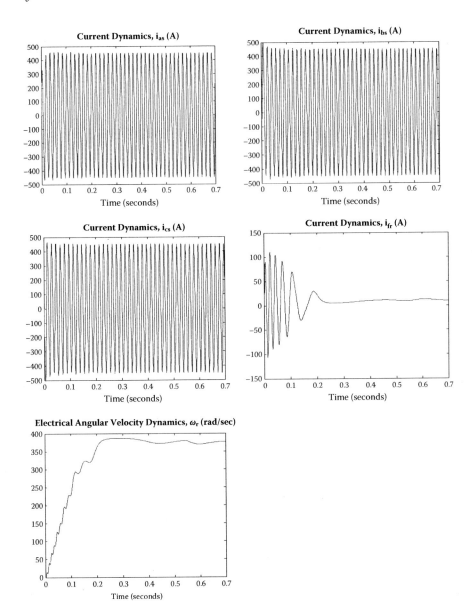

FIGURE 6.39
Transient dynamics of a conventional synchronous motor, $T_{L0} = 1$ N-m and $T_L = 2$ N-m at $t = 0.55$ sec.

starts from stall, and it reaches the synchronous angular velocity ($\omega_e = 377$ rad/sec), that is, $\omega_r = \omega_e$. The load torque varies. We apply

$$T_L = \begin{cases} 1 \text{ N-m}, \ \forall t \in [0 \quad 0.55) \text{ sec} \\ 2 \text{ N-m}, \ \forall t \in [0.55 \quad 0.75] \text{ sec} \end{cases}.$$

The results show that the motor angular velocity is equal to the synchronous angular velocity.

□

We consider three-phase synchronous generators, which should be driven by the prime mover to generate the terminal voltages. Figure 6.38 illustrates the power generation systems. Using Kirchhoff's second law, we have four equations for stator and rotor circuitry

$$0 = -r_s i_{as} + \frac{d\psi_{as}}{dt}, \quad 0 = -r_s i_{bs} + \frac{d\psi_{bs}}{dt}, \quad 0 = -r_s i_{cs} + \frac{d\psi_{cs}}{dt},$$

$$u_{fr} = r_r i_{fr} + \frac{d\psi_{fr}}{dt}. \tag{6.57}$$

Assume the flux linkages are

$$\psi_{as} = -(L_{ls} + \bar{L}_m - L_{\Delta m} \cos 2\theta_r) i_{as} - \left(-\frac{1}{2}\bar{L}_m - L_{\Delta m} \cos 2\left(\theta_r - \frac{1}{3}\pi\right) \right) i_{bs}$$

$$- \left(-\frac{1}{2}\bar{L}_m - L_{\Delta m} \cos 2\left(\theta_r + \frac{1}{3}\pi\right) \right) i_{cs} + L_{md} \sin \theta_r i_{fr},$$

$$\psi_{bs} = -\left(-\frac{1}{2}\bar{L}_m - L_{\Delta m} \cos 2\left(\theta_r - \frac{1}{3}\pi\right) \right) i_{as} - \left(L_{ls} + \bar{L}_m - L_{\Delta m} \cos 2\left(\theta_r - \frac{2}{3}\pi\right) \right) i_{bs}$$

$$- \left(-\frac{1}{2}\bar{L}_m - L_{\Delta m} \cos 2\theta_r \right) i_{cs} + L_{md} \sin\left(\theta_r - \frac{2}{3}\pi\right) i_{fr},$$

$$\psi_{cs} = \left(-\frac{1}{2}\bar{L}_m - L_{\Delta m} \cos 2\left(\theta_r + \frac{1}{3}\pi\right) \right) i_{as} + \left(-\frac{1}{2}\bar{L}_m - L_{\Delta m} \cos 2\theta_r \right) i_{bs}$$

$$+ \left(L_{ls} + \bar{L}_m - L_{\Delta m} \cos 2\left(\theta_r + \frac{2}{3}\pi\right) \right) i_{cs} + L_{md} \sin\left(\theta_r + \frac{2}{3}\pi\right) i_{fr},$$

$$\psi_{fr} = -L_{md} \sin \theta_r i_{as} - L_{md} \sin\left(\theta_r - \frac{2}{3}\pi\right) i_{bs} - L_{md} \sin\left(\theta_r + \frac{2}{3}\pi\right) i_{cs} + (L_{lf} + L_{mf}) i_{fr}.$$

$$\tag{6.58}$$

For round-rotor synchronous machines $L_{\Delta m} = 0$. From (6.57) and (6.58), one finds

$$\frac{di_{as}}{dt} = \frac{1}{L_{ls} + \bar{L}_m}\left(-r_s i_{as} + \frac{d\left(\frac{1}{2}\bar{L}_m i_{bs} + \frac{1}{2}\bar{L}_m i_{cs} + L_{md}\sin\theta_r i_{fr}\right)}{dt}\right),$$

$$\frac{di_{bs}}{dt} = \frac{1}{L_{ls} + \bar{L}_m}\left(-r_s i_{bs} + \frac{d\left(\frac{1}{2}\bar{L}_m i_{as} + \frac{1}{2}\bar{L}_m i_{cs} + L_{md}\sin\left(\theta_r - \frac{2}{3}\pi\right)i_{fr}\right)}{dt}\right),$$

$$\frac{di_{cs}}{dt} = \frac{1}{L_{ls} + \bar{L}_m}\left(-r_s i_{cs} + \frac{d\left(\frac{1}{2}\bar{L}_m i_{as} + \frac{1}{2}\bar{L}_m i_{bs} + L_{md}\sin\left(\theta_r + \frac{2}{3}\pi\right)i_{fr}\right)}{dt}\right),$$

$$\frac{di_{fr}}{dt} = \frac{1}{L_{lf} + L_{mf}}\left(-r_s i_{fr} + \frac{d(L_{md}\sin\theta_r i_{as} + L_{md}\sin(\theta_r - \frac{2}{3}\pi)i_{bs} + L_{md}\sin(\theta_r + \frac{2}{3}\pi)i_{cs})}{dt} + u_{fr}\right)$$

For conventional synchronous generator, Cauchy's form of the circuitry-electromagnetic equations of motion is

$$\frac{di_{as}}{dt} = \frac{1}{\left(2L_{ls} + 3\bar{L}_m\right)\left(3L_{md}^2 L_{ls} - \left(2L_{ls}^2 + 3L_{ls}\bar{L}_m\right)L_{ff}\right)}\Big[-r_s i_{as}\left(3L_{md}^2\bar{L}_m + 4L_{ls}L_{md}^2\right.$$

$$-\left(3\bar{L}_m^2 + 4L_{ls}^2 + 8L_{ls}\bar{L}_m\right)L_{ff} + 2L_{md}^2 L_{ls}\cos 2\theta_r\Big)$$

$$-r_s i_{bs}\left(3L_{md}^2\bar{L}_m + L_{ls}L_{md}^2 - \left(3\bar{L}_m^2 + 2L_{ls}\bar{L}_m\right)L_{ff} + 2L_{md}^2 L_{ls}\cos 2\left(\theta_r - \frac{1}{3}\pi\right)\right)$$

$$-r_s i_{cs}\left(3L_{md}^2\bar{L}_m + L_{ls}L_{md}^2 - \left(3\bar{L}_m^2 + 2L_{ls}\bar{L}_m\right)L_{ff} + 2L_{md}^2 L_{ls}\cos 2\left(\theta_r + \frac{1}{3}\pi\right)\right)$$

$$-\left(6L_{ls}L_{md}\bar{L}_m + 4L_{ls}^2 L_{md}\right)\sin\theta_r(-r_s i_{fr} + u_{fr})$$

$$-\left(6L_{ls}L_{md}^2\bar{L}_m + 4L_{ls}^2 L_{md}^2\right)\left(i_{as}\cos\theta_r + i_{bs}\cos\left(\theta_r - \frac{2}{3}\pi\right)\right.$$

$$+ i_{cs}\cos\left(\theta_r + \frac{2}{3}\pi\right)\Big)\omega_r\sin\theta_r - \left(\left(6L_{md}L_{ls}\bar{L}_m + 4L_{md}L_{ls}^2\right)L_{ff}\right.$$

$$- 6L_{md}^3 L_{ls}\Big)i_{fr}\omega_r\cos\theta_r\Big],$$

$$\frac{di_{bs}}{dt} = \frac{1}{(2L_{ls} + 3\bar{L}_m)\left(3L_{md}^2 L_{ls} - \left(2L_{ls}^2 + 3L_{ls}\bar{L}_m\right)L_{ff}\right)}\left[-r_s i_{as}\left(3L_{md}^2 \bar{L}_m + L_{ls}L_{md}^2\right.\right.$$

$$-\left(3\bar{L}_m^2 + 2L_{ls}\bar{L}_m\right)L_{ff} + 2L_{md}^2 L_{ls}\cos 2\left(\theta_r - \frac{1}{3}\pi\right)\right) - r_s i_{bs}\left(3L_{md}^2 \bar{L}_m + 4L_{ls}L_{md}^2\right.$$

$$-\left(3\bar{L}_m^2 + 4L_{ls}^2 + 8L_{ls}\bar{L}_m\right)L_{ff} + 2L_{md}^2 L_{ls}\cos 2\left(\theta_r - \frac{2}{3}\pi\right)\right)$$

$$-r_s i_{cs}\left(3L_{md}^2 \bar{L}_m + L_{ls}L_{md}^2 - \left(3\bar{L}_m^2 + 2L_{ls}\bar{L}_m\right)L_{ff} + 2L_{md}^2 L_{ls}\cos 2\theta_r\right)$$

$$-\left(6L_{ls}L_{md}\bar{L}_m + 4L_{ls}^2 L_{md}\right)\sin\left(\theta_r - \frac{2}{3}\pi\right)\left(-r_r i_{fr} + u_{fr}\right)$$

$$-\left(6L_{ls}L_{md}^2 \bar{L}_m + 4L_{ls}^2 L_{md}^2\right)\left(i_{as}\cos\theta_r + i_{bs}\cos\left(\theta_r - \frac{2}{3}\pi\right)\right.$$

$$+ i_{cs}\cos\left(\theta_r + \frac{2}{3}\pi\right)\right)\omega_r \sin\left(\theta_r - \frac{2}{3}\pi\right)$$

$$-\left(\left(6L_{md}L_{ls}\bar{L}_m + 4L_{md}L_{ls}^2\right)L_{ff} - 6L_{md}^3 L_{ls}\right)i_{fr}\omega_r \cos\left(\theta_r - \frac{2}{3}\pi\right)\right],$$

$$\frac{di_{cs}}{dt} = \frac{1}{(2L_{ls} + 3\bar{L}_m)\left(3L_{md}^2 L_{ls} - \left(2L_{ls}^2 + 3L_{ls}\bar{L}_m\right)L_{ff}\right)}\left[-r_s i_{as}\left(3L_{md}^2 \bar{L}_m + L_{ls}L_{md}^2\right.\right.$$

$$-\left(3\bar{L}_m^2 + 2L_{ls}\bar{L}_m\right)L_{ff} + 2L_{md}^2 L_{ls}\cos 2\left(\theta_r + \frac{1}{3}\pi\right)\right) - r_s i_{bs}\left(3L_{md}^2 \bar{L}_m + L_{ls}L_{md}^2\right.$$

$$-\left(3\bar{L}_m^2 + 2L_{ls}\bar{L}_m\right)L_{ff} + 2L_{md}^2 L_{ls}\cos 2\theta_r\right) - r_s i_{cs}\left(3L_{md}^2 \bar{L}_m + 4L_{ls}L_{md}^2\right.$$

$$-\left(3\bar{L}_m^2 + 4L_{ls}^2 + 8L_{ls}\bar{L}_m\right)L_{ff} + 2L_{md}^2 L_{ls}\cos 2\left(\theta_r + \frac{2}{3}\pi\right)\right)$$

$$-\left(6L_{ls}L_{md}\bar{L}_m + 4L_{ls}^2 L_{md}\right)\sin\left(\theta_r + \frac{2}{3}\pi\right)\left(-r_r i_{fr} + u_{fr}\right) - \left(6L_{ls}L_{md}^2 \bar{L}_m + 4L_{ls}^2 L_{md}^2\right)$$

$$\times\left(i_{as}\cos\theta_r + i_{bs}\cos\left(\theta_r - \frac{2}{3}\pi\right) + i_{cs}\cos\left(\theta_r + \frac{2}{3}\pi\right)\right)\omega_r \sin\left(\theta_r + \frac{2}{3}\pi\right)$$

$$-\left(\left(6L_{md}L_{ls}\bar{L}_m + 4L_{md}L_{ls}^2\right)L_{ff} - 6L_{md}^3 L_{ls}\right)i_{fr}\omega_r \cos\left(\theta_r + \frac{2}{3}\pi\right)\right], \qquad (6.59)$$

Equation (6.59) yields

$$
\begin{bmatrix} \dfrac{di_{as}}{dt} \\[6pt] \dfrac{di_{bs}}{dt} \\[6pt] \dfrac{di_{cs}}{dt} \\[6pt] \dfrac{di_{fr}}{dt} \end{bmatrix} =
\begin{bmatrix}
-\dfrac{r_s L_{Ds}}{L_{\Sigma s}} & -\dfrac{r_s L_{Ms}}{L_{\Sigma s}} & -\dfrac{r_s L_{Ms}}{L_{\Sigma s}} & 0 \\[8pt]
-\dfrac{r_s L_{Ms}}{L_{\Sigma s}} & -\dfrac{r_s L_{Ds}}{L_{\Sigma s}} & -\dfrac{r_s L_{Ms}}{L_{\Sigma s}} & 0 \\[8pt]
-\dfrac{r_s L_{Ms}}{L_{\Sigma s}} & -\dfrac{r_s L_{Ms}}{L_{\Sigma s}} & -\dfrac{r_s L_{Ds}}{L_{\Sigma s}} & 0 \\[8pt]
0 & 0 & 0 & \dfrac{r_r(2l_{ls}^2+3\overline{L}_m L_{ls})}{L_{\Sigma f}}
\end{bmatrix}
\begin{bmatrix} i_{as} \\[6pt] i_{bs} \\[6pt] i_{cs} \\[6pt] i_{fr} \end{bmatrix}
$$

$$
+\begin{bmatrix}
-\dfrac{2r_s l_{md}^2 L_{ls}}{L_{\Sigma s}}\cos 2\theta_r & -\dfrac{2r_s l_{md}^2 L_{ls}}{L_{\Sigma s}}\cos 2(\theta_r-\tfrac{1}{3}\pi) & -\dfrac{2r_s l_{md}^2 L_{ls}}{L_{\Sigma s}}\cos 2(\theta_r+\tfrac{1}{3}\pi) & \dfrac{r_r L_{Mf}}{L_{\Sigma s}}\sin\theta_r \\[8pt]
-\dfrac{2r_s l_{md}^2 L_{ls}}{L_{\Sigma s}}\cos 2(\theta_r-\tfrac{1}{3}\pi) & -\dfrac{2r_s l_{md}^2 L_{ls}}{L_{\Sigma s}}\cos 2(\theta_r-\tfrac{2}{3}\pi) & -\dfrac{2r_s l_{md}^2 L_{ls}}{L_{\Sigma s}}\cos 2\theta_r & \dfrac{r_r L_{Mf}}{L_{\Sigma s}}\sin(\theta_r-\tfrac{2}{3}\pi) \\[8pt]
-\dfrac{2r_s l_{md}^2 L_{ls}}{L_{\Sigma s}}\cos 2(\theta_r+\tfrac{1}{3}\pi) & -\dfrac{2r_s l_{md}^2 L_{ls}}{L_{\Sigma s}}\cos 2\theta_r & -\dfrac{2r_s l_{md}^2 L_{ls}}{L_{\Sigma s}}\cos 2(\theta_r+\tfrac{2}{3}\pi) & \dfrac{r_r L_{Mf}}{L_{\Sigma s}}\sin(\theta_r+\tfrac{2}{3}\pi) \\[8pt]
\dfrac{2r_r L_{md}L_{ls}}{L_{\Sigma f}}\sin\theta_r & \dfrac{2r_r L_{md}L_{ls}}{L_{\Sigma f}}\sin(\theta_r-\tfrac{2}{3}\pi) & \dfrac{2r_r L_{md}L_{ls}}{L_{\Sigma f}}\sin(\theta_r+\tfrac{2}{3}\pi) & 0
\end{bmatrix}
\begin{bmatrix} i_{as} \\[6pt] i_{bs} \\[6pt] i_{cs} \\[6pt] i_{fr} \end{bmatrix}
$$

$$
+\begin{bmatrix}
-\dfrac{6L_{ls}l_{md}^2\overline{L}_m+4l_{ls}^2 l_{md}^2}{L_{\Sigma s}}\sin\theta_r & -\dfrac{6L_{ls}l_{md}^2\overline{L}_m+4l_{ls}^2 l_{md}^2}{L_{\Sigma s}}\sin\theta_r & -\dfrac{6L_{ls}l_{md}^2\overline{L}_m+4l_{ls}^2 l_{md}^2}{L_{\Sigma s}}\sin\theta_r \\[8pt]
-\dfrac{6L_{ls}l_{md}^2\overline{L}_m+4l_{ls}^2 l_{md}^2}{L_{\Sigma s}}\sin(\theta_r-\tfrac{2}{3}\pi) & -\dfrac{6L_{ls}l_{md}^2\overline{L}_m+4l_{ls}^2 l_{md}^2}{L_{\Sigma s}}\sin(\theta_r-\tfrac{2}{3}\pi) & -\dfrac{6L_{ls}l_{md}^2\overline{L}_m+4l_{ls}^2 l_{md}^2}{L_{\Sigma s}}\sin(\theta_r-\tfrac{2}{3}\pi) \\[8pt]
-\dfrac{6L_{ls}l_{md}^2\overline{L}_m+4l_{ls}^2 l_{md}^2}{L_{\Sigma s}}\sin(\theta_r+\tfrac{2}{3}\pi) & -\dfrac{6L_{ls}l_{md}^2\overline{L}_m+4l_{ls}^2 l_{md}^2}{L_{\Sigma s}}\sin(\theta_r+\tfrac{2}{3}\pi) & -\dfrac{6L_{ls}l_{md}^2\overline{L}_m+4l_{ls}^2 l_{md}^2}{L_{\Sigma s}}\sin(\theta_r+\tfrac{2}{3}\pi) \\[8pt]
-\dfrac{3L_{ls}L_{md}\overline{L}_m+2l_{ls}^2 L_{md}}{L_{\Sigma f}} & -\dfrac{3L_{ls}L_{md}\overline{L}_m+2l_{ls}^2 L_{md}}{L_{\Sigma f}} & -\dfrac{3L_{ls}L_{md}\overline{L}_m+2l_{ls}^2 L_{md}}{L_{\Sigma f}}
\end{bmatrix}
$$

$$
\times\begin{bmatrix} i_{as}\omega_r\cos\theta_r \\[6pt] i_{bs}\omega_r\cos(\theta_r-\tfrac{2}{3}\pi) \\[6pt] i_{cs}\omega_r\cos(\theta_r+\tfrac{2}{3}\pi) \end{bmatrix}
-\begin{bmatrix}
\dfrac{(6L_{md}L_{ls}\overline{L}_m+4L_{md}l_{ls}^2)L_{ff}-6L_{md}^3 L_{ls}}{L_{\Sigma s}}i_{fr}\omega_r\cos\theta_r \\[8pt]
\dfrac{(6L_{md}L_{ls}\overline{L}_m+4L_{md}l_{ls}^2)L_{ff}-6L_{md}^3 L_{ls}}{L_{\Sigma s}}i_{fr}\omega_r\cos(\theta_r-\tfrac{2}{3}\pi) \\[8pt]
\dfrac{(6L_{md}L_{ls}\overline{L}_m+4L_{md}l_{ls}^2)L_{ff}-6L_{md}^3 L_{ls}}{L_{\Sigma s}}i_{fr}\omega_r\cos(\theta_r+\tfrac{2}{3}\pi) \\[8pt]
0
\end{bmatrix}
$$

The *torsional-mechanical* dynamics is

$$
\frac{d\omega_r}{dt}=-\frac{P^2 L_{md}}{4J}i_{fr}\left[i_{as}\cos\theta_r+i_{bs}\cos\left(\theta_r-\frac{2}{3}\pi\right)+i_{cs}\cos\left(\theta_r+\frac{2}{3}\pi\right)\right]-\frac{B_m}{J}\omega_r+\frac{P}{2J}T_{pm},
$$

$$
\frac{d\theta_r}{dt}=\omega_r. \tag{6.60}
$$

The resulting mathematical model for three-phase synchronous generators is given by (6.59) and (6.60). For a specific prime mover, one integrates

the prime mover and synchronous generator dynamics to perform modeling, simulation, and other analysis tasks.

Example 6.14

Examine a two-pole synchronous generator, which was studied in Example 6.13, where the machine parameters were reported. The power generation system is driven by a permanent-magnet DC motor which has the following parameters: $r_{apm} = 0.4$ ohm, $L_{apm} = 0.015$ H, $k_{apm} = 0.15$ V-sec/rad (N-m/A). The armature voltage, applied to the prime mover is $u_{apm} = 100$ V. The voltage supplied to the generator field winding is $u_{fr} = 200$ V.

The differential equations for permanent-magnet DC motors are

$$\frac{di_{apm}}{dt} = -\frac{r_{apm}}{L_{apm}} i_{apm} - \frac{k_{apm}}{L_{apm}} \omega_{rm} + \frac{1}{L_{apm}} u_{apm},$$

$$\frac{d\omega_{rm}}{dt} = \frac{k_a}{J} i_{apm} - \frac{B_m}{J} \omega_{rm} - T_L.$$

The resulting equations of motion to be simulated and analyzed include

$$\frac{di_{apm}}{dt} - = -\frac{r_{apm}}{L_{apm}} i_{apm} - \frac{k_{apm}}{L_{apm}} \omega_{rm} + \frac{1}{L_{apm}} u_{apm},$$

the circuit-electromagnetic generator dynamics (6.59), as well as the *torsional-mechanical* dynamics

$$\frac{d\omega_{rm}}{dt} = \frac{k_{apm}}{J} i_{apm} - \frac{B_m}{J} \omega_{rm} - \frac{PL_{md}}{2J} i_{fr} \left[i_{as} \cos\left(\frac{P}{2}\theta_{rm}\right) + i_{bs} \cos\left(\frac{P}{2}\theta_{rm} - \frac{2}{3}\pi\right) \right.$$

$$\left. + i_{cs} \cos\left(\frac{P}{2}\theta_{rm} + \frac{2}{3}\pi\right) \right],$$

$$\frac{d\theta_{rm}}{dt} = \omega_{rm}.$$

Two MATLAB files (ch6 _ 08.m and ch6 _ 09.m) were developed as documented below

```
tspan = [0 0.8]; y0=[0 0 0 0 0 500 0]';
options = odeset('RelTol',5e-2,'AbsTol',[1e-4 1e-4 1e-4 1e-4 1e-4 1e-4 1e-4]);
[t,y] = ode45('ch6 _ 09',tspan,y0,options);
plot(t,y(:,1));
xlabel('Time [seconds]','FontSize',14);
```

```
title('Prime Mover Current Dynamics, i _ a [A]','FontSize',14); pause;
plot(t,y(:,2));
xlabel('Time [seconds]','FontSize',14);
title('Current Dynamics, i _ a _ s [A]','FontSize',14); pause;
plot(t,y(:,3));
xlabel('Time [seconds]','FontSize',14);
title('Current Dynamics, i _ b _ s [A]','FontSize',14); pause;
plot(t,y(:,4));
xlabel('Time [seconds]','FontSize',14);
title('Current Dynamics, i _ c _ s [A]','FontSize',14); pause;
plot(t,y(:,5));
xlabel('Time [seconds]','FontSize',14);
title('Current Dynamics, i _ f _ r [A]','FontSize',14); pause;
plot(t,y(:,6));
xlabel('Time [seconds]','FontSize',14);
title('Angular Velocity Dynamics, \omega _ r _ m [rad/sec]','FontSize',14);
```

and

```
function yprime=difer(t,y);
% Voltage applied to the prime mover
uapm=100;
% Parameters of the permanent-magnet DC motor
rapm=0.4; Lapm=0.015; kapm=0.15;
% Stator leakage and magnetizing inductances
Lls=0.0001; Lmqs=0.00095; Lmds=0.00095;
% Rotor leakage and magnetizing inductances
Llf=0.0002; Lmf=0.002;
% Mutual inductance
Lmd=0.0004;
% Average value of the magnetizing inductance
Lmb=(Lmqs+Lmds)/3;
% Resistances of the stator and rotor windings
rs=0.25; rr=0.5;
% Equivalent moment of inertia and viscous friction coefficient
J=0.003; Bm=0.0007;
% Number of poles
P=2;
% Voltage applied to the rotor field winding
ufr=200;
% Load resistance
if t<=0.5
Rl=5;
else
Rl=10;
end
wr=2*y(6,:)/P;
% Numerical expressions used
Ld=2*Lls*Lmd^2;
Ldr=2*Lls*Lmd;
Lss=(-3*Lmb^2-4*Lls^2-8*Lls*Lmb)*(Lmf+Llf)+3*Lmd^2*Lmb+4*Lls*Lmd^2;
Ldens=(2*Lls+3*Lmb)*((-2*Lls^2-3*Lmb*Lls)*(Llf+Lmf)+3*Lmd^2*Lls);
Ldenr=(-2*Lls^2-3*Lmb*Lls)*(Llf+Lmf)+3*Lmd^2*Lls;
Lms=(-3*Lmb^2-2*Lls*Lmb)*(Lmf+Llf)+3*Lmd^2*Lmb+Lls*Lmd^2;
Lrr=6*Lls*Lmd*Lmb+4*Lls^2*Lmd;
Llr=-(2*Lls^2+3*Lmb*Lls);
% Variables used
S1=sin(y(7,:));S2=sin(y(7,:)-2*pi/3);S3=sin(y(7,:)+2*pi/3);
Ias=-(rs+Rl)*y(2,:);Ibs=-(rs+Rl)*y(3,:);Ics=-(rs+Rl)*y(4,:);
Ifr=-rr*y(5,:)+ufr;
C1=cos(y(7,:));C2=cos(y(7,:)-2*pi/3);C3=cos(y(7,:)+2*pi/3);
```

```
C12=cos(2*y(7,:));C22=cos(2*y(7,:)-2*pi/3);C32=cos(2*y(7,:)+2*pi/3);
Nsts=(-6*Lmd^2*Lls*Lmb-4*Lmd^2*Lls^2)*wr.*(C1.*y(2,:)+C2.*y(3,:)+C3.*y(4,:));
Nstr=(3*Lmd*Lmb*Lls+2*Lmd*Lls^2)*wr.*(C1.*y(2,:)+C2.*y(3,:)+C3.*y(4,:));
Nct=((6*Lmd*Lls*Lmb+4*Lmd*Lls^2)*(Llf+Lmf)-6*Lmd^3*Lls)*wr.*y(5,:);
Te=P*(Lmd*y(5,:).*(y(2,:).*C1+y(3,:).*C2+y(4,:).*C3))/2;
% Differential equations
yprime=[(-rapm*y(1,:)-kapm*y(6,:)+uapm)/Lapm;...
((Lss+Ld*C12)*Ias+(Lms+Ld*C22)*Ibs+(Lms+Ld*C32)*Ics-Lrr*S1*Ifr+Nsts*S1-Nct*C1)/Ldens;...
((Lms+Ld*C22)*Ias+(Lss+Ld*C32)*Ibs+(Lms+Ld*C12)*Ics-Lrr*S2*Ifr+Nsts*S2-Nct*C2)/Ldens;...
((Lms+Ld*C32)*Ias+(Lms+Ld*C12)*Ibs+(Lss+Ld*C22)*Ics-Lrr*S3*Ifr+Nsts*S3-Nct*C3)/Ldens;...
(-Ldr*S1*Ias-Ldr*S2*Ibs-Ldr*S3*Ics+Llr*Ifr-Nstr)/Ldenr;...
(-Te-P*Bm*y(6,:)+kapm*y(1,:))/J;...
wr];
```

Figure 6.40 illustrates the resulting waveforms for the phase currents, and the terminal voltages are generated as the induced *emfs*. The wye-connected resistive load, inserted in series with the stator phase windings at the specified instances, is

$$R_L = \begin{cases} 5 \text{ ohm, } \forall t \in [0 \quad 0.5) \text{ sec} \\ 10 \text{ ohm, } \forall t \in [0.5 \quad 0.8] \text{ sec} \end{cases}.$$

In the steady-state, the frequency and magnitude of the induced *emfs* depend on the angular velocity, which is a function of the load. One observes that if the load reduces (as R_L increases), the magnitude of phase currents reduces, and ω_{rm} increases. □

Example 6.15 Modeling of Conventional Synchronous Motors in the Rotor Reference Frame

We develop the equations of motion for conventional synchronous machines in the rotor reference frame. One recalls that the rotor rotates with the synchronous angular velocity, and $\omega_r = \omega_e$. Therefore, for synchronous machines the models in the rotor and synchronous reference frames are identical. In the rotor reference frame, the Park transformation, reported in Table 5.1, is

$$\mathbf{u}^r_{qd0s} = \mathbf{K}^r_s \mathbf{u}_{abcs}, \quad \mathbf{i}^r_{qd0s} = \mathbf{K}^r_s \mathbf{i}_{abcs}, \quad \mathbf{\psi}^r_{qd0s} = \mathbf{K}^r_s \mathbf{\psi}_{abcs},$$

$$\mathbf{K}^r_s = \frac{2}{3} \begin{bmatrix} \cos\theta_r & \cos\left(\theta_r - \frac{2}{3}\pi\right) & \cos\left(\theta_r + \frac{2}{3}\pi\right) \\ \sin\theta_r & \sin\left(\theta_r - \frac{2}{3}\pi\right) & \sin\left(\theta_r + \frac{2}{3}\pi\right) \\ \frac{1}{2} & \frac{1}{2} & \frac{1}{2} \end{bmatrix}.$$

One transforms the *abc* variables used to describe the circuitry-electromagnetic dynamics into the *qd0* quantities. For synchronous motors

$$\mathbf{u}_{abcs} = \mathbf{r}_s \mathbf{i}_{abcs} + \frac{d\mathbf{\psi}_{abcs}}{dt},$$

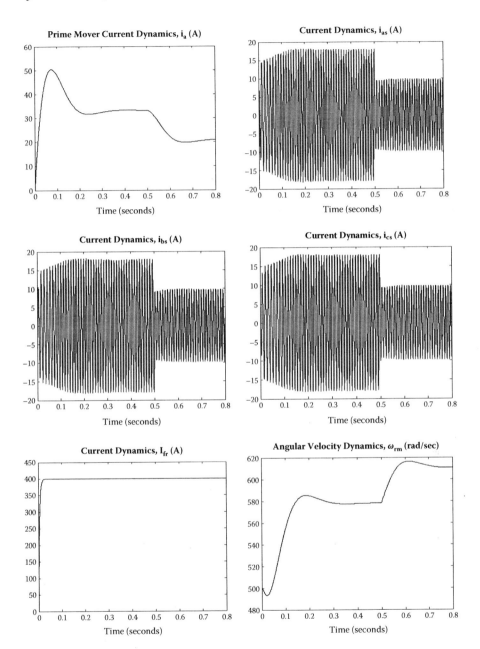

FIGURE 6.40
Transient dynamics of a conventional synchronous generator.

whereas for generators

$$0 = -\mathbf{r}_s \mathbf{i}_{abcs} + \frac{d\boldsymbol{\psi}_{abcs}}{dt}.$$

For motors, the application of the Park transformations results in four differential equations

$$u_{qs}^r = r_s i_{qs}^r + \omega_r \psi_{ds}^r + \frac{d\psi_{qs}^r}{dt}, \quad u_{ds}^r = r_s i_{ds}^r - \omega_r \psi_{qs}^r + \frac{d\psi_{ds}^r}{dt}, \quad u_{0s}^r = r_s i_{0s}^r + \frac{d\psi_{0s}^r}{dt},$$

$$u_{fr} = r_r i_{fr} + \frac{d\psi_{fr}}{dt}.$$

The equations for the flux linkages are

$$\psi_{qs}^r = (L_{ls} + L_{mq})i_{qs}^r, \quad \psi_{ds}^r = (L_{ls} + L_{md})i_{ds}^r + L_{md}i_{fr}, \quad \psi_{0s}^r = L_{ls}i_{0s}^r,$$

$$\psi_{fr} = L_{md}i_{ds}^r + (L_{lf} + L_{mq})i_{fr}.$$

One finds a set of four differential equations

$$u_{qs}^r = r_s i_{qs}^r + (L_{ls} + L_{mq})\frac{di_{qs}^r}{dt} + (L_{ls} + L_{md})i_{ds}^r \omega_r + L_{md}i_{fr}\omega_r,$$

$$u_{ds}^r = r_s i_{ds}^r + (L_{ls} + L_{md})\frac{di_{ds}^r}{dt} - (L_{ls} + L_{mq})i_{qs}^r \omega_r + L_{md}\frac{di_{fr}}{dt},$$

$$u_{0s}^r = r_s i_{0s}^r + L_{ls}\frac{di_{0s}^r}{dt},$$

$$u_{fr} = r_r i_{fr} + L_{md}\frac{di_{ds}^r}{dt} + (L_{lf} + L_{md})\frac{di_{fr}}{dt}. \tag{6.61}$$

The electromagnetic torque is found to be

$$T_e = \frac{3P}{4}\left[L_{mq}i_{qs}^r i_{ds}^r - L_{md}i_{qs}^r \left(i_{ds}^r - i_{fr} \right) \right].$$

Hence, the *torsional-mechanical* dynamics for conventional synchronous motors in the rotor reference frames is described as

$$\frac{d\omega_r}{dt} = \frac{3P^2}{8J}\left[L_{mq}i_{qs}^r i_{ds}^r - L_{md}i_{qs}^r \left(i_{ds}^r - i_{fr} \right) \right] - \frac{B_m}{J}\omega_r - \frac{P}{2J}T_L,$$

$$\frac{d\theta_r}{dt} = \omega_r, \tag{6.62}$$

From (6.61) and (6.62), one obtains the resulting equations of motion.

□

Homework Problems

Problem 6.1

Consider a two-phase permanent-magnet synchronous motor. Let $\psi_{asm} = \psi_m \cos^7\theta_r$ and $\psi_{bsm} = \psi_m \sin^7\theta_r$.

6.1.1. Using Kirchhoff's voltage law, derive the differential equation for the phase current i_{as} and i_{bs}. For example, obtain and report the circuitry-electromagnetic differential equations.

6.1.2. Report the emf_{as} induced (as found in Section 6.1.1). Plot the derived emf_{as} as a function of θ_r at the steady-state operation if $\omega_r = 100$ rad/sec and $\psi_m = 0.1$. Report the MATLAB statement to calculate and plot emf_{as}.

6.1.3. Derive an explicit expression for the electromagnetic torque T_e.

6.1.4. For T_e found, derive the balanced voltage and current sets with the goal to eliminate the torque ripple and current chattering. Report the problems one faces.

Problem 6.2

Consider a two-phase axial topology permanent-magnet synchronous motor. The *effective* fluxes with respect to the *as* and *bs* windings are $B_{as} = B_{Mmax}\sin^7\theta_r$ and $B_{bs} = B_{max}\cos^7\theta_r$. The motor parameters are $N = 100$, $A_{ag} = 0.0001$, $B_{max} = 0.75$, and $N_m = 10$. Assume the phase currents are $i_{as} = i_M\cos\theta_r$ and $i_{bs} = -i_M\sin\theta_r$.

6.2.1. Drive the expression for the electromagnetic torque.

6.2.2. Examine and document how the electromagnetic toque varies as a function of the rotor displacement. Plot torque versus displacement.

6.2.3. Make conclusions how to improve the motor performance and enhance its capabilities. For example, how to maximize the torque and ensure that T_e does not have the ripple.

6.2.4. Document how to use MATLAB to solve problems 6.2.1 through 6.2.3.

Problem 6.3

Simulate and analyze an electromechanical system actuated by a NEMA 23 size, two-phase 1.8° full-step, 5.4 V (*rms*), and 1.4 N-m permanent-magnet stepper motor. The parameters are: $RT = 50$, $r_s = 1.68$ ohm, $L_{ss} = 0.0057$ H, $\psi_m = 0.0064$ V-sec/rad (N-m/A), $B_m = 0.000074$ N-m-sec/rad, and $J = 0.000024$ kg-m^2. Study the motor performance for:

(1) $u_{as} = -u_M \text{sgn}(\sin(RT\theta_{rm}))$ and $u_{bs} = u_M \text{sgn}(\cos(RT\theta_{rm}))$;

(2) $u_{as} = -\sqrt{2}u_M \sin(RT\theta_{rm})$ and $u_{bs} = \sqrt{2}u_M \cos(RT\theta_{rm})$; and

(3) phase voltages u_{as} and u_{bs} with the magnitude u_M are applied as sequences of pulses with different frequency to ensure a step-by-step operation.

Explain why stepper motors can be a favorable solution in some direct drives and *servos* applications. Discuss the possibility to use stepper motors in the open-loop systems without Hall-effect sensors to measure θ_r. Report the challenges one faces using the stepper motors in open-loop configuration and not using the rotor displacement sensor.

References

1. S.J. Chapman, *Electric Machinery Fundamentals*, McGraw-Hill, New York, 1999.
2. A.E. Fitzgerald, C. Kingsley, and S.D. Umans, *Electric Machinery*, McGraw-Hill, New York, 1990.
3. P.C. Krause and O. Wasynczuk, *Electromechanical Motion Devices*, McGraw-Hill, New York, 1989.
4. P.C. Krause, O. Wasynczuk, and S.D. Sudhoff, *Analysis of Electric Machinery*, IEEE Press, New York, 1995.
5. S.E. Lyshevski, *Electromechanical Systems, Electric Machines, and Applied Mechatronics*, CRC Press, Boca Raton, FL, 1999.
6. G.R. Slemon, *Electric Machines and Drives*, Addison-Wesley Publishing Company, Reading, MA, 1992.
7. S.E. Lyshevski, *Engineering and Scientific Computations Using MATLAB®*, Wiley-Interscience, Hoboken, NJ, 2003.
8. *MATLAB R2006b*, CD-ROM, MathWorks, Inc., 2007.

7

Introduction to Control
of Electromechanical Systems and
Proportional-Integral-Derivative
Control Laws

7.1   Electromechanical Systems Dynamics

Electromechanical systems integrate a variety of components (actuators, power converters, sensors, ICs, etc.). As a result of the fast dynamics of ICs and sensors, the overall behavior of electromechanical systems is usually defined by the mechanical dynamics of electromechanical motion devices with the attached kinematics. In conventional and miniscale electromechanical systems, the output behavior depends on the equivalent mass m (for translational motion) or moment of inertia J (for rotational motion) as well as all other parameters.

Various electric machines (actuators, generators, transducers, etc.), their capabilities, and specific performance measures were examined in previous chapters. The basic electromagnetics, energy conversion, efficiency, torque production, and other important features were studied. To perform the analysis, the electromechanical motion devices were modeled, simulated, and examined. The derived mathematical models, in forms of nonlinear differential equations, allow one to accomplish sound quantitative and qualitative analysis. We analyzed various electromechanical systems and their components by numerically and analytically solving the equations of motion. The Newtonian and Lagrangian mechanics, Maxwell's equations, Kirchhoff's laws, energy conversion principles, and other concepts were used. The modeling task was aimed to derive accurate mathematical descriptions with the ultimate objective to assist analysis tasks. Using the derived descriptive modeling and analysis features, one may enhance the system performance through the design in the behavioral domain. Chapters 7 and 8 introduce and cover control topics for electromechanical systems.

Closed-loop systems are designed in order to ensure the best (*achievable*) performance as measured against a spectrum of specifications and requirements. Analog and digital controllers have been derived and implemented for a large class of dynamic systems [1–7]. One performs the application-specific

423

structural design, tailored to the system organization and hardware solutions. Then, the behavioral analysis and design are accomplished. Hardware solutions predefine system performance and capabilities, while control laws affect the system performance and capabilities. For example, for a high-performance ~30 kW electric drive (automotive application), one determines that a permanent-magnet synchronous motor with PWM amplifier will ensure the best performance and capabilities. This electric drive is open-loop stable and operational. However, the tracking control should be ensured by designing a tracking controller. Unsound behavioral design may lead to control laws which may destabilize the stable system (leading to unstable closed-loop system), or controllers which might result in inadequate performance. Efficiency, stability, robustness, and disturbance attenuation are obvious criteria. The need to minimize tracking error and reduce the settling time results in the high feedback gains and discontinuous control laws. However, chattering phenomena and oscillatory dynamics, observed in relay-type and high-gain control laws, result in losses and low efficiency. Therefore, control laws which are sound from a mathematical perspective may not be applicable to various electromechanical systems. The specifications imposed on closed-loop systems are given in the performance (behavioral) domain. For example, the criteria under consideration can be:

- Electromagnetics-centered control soundness and efficiency;
- Stability with the desired stability margins in the full operating envelope;
- Robustness to parameter variations;
- Robustness to structural, kinematical, and environmental changes;
- Tracking accuracy, dynamic and steady-state errors;
- Disturbance (load) and noise attenuation;
- Transient response specifications (settling times, maximum overshoot, etc.).

The systems performance is measured and assessed against multiple criteria (stability, robustness, transient behavior, accuracy, disturbance attenuation, steady-state responses, etc.). The requirements are defined by the specifications imposed in the full operating envelope. Some performance characteristics are assigned and assessed using criteria and metrics in the form of performance functionals (for continuous-time systems) and performance indexes (for discrete-time systems). For example, in the behavioral domain, one can examine and optimize the input-output transient dynamics. Denoting the reference (command) and output variables as $r(t)$ and $y(t)$, the tracking error $e(t) = r(t) - y(t)$ is minimized and controlled by using different control laws $u(t)$. These control laws can be found by applying various performance and stability criteria. The electromechanical system with the reference $r(t)$ and output $y(t)$ is illustrated in Figure 7.1a depicting the input (reference)–output dynamics. In this chapter, systems are studied in the behavioral domain examining and optimizing the steady-state responses and dynamic transients. It is assumed that the optimal design

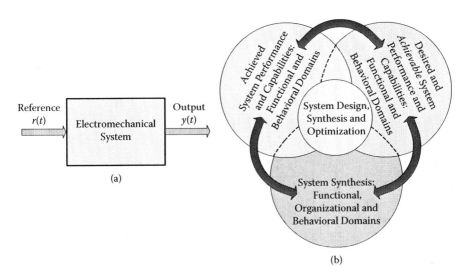

FIGURE 7.1
(a) Electromechanical system with input $r(t)$ and output $y(t)$; (b) Design taxonomy.

in the structural domain was used, and the system organization, advanced matching components (motion devices, power electronics, DSPs, etc.) are utilized. By using the design taxonomy, as reported in Figure 7.1b, the behavioral optimization will be performed.

In Chapter 1, various components of electromechanical system were reported. The high-level system schematics with $r(t)$ = const is shown in Figure 7.2.

The objective is to design control laws to optimize the system performance. The *generic* functional block-diagram of controlled (closed-loop) systems and

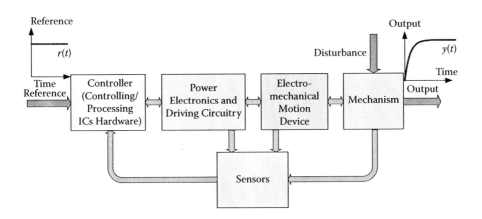

FIGURE 7.2
Functional block-diagram of the closed-loop electromechanical system.

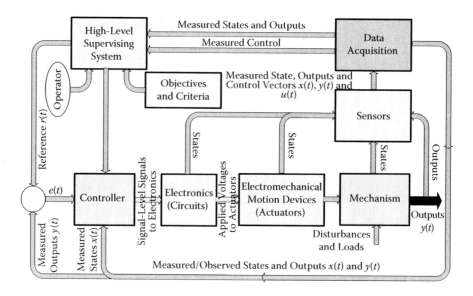

FIGURE 7.3
High-level functional block diagram of closed-loop electromechanical systems.

a possible closed-loop system organization are reported in Figure 7.3. Optimal control laws must be designed, examined, verified, and implemented. We will apply and demonstrate different design methods. To implement these analog and digital controllers, different ICs, microcontrollers, and DSPs are used. These processing and signal conditioning ICs are the components of electromechanical systems.

To optimize system dynamics and attain other performance specifications one can minimize the tracking error and settling time. Steady-state and dynamic behavior can be optimized using different state-space and frequency-domain paradigms. The state-space methods imply the use of the state, output, control, reference, disturbance, and other variables utilizing the nonlinear equations of motion in the time-domain. In contrast, the frequency- and s-domain methods (Laplace and Fourier transforms, transfer function, characteristic equation, pole placement, etc.) are applicable mainly to linear and near-linear systems. The majority of electromechanical systems are nonlinear and cannot be linearized. Therefore, nonlinear methods are emphasized.

**Example 7.1 System Performance and Its Evaluation
               Using Performance Functionals**

Control laws can be designed or evaluated using the performance functionals and indexes, which represent and assess specifications and requirements. Various methods in the analytic design of control laws are covered in Chapter 8. In this chapter we cover the synthesis of PID-type control laws. Some analytic methods are based on the performance functionals minimized that define control laws u, whereas other methods utilize performance functionals to evaluate and assess the system performance.

For example, by using the settling time and the tracking error $e(t) = r(t) - y(t)$, the performance criteria can be expressed as

$$J = \min_{t,e} \int_0^\infty t\,|e|\,dt.$$

The functionals that utilize only e and t, such as

$$J = \min_e \int_0^\infty |e|\,dt, \quad J = \min_e \int_0^\infty e^2 dt, \quad J = \min_e \int_0^\infty (e^2 + e^4 + e^6)dt,$$

may have a limited practicality because other performance requirements are imposed and must be evaluated. Various performance and capabilities features may be integrated and utilized, for example, efficiency, losses, control efforts, transient evolutions, and so on. For example, the control efforts can be assessed using the positive-definite integrands u^{2n} ($n = 1,2,3,\ldots$) or $|u|$, the control rate is evaluated as $(du/dt)^{2n}$ ($n = 1,2,3,\ldots$) or $|du/dt|$, whereas the torque ripple is assessed as T_e^{2n} ($n = 1,2,3,\ldots$) or $|T_e|$. The analysis of the torque ripple leads to the qualitative and quantitative assessment of efficiency, heating, vibration, noise, and so on. The state variables x can be used in the performance functionals to integrate the dissipated energy and other criteria. Furthermore, T_e is given as a nonlinear function of the state variables (currents and angular displacement). As an illustrative example, one may use

$$J = \min_{e,T_e,u} \int_0^\infty (e^6 + T_e^4 + u^2)dt.$$

The synthesis of performance functionals and integrands is covered in Chapter 8. However, the reported basic physics (electromagnetics, mechanics, thermodynamics, vibroacoustics, etc.) were introduced to demonstrate the necessity and possibility to unify the physics and mathematic features. □

A great number of specifications are imposed. Stability, efficiency, robustness, and output dynamics are usually prioritized examining the constraints imposed on the system variables. We consider the output transient dynamics and the evolution envelopes. The output transient response is illustrated in Figure 7.4 for the step reference command $r(t) = $ const. The system is stable because output is bounded and converges to the steady-state value $y_{\text{steady-state}}$, that is,

$$\lim_{t \to \infty} y(t) = y_{\text{steady-state}} \quad \text{and} \quad e(t) \to 0.$$

The output $y(t)$ can be studied within the evolution envelopes. Two evolution envelopes (I and II) are illustrated in Figure 7.4. These envelopes are assigned specifying the desired accuracy, settling time, overshoot, etc. The systems dynamics is within the evolution envelope II, which can represent the best *achievable* performance. The ideal (desired) performance within evolution

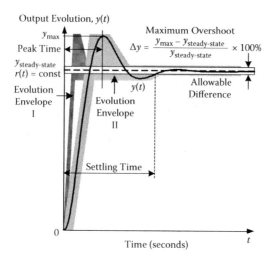

FIGURE 7.4
Output transient response and evolution envelopes, $r(t)$ = const.

envelope I is not achieved. To attain the optimal performance, control laws are used. These control laws change transient behavior ensuring the systems dynamics within an *achievable* evolution envelope, which may or may not meet the specifications imposed. Because of the limits (peak, rated and maximum torque, force, power, voltage, current, and other variables), the designer may not be able to achieve the desired performance while guaranteeing the best *achievable* performance. The system structural redesign (organization, hardware, etc.) may be needed if necessary. The *physical limits* may not be surpassed and/or overcome by any software or controller solutions. The designer may refine the system organization and apply advanced hardware components which may ensure the performance improvements.

The system performance and capabilities in the behavioral domain are examined using the well-defined criteria, for example, settling time, overshoot, accuracy, acceleration capabilities, and others as represented in Figure 7.4. The settling time is the time needed for the system output $y(t)$ to reach and stay within the steady-state value $y_{\text{steady-state}}$. The allowable difference between $y(t)$ and $y_{\text{steady-state}}$ is used to find the settling time. This difference may vary from ~5% to ~0.001% or less. For example, in high-accuracy pointing systems, the required accuracy can be μrad for the overall displacements (repositioning) in the rad range. The settling time is the minimum time after which the system response remains within the desired accuracy taking into account the steady-state value $y_{\text{steady-state}}$ and the command $r(t)$. The maximum overshoot is the difference between the maximum peak value of the systems output $y(t)$ and the steady-state value $y_{\text{steady-state}}$ divided by $y_{\text{steady-state}}$, that is,

$$\Delta y = \frac{y_{\text{max}} - y_{\text{steady-state}}}{y_{\text{steady-state}}} \times 100\%.$$

The rise time is the time needed for the system output $y(t)$ to increase from 10 to 90% of the steady-state value $y_{\text{steady-state}}$ if $r(t) = $ const. The delay time is the time needed for the system output $y(t)$ to reach 50% of $y_{\text{steady-state}}$ if the reference is the step. The peak time is the time required for the system output $y(t)$ to reach the first peak of the overshoot.

In high-performance electromechanical systems, some criteria are stringent. The most common criteria are stability, robustness to parameter variations in the full operating envelope, output dynamics, tracking error, disturbance attenuation, and so on. The transient evolution envelope is examined in the operating envelope. We focus our efforts on the dynamic optimization and behavioral design assuming that the structural design (system organization with the matching components, optimal components selection, measurement accuracy, sensor delays, component efficiency, reliability, passive vibroacoustics optimization, kinematics soundness, etc.) was achieved. However, efficiency, vibroacoustics, and other characteristics can be further improved by designing control laws.

This chapter covers the basic methods to solve the motion control problem designing analog and digital proportional-integral-derivative (PID) controllers. These control laws are widely applied to guarantee stability, attain tracking, ensure disturbance attenuation, guarantee robustness, meet accuracy specifications, and so on. The advanced methods in linear and nonlinear control are reported in Chapter 8 using state-space methods. The PID controllers use the tracking error $e(t)$, whereas the state-space controllers may use both the tracking error $e(t)$ and the state variables $x(t)$.

7.2 Equations of Motion: Electromechanical Systems Dynamics in the State-Space Form and Transfer Functions

This section reports the well-known basics in representation of dynamic systems using state-space equations of motion and transfer functions. The mathematical models of various electromechanical motion devices and power converters were found as given by nonlinear differential equations. These equations of motion are used in order to perform the analysis tasks and design control laws.

Linear systems can be described in the s and z domains using transfer functions $G_{\text{sys}}(s)$ and $G_{\text{sys}}(z)$. The Laplace operator $s = d/dt$ is used for continuous-time systems. The transfer functions can be used to synthesize linear PID-type control laws. There is a very limited class of systems which can be described by linear differential equations. For nonlinear systems, one cannot apply transfer functions and other methods of linear theory. However, many nonlinear electromechanical systems can be controlled using PID controllers ensuring desired performance.

For linear systems, described in the state-space form, we use n states $x \in \mathbb{R}^n$ and m controls $u \in \mathbb{R}^m$. The transient dynamics of linear systems is described by a set of n linear first-order differential equations

$$\frac{dx_1}{dt} = a_{11}x_1 + a_{12}x_2 + \cdots + a_{1n-1}x_{n-1} + a_{1n}x_n + b_{11}u_1$$

$$+ b_{12}u_2 + \cdots + b_{1m-1}u_{m-1} + b_{1m}u_m, \ x_1(t_0) = x_{10},$$

$$\frac{dx_2}{dt} = a_{21}x_1 + a_{22}x_2 + \cdots + a_{2n-1}x_{n-1} + a_{2n}x_n + b_{21}u_1$$

$$+ b_{22}u_2 + \cdots + b_{2m-1}u_{m-1} + b_{2m}u_m, \ x_2(t_0) = x_{20},$$

$$\vdots$$

$$\frac{dx_{n-1}}{dt} = a_{n-11}x_1 + a_{n-12}x_2 + \cdots + a_{n-1n-1}x_{n-1} + a_{n-1n}x_n + b_{n-11}u_1$$

$$+ b_{n-12}u_2 + \cdots + b_{n-1m-1}u_{m-1} + b_{m-1m}u_m, \ x_{n-1}(t_0) = x_{n-10},$$

$$\frac{dx_n}{dt} = a_{n1}x_1 + a_{n2}x_2 + \cdots + a_{nn-1}x_{n-1} + a_{nn}x_n + b_{n1}u_1$$

$$+ b_{n2}u_2 + \cdots + b_{nm-1}u_{m-1} + b_{nm}u_m, \ x_n(t_0) = x_{n0}.$$

In the matrix form, we have

$$\frac{dx}{dt} = \begin{bmatrix} \dfrac{dx_1}{dt} \\ \dfrac{dx_2}{dt} \\ \vdots \\ \dfrac{dx_{n-1}}{dt} \\ \dfrac{dx_n}{dt} \end{bmatrix} = \begin{bmatrix} a_{11} & a_{12} & \cdots & a_{1\,n-1} & a_{1\,n} \\ a_{21} & a_{22} & \cdots & a_{2\,n-1} & a_{2\,n} \\ \vdots & \vdots & \ddots & \vdots & \vdots \\ a_{n-1\,1} & a_{n-1\,2} & \cdots & a_{n-1\,n-1} & a_{n-1\,n} \\ a_{n\,1} & a_{n\,2} & \cdots & a_{n\,n-1} & a_{n\,n} \end{bmatrix} \begin{bmatrix} x_1 \\ x_2 \\ \vdots \\ x_{n-1} \\ x_n \end{bmatrix}$$

$$+ \begin{bmatrix} b_{11} & b_{12} & \cdots & b_{1\,m-1} & b_{1\,m} \\ b_{21} & b_{22} & \cdots & b_{2\,m-1} & b_{2\,m} \\ \vdots & \vdots & \ddots & \vdots & \vdots \\ b_{n-1\,1} & b_{n-1\,2} & \cdots & b_{n-1\,m-1} & b_{n-1\,m} \\ b_{n\,1} & b_{n\,2} & \cdots & b_{n\,m-1} & b_{n\,m} \end{bmatrix} \begin{bmatrix} u_1 \\ u_2 \\ \vdots \\ u_{m-1} \\ u_m \end{bmatrix} = Ax + Bu, \quad x(t_0) = x_0$$

Assuming that matrices $A \in \mathbb{R}^{n \times n}$ and $B \in \mathbb{R}^{n \times m}$ are constant-coefficients (system parameters are constant), we have the characteristic equation as

$$\left| sI - A \right| = 0 \quad \text{or} \quad \left| a_n s^n + a_{n-1} s^{n-1} + \cdots + a_1 s + a_0 \right| = 0.$$

Here, $I \in \mathbb{R}^{n \times n}$ is the identity matrix.

Solving the characteristic equation, one finds the eigenvalues, which are also called the characteristic roots and poles. The system is stable if real parts of all eigenvalues are negative. The stability analysis using the eigenvalues is valid only for linear dynamic systems.

The transfer function

$$G(s) = \frac{Y(s)}{U(s)}$$

can be found using the state-space equations. Consider the linear time-invariant system as described by

$$\frac{dx}{dt} = Ax + Bu, \quad y = Hx.$$

For the output vector $y \in \mathbb{R}^b$, the output equation is $y = Hx$, where $H \in \mathbb{R}^{b \times n}$ is the matrix of the constant coefficients.

The Laplace transform for the state-space

$$\frac{dx}{dt} = Ax + Bu$$

and output $y = Hx$ equations give

$$sX(s) - x(t_0) = AX(s) + BU(s), \quad Y(s) = HX(s).$$

Assuming that the initial conditions are zero, we have

$$X(s) = (sI - A)^{-1} BU(s).$$

Hence,

$$Y(s) = HX(s) = H(sI - A)^{-1} BU(s).$$

The transfer function is found as

$$G(s) = \frac{Y(s)}{U(s)} = H(sI - A)^{-1} B.$$

Assuming that the initial conditions are zero, we apply the Laplace transform to both sides of the n-order differential equation

$$\sum_{i=0}^{n} a_i \frac{d^i y(t)}{dt^i} = \sum_{i=0}^{m} b_i \frac{d^i u(t)}{dt^i}.$$

Taking note of

$$\left(\sum_{i=0}^{n} a_i s^i \right) Y(s) = \left(\sum_{i=0}^{m} b_i s^i \right) U(s)$$

one concludes that the transfer function is

$$G(s) = \frac{Y(s)}{U(s)} = \frac{b_m s^m + b_{m-1} s^{m-1} + \cdots + b_1 s + b_0}{a_n s^n + a_{n-1} s^{n-1} + \cdots + a_1 s + a_0}.$$

By setting the denominator polynomial of the transfer function to zero, one obtains the characteristic equation. The stability of linear time-invariant systems is guaranteed if all characteristic eigenvalues, obtained by solving the characteristic equation

$$\left| a_n s^n + a_{n-1} s^{n-1} + \cdots + a_1 s + a_0 \right| = 0,$$

have negative real parts.

In general, electromechanical systems are described by nonlinear differential equations. In the state-space form we have

$$\dot{x}(t) = F(x, r, d) + B(x)u, \quad y = H(x), \quad u_{min} \le u \le u_{max}, \quad x(t_0) = x_0,$$

where $x \in X \subset \mathbb{R}^n$ is the state vector (displacement, position, velocity, current, voltage, etc.); $u \in U \subset \mathbb{R}^m$ is the bounded control vector (voltage, duty cycle, signal-level voltage to the comparator, etc.); $r \in R \subset \mathbb{R}^b$ and $y \in Y \subset \mathbb{R}^b$ are the measured reference and output vectors; $d \in D \subset \mathbb{R}^v$ is the disturbance vector (load, noise, etc.); $F(\cdot): \mathbb{R}^n \times \mathbb{R}^b \times \mathbb{R}^v \to \mathbb{R}^n$ and $B(\cdot): \mathbb{R}^n \to \mathbb{R}^{n \times m}$ are the nonlinear maps; $H(\cdot): \mathbb{R}^n \to \mathbb{R}^b$ is the nonlinear map defined in the neighborhood of the origin, $H(0) = 0$.

The output equation $y = H(x)$ illustrates that the system output $y(t)$ is a nonlinear function of the state variables $x(t)$. The control bounds are represented as $u_{min} \le u \le u_{max}$.

The majority of electromechanical motion devices are continuous and described by differential equations. For discrete motion devices, or, if digital control laws to be designed, one studies discrete systems which are described by difference equations. For n-dimensional state, m-dimensional control, and b-dimensional output vectors, the electromechanical system states, controls, and outputs variables are

$$
x_k = \begin{bmatrix} x_{k1} \\ x_{k2} \\ \vdots \\ x_{kn-1} \\ x_{kn} \end{bmatrix}, \quad u_k = \begin{bmatrix} u_{k1} \\ u_{k2} \\ \vdots \\ u_{km-1} \\ u_{km} \end{bmatrix} \quad \text{and} \quad y_k = \begin{bmatrix} y_{k1} \\ y_{k2} \\ \vdots \\ y_{kb-1} \\ y_{kb} \end{bmatrix}.
$$

In matrix form, the state-space equations are

$$
x_{k+1} = \begin{bmatrix} x_{k+1,1} \\ x_{k+1,2} \\ \vdots \\ x_{k+1,n-1} \\ x_{k+1,n} \end{bmatrix} = \begin{bmatrix} a_{k11} & a_{k12} & \cdots & a_{k1n-1} & a_{k1n} \\ a_{k21} & a_{k22} & \cdots & a_{k2n-1} & a_{k2n} \\ \vdots & \vdots & \ddots & \vdots & \vdots \\ a_{kn-11} & a_{kn-12} & \cdots & a_{kn-1n-1} & a_{kn-1n} \\ a_{kn1} & a_{kn2} & \cdots & a_{knn-1} & a_{knn} \end{bmatrix} \begin{bmatrix} x_{k1} \\ x_{k2} \\ \vdots \\ x_{kn-1} \\ x_{kn} \end{bmatrix}
$$

$$
+ \begin{bmatrix} b_{k11} & b_{k12} & \cdots & b_{k1m-1} & b_{k1m} \\ b_{k21} & b_{k22} & \cdots & b_{k2m-1} & b_{k2m} \\ \vdots & \vdots & \ddots & \vdots & \vdots \\ b_{kn-11} & b_{kn-12} & \cdots & b_{kn-1m-1} & b_{kn-1m} \\ b_{kn1} & b_{kn2} & \cdots & b_{knm-1} & b_{knm} \end{bmatrix} \begin{bmatrix} u_{k1} \\ u_{k2} \\ \vdots \\ u_{km-1} \\ u_{km} \end{bmatrix}
$$

$$
= A_k x_k + B_k u_k , \; x_{k=k_0} = x_{k_0}.
$$

Here, $A_k \in \mathbb{R}^{n \times n}$ and $B_k \in \mathbb{R}^{n \times m}$ are the matrices of coefficients.

The output equation that integrates the system outputs and states variables is

$$
y_k = H_k x_k ,
$$

where $H_k \in \mathbb{R}^{b \times n}$ is the matrix of the constant coefficients.

The n-order linear difference equation is

$$
\sum_{i=0}^{n} a_i y_{n-i} = \sum_{i=0}^{m} b_i u_{n-i} , \quad n \geq m.
$$

Assuming that the coefficients are time-invariant (constant), using the z-transform and letting the initial conditions to be zero, one has

$$
\left(\sum_{i=0}^{n} a_i z^i \right) Y(z) = \left(\sum_{i=0}^{m} b_i z^i \right) U(z).
$$

Therefore, the transfer function is

$$G(z) = \frac{Y(z)}{U(z)} = \frac{b_m z^m + b_{m-1} z^{m-1} + \cdots + b_1 z + b_0}{a_n z^n + a_{n-1} z^{n-1} + \cdots + a_1 z + a_0}.$$

Nonlinear discrete electromechanical systems are described using nonlinear difference equations

$$x_{k+1} = F(x_k, r_k, d_k) + B(x_k)u_k, \qquad y_k = H(x_k), \quad u_{k\,min} \le u \le u_{k\,max}.$$

This chapter covers the design aspects, as well as introduces analog and digital control for electromechanical systems. We emphasize the application of PID control laws. The linear quadratic regulator problem, Hamilton-Jacobi, state-space, and other advanced concepts applicable to linear and nonlinear electromechanical systems are covered in Chapter 8.

7.3   Analog Control of Electromechanical Systems

7.3.1 Analog Proportional-Integral-Derivative Control Laws

Electromechanical systems can be controlled using PID controllers. The majority of electromechanical motion devices are analog, and they evolve in continuous-time domain. It was documented that PWM power amplifiers are continuous-time systems and described by differential equations.

The simple and effective control laws, utilized for decades in electromechanical systems, are the PID-type controllers. The linear analog PID control law is

$$u(t) = \underbrace{k_p e(t)}_{\text{proportional term}} + \underbrace{k_i \int e(t)dt}_{\text{integral term}} + \underbrace{k_d \frac{de(t)}{dt}}_{\text{derivative term}}, \qquad (7.1)$$

where $e(t)$ is the error between the reference signal and the system output, $e(t) = r(t) - y(t)$; k_p, k_i and k_d are the proportional, integral and derivative feedback gains.

The block diagram of the analog PID control law (7.1) is shown in Figure 7.5. The Laplace operator $s = d/dt$ is used to obtain the corresponding equations in the s-domain. For notation simplicity and illustrative purposes, occasionally, we will "mix" the time- and s-domain notations meaning that s represents the differentiation and $1/s$ is integration. For example, control law (7.1) in s-domain is

$$U(s) = \left(k_p + \frac{k_i}{s} + k_d s \right) E(s),$$

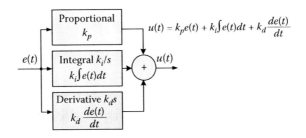

FIGURE 7.5
Analog PID control law.

which gives the transfer function

$$G_{PID}(s) = \frac{U(s)}{E(s)} = \frac{k_d s^2 + k_p s + k_i}{s}.$$

However, for notation simplicity, for example

$$u(t) = k_p e(t) + k_i \int \int \int e(t)dt + k_d \frac{de(t)}{dt},$$

can be written as

$$u(t) = k_p e(t) + k_i \frac{e}{s^3} + k_d \frac{de(t)}{dt}.$$

The reader can easily refine this time- and s-domain notation inconsistency.

The variety of control laws can be obtained utilizing (7.1). Setting k_d equal to zero, the proportional-integral (PI) control law is

$$u(t) = k_p e(t) + k_i \int e(t)dt.$$

Assigning the integral feedback coefficient k_i to be zero, we have the proportional-derivative (PD) control law as given by

$$u(t) = k_p e(t) + k_d \frac{de(t)}{dt}.$$

If $k_i = 0$ and $k_d = 0$, the proportional (P) control law yields as

$$u(t) = k_p e(t).$$

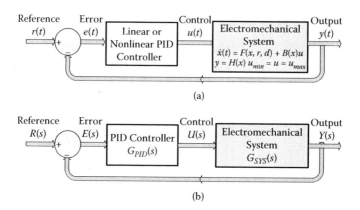

FIGURE 7.6
(a) Time-domain diagram for a nonlinear closed-loop system with a PID controller; (b) s-domain diagram of the linear closed-loop system with an analog PID controller.

Using the Laplace transform, from (7.1), we have

$$U(s) = \left(k_p + \frac{k_i}{s} + k_d s \right) E(s).$$

One finds the transfer function of the analog PID control law as

$$G_{PID}(s) = \frac{U(s)}{E(s)} = \frac{k_d s^2 + k_p s + k_i}{s}.$$

Different linear and nonlinear analog PID-type control laws can be designed and implemented (e.g., proportional, proportional-integral, etc.). The closed-loop electromechanical system with a PID-type control law in the time and s domains are represented in Figures 7.6a and 7.6b. In the time-domain, we use a nonlinear system with control bounds. Using transfer functions for the system and controller, we assume that the system can be linearized and transfer functions are applied as depicted in Figure 7.6b.

If the system output $y(t)$ converges to the bounded reference signal $r(t)$ as time approaches infinity, the tracking of the reference input is accomplished. Ideally, the error vector $e(t)$ approaches zero. Tracking is achieved if

$$e(t) = [r(t) - y(t)] \to 0 \text{ as } t \to \infty.$$

Ideally

$$\lim_{t \to \infty} e(t) = 0,$$

whereas, in practice,

$$\lim_{t \to \infty} |e(t)| \le \varepsilon.$$

In the time domain, the tracking error is $e(t) = r(t) - y(t)$. The Laplace transform of the error signal is $E(s) = R(s) - Y(s)$. For a linear closed-loop system, as given in Figure 7.6b, the Laplace transform of the output $y(t)$ is

$$Y(s) = G_{sys}(s)U(s) = G_{sys}(s)G_{PID}(s)E(s) = G_{sys}(s)G_{PID}(s)[R(s) - Y(s)].$$

The following transfer function of the closed-loop electromechanical systems with a linear PID control law results

$$G(s) = \frac{Y(s)}{R(s)} = \frac{G_{sys}(s)G_{PID}(s)}{1 + G_{sys}(s)G_{PID}(s)}.$$

In the frequency domain, using $s = j\omega$, one obtains

$$G(j\omega) = \frac{Y(j\omega)}{R(j\omega)} = \frac{G_{sys}(j\omega)G_{PID}(j\omega)}{1 + G_{sys}(j\omega)G_{PID}(j\omega)}.$$

The characteristic equation of the linear closed-loop system can be found. The stability can be ensured by the controller $G_{PID}(s)$, while the performance can be refined by adjusting the proportional, integral, and derivative feedback gains. In particular, k_p, k_i and k_d coefficients affect the characteristic equation and closed-loop system performance.

Using the constant factor k, poles at the origin, as well as real and complex-conjugate poles and zeros, one can write

$$G(s) = \frac{k(T_{n1}s+1)(T_{n2}s+1)\cdots\left(T_{n,l-1}^2 s^2 + 2\xi_{n,l-1}T_{n,l-1}s + 1\right)\left(T_{n,l}^2 s^2 + 2\xi_{n,l}T_{n,l}s + 1\right)}{s^M(T_{d1}s+1)(T_{d2}s+1)\cdots\left(T_{d,p-1}^2 s^2 + 2\xi_{d,p-1}T_{d,p-1}s + 1\right)\left(T_{d,p}^2 s^2 + 2\xi_{d,p}T_{d,p}s + 1\right)},$$

where T_i and ζ_i are the time constants and damping coefficients; M is the order of the poles at the origin.

The controller transfer function $G_{PID}(s)$ can be found assigning the "desired" $G(s)$, which "specifies" the location of the poles and zeros, which affect the settling time, stability margins, accuracy, overshoot, and so on. For example, if the PID control law (7.1) is used, the feedback gains k_p, k_i, and k_d can be derived to attain the specific *principal* characteristic eigenvalues because other poles can be located far left in the complex plane (see Figure 7.7.) Many textbooks in feedback control, for example, References 1 through 7, cover the controller design using the pole-centered methods. However, these methods are applicable only for linear systems with no control bounds. In contrast, we

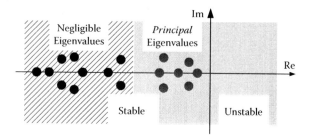

FIGURE 7.7
Eigenvalues in the complex plane.

found that electromechanical systems are nonlinear and control constraints are imposed.

The linear PID control law can be written as

$$u(t) = \underbrace{k_p e(t)}_{\text{proportional}} + \underbrace{\sum_{j=1}^{N_i} k_{i\,j} \frac{e}{s^j}}_{\text{integral}} + \underbrace{\sum_{j=1}^{N_d} k_{d\,j} \frac{d^j e(t)}{dt^j}}_{\text{derivative}}, \tag{7.2}$$

where N_i and N_d are the positive integers; k_{ij} and k_{dj} are the integral and derivative feedback coefficients.

For (7.2), we obtain

$$U(s) = k_p E(s) + \sum_{j=1}^{N_i} k_{ij} \frac{E(s)}{s^j} + \sum_{j=1}^{N_d} k_{dj} s^j E(s)$$

which results in the transfer function $G_{PID}(s)$.

Taking note of $G_{sys}(s)$ and $G_{PID}(s)$, one finds $G(s)$.

Nonlinear PID controllers can be designed and implemented. For example, one may define the nonlinear mappings, and

$$u(t) = \underbrace{\sum_{k=1}^{K_p} k_{p(2k-1)} e^{2k-1}(t)}_{\text{proportional}} + \underbrace{\sum_{j=1}^{N_i} \sum_{k=1}^{K_i} k_{ij(2k-1)} \frac{e^{2k-1}}{s^j}}_{\text{integral}} + \underbrace{\sum_{j=1}^{N_d} \sum_{k=1}^{K_d} k_{dj(2k-1)} \frac{d^j e^{2k-1}(t)}{dt^j}}_{\text{derivative}}, \tag{7.3}$$

where K_p, K_i and K_d are the positive integers; $k_{p(2k-1)}$, $k_{i\,j(2k-1)}$ and $k_{d\,j(2k-1)}$ are the proportional, integral, and derivative feedback coefficients.

In (7.3), integers K_p, K_i and K_d are assigned by the designer defining the power for the tracking error mappings. Setting $N_i = 1$, $N_d = 1$, $K_p = 1$, $K_i = 1$ and $K_d = 1$, we have the PID control law as given by (7.1). Letting $N_i = 2$, $N_d = 1$, $K_p = 3$, $K_i = 2$ and $K_d = 1$, from (7.3), one obtains the nonlinear PID control law as given by

$$u(t) = k_{p1} e(t) + k_{p3} e^3(t) + k_{p5} e^5(t) + k_{i1,1} \frac{e}{s} + k_{i2,1} \frac{e}{s^2} + k_{i1,2} \frac{e^3}{s} + k_{i2,2} \frac{e^3}{s^2} + k_{d1,1} \frac{de(t)}{dt},$$

or, more accurately

$$u(t) = k_{p1}e(t) + k_{p3}e^3(t) + k_{p5}e^5(t) + k_{i1,1}\int e(t)dt + k_{i2,1}\iint e(t)dt$$

$$+ k_{i1,2}\int e^3(t)dt + k_{i2,2}\iint e^3(t)dt + k_{d1,1}\frac{de(t)}{dt}.$$

The control function $u(t)$ is a nonlinear function of $e(t)$. Nonlinear control laws can be applied to improve the system dynamics, enhance stability, ensure robustness, guarantee disturbance attenuation, and so on. The power-series nonlinear PID-type control law is given as

$$u(t) = \underbrace{\sum_{k=1}^{K_p} k_{p(2k-1)} e^{\frac{2k-1}{2a_p+1}}(t)}_{\text{proportional}} + \underbrace{\sum_{j=1}^{N_i}\sum_{k=1}^{K_i} k_{i\,j(2k-1)} \frac{e^{\frac{2k-1}{2a_i+1}}}{s^j}}_{\text{integral}} + \underbrace{\sum_{j=1}^{N_d}\sum_{k=1}^{K_d} k_{d\,j(2k-1)} \frac{d^j e^{\frac{2k-1}{2a_d+1}}(t)}{dt^j}}_{\text{derivative}},$$

(7.4)

where a_p, a_i, and a_d are the nonnegative integers.

The linear PID controller (7.1) results if $N_i = 1$, $N_d = 1$, $K_p = 1$, $K_i = 1$, $K_d = 1$, $a_p = 0$, $a_i = 0$, and $a_d = 0$. Letting $a_p = 2$ and $a_i = 1$, one obtains nonlinear feedback with $e^{1/5}(t)$ or $e^{1/3}(t)$, which ensure the large control signal $u(t)$ for the small values of the tracking error, while reducing $u(t)$ for large $e(t)$. As $e(t) < 0$, the conditional statement and look-up table (implemented by analog ICs, microcontrollers, or DSPs) are used to avoid the complex values. In general, PID-type control law (7.4) may provide one with the optimal performance and high accuracy relaxing the effect of control constraints. In fact, control bounds $u_{min} \leq u \leq u_{max}$ are defined by the physical and hardware limits, while the $u(t)$, as given by (7.1) to (7.4), can exceed the hardware capabilities. One recalls that the duty ratio in power amplifiers or applied voltages (rated voltage) to the motor windings are constrained as

$$d_{Dmin} \leq d_D \leq d_{Dmax}, \quad d_D \in [0\ 1] \quad \text{or} \quad d_D \in [-1\ 1],$$

$$u_{amin} \leq u_a \leq u_{amax} \quad \text{or} \quad u_{Mmin} \leq u_M \leq u_{Mmax}.$$

If nonlinear PID control laws (7.3) or (7.4) are utilized, linear systems methods (transfer functions, eigenvalues, pole-placement, etc.) cannot be applied because the closed-loop system is nonlinear. As a result of the control bounds $u_{min} \leq u \leq u_{max}$, even linear PID control law (7.1) usually leads to saturation in the rated operating envelope. Correspondingly, linear analysis must be applied with a great care.

In electromechanical systems, the controls, states, and outputs are bounded. For any electric machines, actuators, and electromechanical motion devices, the voltages, applied to the windings, are bounded. The duty ratio in PWM power amplifiers is constrained. The bounds on the current, charge, force, torque, power, acceleration, and other physical quantities are imposed.

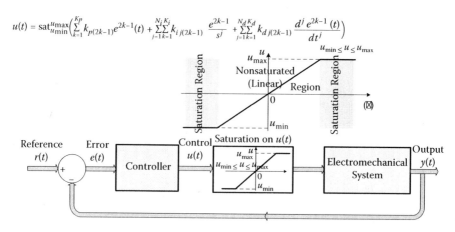

$$u(t) = \mathrm{sat}_{u\min}^{u\max}\left(\sum_{k=1}^{K_p} k_{p(2k-1)} e^{2k-1}(t) + \sum_{j=1}^{N_i}\sum_{k=1}^{K_i} k_{i\,j(2k-1)}\frac{e^{2k-1}}{s^j} + \sum_{j=1}^{N_d}\sum_{k=1}^{K_d} k_{d\,j(2k-1)}\frac{d^j\,e^{2k-1}(t)}{dt^j}\right)$$

FIGURE 7.8
Closed-loop electromechanical system with the bounded control, $u_{\min} \le u \le u_{\max}$.

Mechanical limits are imposed on the maximum angular and linear velocities. These rated and peak (maximum allowed) voltages, currents, velocities, and displacements are specified. Because of the limits imposed, the allowed control is bounded, and the system variables must be within the maximum admissible (rated) set. The closed-loop electromechanical system with a saturated control is shown in Figure 7.8.

The bounded control can be expressed as

$$u(t) = \mathrm{sat}_{u_{\min}}^{u_{\max}}\left(k_p e(t) + k_i \int e(t)dt + k_d \frac{de(t)}{dt}\right), \qquad u_{\min} \le u \le u_{\max}. \qquad (7.5)$$

Thus, the control signal $u(t)$ is bounded between the minimum and maximum values, for example, $u_{\min} \le u \le u_{\max}$, $u_{\min} \le 0$ and $u_{\max} > 0$. In the linear region, the control varies between the maximum $u_{\max}$ and minimum $u_{\min}$ values, and

$$u(t) = k_p e(t) + k_i \int e(t)dt + k_d \frac{de(t)}{dt}.$$

If $k_p e(t) + k_i \int e(t)dt + k_d \frac{de(t)}{dt} > u_{\max}$, the control is bounded, and $u(t) = u_{\max}$.
For $k_p e(t) + k_i \int e(t)dt + k_d \frac{de(t)}{dt} < u_{\min}$, we have $u(t) = u_{\min}$.

Because of control bounds, one must coherently use the control theory even if the electromechanical system itself is described by linear differential equations or transfer functions.

The constrained PID-type control laws with nonlinear mappings are found using (7.3) and (7.4) as

$$u(t) = \text{sat}_{u_{min}}^{u_{max}} \left(\sum_{k=1}^{K_p} k_{p(2k-1)} e^{2k-1}(t) + \sum_{j=1}^{N_i} \sum_{k=1}^{K_i} k_{i\ j(2k-1)} \frac{e^{2k-1}}{s^j} \right.$$

$$\left. + \sum_{j=1}^{N_d} \sum_{k=1}^{K_d} k_{d\ j(2k-1)} \frac{d^j e^{2k-1}(t)}{dt^j} \right), \quad u_{min} \le u \le u_{max},$$

$$u(t) = \text{sat}_{u_{min}}^{u_{max}} \left(\sum_{k=1}^{K_p} k_{p(2k-1)} e^{\frac{2k-1}{2a_p+1}}(t) + \sum_{j=1}^{N_i} \sum_{k=1}^{K_i} k_{i\ j(2k-1)} \frac{e^{\frac{2k-1}{2a_i+1}}}{s^j} \right.$$

$$\left. + \sum_{j=1}^{N_d} \sum_{k=1}^{K_d} k_{d\ j(2k-1)} \frac{d^j e^{\frac{2k-1}{2a_d+1}}(t)}{dt^j} \right), \quad u_{min} \le u \le u_{max}.$$

Example 7.2

Consider one-dimensional motion of a rigid-body mechanical system described by a set of two first-order differential equations

$$\frac{dx_1}{dt} = x_2, \quad \frac{dx_2}{dt} = u,$$

where $x_1(t)$ and $x_2(t)$ are the state variables, in particular, the displacement is $x_1(t)$, and velocity is $x_2(t)$; u is the force or torque to be applied to control the system.

Let the proportional-derivative tracking control law be used. In particular,

$$u(t) = k_p e(t) + k_d \frac{de(t)}{dt}.$$

Hence,

$$G_{PD}(s) = \frac{U(s)}{E(s)} = k_p + k_d s.$$

The transfer function of the open-loop system is

$$G_{sys}(s) = \frac{1}{s^2}.$$

The transfer function of the closed-loop system is

$$G(s) = \frac{Y(s)}{R(s)} = \frac{G_{sys}(s)G_{PD}(s)}{1 + G_{sys}(s)G_{PD}(s)} = \frac{\frac{1}{s^2}(k_p + k_d s)}{1 + \frac{1}{s^2}(k_p + k_d s)} = \frac{k_p + k_d s}{s^2 + k_d s + k_p}.$$

The characteristic equation is $|s^2 + k_d s + k_p| = 0$. One can specify the settling time, which results in the desired characteristic eigenvalues. For example, let the desired poles be -1 and -1. One finds the corresponding feedback gain coefficients to be $k_p = 1$ and $k_d = 2$. The system is stable, and the analytic expressions for $x_1(t)$ and $x_2(t)$ can be found by using the Laplace transform as $r(t)$ and initial conditions are assigned. However, the use of the derivative term results in the sensitivity of the system to the noise and dependence on the waveform of $r(t)$. Furthermore, the control bounds must be integrated.   □

Example 7.3

Consider an electric drive with a permanent-magnet motor. Using Kirchhoff's voltage law and Newton's second law of motion, the differential equations are

$$\frac{di_a}{dt} = -\frac{r_a}{L_a}i_a - \frac{k_a}{L_a}\omega_r + \frac{1}{L_a}u_a \quad \text{and} \quad \frac{d\omega_r}{dt} = \frac{k_a}{J}i_a - \frac{B_m}{J}\omega_r - \frac{1}{J}T_L.$$

The transfer function for an open-loop electric drive (the output is ω_r) is

$$G_{sys}(s) = \frac{Y(s)}{U(s)} = \frac{k_a}{L_a Js^2 + (r_a J + L_a B_m)s + r_a B_m + k_a^2}.$$

The characteristic equation is

$$|L_a Js^2 + (r_a J + L_a B_m)s + r_a B_m + k_a^2| = 0.$$

We have the second-order quadratic equation $as^2 + bs + c = 0$. The solution is

$$s_{1,2} = \frac{-b \pm \sqrt{b^2 - 4ac}}{2a}.$$

The stability of the open-loop system (electric drive) is guaranteed only if the real parts of all characteristic roots are negative. All motor parameters are positive. Hence, $a > 0$, $b > 0$ and $c > 0$. One concludes that for any possible values of a, b, and c, the real parts of the characteristic roots are negative. Thus, an electric drive with a permanent-magnet motor is stable.

Let us examine the derivative tracking control law

$$u_a(t) = k_d \frac{de(t)}{dt}$$

with

$$G_D(s) = k_d s.$$

We obtain the transfer function for the closed-loop system as

$$G(s) = \frac{Y(s)}{R(s)} = \frac{G_{sys}(s)G_{PID}(s)}{1 + G_{sys}(s)G_{PID}(s)} = \frac{k_a k_d s}{L_a J s^2 + (r_a J + L_a B_m + k_a k_d)s + r_a B_m + k_a^2}.$$

The characteristic equation is

$$| L_a J s^2 + (r_a J + L_a B_m + k_a k_d)s + r_a B_m + k_a^2 | = 0.$$

Solving the quadratic equation $as^2 + bs + c = 0$, one has

$$s_{1,2} = \frac{-b \pm \sqrt{b^2 - 4ac}}{2a}.$$

The stability is guaranteed only if the real parts of all characteristic roots are negative. Thus, b must be positive. The system becomes unstable if $b \leq 0$. This can occur if

$$r_a J + L_a B_m + k_a k_d \leq 0.$$

All motor parameters are positive. Hence, the negative value of k_d (positive destabilizing feedback) at which a system becomes unstable is

$$k_d \leq -\frac{r_a J + L_a B_m}{k_a}.$$

One can obtain the characteristics roots using the `roots` command, for example, `Eigenvalues = roots(den_s)`. $\quad\square$

7.3.2 Control of an Electromechanical System with a Permanent-Magnet DC Motor Using Proportional-Integral-Derivative Control Law

One recalls that only a permanent-magnet DC machine under many assumptions can be described by linear differential equations. Consider a servo-system with a permanent-magnet DC motor which actuates a rotating stage (see Figure 7.9).

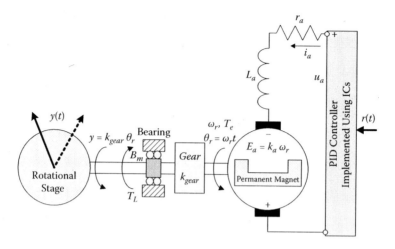

FIGURE 7.9
Schematic diagram of a servo-system with a permanent-magnet DC motor.

This geared motor (with planetary gearhead) is directly attached to the stage. Our goal is to design the control law to attain the desired performance (stability, fast displacement and repositioning of the stage, disturbance attenuation, minimal steady-state tracking error, etc.).

The stage angular displacement is a function of the rotor displacement. Taking into account the gear ratio k_{gear}, one obtains the output equation $y = Hx$ as $y(t) = k_{gear}\theta_r(t)$. To change the angular velocity and displacement, one regulates the voltage applied to the armature winding u_a. The analog PID control law should be designed, and the feedback coefficients must be found. The rated armature voltage for the motor is $\pm u_{max}$ V. The rated (maximum) current is i_{amax}, and the maximum angular velocity is ω_{rmax}. The bounds and parameters must be found. For a DC motor, from the experiments and catalog data we have: $u_{max} = 30$ V ($-30 \le u_a \le 30$ V), $i_{amax} = 0.15$ A, $\omega_{rmax} = 150$ rad/sec, $r_a = 200$ ohm, $L_a = 0.002$ H, $k_a = 0.2$ V-sec/rad, (N-m/A), $J = 0.00000002$ kg-m² and $B_m = 0.00000005$ N-m-sec/rad. The reduction gear ratio is 100:1.

For permanent-magnet DC motors in Chapter 4, we derived the following differential equations

$$\frac{di_a}{dt} = \frac{1}{L_a}(-r_a i_a - k_a \omega_r + u_a),$$

$$\frac{d\omega_r}{dt} = \frac{1}{J}(T_e - T_{viscous} - T_L) = \frac{1}{J}(k_a i_a - B_m \omega_r - T_L),$$

$$\frac{d\theta_r}{dt} = \omega_r.$$

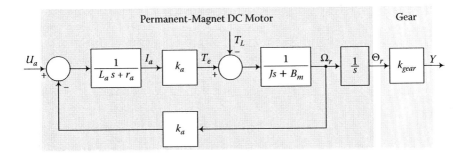

FIGURE 7.10
s-domain diagram of the open-loop system.

Using the Laplace operator $s = d/dt$, one obtains the following equations in the s-domain

$$\left(s + \frac{r_a}{L_a}\right) I_a(s) = -\frac{k_a}{L_a} \Omega_r(s) + \frac{1}{L_a} U_a(s),$$

$$\left(s + \frac{B_m}{J}\right) \Omega_r(s) = \frac{1}{J} k_a I_a(s) - \frac{1}{J} T_L(s),$$

$$s\Theta_r(s) = \Omega_r(s).$$

These equations allow us to obtain the s-domain diagram. In particular, making use of the output equation

$$y(t) = k_{gear}\Theta_r(t), \quad Y(s) = k_{gear}\Theta_r(s),$$

one obtains the s-domain diagram of the open-loop servo-system as documented in Figure 7.10.

The transfer functions of the open-loop system is

$$G_{sys}(s) = \frac{Y(s)}{U_a(s)} = \frac{k_{gear} k_a}{s(L_a J s^2 + (r_a J + L_a B_m)s + r_a B_m + k_a^2)}.$$

Using the linear analog PID control law

$$u_a(t) = k_p e(t) + k_i \int e(t)dt + k_d \frac{de(t)}{dt},$$

we have

$$G_{PID}(s) = \frac{U_a(s)}{E(s)} = \frac{k_d s^2 + k_p s + k_i}{s}.$$

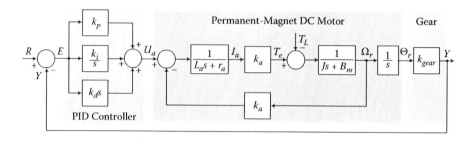

FIGURE 7.11
s-domain diagram of the closed-loop system with an analog PID controller.

The closed-loop s-domain diagram is documented in Figure 7.11. The closed-loop transfer function is

$$G(s) = \frac{Y(s)}{R(s)} = \frac{G_{sys}(s)G_{PID}(s)}{1+G_{sys}(s)G_{PID}(s)}$$

$$= \frac{k_{gear}k_a(k_d s^2 + k_p s + k_i)}{s^2\left(L_a J s^2 + (r_a J + L_a B_m)s + r_a B_m + k_a^2\right) + k_{gear}k_a(k_d s^2 + k_p s + k_i)}$$

$$= \frac{\dfrac{k_d}{k_i}s^2 + \dfrac{k_p}{k_i}s+1}{\dfrac{L_a J}{k_{gear}k_a k_i}s^4 + \dfrac{(r_a J + L_a B_m)}{k_{gear}k_a k_i}s^3 + \dfrac{(r_a B_m + k_a^2 + k_{gear}k_a k_d)}{k_{gear}k_a k_i}s^2 + \dfrac{k_p}{k_i}s+1}.$$

The numerical values of the numerator and denominator coefficients in the transfer function

$$G_{sys}(s) = \frac{Y(s)}{U_a(s)} = \frac{k_{gear}k_a}{s\left(L_a J s^2 + (r_a J + L_a B_m)s + r_a B_m + k_a^2\right)}$$

are found running the following MATLAB statements

```
% System parameters
ra=200; La=0.002; ka=0.2; J=0.00000002; Bm=0.00000005; kgear=0.01;
% Numerator and denominator of the open-loop transfer function
format short e
num _ s=[ka*kgear]; den _ s=[La*J ra*J+La*Bm ra*Bm+ka^2 0];
num _ s, den _ s
```

Using the results displayed in the Command Window

```
num _ s =
   2.0000e-003
den _ s =
   4.0000e-011  4.0001e-006  4.0010e-002  0
```

we conclude that for the open-loop system, the transfer function is

$$G_{sys}(s) = \frac{Y(s)}{U(s)} = \frac{2 \times 10^{-3}}{s(4 \times 10^{-11} s^2 + 4 \times 10^{-6} s + 4 \times 10^{-2})}.$$

The open-loop system is unstable because one of the eigenvalues is at origin. In particular, using the roots command, we have

```
>> Eigenvalues=roots(den _ s)
Eigenvalues =
0
-8.8729e+004
-1.1273e+004
```

To stabilize the servo and attain the desired dynamic performance, control laws must be designed. The characteristic equation of the closed-loop transfer function $G(s)$ with an analog PID control law (7.1) is

$$\frac{L_a J}{k_{gear} k_a k_i} s^4 + \frac{(r_a J + L_a B_m)}{k_{gear} k_a k_i} s^3 + \frac{(r_a B_m + k_a^2 + k_{gear} k_a k_d)}{k_{gear} k_a k_i} s^2 + \frac{k_p}{k_i} s + 1 = 0.$$

The proportional k_p, integral k_i and derivative k_d feedback coefficients of the controller

$$u_a(t) = k_p e(t) + k_i \int e(t) dt + k_d \frac{de(t)}{dt}$$

affect the location of the eigenvalues. Let $k_p = 25000$, $k_i = 250$ and $k_d = 25$. Hence, the PID control law is

$$u_a(t) = 25000 e(t) + 250 \int e(t) dt + 25 \frac{de(t)}{dt}.$$

The characteristic eigenvalues of the closed-loop system are of interest. To derive the eigenvalues, the following MATLAB file is used

```
% System parameters
  ra=200; La=0.002; ka=0.2; J=0.00000002; Bm=0.00000005; kgear=0.01;
% Feedback coefficients
kp=25000; ki=250; kd=25;
```

```
% Denominator of the closed-loop transfer function
den _ c=[(La*J)/(kgear*ka*ki)  (ra*J+La*Bm)/(kgear*ka*ki) ...
(ra*Bm+ka^2+kgear*ka*kd)/(kgear*ka*ki) kp/ki 1];
%         Eigenvalues of the closed-loop system
Eigenvalues _ Closed _ Loop=roots(den _ c)
```

The resulting eigenvalues of the closed-loop system are

```
Eigenvalues _ Closed _ Loop =
-6.6393e+004
-3.3039e+004
-5.6983e+002
-1.0000e-002
```

All four eigenvalues are real. The closed-loop system is stable because the real parts of poles are negative. Because all eigenvalues are real, there should be no overshoot (overshoot is usually an undesirable phenomenon in high-performance positioning systems). The transient dynamics is studied to assess the closed-loop system performance. The following MATLAB file allows the user to simulate the closed-loop electromechanical system

```
% System parameters
ra=200; La=0.002; ka=0.2; J=0.00000002; Bm=0.00000005; kgear=0.01;
% Feedback coefficients
kp=25000; ki=250; kd=25;
ref=1; % reference (command) displacement is 1 rad
% Numerator and denominator of the closed-loop transfer function
num _ c=[kd/ki kp/ki 1];
den _ c=[(La*J)/(kgear*ka*ki)  (ra*J+La*Bm)/(kgear*ka*ki) ...
(ra*Bm+ka^2+kgear*ka*kd)/(kgear*ka*ki) kp/ki 1];
t=0:0.0001:0.02;
u=ref*ones(size(t));
y=lsim(num _ c,den _ c,u,t);
plot(t,y,'-',y,u,':');
title('Angular Displacement, y(t)=0.01\theta _ r, r(t)=1 [rad]','FontSize',14);
xlabel('Time (seconds)','FontSize',14);
ylabel('Output y(t) and Reference r(t)','FontSize',14);
axis([0 0.02,0 1.2]) % axis
```

The closed-loop servo output (angular displacement) and reference $r(t)$ are illustrated in Figure 7.12 if $r(t) = 1$ rad.

The feedback gains significantly affect the stability and dynamics. We reduce the proportional gain k_p and increase the integral feedback k_i. Let $k_p = 2500$, $k_i = 250000$ and $k_d = 25$. We calculate the eigenvalues and simulate the system using the following MATLAB file

```
% System parameters
ra=200; La=0.002; ka=0.2; J=0.00000002; Bm=0.00000005; kgear=0.01;
% Feedback coefficients
kp=2500; ki=250000; kd=25;
ref=1; % reference (command) displacement is 1 rad
% Numerator and denominator of the closed-loop transfer function
num _ c=[kd/ki kp/ki 1];
den _ c=[(La*J)/(kgear*ka*ki)  (ra*J+La*Bm)/(kgear*ka*ki) ...
(ra*Bm+ka^2+kgear*ka*kd)/(kgear*ka*ki) kp/ki 1];
t=0:0.0001:0.2;
u=ref*ones(size(t));
y=lsim(num _ c,den _ c,u,t);
```

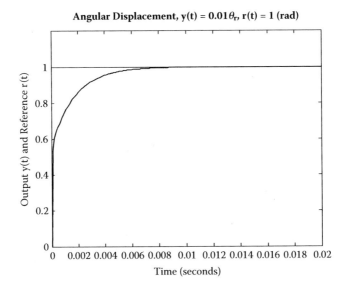

FIGURE 7.12
Dynamics of the closed-loop system with an analog PID controller.

```
plot(t,y,'-',t,u,':');
title('Angular Displacement, y(t)=0.01\theta _ r, r(t)=1 [rad]','FontSize',14);
xlabel('Time (seconds)','FontSize',14);
ylabel('Output y(t) and Reference r(t)','FontSize',14);
axis([0 0.2,0 1.2]) % axis limits
% Denominator of the closed-loop transfer function
den _ c=[(La*J)/(kgear*ka*ki) (ra*J+La*Bm)/(kgear*ka*ki)...
(ra*Bm+ka^2+kgear*ka*kd)/(kgear*ka*ki) kp/ki 1];
% Eigenvalues of the closed-loop system
Eigenvalues _ Closed _ Loop=roots(den _ c)
```

The characteristic eigenvalues are

```
Eigenvalues _ Closed _ Loop =
-6.5869e+004
-3.4079e+004
-2.7719e+001 +6.9284e+001i
-2.7719e+001 -6.9284e+001i
```

The *principal* eigenvalues are complex resulting in the overshoot and longer settling time. The system dynamics for $r(t) = 1$ rad is illustrated in Figure 7.13.

One must soundly design the PID controller and coherently derive the feedback gains. The derivative feedback is not usually used. We let $k_p = 25000$, $k_i = 250$, and $k_d = 0$. The system dynamics is reported in Figure 7.14, and the eigenvalues are real and found to be

```
Eigenvalues _ Closed _ Loop =
-8.8911e+004
-9.6324e+003
-1.4596e+003
-1.0000e-002
```

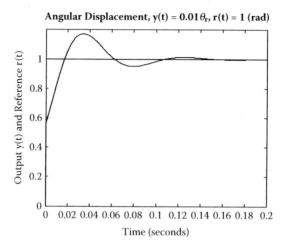

FIGURE 7.13
The output dynamics in the closed-loop system, $k_p = 2500$, $k_i = 250000$ and $k_d = 25$.

Different approaches can be used to simulate electromechanical systems in the MATLAB environment. Simulink was applied in previous chapters. The resulting diagram to perform simulations is documented in Figure 7.15. We use the PID controller block. The system parameters

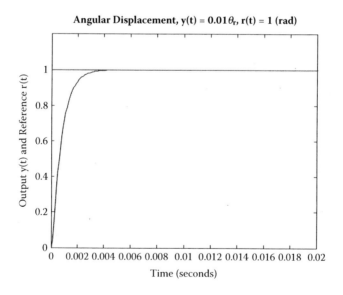

FIGURE 7.14
Dynamics of the output y(t) in the closed-loop system with an analog PID controller, $k_p = 25000$, $k_i = 250$ and $k_d = 0$.

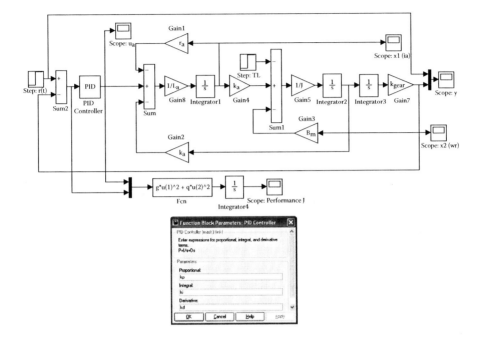

FIGURE 7.15
Simulink diagram to simulate a closed-loop electromechanical system with a linear PID controller (ch7 _ 1.mdl).

and feedback gain coefficients are downloaded in the Command Window. In particular,

```
% System parameters
ra=200; La=0.002; ka=0.2; J=0.00000002; Bm=0.00000005; kgear=0.01;
% Feedback coefficients
kp=25000; ki=250; kd=0;
```

Running the mdl-file, the simulation results become available. To plot the state variables $x_1(t)$ and $x_2(t)$, output angular displacement $y(t) = k_{gear}\theta_r(t)$, and applied voltage $u_a(t)$, which is the control $u(t)$, one types in the Command Window the corresponding statements as reported in Figure 7.16.

Here, $x_1(t)$ denotes the armature current $i_a(t)$, whereas $x_2(t)$ corresponds to the angular velocity $\omega_r(t)$. The transient dynamics of the system variables, as well as the output and voltage evolutions, are documented in Figure 7.16. The analysis of the transients indicates that the settling time is 0.0042 seconds and there is no overshoot. The closed-loop system is stable. However, there are alarming issues to address. The armature current and voltage reach ~102 A and 24750 V, whereas for the motor the peak current is ~0.2 A and the rated voltage is ±30 V. The armature current $i_a(t)$, angular velocity $\omega_r(t)$, and applied voltage $u_a(t)$ significantly exceed the rated values. The saturation $u_{min} \le u \le u_{max}$, $-30 \le u_a \le 30$ V must be integrated.

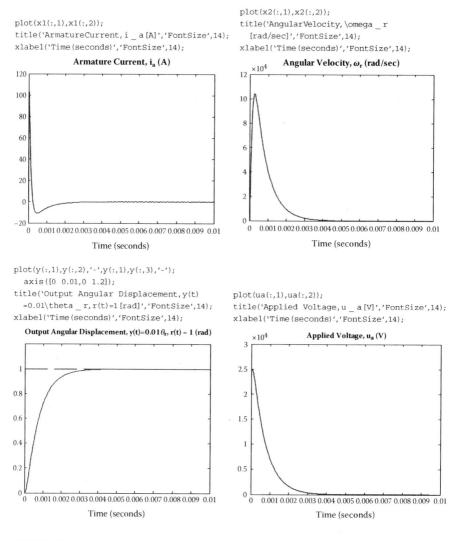

FIGURE 7.16
Dynamics of the closed-loop system with a PID controller, $k_p = 25000$, $k_i = 250$ and $k_d = 0$.

The quadratic performance functional, which can be used to assess the system dynamics, can be given as

$$J = \int_0^\infty \left(qe^2 + gu_a^2 \right) dt.$$

Let $q = 1$ and $g = 0$. The value of J is calculated to be $J = 0.00045$, and the evolution of $J(t)$ is depicted in Figure 7.17. The developed Simulink diagram, as given in Figure 7.15, performs the integration and calculations.

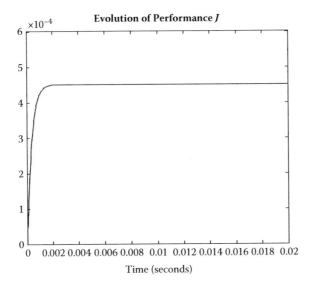

FIGURE 7.17
Performance functional $J = \int_0^\infty (qe^2 + gu_a^2)dt$ evolution if $q = 1$ and $g = 0$.

The bounded PI control law is

$$u_a(t) = \text{sat}_{-30}^{+30}\left(25000e(t) + 250\int e(t)dt \right), \quad -30 \le u_a \le 30 \text{ V.}$$

With these control bounds, the simulation is performed. The Simulink model is built utilizing the Saturation block as illustrated in Figure 7.18.

For the angular displacement $r(t) = 1$ rad, the resulting states and output responses are documented in Figures 7.19 if $T_L = 0$ N-m, $t \in [0\ 0.8)$ sec, and $T_L = 0.02$ N-m, $t \in [0.8\ 1]$ sec. The evolution of the state variables, output, and the bounded voltage are illustrated in Figure 7.19.

The comparison of the simulation results, reported in Figures 7.16 and 7.19, provides one with the evidence that the physical limits and constraints imposed significantly increase the settling time. One observes the effect of the load as T_L is applied at 0.8 sec. The reference (command) input significantly affects the settling time and system behavior. The control bounds, as well as other nonlinearities, must be integrated. For example, the friction, backlash, dead zone, and other nonlinear phenomena affect the closed-loop system performance. Some of those nonlinearities, such as the ready-to-use blocks, are available in the Simulink Library Browser. However, the applicability and validity of the developed simulation tools and various components must be studied.

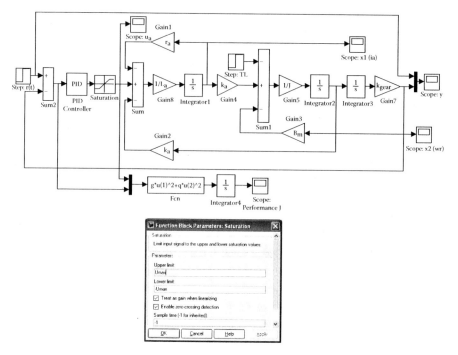

FIGURE 7.18
Simulink diagram of the closed-loop system with saturation (`ch7 _ 2.mdl`).

7.4　Digital Control of Electromechanical Systems

7.4.1　Proportional-Integral-Derivative Digital Control Laws and Transfer Functions

Microcontrollers and DSPs can be utilized to implement control algorithms using analog and digital variables or physical quantities measured by the sensors or observed by the observers. Diagnostics, filtering, data acquisition, and other tasks can be performed using discrete mathematics and digital processing offered by DSPs. Digital control algorithms can be designed, and discrete-time systems are studied in this section.

Continuous-time signals

$$x(t),\ u(t),\ y(t),\ e(t)$$

and others can be sampled with the sampling period T_s, and the continuous- and discrete-time domains are related as $t = kT_s$, where k is the integer. The majority of electromechanical system components (actuators, sensors, etc.) are analog and described by differential equations. One examines the evolution of continuous-time variables. In contrast, discrete-time systems

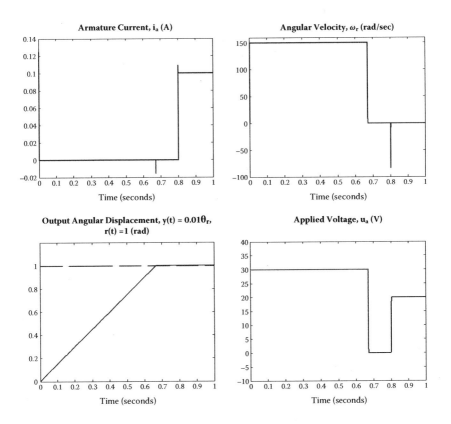

FIGURE 7.19
Dynamics of the closed-loop system with the bounded PID controller ($k_p = 25000$, $k_i = 250$ and $k_d = 0$) if $r(t) = 1$ rad.

are described by difference equations, digital ICs utilize digital quantities, and the evolution of these systems is analyzed in the discrete-time domain. Many systems to be studied are hybrid systems, for example, the systems integrate analog and digital components, devices, and subsystems. To design digital controllers, the differential equations can be discretized, difference equations can be used, or hybrid models are applied.

Example 7.4
For the first-order linear constant-coefficient differential equation

$$\frac{dx}{dt} = -ax(t) + bu(t),$$

we derive the discrete-time model in the form of a difference equation.
  Differential equation

$$\frac{dx}{dt} = -ax(t) + bu(t)$$

is discretized by using $t = kT_s$, yielding

$$\frac{dx}{dt}\bigg|_{t=kT_s} = -ax(kT_s) + bu(kT_s).$$

For a sufficiently small sampling period T_s, the forward rectangular rule (Euler approximation) gives

$$\frac{dx}{dt} \approx \frac{x(t + T_s) - x(t)}{T_s}.$$

Thus,

$$\frac{dx}{dt}\bigg|_{t=kT_s} = \frac{x(kT_s + T_s) - x(kT_s)}{T_s}.$$

Using the forward difference, one obtains

$$\frac{x(kT_s + T_s) - x(kT_s)}{T_s} = -ax(kT_s) + bu(kT_s)$$

We denote $x(t)$ and $u(t)$ at discrete instances t_k and t_{k+1} as

$$x_k = x(t)\big|_{t=kT_s}, \quad x_{k+1} = x(t)\big|_{t=(k+1)T_s} \quad \text{and} \quad u_k = u(t)\big|_{t=kT_s}.$$

Hence, one obtains

$$\frac{x_{k+1} - x_k}{T_s} = -ax_k + bu_k,$$

where

$$x_{k+1} = x[(k+1)T_s], \quad x_k = x(kT_s) \text{ and } u_k = u(kT_s).$$

The following difference equation results

$$x_{k+1} = (1 - aT_s)x_k + bT_s u_k$$

or

$$x_{k+1} = a_k x_k + b_k u_k,$$

where $a_k = (1 - aT_s)$ and $b_k = bT_s$.

This difference equation can be written as

$$x_k = (1 - aT_s)x_{k-1} + bT_s u_{k-1}.$$

From the obtained difference equation, the transfer function results. In particular,

$$G(z) = \frac{X(z)}{U(z)} = \frac{bT_s z^{-1}}{1 - (1 - aT_s)z^{-1}} = \frac{bT_s}{z - (1 - aT_s)}.$$

Thus, the continuous-time system was represented in the discrete-time domain by the difference equation. The z-domain transfer function was derived. □

Hybrid systems integrate analog and digital components as shown in Figure 7.20. Nonlinear and linear systems with digital controllers, hybrid circuits (including A/D and D/A converters, data hold circuits, etc.), power electronics, and analog electromechanical motion devices are represented by Figures 7.20a and 7.20b, respectively.

Assume that the electromechanical motion device dynamics is described by linear constant-coefficient (time-invariant) differential equations. The closed-loop system is documented in Figure 7.20b using the transfer function for the electronics-actuator-mechanism system $G_{sys}(s)$, data hold circuit $G_H(s)$, and digital controller $G_C(z)$. To convert the discrete-time signals from microcontrollers or DSPs to piecewise continuous signals to drive transistors in PWM amplifiers, distinct data hold circuits are used. Zero- and first-order data hold circuits are usually implemented to avoid the complexity and time delay associated with the application of high-order data hold circuits. The N-order data hold circuit with the zero-order data hold is documented in Figure 7.21.

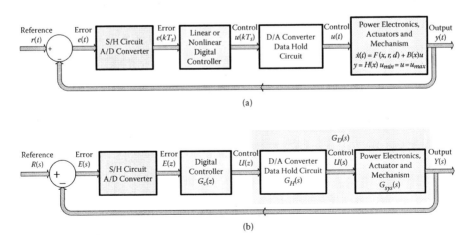

(a)

(b)

FIGURE 7.20
Block diagrams of nonlinear and linear hybrid systems with digital controllers.

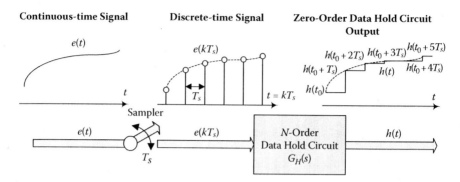

FIGURE 7.21
Sampler and N-order data hold circuit with zero-order data hold.

For the zero-order data hold circuit, the piecewise continuous data hold output is

$$h(t) = \sum_{k=0}^{\infty} e\left(kT_s\right)[1(t - kT_s) - 1(t - (k+1)T_s)].$$

The output of the zero-order data hold is the piecewise continuous signal which is equal to the last sampled value (the value of the continuous-time signal at $t = kT_s$) until the next sampled value is available. This feature is illustrated in Figure 7.21 plotting $e(kT_s)$. We have

$$h(kT_s + t) = h(kT_s) = e(t)\big|_{t=kT_s} \qquad \text{for } 0 \le t < T_s.$$

The Laplace transforms

$$L[1(t)] = \frac{1}{s} \text{ and } L[1(t - kT_s)] = \frac{e^{-kT_s s}}{s}$$

yield

$$L[h(t)] = \sum_{k=0}^{\infty} e(kT_s) \frac{e^{-kT_s s} - e^{-(k+1)T_s s}}{s} = \frac{1 - e^{-T_s s}}{s} \sum_{k=0}^{\infty} e(kT_s)e^{-kT_s s}.$$

Using the Laplace transform for the signal $e(kT_s)$, as given by

$$E_{sampled}(s) = \sum_{k=0}^{\infty} e(kT_s)e^{-kT_s s},$$

one obtains the transfer function of the zero-order data hold circuit applying the following relation

$$L[h(t)] = \frac{1 - e^{-T_s s}}{s} \sum_{k=0}^{\infty} e(kT_s)e^{-kT_s s}.$$

The transfer function of the zero-order data hold is

$$G_H(s) = \frac{1 - e^{-T_s s}}{s}$$

The first-order data hold, which can be used to perform the direct linear extrapolation, is expressed in the time domain as

$$h(t) = 1(t) + \frac{t}{T_s} 1(t) - \frac{t - T_s}{T_s} 1(t - T_s) - 1(t - T_s).$$

Hence, the transfer function is

$$G_H(s) = \frac{1}{s} + \frac{1}{T_s s^2} - \frac{1}{T_s s^2} e^{-T_s s} - \frac{1}{s} e^{-T_s s} = (1 - e^{-T_s s}) \frac{T_s s + 1}{T_s s^2}.$$

The dynamic system $G_{sys}(s)$ with the data hold circuit $G_H(s)$ is represented by the transfer function

$$G_D(s) = G_H(s) G_{sys}(s).$$

Having derived $G_{sys}(s)$, for the chosen $G_H(s)$, one obtains $G_D(s)$ with the corresponding $G_D(z)$. Table 7.1 reports s- and z-domain transforms for distinct continuous- and discrete-time signals.

Example 7.5
We derive the z-domain representations for digital proportional, integral and derivative terms of the PID control law

$$u(t) = k_p e(t) + k_i \int e(t) dt + k_d \frac{de(t)}{dt}.$$

One recalls that the transfer function of an analog PID control law is

$$G_{PID}(s) = \frac{U(s)}{E(s)} = \frac{k_d s^2 + k_p s + k_i}{s}.$$

For the proportional control law, one has

$$u_p(t) = k_p e(t) \quad \text{and} \quad G_p(s) = \frac{U_p(s)}{E(s)} = k_p.$$

Thus, the proportional digital control law is

$$u_p(kT_s) = k_p e(kT_s) \quad \text{and} \quad G_p(z) = \frac{U_p(z)}{E(z)} = k_p.$$

TABLE 7.1

Signals and Their s- and z-Transform

Laplace Transform $X(s)$	Time-Domain Signal $x(t)$	Time-Domain Signal $x(kT_s)$	z-Transform $X(z)$
$\dfrac{1}{s}$	Unit step $1(t)$	$1(kT_s)$	$\dfrac{1}{1-z^{-1}} = \dfrac{z}{z-1}$
$\dfrac{1}{s^2}$	$t1(t)$	$kT_s1(kT_s)$	$\dfrac{T_s z^{-1}}{(1-z^{-1})^2} = \dfrac{T_s z}{(z-1)^2}$
$\dfrac{2}{s^3}$	$t^2 1(t)$	$(kT_s)^2 1(kT_s)$	$\dfrac{T_s^2 z^{-1}(1+z^{-1})}{(1-z^{-1})^3}$
$\dfrac{6}{s^4}$	$t^3 1(t)$	$(kT_s)^3 1(kT_s)$	$\dfrac{T_s^3 z^{-1}(1+4z^{-1}+z^{-2})}{(1-z^{-1})^4}$
$\dfrac{24}{s^5}$	$t^4 1(t)$	$(kT_s)^4 1(kT_s)$	$\dfrac{T_s^4 z^{-1}(1+11z^{-1}+11z^{-2}+z^{-3})}{(1-z^{-1})^5}$
$\dfrac{1}{s+a}$	$e^{-at} 1(t)$	$e^{-akT_s} 1(kT_s)$	$\dfrac{1}{1-e^{-aT_s}z^{-1}}$
$\dfrac{a}{s(s+a)}$	$(1-e^{-at})1(t)$	$(1-e^{-akT_s})1(kT_s)$	$\dfrac{(1-e^{-aT_s})z^{-1}}{(1-z^{-1})(1-e^{-aT_s}z^{-1})}$
$\dfrac{b-a}{(s+a)(s+b)}$	$(e^{-at}-e^{-bt})1(t)$	$(e^{-akT_s}-e^{-bkT_s})1(kT_s)$	$\dfrac{(e^{-aT_s}-e^{-bT_s})z^{-1}}{(1-e^{-aT_s}z^{-1})(1-e^{-bT_s}z^{-1})}$
$\dfrac{1}{(s+a)^2}$	$te^{-at}1(t)$	$kT_s e^{-akT_s}1(kT_s)$	$\dfrac{T_s e^{-aT_s}z^{-1}}{(1-e^{-aT_s}z^{-1})^2}$
$\dfrac{s}{(s+a)^2}$	$(1-at)e^{-at}1(t)$	$(1-akT_s)e^{-akT_s}1(kT_s)$	$\dfrac{1-(1+aT_s)e^{-aT_s}z^{-1}}{(1-e^{-aT_s}z^{-1})^2}$
$\dfrac{\omega_0}{s^2+\omega_0^2}$	$\sin(\omega_0 t)1(t)$	$\sin(\omega_0 kT_s)1(kT_s)$	$\dfrac{z^{-1}\sin(\omega_0 T_s)}{1-2z^{-1}\cos(\omega_0 T_s)+z^{-2}}$
$\dfrac{s}{s^2+\omega_0^2}$	$\cos(\omega_0 t)1(t)$	$\cos(\omega_0 kT_s)1(kT_s)$	$\dfrac{1-z^{-1}\cos(\omega_0 T_s)}{1-2z^{-1}\cos(\omega_0 T_s)+z^{-2}}$
$\dfrac{\omega_0}{(s+a)^2+\omega_0^2}$	$e^{-at}\sin(\omega_0 t)1(t)$	$e^{-akT_s}\sin(\omega_0 kT_s)1(kT_s)$	$\dfrac{e^{-akT_s}z^{-1}\sin(\omega_0 T_s)}{1-2e^{-aT_s}z^{-1}\cos(\omega_0 T_s)+e^{-2aT_s}z^{-2}}$
$\dfrac{s+a}{(s+a)^2+\omega_0^2}$	$e^{-at}\cos(\omega_0 t)1(t)$	$e^{-akT_s}\cos(\omega_0 kT_s)1(kT_s)$	$\dfrac{1-e^{-aT_s}z^{-1}\cos(\omega_0 T_s)}{1-2e^{-aT_s}z^{-1}\cos(\omega_0 T_s)+e^{-2aT_s}z^{-2}}$

The integral

$$u_i(t) = k_i \int e(t) dt$$

and derivative

$$u_d(t) = k_d \frac{de(t)}{dt}$$

terms, with transfer functions

$$G_i(s) = \frac{U_i(s)}{E(s)} = \frac{k_i}{s} \text{ and } G_d(s) = \frac{U_d(s)}{E(s)} = k_d s,$$

can be discretized and represented in the z-domain. Using the z-transform, reported in Table 7.1, for the integral part, using the Euler approximation, the transfer function is

$$G_i(z) = \frac{U_i(z)}{E(z)} = \frac{T_s}{1 - z^{-1}} = \frac{T_s z}{z - 1}.$$

To find the derivative term, using the trapezoidal approximation the first difference results, and

$$G_d(z) = \frac{U_d(z)}{E(z)} = \frac{1 - z^{-1}}{T_s} = \frac{z - 1}{T_s z}.$$

Performing the summation of the derived terms, PI, PD, or PID control laws result. □

There exists a great variety of analog PID-type controllers with the corresponding transfer functions $G_{PID}(s)$. For a PID control law

$$u(t) = k_p e(t) + k_i \int e(t) dt + k_d \frac{de(t)}{dt}$$

with

$$G_{PID}(s) = \frac{U(s)}{E(s)} = \frac{k_d s^2 + k_p s + k_i}{s},$$

one finds the z-domain representation of the control signal $U(z)$ and the transfer functions $G_{PID}(z)$. In the *error* form (the error signal is commonly used to calculate the control), the following expressions result

$$U(z) = \left(k_{dp} + \frac{k_{di}}{1 - z^{-1}} + k_{dd}(1 - z^{-1}) \right) E(z)$$

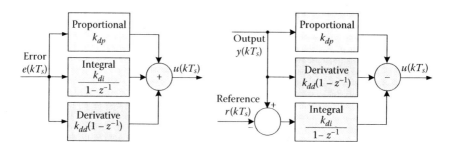

FIGURE 7.22
Error and *reference-output* forms of the digital PID control law.

and

$$G_{PID}(z) = \frac{U(z)}{E(z)} = k_{dp} + \frac{k_{di}}{1 - z^{-1}} + k_{dd}(1 - z^{-1}).$$

We have

$$G_{PID}(z) = \frac{(k_{dp} + k_{di} + k_{dd})z^2 - (k_{dp} + 2k_{dd})z + k_{dd}}{z^2 - z}.$$

The *reference-output* form of the digital PID control law is

$$U(z) = -k_{dp}Y(z) - k_{di}\frac{Y(z) - R(z)}{1 - z^{-1}} - k_{dd}(1 - z^{-1})Y(z).$$

For the *error* and *reference-output* forms, the z-domain block diagrams of the digital PID control laws are illustrated in Figure 7.22.

The feedback gains k_{dp}, k_{di} and k_{dd} of the digital control laws are related to the proportional, integral, and derivative coefficients of the analog PID controller (k_p, k_i and k_d) as well as the sampling period T_s. The relationships between k_{dp}, k_{di}, k_{dd} and k_p, k_i, k_d can be obtained utilizing various analytical and numerical approaches, for example, applying specific formulas that relate z and s. Approximating the integral term by the trapezoidal summation and derivative term by a two-point difference yields

$$k_{dp} = k_p - \frac{1}{2}k_{di}, \quad k_{di} = k_i T_s \quad \text{and} \quad k_{dd} = k_d/T_s.$$

Approximating the integration (rectangular, trapezoidal Tustin, bilinear, etc.) and differentiation (Euler, Taylor, backward difference, etc.), one may obtain other $G_{PID}(z)$ and expressions for feedback gains.

Using microcontrollers and DSPs, the PID control law can be implemented as

$$u(kT_s) = \underbrace{k_{dp}e(kT_s)}_{\text{Proportional}} + \underbrace{\frac{1}{2}k_i T_s \sum_{i=1}^{k}[e((i-1)T_s) + e(iT_s)]}_{\text{Integral}} + \underbrace{\frac{k_d}{T_s}[e(kT_s) - e((k-1)T_s)]}_{\text{Derivative}}.$$

To find the transfer function for systems and controllers in the z-domain, the Tustin approximation is commonly applied to the transfer functions in the s-domain. In particular, from $z = e^{sT_s}$, we have

$$s = \frac{1}{T_s} \ln(z).$$

The series expansion of $\ln(z)$ is

$$\ln(z) = 2\left[\frac{z-1}{z+1} + \frac{1}{3}\left(\frac{z-1}{z+1}\right)^3 + \frac{1}{5}\left(\frac{z-1}{z+1}\right)^5 + \cdots \right], \quad z > 0.$$

By truncating this series expansion for $\ln(z)$, one obtains the Tustin approximation

$$\ln(z) \approx 2\frac{z-1}{z+1} = 2\frac{1-z^{-1}}{1+z^{-1}}.$$

Thus, we have

$$s = \frac{1}{T_s} \ln(z) \approx \frac{2}{T_s}\frac{z-1}{z+1} = \frac{2}{T_s}\frac{1-z^{-1}}{1+z^{-1}}.$$

Example 7.6
Using a linear PID control law, we derive the expression for $G_{PID}(z)$ applying the Tustin approximation. From

$$G_{PID}(s) = \frac{U(s)}{E(s)} = \frac{k_d s^2 + k_p s + k_i}{s},$$

by using

$$s \approx \frac{2}{T_s}\frac{1-z^{-1}}{1+z^{-1}}$$

we have

$$G_{PID}(z) = \frac{U(z)}{E(z)}$$

$$= \frac{k_d \left(\dfrac{2}{T_s}\dfrac{1-z^{-1}}{1+z^{-1}}\right)^2 + k_p \dfrac{2}{T_s}\dfrac{1-z^{-1}}{1+z^{-1}} + k_i}{\dfrac{2}{T_s}\dfrac{1-z^{-1}}{1+z^{-1}}}$$

$$= \frac{\left(2k_p T_s + k_i T_s^2 + 4k_d\right) + \left(2k_i T_s^2 - 8k_d\right)z^{-1} + \left(-2k_p T_s + k_i T_s^2 + 4k_d\right)z^{-2}}{2T_s(1-z^{-2})}.$$

Thus,

$$U(z) - U(z)z^{-2} = k_{e0}E(z) + k_{e1}E(z)z^{-1} + k_{e2}E(z)z^{-2},$$

where

$$k_{e0} = k_p + \frac{1}{2}k_i T_s + 2\frac{k_d}{T_s}, \quad k_{e1} = k_i T_s - 4\frac{k_d}{T_s} \quad \text{and} \quad k_{e2} = -k_p + \frac{1}{2}k_i T_s + 2\frac{k_d}{T_s}.$$

The expression to implement the digital control law is

$$u(k) = u(k-2) + k_{e0}e(k) + k_{e1}e(k-1) + k_{e2}e(k-2). \qquad \square$$

Thus, to implement a digital control law, one should use

$$e(k), \quad e(k-1), \quad e(k-2) \quad \text{and} \quad u(k-2).$$

The closed-loop system with a digital control law $G_C(z)$ is illustrated in Figure 7.20b. The transfer function of the closed-loop system is

$$G(z) = \frac{Y(z)}{R(z)} = \frac{G_C(z)G_D(z)}{1 + G_C(z)G_D(z)}.$$

For a digital PID control law, one has

$$G(z) = \frac{Y(z)}{R(z)} = \frac{G_{PID}(z)G_D(z)}{1 + G_{PID}(z)G_D(z)}.$$

The analysis of linear discrete-time systems is straightforward by applying the methods of linear control theory. The application of linear theory and MATLAB are covered in the following section.

7.4.2 Digital Electromechanical Servosystem with a Permanent-Magnet DC Motor

Consider a pointing system actuated by a permanent-magnet DC motor. For this system, analog control was examined in Section 7.3.2. Our goal is to study the digital PID control laws and analyze the system behavior. The objectives are to guarantee stability, attain the fast displacement (rapid repositioning), minimize tracking error, and so on.

Three differential equations were derived to describe the evolution of the open-loop system. In particular, the differential equations are

$$\frac{di_a}{dt} = \frac{1}{L_a}(-r_a i_a - k_a \omega_r + u_a), \quad \frac{d\omega_r}{dt} = \frac{1}{J}(k_a i_a - B_m \omega_r - T_L), \quad \frac{d\theta_r}{dt} = \omega_r,$$

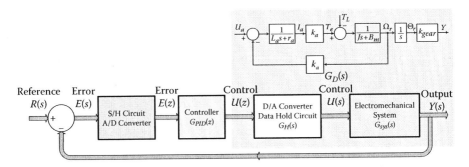

FIGURE 7.23
Block diagrams of the closed-loop systems with the digital PID controller.

and the output equation is $y(t) = k_{gear}\theta_r$. The block diagram of the closed-loop system with a digital PID controller, A/D and D/A converters, and data hold circuit is documented in Figure 7.23.

The transfer function of the open-loop system (permanent-magnet DC motor with a gearhead) is

$$G_{sys}(s) = \frac{Y(s)}{U(s)} = \frac{k_{gear}k_a}{s\left(L_aJs^2 + (r_aJ + L_aB_m)s + r_aB_m + k_a^2\right)}.$$

The transfer function of the zero-order data hold is

$$G_H(s) = \frac{1 - e^{-T_ss}}{s}.$$

One obtains

$$G_D(s) = G_H(s)G_{sys}(s) = \frac{1 - e^{-T_ss}}{s} \frac{k_{gear}k_a}{s\left(L_aJs^2 + (r_aJ + L_aB_m)s + r_aB_m + k_a^2\right)}.$$

The DC motor parameters are: $r_a = 200$ ohm, $L_a = 0.002$ H, $k_a = 0.2$ V-sec/rad (N-m/A), $J = 0.00000002$ kg-m^2 and $B_m = 0.00000005$ N-m-sec/rad.

The transfer function in the z-domain $G_D(z)$ is found from $G_D(s)$ by using the c2dm command. The filter command is used to simulate the dynamics. The following MATLAB file was developed to discretize the system, perform simulations, and plot the dynamic evolution.

```
% System parameters
ra=200; La=0.002; ka=0.2; J=0.00000002; Bm=0.00000005; kgear=0.01;
% Numerator and denominator of the open-loop transfer function
format short e
num_s=[ka*kgear]; den_s=[La*J ra*J+La*Bm ra*Bm+ka^2 0];
```

```
num _ s, den _ s
pause;
% Numerator and denominator of GD(z) with zero-order data hold
Ts=0.0002;          % Sampling time (sampling period) Ts
[num _ dz,den _ dz]=c2dm(num _ s,den _ s,Ts,'zoh');
num _ dz, den _ dz
pause;
% Feedback coefficient gains of the analog PID controller
kp=25000; ki=250; kd=0.25;
% Feedback coefficient gains of the digital PID controller
kdi=ki*Ts; kdp=kp-kdi/2; kdd=kd/Ts;
% Numerator and denominator of the transfer function of the PID controller
num _ pidz=[(kdp+kdi+kdd) -(kdp+2*kdd) kdd]; den _ pidz=[1 -1 0];
num _ pidz, den _ pidz
pause;
% Numerator and denominator of the closed-loop transfer function G(z)
num _ z=conv(num _ pidz,num _ dz);
den _ z=conv(den _ pidz,den _ dz)+conv(num _ pidz,num _ dz);
num _ z, den _ z
pause;
% Samples, t=k*Ts
k _ final=20; k=0:1:k _ final;
% Reference input r(t)=1 rad
ref=1; % Reference (command) input is 1 rad
r=ref*ones(1,k _ final+1);
% Modeling of the servo-system output y(k)
y=filter(num _ z,den _ z,r);
% Plotting statement
plot(k,y,'o',k,y,'--',k,r,':');
title('Angular Displacement, y(t)=0.01\theta _ r, r(t)=1 [rad]','FontSize',14);
xlabel('Discrete Time k, Continuous Time t=kT _ s [seconds]','FontSize',14);
ylabel('Output y(k) and Reference r(k)','FontSize',14);
axis([0 20,0 1.2]) % Axis limits
```

Having found the numerator and denominator of $G_{sys}(s)$ to be

```
num _ s =
2.0000e-003
den _ s =
4.0000e-011    4.0001e-006    4.0010e-002        0
```

the following transfer function for the open-loop system yields

$$G_{sys}(s) = \frac{Y(s)}{U(s)} = \frac{2 \times 10^{-3}}{s(4 \times 10^{-11} s^2 + 4 \times 10^{-6} s + 4 \times 10^{-2})}.$$

The sampling time is assigned to be 0.0002 sec, $T_s = 0.0002$ sec. The transfer function $G_D(z)$ in the z-domain is

$$G_D(z) = \frac{5.53 \times 10^{-6} z^2 + 3.41 \times 10^{-6} z + 8.6 \times 10^{-9}}{z^3 - 1.1 z^2 + 0.105 z - 2.06 \times 10^{-9}},$$

which is found using the numerical results obtained

```
num _ dz =
0      5.5328e-006    3.4072e-006    8.6022e-009
den _ dz =
1.0000e+000    -1.1049e+000    1.0491e-001    -2.0601e-009
```

The transfer function of the digital PID controller is

$$G_{PID}(z) = \frac{(k_{dp} + k_{di} + k_{dd})z^2 - (k_{dp} + 2k_{dd})z + k_{dp}}{z^2 - z}, \quad k_{dp} = k_p - \tfrac{1}{2}k_{di},$$

$$k_{di} = k_i T_s, \quad k_{dd} = k_d / T_s.$$

Assume the feedback gains of the analog PID controller are $k_p = 25000$, $k_i = 250$, and $k_d = 0.25$. The feedback coefficients of the digital controller are found using the equations

$$k_{dp} = k_p - \frac{1}{2}k_{di}, \quad k_{di} = k_i T_s, \quad \text{and} \quad k_{dd} = k_d / T_s.$$

The numerator and denominator of the transfer function $G_{PID}(z)$ are found as

```
num _ pidz =
2.6250e+004        -2.7500e+004        1.2500e+003
den _ pidz =
1               -1                  0
```

Hence,

$$G_{PID}(z) = \frac{2.63 \times 10^4 z^2 - 2.75 \times 10^4 z + 1.25 \times 10^3}{z^2 - z}.$$

The transfer function of the closed-loop system is

$$G(z) = \frac{Y(z)}{R(z)} = \frac{G_{PID}(z)G_D(z)}{1 + G_{PID}(z)G_D(z)}.$$

The following numerical results are found:

```
num _ z =
0 1.4524e-001 -6.2713e-002 -8.6556e-002 4.0225e-003      1.0753e-005
den _ z =
1.0000e+000 -1.9597e+000 1.1471e+000 -1.9147e-001 4.0225e-003 1.0753e-005
```

Thus, we have

$$G(z) = \frac{0.145z^4 - 0.063z^3 - 0.087z^2 + 0.004z + 1.07 \times 10^{-5}}{z^5 - 1.96z^4 + 1.15z^3 - 0.19z^2 + 0.004z + 1.07 \times 10^{-5}}.$$

The output dynamics for the reference input $r(kT_s) = 1$ rad, $k \geq 0$ is shown in Figure 7.24. The settling time is $k_{settling}T_s = 15 \times 0.0002 = 0.003$ sec, and there is no overshoot.

The sampling time, affected by the microcontroller or DSP capabilities, significantly affects the system dynamics. We increase the sampling time

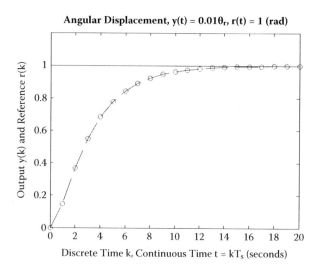

FIGURE 7.24
Output dynamics of the system with a digital PID controller, $T_s = 0.0002$ sec.

to be $T_s = 0.001$ sec. Using the MATLAB file reported, for the sampling time $T_s = 0.001$ sec, we have

```
num _ z =
0 1.1360e+000 -1.0211e+000 -1.1617e-001 1.2492e-003 2.6056e-010
den _ z =
1.0000e+000 -8.6401e-001 -2.1039e-002 -1.1619e-001 1.2492e-003 2.6056e-010
```

That is, the closed-loop transfer function is found to be

$$G(z) = \frac{1.14z^4 - 1.02z^3 - 0.12z^2 + 0.0012z + 2.61 \times 10^{-10}}{z^5 - 0.86z^4 - 0.021z^3 - 0.12z^2 + 0.0012z + 2.61 \times 10^{-10}}.$$

The output of the servosystem $y(kT_s)$ is plotted in Figure 7.25 for $T_s = 0.001$ sec and $T_s = 0.0015$ sec.

If $T_s = 0.001$ sec, the settling time is $k_{settling}T_s = 5 \times 0.001 = 0.005$ sec, and the overshoot is ~14%. For $T_s = 0.0015$ sec, as documented in Figure 7.25, the overshoot is 77%, and the settling time $k_{settling}T_s = 10 \times 0.0015 = 0.015$ sec. If $T_s = 0.0018$ sec, the closed-loop system becomes unstable. Thus, the sampling time significantly affects the closed-loop system performance and stabilities. The sampling time is defined by the microcontroller or DSP used. For "high" $T_{s'}$ one must refine the feedback coefficients k_{dp}, k_{di} and k_{dd} to ensure the stability and desired dynamic responses within the desired settling time, overshoot, accuracy, etc. As was reported, the control bounds $u_{min} \leq u \leq u_{max}$ must be integrated in the analysis. The stability analysis of the closed-loop system performed without system nonlinearities do not provide sound and accurate results. Neglecting nonlinearities, presumable stable systems may

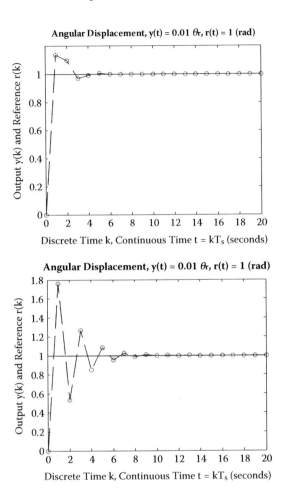

FIGURE 7.25
Output of a servo with a digital PID controller, $T_s = 0.001$ sec and $T_s = 0.0015$ sec.

become unstable as nonlinearities are integrated, while unstable systems can be stable as one integrates the existing nonlinearities and constraints. Non-linearities, as inherent hardware features, must be integrated in the analysis to ensure soundness, accuracy, and coherence. Using the `filter` command, simulations were performed assuming that the system is linear and no con-straints are imposed. One must carry out the nonlinear simulations recall-ing that $-30 \leq u_a \leq 30$ V. Various MATLAB built-in commands and Simulink components from block libraries can be effectively utilized.

The analytical results are important in addition to numerical solutions that were obtained in the MATLAB environment. Assuming that the system does not reach control constraints and the system is linear, we use the transfer functions

$$G_{sys}(s) \quad \text{and} \quad G_H(s)$$

to find

$$G_D(s) = G_H(s)G_{sys}(s).$$

For the zero-order data hold circuit

$$G_H(s) = \frac{1 - e^{-T_s s}}{s}.$$

Thus,

$$G_D(s) = G_H(s)G_{sys}(s) = \frac{1 - e^{-T_s s}}{s} \frac{k_{gear}k_a}{s\left(L_a J s^2 + (r_a J + L_a B_m)s + r_a B_m + k_a^2\right)}.$$

Assume that the eigenvalues are real and distinct. We have

$$G_D(s) = G_H(s)G_{sys}(s) = \frac{1 - e^{-T_s s}}{s} \frac{k_D}{s(T_1 s + 1)(T_2 s + 1)},$$

where $k_D = k_{gear}k_a$; T_1 and T_2 are the time constants that correspond to the second-order polynomial

$$L_a J s^2 + (r_a J + L_a B_m)s + r_a B_m + k_a^2.$$

The partial fraction expansion is performed letting $T_1 \neq T_2$. This situation is the most common in practice. Assuming that the eigenvalues

$$-\frac{1}{T_1} \quad \text{and} \quad -\frac{1}{T_2}$$

are distinct (nonrepeated), from Heaviside's expansion formula, we have

$$\frac{k_D}{s^2(T_1 s + 1)(T_2 s + 1)} = \frac{c_1}{s} + \frac{c_2}{s^2} + \frac{c_3}{T_1 s + 1} + \frac{c_4}{T_2 s + 1},$$

where c_1, c_2, c_3 and c_4 are the unknown coefficients which are derived as

$$c_1 = \frac{d}{ds}\left(\frac{k_D s^2}{s^2(T_1 s + 1)(T_2 s + 1)}\right)\bigg|_{s=0} = \frac{-k_D(2T_1 T_2 s + T_1 + T_2)}{(T_1 T_2 s^2 + (T_1 + T_2)s + 1)}\bigg|_{s=0} = -k_D(T_1 + T_2),$$

$$c_2 = \frac{k_D}{(T_1 s + 1)(T_2 s + 1)}\bigg|_{s=0} = k_D,$$

$$c_3 = \frac{k_D}{s^2(T_2 s + 1)}\bigg|_{s=-\frac{1}{T_1}} = \frac{k_D T_1^3}{T_1 - T_2} \quad \text{and}$$

$$c_4 = \frac{k_D}{s^2(T_1 s + 1)}\bigg|_{s=-\frac{1}{T_2}} = \frac{k_D T_2^3}{T_2 - T_1}.$$

One finds

$$G_D(z) = Z[G_D(s)] = Z[G_H(s)G_{sys}(s)] = \frac{z-1}{z} Z\left[\frac{G_{sys}(s)}{s}\right].$$

Using the z-transform table, we have the expression for $G_D(z)$ as

$$G_D(z) = \frac{z-1}{z} Z\left[\frac{k_D}{s^2(T_1 s+1)(T_2 s+1)}\right]$$

$$= k_D \frac{z-1}{z} Z\left[-\frac{T_1+T_2}{s} + \frac{1}{s^2} + \frac{T_1^3}{T_1-T_2}\left(\frac{1}{T_1 s+1}\right) + \frac{T_2^3}{T_2-T_1}\frac{1}{(T_2 s+1)}\right]$$

$$= k_D \frac{z-1}{z}\left(-(T_1+T_2)\frac{z}{z-1} + \frac{T_s z}{(z-1)^2} + \frac{T_1^2}{T_1-T_2}\frac{z}{z-e^{-\frac{T_s}{\times T_1}}}\right.$$

$$\left. + \frac{T_2^2}{T_2-T_1}\frac{z}{z-e^{-\frac{T_s}{\times T_2}}}\right).$$

From

$$G_D(z) = k_D\left(-T_1-T_2+T_s\frac{1}{z-1} + \frac{T_1^2}{T_1-T_2}\frac{z-1}{z-e^{-\frac{T_s}{\times T_1}}} + \frac{T_2^2}{T_2-T_1}\frac{z-1}{z-e^{-\frac{T_s}{\times T_2}}}\right)$$

and

$$G_{PID}(z) = \frac{(k_{dp}+k_{di}+k_{dd})z^2 - (k_{dp}+2k_{dd})z + k_{dd}}{z^2-z},$$

the transfer function of the closed-loop system results as

$$G(z) = \frac{Y(z)}{R(z)} = \frac{G_{PID}(z)G_D(z)}{1+G_{PID}(z)G_D(z)}.$$

The analytic solution $y(k)$ can be derived for different system parameters, distinct feedback gains (which vary to attain the desired dynamics), various sampling time T_s, distinct waveforms of the reference inputs $r(t)$, etc. Thus, in addition to numerical results, analytic studies can be carried out. Unfortunately, due to the fact that systems are nonlinear and variables are bounded, the results obtained assuming that the system dynamics is described by linear differential or difference equations may not be valid in the full operating envelope. Therefore, the designer performs numerical analysis integrating system nonlinearities.

Homework Problems

Problem 7.1

Why should one control electromechanical systems?

Problem 7.2

List the specifications imposed on systems in the behavioral domain.

Problem 7.3

Explain the differences between bounded and unbounded control laws. Provide examples. Explain how control bounds influence the system performance and capabilities.

Problem 7.4

What are the challenges in the design of bounded control laws? How may the designer approach and solve control problems for electromechanical systems?

Problem 7.5
Let the control law be expressed as

$$u = k_{p1}e + k_{p2}\,|e^2| + k_{p3}e^3 + k_i \int edt.$$

Elaborate the feedback terms used. Make the conclusion if this control law can be applied. Explain why one should study the bounded control law

$$u = \mathbf{sat}\left(k_{p1}e + k_{p2}\,|e^2| + k_{p3}e^3 + k_i \int edt \right)$$

to examine the system performance. Propose the terms to improve system performance. Also provide the terms that degrade the system performance and lead to unstable closed-loop systems.

Problem 7.6
Let the performance functional be given as

$$J = \min_{t,x,e} \int_0^\infty (x^2 + e^4 + t\,|e|)dt.$$

Explain what performance is specified and how. Justify the results. If strict specifications should be imposed on the settling time and tracking error, propose the additional integrands in the performance functional to be used.

Problem 7.7

Let the system is modeled as

$$\frac{d\omega}{dt} = -\omega + 100u.$$

The control is bounded as $-10 \le u \le 10$ V. Propose the bounded PID control law and simulate the closed-loop system in Simulink. Let the desired speed be 200 rad/sec. Find the feedback coefficients by varying the feedback gains. Report the simulation results for the closed-loop system.

Problem 7.8

Consider a pointing system actuated by a geared permanent-magnet DC motor. The angular displacement of the pointing stage is $y(t) = k_{gear}\theta_r$. The motor data is: $u_{max} = 24$ V ($-24 \le u_a \le 24$ V), $i_{amax} = 10$ A, $\omega_{rmax} = 240$ rad/sec, $r_a = 1$ ohm, $L_a = 0.005$ H, $k_a = 0.1$ V-sec/rad (N-m/A), $J = 0.0005$ kg-m^2, and $B_m = 0.0005$ N-m-sec/rad. The reduction gear ratio is 10:1. Design and analyze

- Not bounded linear and nonlinear PID control laws;
- Bounded PID control laws.

For different control laws and feedback coefficients, study the transient dynamics for $r(t) = 0.1$ rad and $r(t) = 1$ rad. Simulations must be performed by developing the MATLAB (Simulink) files.

References

1. R.C. Dorf and R.H. Bishop, *Modern Control Systems*, Addison-Wesley Publishing Company, Reading, MA, 1995.
2. J.F. Franklin, J.D. Powell, and A. Emami-Naeini, *Feedback Control of Dynamic Systems*, Addison-Wesley Publishing Company, Reading, MA, 1994.
3. B.C. Kuo, *Automatic Control Systems*, Prentice Hall, Englewood Cliffs, NJ, 1995.
4. S.E. Lyshevski, *Control Systems Theory with Engineering Applications*, Birkhauser, Boston, MA, 2000.
5. K. Ogata, *Discrete-Time Control Systems*, Prentice-Hall, Upper Saddle River, NJ, 1995.
6. K. Ogata, *Modern Control Engineering*, Prentice-Hall, Upper Saddle River, NJ, 1997.
7. C.L. Phillips and R.D. Harbor, *Feedback Control Systems*, Prentice Hall, Englewood Cliffs, NJ, 1996.

8

Advanced Control of Electromechanical Systems

Chapters 1 and 7 established and emphasized the need for control of electromechanical systems. Linear, nonlinear and bounded PID control laws were examined in Chapter 7 to optimize performance with the goal to attain the desired systems capabilities. The differences between the desired, specified, and *achievable* performance and capabilities are evident. Through the optimal design (structural and behavioral) and hardware-software codesign, the *achievable* performance of closed-loop electromechanical systems results. The desired requirements and specifications may not be met for the specific solutions, and re-design should be performed if needed. As will be discussed, the *achievable* performance and capabilities may be ensured through a coherent design and optimization. Advanced solutions may significantly complicate hardware and software increasing the overall system complexity, which leads to the obvious drawbacks. References 1 through 11 report the fundamentals of linear and nonlinear designs for continuous-time and discrete systems.

Advanced control laws can be designed with the goal to improve the system performance [7, 8]. To implement PID control laws, only the tracking error $e(t)$ must be measured or estimated. The $e(t)$ usually is directly accessible or can be derived using $y(t)$ and $r(t)$. The advanced control systems design is an important task for high-performance systems for which multiobjective optimization is studied. The achievable performance capabilities, largely defined by the hardware, can be improved utilizing sound control laws. The departure from PID controllers may result in substantial requirements for advanced hardware such as sensors, ICs, DSPs, and so on. For example, if the state variables $x(t)$ are used, these $x(t)$ must be measured and utilized in deriving the control $u(t)$ in real time. For some systems, one may achieve a substantial performance improvement as advanced control laws are applied. For other systems, a linear PI controller may guarantee the desired minimal hardware and software complexities while ensuring near-optimal performance. The physical limits, constraints, and bounds must be coherently examined. For many open-loop stable and unstable systems, PID controllers have been successfully utilized for decades. Because of inherent physical limits and increase of the overall system complexity, advanced controllers may ensure a quite moderate improvement. For conventional electromechanical systems, PID-centered controllers have found to be effective, valuable, and sufficient. The advanced methods in the design of control laws are applied if the strengthened specifications are imposed. These specifications should be supported and substantiated by the hardware solutions. For multiobjective (accuracy, efficiency maximization, disturbance attenuation, vibration and noise minimization, etc.) control and optimization, the advanced concepts are applied. Advanced control laws, which should be implementable, are deployed in very-high-accuracy pointing systems, advanced multidegree-of-freedom robots, high-precision positioning systems, sound systems, and so on. This chapter reports possible design methods if one faces the need for advanced control.

8.1 Hamilton-Jacobi Theory and Optimal Control of Electromechanical Systems

We design control algorithms by minimizing performance functionals that quantitatively and qualitatively describe the system performance and capabilities. The following system variables are utilized: states x, outputs y, tracking errors e, and control u. Considering multi-input/multi-output systems, one has

$$x \in \mathbb{R}^n, \quad y \in \mathbb{R}^b, \quad e \in \mathbb{R}^b \quad \text{and} \quad u \in \mathbb{R}^m.$$

The variables tuple (x,e,u) or quadruple (x,y,e,u) describe, specify, quantify, and qualify the system performance and capabilities. The performance functional can be maximized or minimized.

A general problem formulation in design of control laws using the Hamilton-Jacobi concept is formulated thus: find the *admissible* (bounded or unbounded) time-invariant or time-varying control law as a nonlinear function of error $e(t)$ and state $x(t)$ vectors

$$u = \phi(t,e,x), \tag{8.1}$$

minimizing the performance functional

$$J(x(\cdot), e(\cdot), u(\cdot)) = \int_{t_0}^{t_f} W_{xeu}(x, e, u)\, dt, \tag{8.2}$$

subject to the system dynamics and constraints.

In (8.2),

$$W_{xeu}(\cdot){:}\mathbb{R}^n \times \mathbb{R}^b \times \mathbb{R}^m \to \mathbb{R}_{\geq 0}$$

is the positive-definite, continuous, and differentiable integrand function synthesized by the designer; t_0 and t_f are the initial and final time, which define the time horizon.

The performance functional (8.2) plays a very important role. One may use the squired $e(t)$ and $u(t)$ obtaining the quadratic functional

$$J = \int_{t_0}^{t_f} (e^2 + u^2)\, dt,$$

where $W_{eu}(e,u) = e^2 + u^2$. This performance functional is positive-definite, and the performance integrands are differentiable ensuring the analytic solutions. A positive-definite $W_{eu}(e,u) = |e| + |u|$ may not be used with an ease to analytically synthesize control laws applying the Hamilton-Jacobi and many other concepts, although discontinuous integrands $W_{xeu}(\cdot)$ can be applied in search, parametric, and numeric optimization methods.

One can design and minimize a great variety of functionals in an attempt to specify and achieve better performance. For example, nonquadratic functionals

$$J = \int_{t_0}^{t_f} (e^8 + u^6)dt, \quad J = \int_{t_0}^{t_f} (|e|e^2 + |u|u^2)dt, \quad \text{or} \quad J = \int_{t_0}^{t_f} (|e|e^4 + |u|u^6)dt$$

may be used with differentiable $W_{eu}(\cdot)$. The quadratic functionals, such as,

$$J = \int_{t_0}^{t_f} (x^2 + e^2 + u^2)dt,$$

are commonly used. These quadratic functionals ease the analytic design ensuring that the problem is solvable. In general, the application of nonquadratic integrands results in mathematical complexity. However, the system performance may be improved if sound nonlinear functionals are used. It will be documented that the analytic design of constrained control laws is centered on the use of nonquadratic functionals.

As a performance functional (8.2) is chosen by synthesizing the integrand functions, the design centers on the minimization or maximization problem using the Hamilton-Jacobi concept, dynamic programming, maximum principle, variational calculus, nonlinear programming, or other concepts. The ultimate objective is to design control laws (8.1) to attain optimal system performance, for example, stability, robustness, accuracy, and so on.

Assume the electromechanical system dynamics is described by nonlinear differential equations

$$x(t) = F(x) + B(x)u, \quad x(t_0) = x_0. \tag{8.3}$$

To find an optimal control, the necessary conditions for optimality must be studied.

For

$$J = \int_{t_0}^{t_f} W_{xu}(x,u)dt,$$

the Hamiltonian function is

$$H\left(x, u, \frac{\partial V}{\partial x}\right) = W_{xu}(x,u) + \left(\frac{\partial V}{\partial x}\right)^T (F(x) + B(x)u), \tag{8.4}$$

where $V(\cdot):\mathbb{R}^n \to \mathbb{R}_{\geq 0}$ is the continuous and differentiable return function, $V(0) = 0$.

Control law (8.1) can be found using the following first-order necessary condition for optimality

$$\frac{\partial H\left(x, u, \frac{\partial V}{\partial x}\right)}{\partial u} = 0. \tag{8.5}$$

The control function $u(\cdot):[t_0, t_f) \to \mathbb{R}^m$ is obtained from (8.5) as the performance functional (8.2) is defined. Using the quadratic performance functional

$$J = \frac{1}{2}\int_{t_0}^{t_f}(x^T Q x + u^T G u)dt, \quad Q \in \mathbb{R}^{n \times n}, \quad Q \geq 0, \quad G \in \mathbb{R}^{m \times m}, \quad G > 0, \tag{8.6}$$

from (8.4) and (8.5) one finds

$$\frac{\partial H}{\partial u} = u^T G + \left(\frac{\partial V}{\partial x}\right)^T B(x).$$

Hence, the control law is

$$u = -G^{-1} B^T(x)\frac{\partial V}{\partial x}. \tag{8.7}$$

The second-order necessary conditions for optimality

$$\frac{\partial^2 H\left(x, u, \frac{\partial V}{\partial x}\right)}{\partial u \times \partial u^T} > 0 \tag{8.8}$$

is satisfied because

$$\frac{\partial^2 H}{\partial u \times \partial u^T} = G > 0.$$

Example 8.1
Consider a moving object assuming that the applied force F_a is a control variable u. Assume the velocity v is the state variable x. From

$$\dot{x}(t) = \frac{1}{m}\sum F,$$

taking note of the viscous friction, the dynamics is

$$\dot{x}(t) = \frac{1}{m}\sum F = \frac{1}{m}(F_a - B_m x).$$

Using the state and control variables, one has the first-order differential equation which describes the input-output dynamics. In particular,

$$x(t) = ax + bu,$$

where

$$a = -\frac{B_m}{m} \quad \text{and} \quad b = \frac{1}{m}.$$

For a first-order system, we minimize the quadratic functional (8.6) with the weighting coefficients q and g. The functional

$$J = \frac{1}{2} \int_{t_0}^{t_f} (qx^2 + gu^2) dt, \quad q \geq 0, \quad g > 0,$$

yields the Hamiltonian function (8.4) as

$$H\left(x, u, \frac{\partial V}{\partial x}\right) = \underbrace{\frac{1}{2}(qx^2 + gu^2)}_{\text{Performance Functional}} + \frac{\partial V}{\partial x} \underbrace{\frac{dx}{dt}}_{\substack{\text{System Dynamics} \\ \dot{x}(t) = ax + bu}} = \frac{1}{2}(qx^2 + gu^2) + \frac{\partial V}{\partial x}(ax + bu).$$

$$J(x,u) = \int_{t_0}^{t_f} (qx^2 + gu^2) dt$$

The Hamiltonian function $H(\cdot)$ is minimized using the first-order necessary condition for optimality (8.5). From (8.5), one obtains

$$gu + \frac{\partial V}{\partial x} b = 0.$$

From this expression, the control law is found to be

$$u = -\frac{b}{g} \frac{\partial V}{\partial x} = -g^{-1}b \frac{\partial V}{\partial x}.$$

Let the continuous and differentiable return function $V(x)$ be given in the quadratic form, as,

$$V(x) = \frac{1}{2} kx^2,$$

where k is the positive-definite unknown coefficient.
Thus, from

$$u = -g^{-1}b \frac{\partial V}{\partial x},$$

the control law is given as

$$u = -g^{-1}bkx, \quad k > 0.$$

The unknown coefficient k is found by solving the Riccati differential equation

$$-dk/dt = q + 2ak - g^{-1}b^2k^2$$

or the quadratic algebraic equation

$$-q-2ak + g^{-1}b^2k^2 = 0$$

as will be illustrated later.

Making use of the system dynamics

$$\dot{x}(t) = ax + bu$$

with control law $u = -g^{-1}bkx$, the closed-loop system is expressed as

$$x(t) = (a - g^{-1}b^2k)x.$$

The closed-loop system is stable if $(a-g^{-1}b^2k) < 0$. This condition for stability is guaranteed because $a < 0$, $g > 0$, and $k > 0$.

The second-order necessary conditions for optimality (8.8) is guaranteed because

$$\frac{\partial^2 H}{\partial u^2} = g > 0.$$

The electromechanical systems are nonlinear. For example, if the object is attached to a nonlinear spring with the restoring force $k_s x^3$, we have

$$x(t) = \frac{1}{m}\sum F = \frac{1}{m}(F_a - B_m x - k_s x^3).$$

Minimizing the quadratic functional, the control law is found to be

$$u = -g^{-1}bkx, \quad k > 0.$$

The closed-loop system dynamics is

$$\dot{x}(t) = ax - \frac{1}{m}k_s x^3 - g^{-1}b^2kx.$$

This system is stable, which can be easily proven by applying the Lyapunov stability theory covered in Section 8.9. However, in general, one may not be able to utilize the quadratic return function $V(x)$. The nonquadratic $V(x)$ results in the nonlinear control law.

□

8.2 Stabilization Problem for Linear Electromechanical Systems

A linear control law was synthesized in Example 8.1 for the system dynamics as described by the first-order linear and nonlinear differential equation. Consider a linear time-invariant electromechanical system described by the linear differential equations

$$x(t) = Ax + Bu, \qquad x(t_0) = x_0, \tag{8.9}$$

where $A \in \mathbb{R}^{n \times n}$ and $B \in \mathbb{R}^{n \times m}$ are the constant-coefficient matrices.

Using the quadratic integrands, the quadratic performance functional is

$$J = \frac{1}{2} \int_{t_0}^{t_f} (x^T Q x + u^T G u) dt, \qquad Q \geq 0, \quad G > 0, \tag{8.10}$$

where $Q \in \mathbb{R}^{n \times n}$ is the positive semidefinite constant-coefficient weighting matrix; $G \in \mathbb{R}^{m \times m}$ is the positive-definite constant-coefficient weighting matrix.

From (8.9) and (8.10), the Hamiltonian function is found to be

$$H\left(x, u, \frac{\partial V}{\partial x}\right) = \frac{1}{2}\left(x^T Q x + u^T G u\right) + \left(\frac{\partial V}{\partial x}\right)^T (Ax + Bu). \tag{8.11}$$

The Hamilton-Jacobi functional equation is

$$-\frac{\partial V}{\partial t} = \min_{u} \left[\frac{1}{2}(x^T Q x + u^T G u) + \left(\frac{\partial V}{\partial x}\right)^T (Ax + Bu) \right]. \tag{8.12}$$

The derivative of the Hamiltonian H exists, and the control function $u(\cdot){:}[t_0, t_f) \to \mathbb{R}^m$ is found by using the first-order necessary condition for optimality (8.5). From

$$\frac{\partial H}{\partial u} = u^T G + \left(\frac{\partial V}{\partial x}\right)^T B,$$

one finds

$$u = -G^{-1} B^T \frac{\partial V}{\partial x}. \tag{8.13}$$

This control law is found minimizing the quadratic performance functional (8.10). The second-order necessary condition for optimality (8.8) is guaranteed. In fact, the weighting matrix G is positive-definite, yielding

$$\frac{\partial^2 H}{\partial u \times \partial u^T} = G > 0.$$

Substituting control law (8.13) in (8.12), we obtain the partial differential equation

$$-\frac{\partial V}{\partial t} = \frac{1}{2}\left(x^T Q x + \left(\frac{\partial V}{\partial x}\right)^T BG^{-1}B^T \frac{\partial V}{\partial x}\right) + \left(\frac{\partial V}{\partial x}\right)^T Ax - \left(\frac{\partial V}{\partial x}\right)^T BG^{-1}B^T \frac{\partial V}{\partial x}$$

$$= \frac{1}{2}x^T Q x + \left(\frac{\partial V}{\partial x}\right)^T Ax - \frac{1}{2}\left(\frac{\partial V}{\partial x}\right)^T BG^{-1}B^T \frac{\partial V}{\partial x}. \tag{8.14}$$

The solution of (8.14) is satisfied by the quadratic return function

$$V(x) = \frac{1}{2}x^T K(t)x, \tag{8.15}$$

where $K \in \mathbb{R}^{n \times n}$ is the symmetric matrix, $K = K^T$.
The matrix

$$K = \begin{bmatrix} k_{11} & k_{12} & \cdots & k_{1n-1} & k_{1n} \\ k_{21} & k_{22} & \cdots & k_{2n-1} & k_{2n} \\ \vdots & \vdots & \ddots & \vdots & \vdots \\ k_{n-11} & k_{n-12} & \cdots & k_{n-1n-1} & k_{n-1n} \\ k_{n1} & k_{n2} & \cdots & k_{nn-1} & k_{nn} \end{bmatrix}, \quad k_{ij} = k_{ji}$$

must be positive-definite because positive semidefinite and positive-definite constant-coefficient weighting matrices Q and G are used in the quadratic performance functional (8.10) yielding $J > 0$. The positive definiteness of the quadratic return function $V(x)$ can be verified using the Sylvester criterion.

From (8.15) and applying the matrix identity

$$x^T KAx = \frac{1}{2}x^T(A^T K + KA)x$$

in (8.14), one obtains

$$-\frac{\partial\left(\frac{1}{2}x^T Kx\right)}{\partial t} = \frac{1}{2}x^T Q x + \frac{1}{2}x^T A^T Kx + \frac{1}{2}x^T KAx - \frac{1}{2}x^T KBG^{-1}B^T Kx. \tag{8.16}$$

Using the boundary conditions

$$V(t_f, x) = \frac{1}{2}x^T K(t_f)x = \frac{1}{2}x^T K_f x, \tag{8.17}$$

the following nonlinear differential equation, called the Riccati equation, must be solved to find the unknown symmetric matrix K

$$-\dot{K} = Q + A^T K + K^T A - K^T B G^{-1} B^T K, \quad K(t_f) = K_f. \tag{8.18}$$

From (8.13) and (8.15), the control law is

$$u = -G^{-1} B^T K x. \tag{8.19}$$

The feedback gain matrix K_F results, and $K_F = G^{-1}B^T K$. Using (8.9) and (8.19), we have the closed-loop system as

$$x(t) = Ax + Bu = Ax - BG^{-1}B^T Kx = (A - BG^{-1}B^T K)x = (A - BK_F)x. \tag{8.20}$$

The eigenvalues of the matrix

$$(A - BG^{-1}B^T K) = (A - BK_F) \in \mathbb{R}^{n \times n}$$

have negative real parts ensuring stability.

If in functional (8.10) $t_f = \infty$, the matrix K can be found by solving the non-linear algebraic equation

$$0 = -Q - A^T K - K^T A + K^T B G^{-1} B^T K. \tag{8.21}$$

To solve (8.21), the Riccati equation, the solver `lqr` is available in MATLAB. In particular,

```
>> help lqr
LQR Linear-quadratic regulator design for state space systems.
  [K,S,E] = LQR(SYS,Q,R,N) calculates the optimal gain matrix K such that:
    For a continuous-time state-space model SYS, the state-feedback law u = -Kx
    minimizes the cost function
        J = Integral {x'Qx + u'Ru + 2*x'Nu} dt
  subject to the system dynamics dx/dt = Ax + Bu
  For a discrete-time state-space model SYS, u[n] = -Kx[n] minimizes
        J = Sum {x'Qx + u'Ru + 2*x'Nu}
  subject to x[n+1] = Ax[n] + Bu[n].
  The matrix N is set to zero when omitted. Also returned are the solution S of
    the associated algebraic Riccati equation and the closed-loop
    eigenvalues E = EIG(A-B*K).
  [K,S,E] = LQR(A,B,Q,R,N) is an equivalent syntax for continuous-time
    models with dynamics dx/dt = Ax + Bu
    See also lqry, lqgreg, lqg, dlqr, care, dare.
```

Using the `lqr` command, one finds the feedback gain matrix K_F, matrix K, and eigenvalues of the closed-loop system.

Example 8.2
Consider the system studied in Example 8.1, assuming that $m = 1$ and $B_m = 0$ (the viscous friction is neglected). That is, the differential equation is

$$\frac{dx}{dt} = u.$$

In the model

$$x(t) = Ax + Bu,$$

the matrices A and B are $A = [0]$ and $B = [1]$.
   Let

$$J = \frac{1}{2} \int_{t_0}^{t_f} (qx^2 + gu^2)dt, \quad q = 1, \quad \text{and} \quad g = 1.$$

Hence, $Q = [1]$ and $G = [1]$.
   The unknown k of the quadratic return function (8.15)

$$V(x) = \frac{1}{2}kx^2$$

is found by solving (8.18) which is expressed as

$$-\dot{k}(t) = 1 - k^2(t), \quad k(t_f) = 0.$$

The solution of this nonlinear differential equation is

$$k(t) = \frac{1 - e^{-2(t_f - t)}}{1 + e^{-2(t_f - t)}}.$$

A control law, which guarantees the minimum of the quadratic functional

$$J = \frac{1}{2} \int_0^{t_f} (x^2 + u^2)dt$$

subject to the system dynamics

$$\frac{dx}{dt} = u,$$

is obtained using (8.19) as

$$u = -k(t)x = -\frac{1-e^{-2(t_f-t)}}{1+e^{-2(t_f-t)}}x.$$

By applying the `lqr` MATLAB command, for $t_f = \infty$, one finds the feedback gain, return function coefficient k, as well as the eigenvalue. Minimizing the functional

$$J = \frac{1}{2}\int_0^\infty (x^2 + u^2)dt$$

using the statement

```
[K _ feedback,K,Eigenvalues]  =  lqr(0,1,1,1,0)
```

we have the following numerical results

```
K _ feedback =              1
K =                         1
Eigenvalues =               -1
```

One concludes that the control law is $u = -x$, and the eigenvalue (pole) is -1. The pole has a negative real part, and the closed-loop system is stable. The numerical results correspond to the analytic solution found. □

Example 8.3
Consider one-dimensional motion of a rigid-body mechanical system described by a set of two first-order differential equations. The state variables are the displacement $x_1(t)$ and velocity $x_2(t)$. Neglecting the viscous friction, one has

$$\frac{dx_1}{dt} = x_2, \quad \frac{dx_2}{dt} = u,$$

where u represents the force or torque to be applied to control the system.

Using the state-space notations (8.9), we have the matrix differential equation

$$\dot{x}(t) = Ax + Bu, \quad \begin{bmatrix} \dot{x}_1 \\ \dot{x}_2 \end{bmatrix} = \begin{bmatrix} 0 & 1 \\ 0 & 0 \end{bmatrix}\begin{bmatrix} x_1 \\ x_2 \end{bmatrix} + \begin{bmatrix} 0 \\ 1 \end{bmatrix}u, \quad A = \begin{bmatrix} 0 & 1 \\ 0 & 0 \end{bmatrix}, \quad B = \begin{bmatrix} 0 \\ 1 \end{bmatrix}.$$

The quadratic functional (8.10) is

$$J = \frac{1}{2}\int_0^\infty \left(q_{11}x_1^2 + q_{22}x_2^2 + gu^2\right)dt, \quad q_{11} \ge 0, \quad q_{22} \ge 0 \quad \text{and} \quad g > 0$$

or

$$J = \frac{1}{2} \int_0^\infty \left(\begin{bmatrix} x_1 & x_2 \end{bmatrix} \begin{bmatrix} q_{11} & 0 \\ 0 & q_{22} \end{bmatrix} \begin{bmatrix} x_1 \\ x_2 \end{bmatrix} + uGu \right) dt, \quad Q = \begin{bmatrix} q_{11} & 0 \\ 0 & q_{22} \end{bmatrix}, \quad G = g.$$

Using the quadratic return function (8.15)

$$V(x) = \frac{1}{2} k_{11} x_1^2 + k_{12} x_1 x_2 + \frac{1}{2} k_{22} x_2^2 = \frac{1}{2} \begin{bmatrix} x_1 & x_2 \end{bmatrix} \begin{bmatrix} k_{11} & k_{12} \\ k_{21} & k_{22} \end{bmatrix} \begin{bmatrix} x_1 \\ x_2 \end{bmatrix}, \quad k_{12} = k_{21},$$

the control law (8.19) is

$$u = -G^{-1} B^T K x = -g^{-1} \begin{bmatrix} 0 & 1 \end{bmatrix} \begin{bmatrix} k_{11} & k_{12} \\ k_{21} & k_{22} \end{bmatrix} \begin{bmatrix} x_1 \\ x_2 \end{bmatrix} = -\frac{1}{g} (k_{21} x_1 + k_{22} x_2).$$

The unknown matrix

$$K = \begin{bmatrix} k_{11} & k_{12} \\ k_{21} & k_{22} \end{bmatrix}, \quad k_{12} = k_{21}$$

is found by solving the matrix Riccati equation (8.21) which is

$$-Q - A^T K - K^T A + K^T B G^{-1} B^T K =$$

$$-\begin{bmatrix} q_{11} & 0 \\ 0 & q_{22} \end{bmatrix} - \begin{bmatrix} 0 & 0 \\ 1 & 0 \end{bmatrix} \begin{bmatrix} k_{11} & k_{12} \\ k_{21} & k_{22} \end{bmatrix} - \begin{bmatrix} k_{11} & k_{21} \\ k_{12} & k_{22} \end{bmatrix} \begin{bmatrix} 0 & 1 \\ 0 & 0 \end{bmatrix}$$

$$+ \begin{bmatrix} k_{11} & k_{21} \\ k_{12} & k_{22} \end{bmatrix} \begin{bmatrix} 0 \\ 1 \end{bmatrix} g^{-1} \begin{bmatrix} 0 & 1 \end{bmatrix} \begin{bmatrix} k_{11} & k_{12} \\ k_{21} & k_{22} \end{bmatrix} = \begin{bmatrix} 0 & 0 \\ 0 & 0 \end{bmatrix}.$$

Three algebraic equations to be solved are

$$\frac{k_{12}^2}{g} - q_{11} = 0, \quad -k_{11} + \frac{k_{12} k_{22}}{g} = 0 \quad \text{and} \quad -2k_{12} + \frac{k_{22}^2}{g} - q_{22} = 0.$$

The solution is

$$k_{12} = k_{21} = \pm\sqrt{q_{11} g}, \quad k_{22} = \pm\sqrt{g(q_{22} + 2k_{12})} \quad \text{and} \quad k_{11} = \frac{k_{12} k_{22}}{g}.$$

The performance functional

$$J = \frac{1}{2} \int_0^\infty \left(q_{11} x_1^2 + q_{22} x_2^2 + g u^2 \right) dt$$

is positive-definite because the quadratic terms are used, and $q_{11} \geq 0$, $q_{22} \geq 0$ and $g > 0$. Hence,

$$k_{11} = \sqrt{q_{11}\left(q_{22} + 2\sqrt{q_{11}g}\right)}, \quad k_{12} = k_{21} = \sqrt{q_{11}g} \quad \text{and} \quad k_{22} = \sqrt{g\left(q_{22} + 2\sqrt{q_{11}g}\right)}.$$

The control law is

$$u = -\frac{1}{g}\left(\sqrt{q_{11}g}\, x_1 + \sqrt{g\left(q_{22} + 2\sqrt{q_{11}g}\right)}\, x_2\right) = -\sqrt{\frac{q_{11}}{g}}\, x_1 - \sqrt{\frac{q_{22} + 2\sqrt{q_{11}g}}{g}}\, x_2.$$

Having obtained the analytic solution in the symbolic form, we calculate the feedback gains and eigenvalues applying the `lqr` command. Let $q_{11} = 100$, $q_{22} = 10$, and $g = 1$. We have

```
A=[0 1;0 0],B=[0;1],Q=[100 0;0 10],G=[1],
[K _ feedback,K,Eigenvalues]= lqr(A,B,Q,G)
```

with resulting

```
A =
        0        1
        0        0
B =
        0
        1
Q =
      100        0
        0       10
G =
        1
K _ feedback =
    10.0000    5.4772
K =
    54.7723   10.0000
    10.0000    5.4772
Eigenvalues =
    -2.7386 + 1.5811i
    -2.7386 - 1.5811i
```

Hence, we have

$$K = \begin{bmatrix} k_{11} & k_{12} \\ k_{21} & k_{22} \end{bmatrix} = \begin{bmatrix} 54.77 & 10 \\ 10 & 5.48 \end{bmatrix}, \quad k_{11} = 54.77, \quad k_{12} = k_{21} = 10, \quad k_{22} = 5.48.$$

The control law is found as

$$u = -10x_1 - 5.48x_2.$$

The stability of the closed-loop system, given as

$$\frac{dx_1}{dt} = x_2, \quad \frac{dx_2}{dt} = -10x_1 - 5.48x_2,$$

is guaranteed. The eigenvalues have negative real parts, and the complex poles found are $-2.74 \pm 1.58i$.

$\square$

Example 8.4

Consider the system described by the following state-space equation (8.9)

$$x = Ax + Bu = \begin{bmatrix} -10 & 0 & -20 & 0 \\ 0 & -10 & -10 & 0 \\ 10 & 5 & -1 & 0 \\ 0 & 0 & 1 & 0 \end{bmatrix} \begin{bmatrix} x_1 \\ x_2 \\ x_3 \\ x_4 \end{bmatrix} + \begin{bmatrix} 10 & 0 \\ 0 & 10 \\ 0 & 0 \\ 0 & 0 \end{bmatrix} \begin{bmatrix} u_1 \\ u_2 \end{bmatrix}.$$

The output is x_4 and $y = x_4$. Hence, the output equation is

$$y = \begin{bmatrix} 0 & 0 & 0 & 1 \end{bmatrix} \begin{bmatrix} x_1 \\ x_2 \\ x_3 \\ x_4 \end{bmatrix} + \begin{bmatrix} 0 & 0 \end{bmatrix} \begin{bmatrix} u_1 \\ u_2 \end{bmatrix}.$$

The quadratic performance functional (8.10) is

$$J = \frac{1}{2} \int_0^\infty (x^T Q x + u^T G u) dt$$

$$= \frac{1}{2} \int_0^\infty \left(\begin{bmatrix} x_1 & x_2 & x_3 & x_4 \end{bmatrix} \begin{bmatrix} 0.05 & 0 & 0 & 0 \\ 0 & 0.1 & 0 & 0 \\ 0 & 0 & 0.01 & 0 \\ 0 & 0 & 0 & 1 \end{bmatrix} \begin{bmatrix} x_1 \\ x_2 \\ x_3 \\ x_4 \end{bmatrix} \right.$$

$$\left. + \begin{bmatrix} u_1 & u_2 \end{bmatrix} \begin{bmatrix} 0.001 & 0 \\ 0 & 0.001 \end{bmatrix} \begin{bmatrix} u_1 \\ u_2 \end{bmatrix} \right) dt$$

$$= \frac{1}{2} \int_0^\infty \left(0.05x_1^2 + 0.1x_2^2 + 0.01x_3^2 + x_4^2 + 0.001u_1^2 + 0.001u_2^2 \right) dt.$$

The MATLAB m-file to perform the controller design and simulate the system is

```
echo off; clear all; format short e;
% Constant-coefficient matrices A, B, C and D of system
A=[-10 0 -20 0;0 -10 -10 0;10 5 -1 0;0 0 1 0];
disp('eigenvalues _ A'); disp(eig(A)); % Eigenvalues of matrix A
B=[10 0;0 10;0 0;0 0];
C=[0 0 0 1]; D=[0 0 0 0];
% Weighting matrices Q and G
Q=[0.05 0 0 0;0 0.1 0 0;0 0 0.01 0;0 0 0 1];
G=[0.001 0;0 0.001];
% Feedback and return function coefficients, eigenvalues
[K _ feedback,K,Eigenvalues]=lqr(A,B,Q,G);
disp('K _ feedback'); disp(K _ feedback);
disp('K'); disp(K);
disp('eigenvalues A-BK _ feedback'); disp(Eigenvalues);
% Closed-loop system
A _ closed _ loop=A-B*K _ feedback;
% Dynamics
t=0:0.002:1;
uu=[0*ones(max(size(t)),4)]; % Applied inputs
x0=[20 10 -10 -20]; % Initial conditions
[y,x]=lsim(A _ closed _ loop,B*K _ feedback,C,D,uu,t,x0);
plot(t,x); title('System Dynamics, x _ 1, x _ 2, x _ 3, x _ 4', 'FontSize',14);
xlabel('Time [seconds]','FontSize',14); pause;
plot(t,y); pause;
plot(t,x(:,1),'-',t,x(:,2),'-',t,x(:,3),'-',t,x(:,4),'-'); pause;
plot(t,x(:,1),'-'); pause; plot(t,x(:,2),'-'); pause;
plot(t,x(:,3),'-'); pause; plot(t,x(:,4),'-'); pause;
disp('End')
```

The feedback gain matrix K_F, the return function matrix K, and the eigenvalues of the closed-loop system

$$(A - BG^{-1}B^T K) = (A - BK_F)$$

are given below

```
eigenvalues _ A

0
-1.0000e+001
-5.5000e+000          +1.5158e+001i
-5.5000e+000          -1.5158e+001i


K _ feedback

6.7782e+000    2.1467e-001    4.7732e+000    2.9687e+001
2.1467e-001    9.1198e+000    1.4555e+000    1.0895e+001

K

6.7782e-004    2.1467e-005    4.7732e-004    2.9687e-003
2.1467e-005    9.1198e-004    1.4555e-004    1.0895e-003
```

| 4.7732e-004 | 1.4555e-004 | 4.8719e-003 | 2.3325e-002 |
| 2.9687e-003 | 1.0895e-003 | 2.3325e-002 | 2.5115e-001 |

eigenvalues A-BK_ feedback

-1.0002e+002	
-6.8253e+001	
-5.8523e+000	+3.8928e+000i
-5.8523e+000	-3.8928e+000i

Having found matrices K and K_F , we obtain the control law as

$$
u = \begin{bmatrix} u_1 \\ u_2 \end{bmatrix} = -K_F x = - \begin{bmatrix} 6.78 & 0.21 & 4.77 & 29.7 \\ 0.21 & 9.12 & 1.46 & 10.9 \end{bmatrix} \begin{bmatrix} x_1 \\ x_2 \\ x_3 \\ x_4 \end{bmatrix}.
$$

The dynamics of the closed-loop system states, if initial conditions are set to be $x_{10} = 20$, $x_{20} = 10$, $x_{30} = -10$, $x_{40} = -20$, are plotted in Figure 8.1. The state variables and output ($y = x_4$) converge to the steady-state values. The closed-loop system is stable. In this section and example, the stabilization problem is considered. The reference $r(t)$ and tracking error $e(t)$ are not used.

□

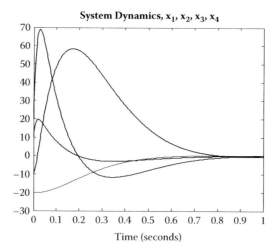

FIGURE 8.1
Dynamics of the state variables.

8.3 Tracking Control of Linear Electromechanical Systems

The stabilization problem and design of stabilizing control laws using the Hamilton-Jacobi theory were covered in Section 8.2. The tracking control laws must be synthesized using the tracking error $e(t) = r(t) - y(t)$.

The tracking control problem can be formulated as: for electromechanical systems, modeled as (8.9)

$$\dot{x}(t) = Ax + Bu$$

with the output equation $y(t) = Hx(t)$, synthesize the tracking optimal control law $u = \phi(e,x)$ by minimizing the performance functional.

Using the output equation $y(t) = Hx(t)$, for multivariable systems we have

$$e(t) = Nr(t) - y(t) = Nr(t) - Hx(t),$$

$$e(t) = \begin{bmatrix} n_{11} & 0 & \cdots & 0 & 0 \\ 0 & n_{22} & \cdots & 0 & 0 \\ \vdots & \vdots & \ddots & \vdots & \vdots \\ 0 & 0 & \cdots & n_{b-1b-1} & 0 \\ 0 & 0 & \cdots & 0 & n_{bb} \end{bmatrix} \begin{bmatrix} r_1 \\ r_2 \\ \vdots \\ r_{b-1} \\ r_b \end{bmatrix}$$

$$- \begin{bmatrix} h_{11} & h_{12} & \cdots & h_{1n-1} & h_{1n} \\ h_{21} & h_{22} & \cdots & h_{2n-1} & h_{2n} \\ \vdots & \vdots & \ddots & \vdots & \vdots \\ h_{b-11} & h_{b-12} & \cdots & h_{b-1n-1} & h_{b-1n} \\ h_{b1} & h_{b2} & \cdots & h_{bn-1} & h_{bn} \end{bmatrix} \begin{bmatrix} x_1 \\ x_2 \\ \vdots \\ x_{n-1} \\ x_n \end{bmatrix},$$

where $N \in \mathbb{R}^{b \times b}$ and $H \in \mathbb{R}^{b \times n}$ are the constant-coefficient matrices.

Denoting

$$e(t) = \dot{x}^{ref}(t),$$

consider the dynamics of the *exogeneous* system

$$x^{ref}(t) = Nr - y = Nr - Hx. \tag{8.22}$$

The system equations of motion (8.9) and (8.22) are

$$\dot{x}(t) = Ax + Bu, \quad y = Hx, \quad x_0(t_0) = x_0, \quad \dot{x}^{ref}(t) = Nr - y = Nr - Hx.$$

Hence, we have

$$\dot{x}_{\Sigma}(t) = A_{\Sigma} x_{\Sigma} + B_{\Sigma} u + N_{\Sigma} r, \quad y = Hx, \quad x_{\Sigma 0}(t_0) = x_{\Sigma 0}, \tag{8.23}$$

where

$$x_{\Sigma} = \begin{bmatrix} x \\ x^{ref} \end{bmatrix} \in \mathbb{R}^c (c=n+b); \; A_{\Sigma} = \begin{bmatrix} A & 0 \\ -H & 0 \end{bmatrix} \in \mathbb{R}^{c \times c}, B_{\Sigma} = \begin{bmatrix} B \\ 0 \end{bmatrix} \in \mathbb{R}^{c \times m}, \text{and } N_{\Sigma} = \begin{bmatrix} 0 \\ N \end{bmatrix} \in \mathbb{R}^{c \times b}.$$

The quadratic performance functional is

$$J = \frac{1}{2} \int_{t_0}^{t_f} \left(\begin{bmatrix} x \\ x^{ref} \end{bmatrix}^T Q \begin{bmatrix} x \\ x^{ref} \end{bmatrix} + u^T G u \right) dt. \tag{8.24}$$

From (8.23) and (8.24), the Hamiltonian function is

$$H\left(x_{\Sigma}, u, r, \frac{\partial V}{\partial x_{\Sigma}} \right) = \frac{1}{2} \left(x_{\Sigma}^T Q x_{\Sigma} + u^T G u \right) + \left(\frac{\partial V}{\partial x_{\Sigma}} \right)^T (A_{\Sigma} x_{\Sigma} + B_{\Sigma} u + N_{\Sigma} r). \tag{8.25}$$

Using the first-order necessary condition for optimality (8.5), from (8.25), one finds

$$\frac{\partial H}{\partial u} = u^T G + \left(\frac{\partial V}{\partial x_{\Sigma}} \right)^T B_{\Sigma}.$$

Thus, the control law is

$$u = -G^{-1} B_{\Sigma}^T \frac{\partial V(x_{\Sigma})}{\partial x_{\Sigma}} = -G^{-1} \begin{bmatrix} B \\ 0 \end{bmatrix}^T \frac{\partial V\left(\begin{bmatrix} x \\ x^{ref} \end{bmatrix} \right)}{\partial \begin{bmatrix} x \\ x^{ref} \end{bmatrix}}. \tag{8.26}$$

The solution of the Hamilton-Jacobi-Bellman partial differential equation

$$-\frac{\partial V}{\partial t} = \frac{1}{2} x_{\Sigma}^T Q x_{\Sigma} + \left(\frac{\partial V}{\partial x_{\Sigma}} \right)^T A x_{\Sigma} - \frac{1}{2} \left(\frac{\partial V}{\partial x_{\Sigma}} \right)^T B_{\Sigma} G^{-1} B_{\Sigma}^T \frac{\partial V}{\partial x_{\Sigma}} \tag{8.27}$$

is satisfied by the quadratic return function

$$V(x_{\Sigma}) = \frac{1}{2} x_{\Sigma}^T K(t) x_{\Sigma}. \tag{8.28}$$

From (8.27) and (8.28), the Riccati equation to obtain the unknown symmetric matrix K is

$$-\dot{K} = Q + A_\Sigma^T K + K^T A_\Sigma - K^T B_\Sigma G^{-1} B_\Sigma^T K, \quad K(t_f) = K_f. \tag{8.29}$$

The control law is found from (8.26) and (8.28) as

$$u = -G^{-1} B_\Sigma^T K x_\Sigma = -G^{-1} \begin{bmatrix} B \\ 0 \end{bmatrix}^T K \begin{bmatrix} x \\ x^{ref} \end{bmatrix}. \tag{8.30}$$

Recalling that

$$x^{ref}(t) = e(t),$$

one has

$$x^{ref}(t) = \int e(t)dt.$$

Hence, we have a control law with the state feedback and integral term for $e(t)$. In particular,

$$u(t) = -G^{-1} B_\Sigma^T K x_\Sigma(t) = -G^{-1} \begin{bmatrix} B \\ 0 \end{bmatrix}^T K \begin{bmatrix} x(t) \\ \int e(t)dt \end{bmatrix}. \tag{8.31}$$

This control law usually does not ensure suitable performance because there is no proportional term for the tracking error $e(t)$. Other design concepts are of particular interest.

8.4 *State Transformation* Method and Tracking Control

The tracking control problem is solved designing the proportional-integral control laws using the *state transformation* method. We define the tracking error vector as

$$e(t) = Nr(t) - y(t) = Nr(t) - Hx^{sys}(t).$$

For linear systems

$$x^{sys} = A^{sys} x^{sys} + B^{sys} u, \tag{8.32}$$

with the output equation $y(t) = Hx^{sys}(t)$, we have

$$e(t) = Nr(t) - y(t) = Nr(t) - Hx^{sys}(t) = Nr(t) - HA^{sys}x^{sys} - HB^{sys}u. \quad (8.33)$$

From (8.32) and (8.33), applying the expanded state vector

$$x(t) = \begin{bmatrix} x^{sys}(t) \\ e(t) \end{bmatrix},$$

one finds

$$\dot{x}(t) = \begin{bmatrix} \dot{x}^{sys}(t) \\ \dot{e}(t) \end{bmatrix} = \begin{bmatrix} A^{sys} & 0 \\ -HA^{sys} & 0 \end{bmatrix}\begin{bmatrix} x^{sys} \\ e \end{bmatrix} + \begin{bmatrix} B^{sys} \\ -HB^{sys} \end{bmatrix}u + \begin{bmatrix} 0 \\ N \end{bmatrix}\dot{r} = Ax + Bu + \begin{bmatrix} 0 \\ N \end{bmatrix}\dot{r}.$$

$$(8.34)$$

The *space transformation* method utilizes the z and v vectors as defined by

$$z = \begin{bmatrix} x \\ u \end{bmatrix} \quad \text{and} \quad v = u. \quad (8.35)$$

Here, $z \in \mathbb{R}^{c+m}$ and $v \in \mathbb{R}^m$ are the *state transformation* variable. Using z and v, one obtains the system model as

$$\dot{z}(t) = \begin{bmatrix} A & B \\ 0 & 0 \end{bmatrix}z + \begin{bmatrix} 0 \\ I \end{bmatrix}v = A_z z + B_z v, \quad y = Hx^{sys}, \quad z(t_0) = z_0, \quad (8.36)$$

where $A_z \in \mathbb{R}^{(c+m)\times(c+m)}$ and $B_z \in \mathbb{R}^{(c+m)\times m}$ are the constant-coefficient matrices. Minimizing the quadratic functional

$$J = \frac{1}{2}\int_{t_0}^{t_f}(z^T Q_z z + v^T G_z v)dt, \quad Q_z \in \mathbb{R}^{(c+m)\times(c+m)}, Q_z \geq 0, G \in \mathbb{R}^{m\times m}, G > 0, \quad (8.37)$$

the application of the first-order necessary condition for optimality (8.5) yields

$$v = -G_z^{-1}B_z^T \frac{\partial V}{\partial z}. \quad (8.38)$$

The solution of the Hamilton-Jacobi-Bellman partial differential equation

$$-\frac{\partial V}{\partial t} = \frac{1}{2}z^T Q_z z + \left(\frac{\partial V}{\partial z}\right)^T A_z z - \frac{1}{2}\left(\frac{\partial V}{\partial z}\right)^T B_z G_z^{-1}B_z^T \frac{\partial V}{\partial z} \quad (8.39)$$

is satisfied by the continuous and differentiable quadratic return function

$$V(z) = \frac{1}{2} z^T K(t) z. \tag{8.40}$$

Using (8.38) and (8.40), one obtains the control function as

$$v = -G_z^{-1} B_z^T K z. \tag{8.41}$$

From (8.39), the Riccati equation to find the unknown matrix $K \in \mathbb{R}^{(c+m) \times (c+m)}$ is

$$-\dot{K} = K A_z + A_z^T K - K B_z G_z^{-1} B_z^T K + Q_z, \quad K(t_f) = K_f.$$

From (8.41) and (8.35), one has

$$u(t) = -G_z^{-1} B_z^T K z = -G_z^{-1} \begin{bmatrix} 0 \\ I \end{bmatrix}^T \begin{bmatrix} K_{11} & K_{21}^T \\ K_{21} & K_{22} \end{bmatrix} \begin{bmatrix} x \\ u \end{bmatrix} = -G_z^{-1} K_{21} x - G_z^{-1} K_{22} u = K_{f1} x + K_{f2} u.$$

$$\tag{8.42}$$

Using (8.34) $x(t) = Ax + Bu$, we have $u = B^{-1}(x(t) - Ax)$. Thus,

$$u = B^{-1}(x(t) - Ax) = (B^T B)^{-1} B^T (x(t) - Ax). \tag{8.43}$$

Applying (8.43) in (8.42), one obtains

$$u(t) = K_{f1} x + K_{f2} u = K_{f1} x + K_{f2} (B^T B)^{-1} B^T (x(t) - Ax)$$

$$= [K_{f1} - K_{f2} (B^T B)^{-1} B^T A] x(t) + K_{f2} (B^T B)^{-1} B^T x(t)$$

$$= (K_{f1} - K_{F1} A) x(t) + K_{F1} x(t) = K_{F2} x(t) + K_{F1} x(t). \tag{8.44}$$

Hence, the control law is derived as

$$u(t) = K_{F1} x(t) - K_{F1} x_0 + \int K_{F2} x(\tau) d\tau + u_0. \tag{8.45}$$

The designed controller (8.45) is the proportional-integral tracking control law with state feedback because

$$x(t) = \begin{bmatrix} x^{sys}(t) \\ e(t) \end{bmatrix}.$$

We utilize the states $x^{sys}(t)$ and tracking error $e(t)$. To implement control law (8.45), these $x^{sys}(t)$ and $e(t)$ must be available.

For nonlinear electromechanical systems, the proposed procedure can be straightforwardly used to derive control laws. In particular, by using

$$v = -G_z^{-1} B_z^T (z) \frac{\partial V}{\partial z},$$
(8.46)

we obtain the proportional-integral control law $u(t)$.

Example 8.5

Design the tracking control law for a system with a PZT actuator controlled by changing the applied voltage u. The second-order equation of motion for a PZT actuator is

$$m \frac{d^2 y}{dt^2} + b \frac{dy}{dt} + k_e y = k_e d_e u,$$

where y is the actuator displacement (output).

Hence, we have

$$\frac{dy}{dt} = v,$$

$$\frac{dv}{dt} = -\frac{k_e}{m} y - \frac{b}{m} v + \frac{k_e d_e}{m} u.$$

The reference input is $r(t)$, and the tracking error is $e(t) = Nr(t) - y(t)$, $N = 1$. We have the system (8.32) and (8.33)

$$x^{sys} = A^{sys} x^{sys} + B^{sys} u, \quad e = Nr - HA^{sys} x^{sys} - HB^{sys} u,$$

where

$$x^{sys} = \begin{bmatrix} y \\ v \end{bmatrix}, \quad A^{sys} = \begin{bmatrix} 0 & 1 \\ -\dfrac{k_e}{m} & -\dfrac{b}{m} \end{bmatrix}, \quad B^{sys} = \begin{bmatrix} 0 \\ \dfrac{k_e d_e}{m} \end{bmatrix} \quad \text{and} \quad H = \begin{bmatrix} 1 & 0 \end{bmatrix}.$$

One obtains (8.34) as

$$\dot{x} = \begin{bmatrix} \dot{x}^{sys} \\ \dot{e} \end{bmatrix} = \begin{bmatrix} A^{sys} & 0 \\ -HA^{sys} & 0 \end{bmatrix} \begin{bmatrix} x^{sys} \\ e \end{bmatrix} + \begin{bmatrix} B^{sys} \\ -HB^{sys} \end{bmatrix} u + \begin{bmatrix} 0 \\ N \end{bmatrix} \dot{r}, \quad y = \begin{bmatrix} H & 0 \end{bmatrix} \begin{bmatrix} x^{sys} \\ e \end{bmatrix}.$$

We apply the *state transformation* method. From (8.35), one has

$$z = \begin{bmatrix} x^{sys} \\ e \\ u \end{bmatrix}.$$

Therefore, the control function (8.41) is

$$u(t) = -K_f z(t) = -K_f \begin{bmatrix} y(t) \\ v(t) \\ e(t) \\ u(t) \end{bmatrix}.$$

The proportional-integral tracking control law (8.45) is derived as

$$u(t) = K_{F1} \begin{bmatrix} y(t) \\ v(t) \\ e(t) \end{bmatrix} + \int K_{F2} \begin{bmatrix} y(\tau) \\ v(\tau) \\ e(\tau) \end{bmatrix} d\tau.$$

Using the actuator parameters $k_e = 3000$, $b = 1$, $d_e = 0.000001$ and $m = 0.02$, the feedback gains are found by using the lqr MATLAB command. The weighting matrices of the quadratic functional (8.37) are

$$Q_z = \begin{bmatrix} 1 & 0 & 0 & 0 \\ 0 & 1 & 0 & 0 \\ 0 & 0 & 1 \times 10^{10} & 0 \\ 0 & 0 & 0 & 1 \end{bmatrix}$$

and $G_z = 10$. The closed-loop actuator dynamics is documented in Figure 8.2. The settling time is 0.022 sec, and the tracking error converges to zero.

The differential equation, which models the PZT actuators, can be given as

$$m\frac{d^2y}{dt^2} + b\frac{dy}{dt} + k_e y = k_e(d_e u - z_h),$$

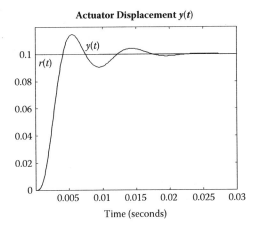

Actuator Displacement $y(t)$

FIGURE 8.2
Closed-loop system dynamics if $r(t) = 0.1$.

may be integrated with the hysteresis model

$$\dot{z}_h = \alpha d_e u - \beta |\dot{u}| z_h - \gamma \dot{u} |z_h|,$$

where α and β are the constants.

Thus, we have the state-space nonlinear model of PZT actuators as

$$\frac{dy}{dt} = v,$$

$$\frac{dv}{dt} = -\frac{k_e}{m} y - \frac{b}{m} v - \frac{k_e}{m} z_h + \frac{kd_e}{m} u,$$

$$\frac{dz_h}{dt} = -\beta |\dot{u}| z_h + \alpha d_e \dot{u} - \gamma |z_h| \dot{u}.$$

The tracking control law can be designed for this nonlinear actuator. □

8.5   Time-Optimal Control of Electromechanical Systems

For dynamic systems, time-optimal control laws can be designed using functionals

$$J = \frac{1}{2} \int_{t_0}^{t_f} W(x) dt \quad \text{or} \quad J = \int_{t_0}^{t_f} 1 dt.$$

For nonlinear systems (8.3), taking note of the Hamilton-Jacobi equation

$$-\frac{\partial V}{\partial t} = \min_{-1 \le u \le 1} \left[1 + \left(\frac{\partial V}{\partial x} \right)^T (F(x) + B(x)u) \right] \tag{8.47}$$

and using the first-order necessary condition for optimality (8.5), the relay-type control law is

$$u = -\operatorname{sgn}\left(B^T(x) \frac{\partial V}{\partial x} \right), \quad -1 \le u \le 1. \tag{8.48}$$

Control law (8.48) cannot be applied to systems because of the chattering phenomena, switching, losses, and other undesirable effects. Therefore, relay-type control laws with dead zone

$$u = -\operatorname{sgn}\left(B(x)^T \frac{\partial V}{\partial x} \right)\bigg|_{\text{dead zone}}, \quad -1 \le u \le 1, \tag{8.49}$$

may be considered as a possible candidate.

Example 8.6

Synthesize the time-optimal control law for the system described by the following differential equations

$$x_1(t) = x_1^5 u_1 + x_2^7, \quad -1 \le u_1 \le 1,$$
$$x_2(t) = x_1^3 x_2^5 u_2, \quad -1 \le u_2 \le 1.$$

The performance functional is

$$J = \int_{t_0}^{t_f} 1 dt.$$

The Hamilton-Jacobi equation is

$$-\frac{\partial V}{\partial t} = \min_{u \in U} \left[1 + \left(\frac{\partial V}{\partial x} \right)^T (F(x) + B(x)u) \right]$$

$$= \min_{\substack{-1 \le u_1 \le 1 \\ -1 \le u_2 \le 1}} \left[1 + \frac{\partial V}{\partial x_1} (x_1^5 u_1 + x_2^7) + \frac{\partial V}{\partial x_2} x_1^3 x_2^5 u_2 \right].$$

From the first-order necessary condition for optimality (8.5), an optimal control law (8.48) is

$$u_1 = -\operatorname{sgn}\left(x_1^5 \frac{\partial V}{\partial x_1} \right) \quad \text{and} \quad u_2 = -\operatorname{sgn}\left(x_1^3 x_2^5 \frac{\partial V}{\partial x_2} \right).$$

□

Example 8.7

We can design an optimal relay-type control law for a system studied in Example 8.3. In particular, the equations of motion are

$$\dot{x}_1(t) = x_2, \quad \dot{x}_2(t) = u, \quad -1 \le u \le 1.$$

The control takes values $u = 1$ and $u = -1$.
If $u = 1$, from $x_1(t) = x_2$, $x_2(t) = 1$, one has

$$\frac{dx_2}{dx_1} = \frac{1}{x_2}.$$

The integration gives $x_2^2 = 2x_1 + c_1$.
If $u = -1$, from $x_1(t) = x_2$, $x_2(t) = -1$, we obtain

$$\frac{dx_2}{dx_1} = -\frac{1}{x_2}.$$

The integration yields $x_2^2 = -2x_1 + c_2$.

As a result of the switching ($u = 1$ or $u = -1$), the switching curve is derived as a function of the state variables. The comparison of $x_2^2 = 2x_1 + c_1$ and $x_2^2 = -2x_1 + c_2$ gives the switching curve as

$$-x_2^2 - 2x_1 \, \mathrm{sgn}(x_2) = 0 \quad \text{or} \quad -x_1 - \frac{1}{2}x_2^2 \, |x_2| = 0.$$

The control takes the values $u = 1$ and $u = -1$. Making use of the derived expression for the switching curve, one finds the time-optimal (relay) control law as

$$u = -\mathrm{sgn}\left(x_1 + \frac{1}{2}x_2^2 \, |x_2|\right), \quad -1 \le u \le 1.$$

We found an optimal control law using the calculus of variations performing analysis of the solutions of the differential equations with the relay control law.

The Hamilton-Jacobi theory is applied. We minimize the functional

$$J = \int_{t_0}^{t_f} 1 dt.$$

From the Hamilton-Jacobi equation

$$-\frac{\partial V}{\partial t} = \min_{-1 \le u \le 1}\left[1 + \frac{\partial V}{\partial x_1}x_2 + \frac{\partial V}{\partial x_2}u\right],$$

an optimal control law derived using (8.5) is

$$u = -\mathrm{sgn}\left(\frac{\partial V}{\partial x_2}\right).$$

The solution of the Hamilton-Jacobi partial differential equation is given by the return function

$$V(x_1, x_2) = k_{11}x_1^2 + k_{12}x_1x_2 + k_{22}x_2^3 \, |x_2|.$$

That is, the nonquadratic, continuous, and differentiable return function is used. The control law is

$$u = -\mathrm{sgn}\left(x_1 + \frac{1}{2}x_2^2 \, |x_2|\right).$$

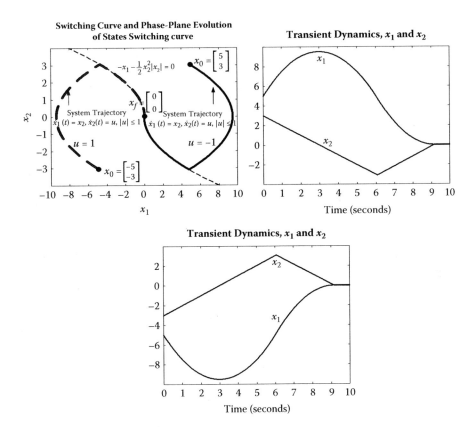

FIGURE 8.3
System evolution with time-optimal control (switching) control.

The time-optimal design using the Hamilton-Jacobi theory corresponds to the results obtained using the calculus of variations. The transient dynamics are analyzed. The switching curve, the phase-plane evolution of the variables, and the transient behavior for different initial conditions are documented in Figure 8.3.

The application of the return function

$$V(x_1, x_2) = \frac{1}{2} k_{11} x_1^2 + k_{12} x_1 x_2 + \frac{1}{2} k_{22} x_2^2$$

results in the control law

$$u = -\operatorname{sgn}(x_1 + x_2)$$

which also stabilizes the closed-loop system. Introducing the switching surface, which will be used in Section 8.6, as

$$v(x_1, x_2) = x_1 + x_2,$$

one has

$$u = \begin{cases} 1 & \text{if } v(x_1, x_2) < 0 \\ -1 & \text{if } v(x_1, x_2) > 0 \end{cases}.$$

$\square$

8.6   Sliding Mode Control

Time-optimal control results in relay-type controllers with switching surfaces. The undesirable phenomena (switching, chattering, ripple, etc.) usually lead to unacceptable performance. Sliding mode control has direct analogies to the time-optimal control. Soft- and hard-switching sliding mode control laws can be synthesized. Sliding mode soft switching control may provide good performance, and the chattering phenomena (typical for relay-type and hard switching sliding mode control laws) is eliminated. To design control laws, we model the states and errors dynamics as

$$x(t) = F(x)x + B(x)u, \quad u_{\min} \le u(t, x, e) \le u_{\max}, \quad u_{\min} < 0, \quad u_{\max} > 0, \quad (8.50)$$

$$\dot{e}(t) = N\dot{r}(t) - H\dot{x}.$$

The smooth sliding manifold is given as

$$M = \{(t, x, e) \in \mathbb{R}_{\ge 0} \times X \times E \mid v(t, x, e) = 0\}$$

$$= \bigcap_{j=1}^{m} \{(t, x, e) \in \mathbb{R}_{\ge 0} \times X \times E \mid v_j(t, x, e) = 0\}, \quad (8.51)$$

where the time-varying nonlinear switching surface is

$$v(t, x, e) = K_{vxe}(t, x, e) = 0 \quad \text{or}$$

$$v(t, x, e) = [K_{vx}(t) \ \ K_{ve}(t)] \begin{bmatrix} x(t) \\ e(t) \end{bmatrix} = K_{vx}(t)x(t) + K_{ve}(t)e(t) = 0,$$

$$\begin{bmatrix} v_1(t, x, e) \\ \vdots \\ v_m(t, x, e) \end{bmatrix} = \begin{bmatrix} k_{vx11}(t) & \cdots & k_{vx1n}(t) & k_{ve11}(t) & \cdots & k_{ve1b}(t) \\ \vdots & \vdots & \vdots & \vdots & \vdots & \vdots \\ k_{vxm1}(t) & \cdots & k_{vxmn}(t) & k_{vem1}(t) & \cdots & k_{vemb}(t) \end{bmatrix} \begin{bmatrix} x_1(t) \\ \vdots \\ x_n(t) \\ e_1(t) \\ \vdots \\ e_b(t) \end{bmatrix} = 0.$$

The soft switching control law is given as

$$u(t, x, e) = -G\phi(v), \quad u_{\min} \leq u(t,x,e) \leq u_{\max}, \quad G \in \mathbb{R}^{m \times m}, \quad G > 0, \quad (8.52)$$

where ϕ is the continuous real-analytic function of class $C^\in (\in \geq 1)$, for example, tanh and erf.

In contrast, the discontinuous (hard-switching) tracking control laws with constant and varying gains are

$$u(t, x, e) = -G \operatorname{sgn}(v), \quad G \in \mathbb{R}^{m \times m}, \quad G > 0,$$

or

$$u(t, x, e) = -G(t, x, e) \operatorname{sgn}(v), \quad G(\cdot): \mathbb{R}_{\geq 0} \times \mathbb{R}^n \times \mathbb{R}^b \to \mathbb{R}^{m \times m}. \quad (8.53)$$

The simplest hard-switching tracking control law (8.53) is

$$u(t, x, e) = \begin{cases} u_{\max}, & \forall v(t, x, e) > 0 \\ 0, & \forall v(t, x, e) = 0, \quad u_{\min} \leq u(t,x,e) \leq u_{\max}, \quad (8.54) \\ u_{\min}, & \forall v(t, x, e) < 0 \end{cases}$$

and a polyhedron in the control space with 2^m vertexes results.

Example 8.8 Control of Permanent-Magnet Synchronous Motors in the Rotor Reference Frame

We design a sliding mode control law for a system with a permanent-magnet synchronous motor. The 40 V motor parameters are:

$$r_s(\cdot) \in [0.5_{T=20^\circ C} \quad 0.75_{T=130^\circ C}] \text{ ohm,}$$
$$L_{ss}(\cdot) \in [0.009 \quad 0.01] \text{ H,}$$
$$L_{ls} = 0.001 \text{ H,}$$
$$\psi_m(\cdot) \in [0.069_{T=20^\circ C} \quad 0.055_{T=130^\circ C}] \text{ V-sec/rad or N-m/A,}$$
$$B_m = 0.000013 \text{ N-m-sec/rad, and}$$
$$J(\cdot) \in [0.0001 \quad 0.0003] \text{ kg-m}^2.$$

As found in Section 6.3.3, in the rotor and synchronous reference frames, permanent-magnet synchronous motors are described by equations (6.33). In particular,

$$\frac{di_{qs}^r}{dt} = -\frac{r_s}{L_{ss}} i_{qs}^r - \frac{\psi_m}{L_{ss}} \omega_r - i_{ds}^r \omega_r + \frac{1}{L_{ss}} u_{qs}^r, \quad \frac{di_{ds}^r}{dt} = -\frac{r_s}{L_{ss}} i_{ds}^r + i_{qs}^r \omega_r + \frac{1}{L_{ss}} u_{ds}^r,$$

$$\frac{di_{0s}^r}{dt} = -\frac{r_s}{L_{ls}} i_{0s}^r + \frac{1}{L_{ls}} u_{0s}^r, \quad \frac{d\omega_r}{dt} = \frac{3P^2 \psi_m}{8J} i_{qs}^r - \frac{B_m}{J} \omega_r - \frac{P}{2J} T_L.$$

The state and control variables are the *quadrature, direct,* and *zero* axis components of currents and voltages, as well as the angular velocity. We have

$$x = \begin{bmatrix} i_{qs}^r \\ i_{ds}^r \\ i_{0s}^r \\ \omega_r \end{bmatrix} \quad \text{and} \quad u = \begin{bmatrix} u_{qs}^r \\ u_{ds}^r \end{bmatrix}.$$

The *qd0* voltage and current components have a dc form.

One can study the stability using the Lyapunov stability theory covered in Section 8.9 (see Example 8.17). For the uncontrolled motor, $u_{qs}^r = u_{ds}^r = u_{0s}^r = 0$. The total derivative of the positive-definite quadratic function

$$V\left(i_{qs}^r, i_{ds}^r, i_{0s}^r, \omega_r\right) = \frac{1}{2} i_{qs}^{r\,2} + \frac{1}{2} i_{ds}^{r\,2} + \frac{1}{2} i_{0s}^{r\,2} + \frac{1}{2} \omega_r^2$$

is

$$\frac{dV\left(i_{qs}^r, i_{ds}^r, i_{0s}^r, \omega_r\right)}{dt} = -\frac{r_s}{L_{ss}} i_{qs}^{r\,2} - \frac{r_s}{L_{ss}} i_{ds}^{r\,2} - \frac{r_s}{L_{ls}} i_{0s}^{r\,2} - \frac{\psi_m(8J - 3P^2 L_{ss})}{8JL_{ss}} i_{qs}^r \omega_r - \frac{B_m}{J} \omega_r^2.$$

The motor parameters are time-varying, and

$$r_s(\cdot) \in [r_{s\min} \quad r_{s\max}], L_{ss}(\cdot) \in [L_{ss\min} \quad L_{ss\max}], \psi_m(\cdot) \in [\psi_{m\min} \quad \psi_{m\max}], J(\cdot) \in [J_{\min} \quad J_{\max}].$$

However, $r_s > 0$, $L_{ss} > 0$, $\psi_m > 0$, and $J > 0$. Hence, the open-loop system is uniformly robustly asymptotically stable in the large because the total derivative of a positive-definite function

$$V\left(i_{qs}^r, i_{ds}^r, i_{0s}^r, \omega_r\right)$$

is negative. Thus, it is not necessary to apply the linearizing feedback to transform a nonlinear motor model into a linear one by canceling the beneficial stabilizing inherent nonlinearities $-i_{ds}^r \omega_r$ and $i_{qs}^r \omega_r$ which are the *back emfs.* Furthermore, as a result of the bounds imposed on the voltages, as well as variations of L_{ss}, one cannot ensure the feedback linearization.

From the electric machinery standpoints, to attain the balanced voltage set, one varies only u_{qs}^r whereas $u_{ds}^r = 0$. Thus, one is unable to linearize the electric motor even if needed.

Our goal is to design a soft-switching control law. The applied phase voltages are bounded by ±40 V. The time-invariant linear and nonlinear *switching surfaces* of stabilizing control laws are obtained as functions of the state variables i_{qs}^r, i_{ds}^r and ω_r. We have

$$v\left(i_{qs}^r, i_{ds}^r, \omega_r\right) = -0.00049 i_{qs}^r - 0.00049 i_{ds}^r - 0.0014\omega_r = 0 \text{ (linear } switching \text{ } surface),$$

$$v\left(i_{qs}^r, i_{ds}^r, \omega_r\right) = -0.00049 i_{qs}^r - 0.00049 i_{ds}^r - 0.000017\omega_r$$

$$-0.000025\omega_r |\omega_r| = 0 \text{ (nonlinear } switching \text{ } surface).$$

It was emphasized that only *quadrature* voltage component is regulated. We denote $u = u_{qs}^r$. A discontinuous hard-switching stabilizing control law is

$$u = \text{sgn}_{-40}^{+40} v\left(i_{qs}^r, i_{ds}^r, \omega_r\right) = \begin{vmatrix} +40, & v\left(i_{qs}^r, i_{ds}^r, \omega_r\right) > 0 \\ 0, & v\left(i_{qs}^r, i_{ds}^r, \omega_r\right) = 0. \\ -40, & v\left(i_{qs}^r, i_{ds}^r, \omega_r\right) < 0 \end{vmatrix}$$

To avoid the singularity, this discontinuous algorithm is *regularized* as

$$u = 40 \frac{v(i_{qs}^r, i_{ds}^r, \omega_r)}{|v(i_{qs}^r, i_{ds}^r, \omega_r)| + \varepsilon}, \quad \varepsilon = 0.0005.$$

A soft-switching stabilizing control law is designed by making use of linear and nonlinear *switching surfaces*. In particular,

$$u = 40 \tanh v\left(i_{qs}^r, i_{ds}^r, \omega_r\right),$$

where $v(i_{qs}^r, i_{ds}^r, \omega_r)$ is the linear or nonlinear *switching surfaces*.

Numerical simulations are performed to study the state evolutions due to initial conditions, references (assigned angular velocity), and disturbances (T_L variations). The three-dimensional plot for the evolution of states i_{qs}^r, i_{ds}^r and ω_r is plotted in Figure 8.4. It is evident that the system evolves to the origin.

The tracking control law is synthesized within the nonlinear time-invariant *switching surface* as

$$v\left(i_{qs}^r, i_{ds}^r, \omega_r, e\right) = -0.0005 i_{qs}^r - 0.0005 i_{ds}^r - 0.00003\omega_r + 0.0015e + 0.0001e^3 = 0.$$

A soft-switching tracking control law is

$$u = 40 \tanh v\left(i_{qs}^r, i_{ds}^r, \omega_r, e\right).$$

Figure 8.5 illustrates the dynamics of a closed-loop system with the designed tracking control law. The reference angular velocity is 200 rad/sec.

The simulation results illustrate that the settling time is 0.039 sec. Good dynamic performance of the synthesized system is evident, and the tracking error approaches zero. As a result of soft switching, the singularity and sensitivity problems are avoided, robustness and stability are improved, and the

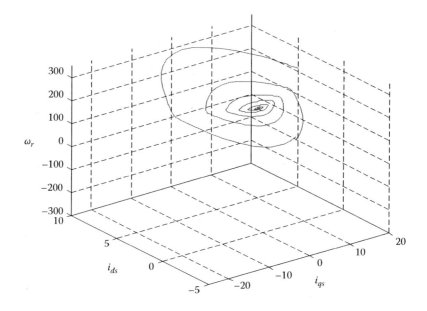

FIGURE 8.4
Three-dimensional state evolution due to initial conditions $\begin{bmatrix} i^r_{qs}(0) \\ i^r_{ds}(0) \\ \omega_r(0) \end{bmatrix} = \begin{bmatrix} 20 \\ 5 \\ 200 \end{bmatrix}$.

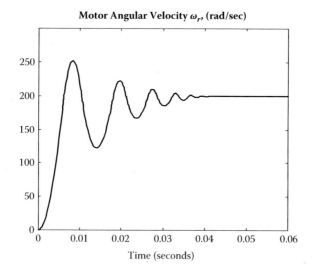

FIGURE 8.5
Transient dynamic of the motor angular velocity, $r = 200$ rad/sec.

chattering (high frequency switching) is eliminated. However, one recalls the drawbacks associated with the use of the *arbitrary* reference frame.

Feedback linearization and control of permanent-magnet synchronous motors. The *linearizability* condition is formally (mathematically) met. In fact, for a model

$$\frac{di_{qs}^r}{dt} = -\frac{r_s}{L_{ss}} i_{qs}^r - \frac{\psi_m}{L_{ss}} \omega_r - i_{ds}^r \omega_r + \frac{1}{L_{ss}} u_{qs}^r, \qquad \frac{di_{ds}^r}{dt} = -\frac{r_s}{L_{ss}} i_{ds}^r + i_{qs}^r \omega_r + \frac{1}{L_{ss}} u_{ds}^r,$$

$$\frac{d\omega_r}{dt} = \frac{3P^2 \psi_m}{8J} i_{qs}^r - \frac{B_m}{J} \omega_r,$$

the controller can be represented as

$$u_{qs}^r = u_{qs}^{r\ lin} + u_{qs}^{r\ cont} \quad \text{and} \quad u_{ds}^r = u_{ds}^{r\ lin} + u_{ds}^{r\ cont}.$$

With the linearizing feedback

$$u_{qs}^{r\ lin} = \left(L_{ls} + \frac{3}{2} \bar{L}_m \right) i_{ds}^r \omega_r \quad \text{and} \quad u_{ds}^{r\ lin} = -\left(L_{ls} + \frac{3}{2} \bar{L}_m \right) i_{qs}^r \omega_r,$$

the mathematical model is linearized to

$$\frac{di_{qs}^r}{dt} = -\frac{r_s}{L_{ls} + \frac{3}{2} \bar{L}_m} i_{qs}^r - \frac{\psi_m}{L_{ls} + \frac{3}{2} \bar{L}_m} \omega_r + \frac{1}{L_{ls} + \frac{3}{2} \bar{L}_m} u_{qs}^r,$$

$$\frac{di_{ds}^r}{dt} = -\frac{r_s}{L_{ls} + \frac{3}{2} \bar{L}_m} i_{ds}^r + \frac{1}{L_{ls} + \frac{3}{2} \bar{L}_m} u_{ds}^r,$$

$$\frac{d\omega_r}{dt} = \frac{3P^2 \psi_m}{8J} i_{qs}^r - \frac{B_m}{J} \omega_r - \frac{P}{2J} T_L.$$

For the linearized model the eigenvalues of the characteristic equation are found to be

$$\lambda_1 = -\frac{\left(2B_m L_{ss} + 2r_s J - \sqrt{4B_m^2 L_{ss}^2 - 8B_m L_{ss} r_s J + 4r_s^2 J^2 - 6L_{ss} J^3 \psi_m^2 P^2} \right)}{4L_{ss} J},$$

$$\lambda_2 = -\frac{\left(2B_m L_{ss} + 2r_s J + \sqrt{4B_m^2 L_{ss}^2 - 8B_m L_{ss} r_s J + 4r_s^2 J^2 - 6L_{ss} J^3 \psi_m^2 P^2} \right)}{4L_{ss} J} \quad \text{and}$$

$$\lambda_3 = -\frac{r_s}{L_{ss}}.$$

The real parts of these eigenvalues are negative. Thus, the stability is guaranteed.

For the resulting linear system, one can design control laws. For example, applying the proportional stabilizing controller

$$u_{qs}^{r\ cont} = -[k_{iq}\ \ k_{id}\ \ k_\omega]\begin{bmatrix} i_{qs}^r \\ i_{ds}^r \\ \omega_r \end{bmatrix} \quad \text{and} \quad u_{ds}^{r\ cont} = 0,$$

the eigenvalues for the closed-loop system are

$$\lambda_1 = -\frac{\left(2k_{iq}J + 2B_m L_{ss} + 2r_s J - \sqrt{4k_{iq}^2 J^2 - 8k_{iq} J B_m L_{ss} + 8k_{iq} J^2 r_s + 4B_m^2 L_{ss}^2 - 8B_m L_{ss} r_s J + 4r_s^2 J^2 - 6L_{ss} J^3 \psi_m P^2 \left(\psi_m + k_\omega\right)}\right)}{4L_{ss} J},$$

$$\lambda_2 = -\frac{\left(2k_{iq}J + 2B_m L_{ss} + 2r_s J + \sqrt{4k_{iq}^2 J^2 - 8k_{iq} J B_m L_{ss} + 8k_{iq} J^2 r_s + 4B_m^2 L_{ss}^2 - 8B_m L_{ss} r_s J + 4r_s^2 J^2 - 6L_{ss} J^3 \psi_m P^2 \left(\psi_m + k_\omega\right)}\right)}{4L_{ss} J},$$

$$\lambda_3 = -\frac{r_s}{L_{ss}}.$$

We derived the expressions for the eigenvalues. One can tentatively conclude that stability, specified transient quantities, dynamic performance, and desired capabilities can be attained. In pole-placement design, the specification on the optimum (desired) transient responses in terms of system models and feedback coefficients is equivalent to the specification imposed on desired transfer function of the closed-loop system. The desired eigenvalues can be specified by the designer (assuming that there is no state and control bounds, model is accurate, system is linear, parameters are constant, etc.), and these eigenvalues can be used to find the corresponding feedback gains. However, there is no guarantee that these eigenvalues can be achieved. Most electromechanical systems cannot be linearized (due to saturation, secondary phenomena, simplifications, parameter variations, etc.) or should not be linearized due to the attempt to cancel beneficial nonlinearities, which stabilize the system. One can try to mathematically perform the linearization assuming a number of unsound assumptions. Furthermore, the pole-placement concept, while theoretically guarantees the desired location of the characteristic eigenvalues, can lead to positive feedback coefficients and very sensitive closed-loop systems as a result of the parameter variations. Hence, the stability, robustness to parameter variations, and system performance must be examined.

The feedback linearization does not reduce the complexity of the corresponding analysis and design. Even from mathematical standpoints, the simplification and the *"optimal"* performance would achieve in expense of large control efforts required due to the linearizing feedback. This leads to saturation. In general, there is no need to linearize the majority of electromechanical system because the open-loop system (drives) are uniformly asymptotically stable as illustrated by using the Lyapunov stability theory. There are no challenges to design control laws for *servos* with one pole at the origin recalling that

$$\frac{d\theta_r}{dt} = \omega_r.$$

The most critical problem is that the linearizing feedback

$$u_{ds}^{r \; lin} = -\left(L_{ls} + \frac{3}{2}\bar{L}_m \right) i_{qs}^r \omega_r$$

cannot be applied as a result of the need for the balanced operating conditions. In fact, to guarantee the balanced operation one has

$$u_{ds}^r = 0 \quad \text{and} \quad u_{0s}^r = 0.$$

That is, nonlinear linearizing feedback

$$u_{ds}^{lin} = -\left(L_{ls} + \frac{3}{2}\bar{L}_m \right) i_{qs}^r \omega_r$$

cannot be utilized. Hence, the feedback linearizing controllers cannot be used to control synchronous motors as well as the majority of (if not all) other electric machines, actuators, and systems. We report sound methods to solve the motion control problem, methods that do not entail the applied voltages to the saturation limits to cancel beneficial nonlinearities, and methods that do not lead to unbalanced motor operation resulting in completely unacceptable performance. □

8.7 Constrained Control of Nonlinear Electromechanical Systems

In general, electromechanical systems are modeled by nonlinear differential equations and saturation must be examined. Our goal is to minimize the functional

$$J = \int_{t_0}^{t_f} W_{xu}(x, u) \, dt, \tag{8.55}$$

subject to the system dynamics described by nonlinear differential equations

$$x(t) = F(x) + B(x)u, \quad u_{min} \leq u \leq u_{max}, \; u_{min} < 0, \; u_{max} > 0, \; x(t_0) = x_0. \tag{8.56}$$

The positive-definite, continuous, and differentiable integrand function

$$W_{xu}(\cdot): \mathbb{R}^n \times \mathbb{R}^m \to \mathbb{R}_{\geq 0}$$

is used. In (8.56),

$$F(\cdot):\mathbb{R}_{\geq 0} \times \mathbb{R}^n \to \mathbb{R}^n \quad \text{and} \quad B(\cdot):\mathbb{R}_{\geq 0} \times \mathbb{R}^n \to \mathbb{R}^{n \times m}$$

are continuous and Lipschitz.

The Hamiltonian is given by

$$H\left(x, u, \frac{\partial V}{\partial x}\right) = W_{xu}(x, u) + \left(\frac{\partial V}{\partial x}\right)^T (F(x) + B(x)u). \tag{8.57}$$

The first- and second-order necessary conditions for optimality are given by (8.5) and (8.8). Using the first-order necessary condition for optimality, one derives the control function

$$u(\cdot):[t_0, t_f) \to \mathbb{R}^m,$$

which minimizes functional (8.55). Constrained optimization of electromechanical systems is a topic of great practical interest. We consider the systems modeled by nonlinear differential equations

$$\dot{x}^{sys}(t) = F_s(x^{sys}) + B_s(x^{sys})u^{2w+1}, \quad y = H(x^{sys}), \quad u_{min} \leq u \leq u_{max},$$

$$x^{sys}(t_0) = x_0^{sys}, \tag{8.58}$$

where $x^{sys} \in X_s$ is the state vector; $u \in U$ is the vector of control inputs; $y \in Y$ is the measured output; w is the nonnegative integer.

Using the Hamilton-Jacobi theory, the bounded control laws can be synthesized for continuous-time systems (8.58). To design the tracking control law, we integrate the system and *exogenous* dynamics, for example, we have

$$\dot{x}^{sys}(t) = F_s(x^{sys}) + B_s(x^{sys})u^{2w+1}, \quad y = H(x^{sys}), \quad u_{min} \leq u \leq u_{max},$$

$$x^{sys}(t_0) = x_0^{sys}, \quad \dot{x}^{ref}(t) = Nr - y = Nr - H(x^{sys}). \tag{8.59}$$

Using the state vector

$$x = \begin{bmatrix} x^{sys} \\ x^{ref} \end{bmatrix} \in X,$$

from (8.59), one obtains

$$\dot{x}(t) = F(x, r) + B(x)u^{2w+1}, \quad u_{min} \leq u \leq u_{max}, \quad x(t_0) = x_0,$$

$$F(x, r) = \begin{bmatrix} F_s(x^{sys}) \\ -H(x^{sys}) \end{bmatrix} + \begin{bmatrix} 0 \\ N \end{bmatrix} r, \quad B(x) = \begin{bmatrix} B_s(x^{sys}) \\ 0 \end{bmatrix}. \tag{8.60}$$

We describe the control bounds by a bounded, integrable, one-to-one, globally Lipschitz, vector-valued continuous function ϕ. Our goal is to analytically design the bounded admissible state-feedback control law in the form $u = \phi(x)$. As ϕ one may apply the algebraic and transcendental (exponential, hyperbolic, logarithmic, trigonometric) continuously differentiable, integrable, one-to-one functions. For example, the odd one-to-one integrable function tanh with domain $(-\infty, +\infty)$ describes the control bounds. This function has the corresponding inverse function $\tanh^{-1}$ with range $(-\infty, +\infty)$.

The performance functional to be minimized is

$$J = \int_{t_0}^{\infty} [W_x(x) + W_u(u)]dt = \int_{t_0}^{\infty} \left[W_x(x) + (2w+1) \int (\phi^{-1}(u))^T G^{-1}\mathrm{diag}(u^{2w})du \right] dt,$$

(8.61)

where $G^{-1} \in \mathbb{R}^{m \times m}$ is the positive-definite diagonal matrix, $G^{-1} > 0$.

Performance integrands W_x and W_u are real-valued, positive-definite, and continuously differentiable integrand functions. Using the properties of ϕ one concludes that inverse function ϕ^{-1} is integrable. Hence, the integral

$$\int (\phi^{-1}(u))^T G^{-1}\mathrm{diag}(u^{2w})du$$

exists.

Example 8.9

Synthesize the performance functional to design a bounded control law for the following system

$$\frac{dx}{dt} = ax + bu^3, \quad u_{\min} \leq u \leq u_{\max}.$$

Using the performance integrand

$$W_u(u) = (2w+1) \int (\phi^{-1}(u))^T G^{-1}\mathrm{diag}(u^{2w})du,$$

and applying the integrable tanh function, one has the following positive-definite integrand

$$W_u(u) = 3 \int \tanh^{-1} u G^{-1} u^2 du.$$

For $G^{-1} = 1/3$, we have

$$W_u(u) = 3 \int \tanh^{-1} u G^{-1} u^2 du = \frac{1}{3} u^3 \tanh^{-1} u + \frac{1}{6} u^2 + \frac{1}{6} \ln(1 - u^2).$$

The hyperbolic tangent can used to describe the saturation effect. In general, one has

$$W_u(u) = (2w+1)\int u^{2w} \tanh^{-1}\frac{u}{k}\,du = u^{2w+1}\tanh^{-1}\frac{u}{k} - k\int \frac{u^{2w+1}}{k^2-u^2}\,du.$$

$\square$

First- and second-order necessary conditions for optimality (8.5) and (8.8) that the control guarantees a minimum to the Hamiltonian

$$H = W_x(x) + (2w+1)\int (\phi^{-1}(u))^T G^{-1}\mathrm{diag}(u^{2w})\,du + \frac{\partial V(x)}{\partial x}^T \left[F(x,r) + B(x)u^{2w+1} \right]$$

(8.62)

are

$$\frac{\partial H}{\partial u} = 0 \quad \text{and} \quad \frac{\partial^2 H}{\partial u \times \partial u^T} > 0.$$

The positive-definite return function is

$$V(x_0) = \inf_{u\in U} J(x_0, u) = \inf J(x_0, \phi(\cdot)) \geq 0.$$

The function

$$V(x),\ V(\cdot):\mathbb{R}^c \to \mathbb{R}_{\geq 0}$$

is found using the Hamilton-Jacobi-Bellman equation

$$-\frac{\partial V}{\partial t} = \min_{u\in U}\left\{ W_x(x) + (2w+1)\int (\phi^{-1}(u))^T G^{-1}\mathrm{diag}(u^{2w})\,du \right.$$

$$\left. + \frac{\partial V(x)}{\partial x}^T \left[F(x,r) + B(x)u^{2w+1} \right] \right\}.$$

(8.63)

The control law is derived by minimizing nonquadratic functional (8.61). The first-order necessary condition (8.5) leads us to a bounded control law

$$u = -\phi\left(GB^T(x)\frac{\partial V(x)}{\partial x} \right), \quad u\in U.$$

(8.64)

The second-order necessary condition for optimality is met because the matrix G^{-1} is positive-definite. Hence, a unique, bounded, real-analytic and continuous control law is designed. The solution of the functional equation should be found using nonquadratic return functions. To obtain $V(x)$, the performance functional is evaluated at the allowed values of the states and control. Linear and nonlinear functionals admit the final values. The minimum value of the nonquadratic functional (8.61) is given in a power-series form

$$J_{min} = \sum_{i=1}^{\eta} v(x_0)^{2i}$$

where η is the integer, $\eta = 1,2,3,....$, The solution of the partial differential equation (8.63) with (8.64) is satisfied by a continuously differentiable positive-definite return function

$$V(x) = \sum_{i=1}^{\eta} \frac{1}{2i}(x^i)^T K_i x^i, \quad K_i \in \mathbb{R}^{c \times c}, \tag{8.65}$$

where matrices K_i are found by solving the Hamilton-Jacobi equation.

From (8.64) and (8.65), the nonlinear bounded control law is given as

$$u = -\phi\left(GB^T(x) \sum_{i=1}^{\eta} \text{diag}[x^{i-1}]K_i x^i \right), \text{diag}[x^{i-1}] = \begin{bmatrix} x_1^{i-1} & 0 & \cdots & 0 & 0 \\ 0 & x_2^{i-1} & \cdots & 0 & 0 \\ \vdots & \vdots & \ddots & \vdots & \vdots \\ 0 & 0 & \cdots & x_{c-1}^{i-1} & 0 \\ 0 & 0 & \cdots & 0 & x_c^{i-1} \end{bmatrix}.$$

$$\tag{8.66}$$

The constraints on u are a result of the hardware (solid-state devices, ICs, electromechanical motion devices, etc.) limits. Those constraints are integrated in the control law design to ensure the optimal dynamics in order to meet the *achievable* system performance ability. One does not need to "implement" those bounds by means of additional software or hardware. The control law to be implemented by the analog or digital controller is

$$GB^T(x)\frac{\partial V(x)}{\partial x} \quad \text{or} \quad GB^T(x)\sum_{i=1}^{\eta} \text{diag}[x^{i-1}]K_i x^i.$$

That is, the nonlinearity ϕ is the existing inherent constraint. However, the control law design complies with the system nonlinearities, and u is found as a nonlinear function of x and e.

Example 8.10

Consider a dynamic system with constraints on control. In particular,

$$x(t) = ax + bu, \quad -1 \leq u \leq 1, \quad u \in U.$$

Using the hyperbolic tangent function to describe the saturation on u, the performance functional (8.61) is

$$J = \int_0^{\infty} \left(qx^2 + g\int \tanh^{-1} u du \right) dt, \quad q \geq 0, \quad g > 0.$$

The Hamiltonian function (8.62) is

$$H = qx^2 + g \int \tanh^{-1} u \, du + \frac{\partial V}{\partial x}(ax + bu).$$

The first-order necessary condition for optimality (8.5) results in

$$g \tanh^{-1} u + \frac{\partial V}{\partial x} b = 0.$$

Hence, the bounded control law is

$$u = -\tanh\left(g^{-1}bkx \frac{\partial V}{\partial x}\right), \quad u \in U.$$

The solution of the Hamilton-Jacobi-Bellman equation (8.63) is found using (8.65). Approximating the solution by the quadratic return function

$$V(x) = \frac{1}{2} kx^2,$$

the bounded control law is found to be

$$u = -\tanh(g^{-1}bkx), \quad k > 0.$$

However, the solution of the Hamilton-Jacobi-Bellman partial differential equation (8.63) should be approximated in $x \in X$ and $u \in U$. From (8.65), using

$$V(x) = \frac{1}{2} k_1 x^2 + \frac{1}{4} k_2 x^4 + \frac{1}{6} k_3 x^6,$$

one obtains

$$u = -\tanh[g^{-1}bkx(k_1 x + k_2 x^3 + k_3 x^5)].$$

$\square$

8.8 Optimization of Systems Using Nonquadratic Performance Functionals

The Hamilton-Jacobi theory, maximum principle, dynamic programming, and Lyapunov concept provide the designer with a general setup to solve linear and nonlinear optimal control problems for electromechanical systems. The general results can be derived using different performance functionals. In particular, quadratic and nonquadratic integrands have been applied. These functionals lead to solution of optimization problems. The importance of synthesis of performance functionals lay on the matter that the control laws are predefined by the functionals used. Furthermore, the closed-loop system performance (stability, robustness, settling time, overshoot, evolution of the state, output and control variables) is defined by the performance integrands applied.

The closed-loop system performance is optimal (with respect to the minimizing functional), and stability margins are assigned in the specific sense as implied by the performance functionals. The performance integrands, which allow one to measure system performance as well as design bounded control laws, were reported in Section 8.7. The system optimality and performance depend to a large extent on the specifications imposed (desired steady-state and dynamic performance, such as settling and rise time, accuracy, steady-state error, overshoot, bandwidth, etc.) and the inherent system capabilities, which include state bounds, control constraints, and so on.

As documented in Section 8.7, to design the *admissible* (bounded) control laws, we minimize

$$J = \int_{t_0}^{t_f} \left[\frac{1}{2} x^T Q x + \int (\phi^{-1}(u))^T G^{-1} du \right] dt$$

or similar functionals using the bounded, integrable, one-to-one, real-analytic globally Lipschitz continuous function

$$\phi(\cdot) : \mathbb{R}^n \to \mathbb{R}^m, \ \phi \in U \subset \mathbb{R}^m.$$

For linear (8.9) and nonlinear (8.3) systems with $u_{min} \leq u \leq u_{max}$, $u \in U$, the minimization of the nonquadratic functional

$$J = \int_{t_0}^{t_f} \left[\frac{1}{2} x^T Q x + \int (\phi^{-1}(u))^T G du \right] dt \qquad (8.67)$$

gives

$$-\frac{\partial V}{\partial t} = \min_{u \in U} \left\{ \frac{1}{2} x^T Q x + \int (\phi^{-1}(u))^T G du + \frac{\partial V}{\partial x}^T (Ax + Bu) \right\} \quad \text{for} \quad \dot{x}(t) = Ax + Bu,$$

and

$$-\frac{\partial V}{\partial t} = \min_{u \in U} \left\{ \frac{1}{2} x^T Q x + \int (\phi^{-1}(u))^T G du + \frac{\partial V}{\partial x}^T [F(x) + B(x)u] \right\} \text{for} \ \dot{x}(t) = F(x) + B(x)u.$$

Using the first-order necessary condition for optimality (8.5), the *admissible* control laws are

$$u = -\phi \left(G^{-1} B^T \frac{\partial V(x)}{\partial x} \right), \quad u \in U, \qquad (8.68)$$

and

$$u = -\phi\left(G^{-1}B^T(x)\frac{\partial V(x)}{\partial x}\right), \quad u \in U. \tag{8.69}$$

As documented, for linear and nonlinear systems, bounded control laws (8.68) and (8.69) can be straightforwardly designed. We further concentrate on the synthesis of performance functionals and design of control laws. Consider electromechanical systems modeled by linear or nonlinear differential equations. We apply the following performance functional

$$J = \int_{t_0}^{t_f} \frac{1}{2}\left[\omega(x)^T Q\omega(x) + \dot\omega(x)^T P\dot\omega(x)\right]dt, \quad Q \geq 0, \quad P > 0, \tag{8.70}$$

where $\omega(\cdot):\mathbb{R}^n \to \mathbb{R}_{\geq 0}$ is the differentiable real-analytic continuous function; $Q \in \mathbb{R}^{n \times n}$ and $P \in \mathbb{R}^{n \times n}$ are the positive-definite diagonal weighting matrices.

Using (8.70), the system transient performance and stability are specified by two integrands, in particular,

$$\omega(x)^T Q\omega(x) \text{ and } \dot\omega(x)^T P\dot\omega(x).$$

These integrands are given as the nonlinear functions of the states and the rate of the variables changes. The performance functional (8.70) depends on the system dynamics (states and control variables), control efforts, energy, and so on. For linear systems described by (8.9), we have

$$\dot\omega(x) = \frac{\partial \omega}{\partial x}\dot{x} = \frac{\partial \omega}{\partial x}(Ax + Bu).$$

Using (8.70), one has the following functional

$$
\begin{aligned}
J &= \int_{t_0}^{t_f} \frac{1}{2}\left[\omega(x)^T Q\omega(x) + \dot{x}^T \frac{\partial \omega}{\partial x}^T P \frac{\partial \omega}{\partial x}\dot{x}\right]dt \\
&= \int_{t_0}^{t_f} \frac{1}{2}\left[\omega(x)^T Q\omega(x) + (Ax + Bu)^T \frac{\partial \omega}{\partial x}^T P \frac{\partial \omega}{\partial x}(Ax + Bu)\right]dt. \tag{8.71}
\end{aligned}
$$

In (8.71), the $\omega(x)$ is the differentiable and integrable real-valued continuous function. For example, one may use

- $\omega(x) = x$, which leads to the quadratic integrand functions;
- $\omega(x) = x^3$, $\omega(x) = \tanh(x)$, or $\omega(x) = e^{-x}$, which result in nonquadratic functionals.

For linear systems (8.9) and performance functional (8.70), the Hamiltonian function is

$$H\left(x, u, \frac{\partial V}{\partial x}\right) = \frac{1}{2}\omega(x)^T Q\omega(x) + \frac{1}{2}\dot\omega(x)^T P\dot\omega(x) + \frac{\partial V}{\partial x}^T (Ax + Bu). \quad (8.72)$$

The application of the first-order condition for optimality (8.5) gives the following *optimal* control law

$$u = -\left(B^T \frac{\partial\omega}{\partial x}^T P \frac{\partial\omega}{\partial x} B\right)^{-1} B^T\left(\frac{\partial\omega}{\partial x}^T P \frac{\partial\omega}{\partial x} Ax + \frac{\partial V}{\partial x}\right). \quad (8.73)$$

The second-order condition for optimality (8.8) is guaranteed. In particular, from (8.72), one has

$$\frac{\partial^2 H}{\partial u \times \partial u^T} = B^T \frac{\partial\omega^T}{\partial x} P \frac{\partial\omega}{\partial x} B > 0$$

because $\omega(x)$ is chosen such that

$$\frac{\partial\omega}{\partial x} B$$

has a full rank, and $P > 0$.

The synthesis of the performance integrands

$$\omega(x)^T Q\omega(x) \quad \text{and} \quad \dot\omega(x)^T P\dot\omega(x)$$

results in devising and utilizing integrable and differentiable function $\omega(x)$. For example, applying $\omega(x) = x$, from (8.70) we have

$$J = \int_{t_0}^{t_f} \frac{1}{2}[x^T Qx + (Ax + Bu)^T P(Ax + Bu)]dt. \quad (8.74)$$

Using (8.74), one obtains the functional equation

$$-\frac{\partial V}{\partial t} = \min_u \left\{\frac{1}{2}[x^T Qx + (Ax + Bu)^T P(Ax + Bu)] + \frac{\partial V^T}{\partial x}(Ax + Bu)\right\}. \quad (8.75)$$

The control law is found by using the first-order necessary condition for optimality (8.5). From

$$H\left(x, u, \frac{\partial V}{\partial x}\right) = \frac{1}{2}[x^T Qx + (Ax + Bu)^T P(Ax + Bu)] + \frac{\partial V^T}{\partial x}(Ax + Bu), \quad (8.76)$$

using

$$\frac{\partial H}{\partial u} = 0,$$

one has

$$u = -(B^T PB)^{-1} B^T \left(PAx + \frac{\partial V}{\partial x} \right).$$ (8.77)

The solution of the functional equation (8.75) is given by the quadratic return function (8.15)

$$V = \frac{1}{2} x^T K x.$$

From (8.77), we have the following linear control law

$$u = -(B^T PB)^{-1} B^T (PA + K)x.$$ (8.78)

Using (8.75) and (8.78), the expression for the unknown symmetric matrix $K \in \mathbb{R}^{n \times n}$ is found. In particular, the symmetric K is obtained by solving the following nonlinear differential equation

$$-\dot{K} = Q - KB(B^T PB)^{-1} B^T K, \quad K(t_f) = K_f.$$ (8.79)

The control law designed different compared with the conventional linear control law (8.19). Furthermore, the equations to compute matrix K are different; see equations (8.18) and (8.79). The conventional and reported results are compared in Table 8.1.

TABLE 8.1

Performance Functional and Control Laws Comparison for Linear Systems $x(t) = Ax + Bu$

Performance Functionals	Quadratic (8.10): $J = \int_{t_0}^{t_f} \frac{1}{2}(x^T Qx + u^T Gu)dt$
	Quadratic generalized (8.70): $J = \int_{t_0}^{t_f} \frac{1}{2}[\omega(x)^T Q\omega(x) + \dot{\omega}(x)^T P\dot{\omega}(x)]dt$
Controls	Linear (8.19): $u = -G^{-1} B^T Kx$
	Nonlinear (8.73): $u = -\left(B^T \frac{\partial \omega}{\partial x}^T P \frac{\partial \omega}{\partial x} B \right)^{-1} B^T \left(\frac{\partial \omega}{\partial x}^T P \frac{\partial \omega}{\partial x} Ax + \frac{\partial V}{\partial x} \right)$
	or linear (8.78): $u = -(B^T PB)^{-1} B^T (PA + K)x$ for $\omega(x) = x$
Riccati Equations	Quadratic (8.18): $-K = Q + A^T K + K^T A - K^T BG^{-1} B^T K$
	Quadratic generalized (8.79): $-\dot{K} = Q - KB(B^T PB)^{-1} B^T K$ for $\omega(x) = x$

We study nonlinear system dynamic as given by (8.3). The performance functional is

$$J = \int_{t_0}^{t_f} \frac{1}{2}\left[\omega(x)^T Q\omega(x) + [F(x)+B(x)u]^T \frac{\partial \omega}{\partial x}^T P \frac{\partial \omega}{\partial x}[F(x)+B(x)u] \right] dt. \quad (8.80)$$

For system (8.3) and performance functional (8.80), we have the following Hamiltonian function

$$H\left(x, u, \frac{\partial V}{\partial x}\right) = \frac{1}{2}\omega(x)^T Q\omega(x)$$

$$+ \frac{1}{2}[F(x)+B(x)u]^T \frac{\partial \omega}{\partial x}^T P \frac{\partial \omega}{\partial x}[F(x)+B(x)u]$$

$$+ \frac{\partial V}{\partial x}^T [F(x)+B(x)u]. \quad (8.81)$$

The positive-definite return function $V(\cdot):\mathbb{R}^n \to \mathbb{R}_{\geq 0}$ satisfies the following differential equation

$$-\frac{\partial V}{\partial t} = \min_u \left\{ \frac{1}{2}\omega(x)^T Q\omega(x) + \frac{1}{2}[F(x)+B(x)u]^T \frac{\partial \omega}{\partial x}^T P \frac{\partial \omega}{\partial x}[F(x)+B(x)u] \right.$$

$$\left. + \frac{\partial V}{\partial x}^T [F(x)+B(x)u] \right\}. \quad (8.82)$$

From (8.82), using (8.5), one finds the control law as

$$u = -\left(B^T(x)\frac{\partial \omega}{\partial x}^T P \frac{\partial \omega}{\partial x}B(x) \right)^{-1} B^T(x)\left(\frac{\partial \omega}{\partial x}^T P \frac{\partial \omega}{\partial x}F(x) + \frac{\partial V}{\partial x} \right). \quad (8.83)$$

Control law (8.83) is an *optimal* control, and the second-order necessary condition for optimality is met because

$$\frac{\partial^2 H}{\partial u \times \partial u^T} = B^T(x)\frac{\partial \omega}{\partial x}^T P \frac{\partial \omega}{\partial x}B(x) > 0.$$

The solution of the partial differential equation (8.82) is approximated by the return function

$$V(x) = \sum_{i=1}^{\eta} \frac{1}{2i} (x^i)^T K_i x^i, \qquad K_i \in \mathbb{R}^{n \times n}. \tag{8.84}$$

Example 8.11

For the first-order system studied in Example 8.1

$$\frac{dx}{dt} = ax + bu,$$

the generalized quadratic performance functional is synthesized assigning $\omega(x) = x$. From (8.71), we have

$$J = \int_{t_0}^{\infty} \frac{1}{2} \left[Q\omega(x)^2 + P \left(\frac{\partial \omega}{\partial x} \right)^2 (a^2 x^2 + 2abxu + b^2 u^2) \right] dt$$

$$= \int_{t_0}^{\infty} \frac{1}{2} (x^2 + a^2 x^2 + 2abxu + b^2 u^2) dt,$$

where we let $Q = 1$ and $P = 1$.

Using the quadratic return function

$$V = \frac{1}{2} kx^2,$$

the linear control law (8.78) is

$$u = -\frac{1}{b}(a + k)x.$$

Solving the differential equation (8.79)

$$-k = 1 - k^2,$$

we obtain $k = 1$.

Thus, the control law is

$$u = -\frac{1}{b}(a + 1)x.$$

The closed-loop system is stable and evolves as

$$\frac{dx}{dt} = -x.$$

$\square$

Example 8.12
For the system

$$\frac{dx}{dt} = ax + bu,$$

minimizing the performance functional (8.71) one finds

$$J = \int_{t_0}^{\infty} \frac{1}{2} \left[Q\omega(x)^2 + P\left(\frac{\partial \omega}{\partial x}\right)^2 (a^2 x^2 + 2abxu + b^2 u^2) \right] dt.$$

The nonquadratic integrands are designed using $\omega(x) = \tanh(x)$. Let $Q = 1$ and $P = 1$. One finds the nonquadratic functional

$$J = \int_{t_0}^{\infty} \frac{1}{2} [\tanh^2 x + \text{sech}^4 x(a^2 x^2 + 2abxu + b^2 u^2)] dt.$$

This performance functional with the synthesized integrands can be straightforwardly examined. For $x \ll 1$, $\tanh^2 x \approx x^2$ and $\text{sech}^4 x \approx 1$. That is, if $x \ll 1$, the performance functional can be expressed as

$$J \approx \int_{t_0}^{\infty} \frac{1}{2} [x^2 + a^2 x^2 + 2abxu + b^2 u^2] dt.$$

This generalized quadratic-like functional was used in Example 8.11 when we let $\omega(x) = x$. However, this conclusion is accurate for $x \ll 1$.

If $x \gg 1$, we have $\tanh^2 x \approx 1$ and $\text{sech}^4 x \approx 0$. Hence, for $x \gg 1$, the performance functional is

$$J \approx \frac{1}{2} \int_{t_0}^{\infty} dt$$

This performance functional is commonly used to solve the time-optimal (minimum time) problem as reported in Section 8.5.

The first-order necessary condition for optimality (8.5) is applied. One finds a control law (8.73) as

$$u = -\frac{a}{b} x - \frac{1}{b \, \text{sech}^4 x} \frac{\partial V}{\partial x}.$$

The functional equation to be solved is

$$-\frac{\partial V}{\partial t} = \frac{1}{2} \tanh^2 x - \frac{1}{2 \, \text{sech}^4 x} \frac{\partial^2 V}{\partial x^2}.$$

In $x \in X$, one approximates continuous functions $\tanh^2 x$ and $\text{sech}^4 x$. The quadratic and nonquadratic return functions are used to solve the Hamilton-Jacobi differential equation. Letting

$$V = \frac{1}{2} kx^2,$$

we have

$$u = -\frac{a}{b} x - \frac{1}{b\,\text{sech}^4 x} kx.$$

The closed-loop systems evolves as

$$\frac{dx}{dt} = -\frac{k}{\text{sech}^4 x} x.$$

If $x \ll 1$, $\text{sech}^4 x \approx 1$, and thus,

$$u \approx -\frac{a+k}{b} x.$$

The system dynamics is

$$\frac{dx}{dt} = -kx.$$

For $x \gg 1$, we have the nonlinear control law which can be typified as a high feedback gain control.

Solving

$$-\frac{\partial V}{\partial t} = \frac{1}{2} \tanh^2 x - \frac{1}{2\,\text{sech}^4 x} \frac{\partial^2 V}{\partial x^2}, \quad V = \frac{1}{2} kx^2$$

we found $k = 1$.

Although the quadratic return function

$$V = \frac{1}{2} kx^2$$

may approximate the solution of nonlinear functional partial differential equation in the specified $x \in X$, in general, nonquadratic return functions must be used. Letting

$$V = \frac{1}{2} k_1 x^2 + \frac{1}{4} k_2 x^4,$$

one has

$$u = -\frac{a}{b} x - \frac{1}{b\,\text{sech}^4 x} (k_1 x + k_2 x^3).$$

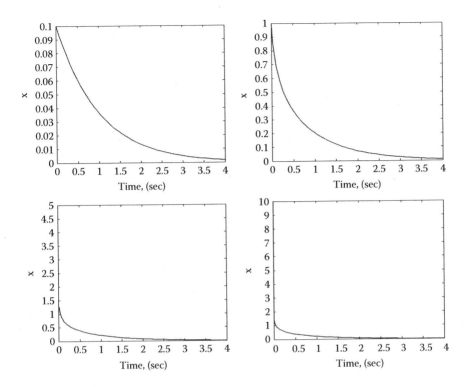

FIGURE 8.6
Transient dynamics of the closed-loop system.

The solution of

$$-\frac{\partial V}{\partial t} = \frac{1}{2}\tanh^2 x - \frac{1}{2\operatorname{sech}^4 x}\frac{\partial^2 V}{\partial x^2}$$

gives $k_1 = 1$ and $k_2 = 0.5$. The transient dynamics of the closed-loop system for four different initial conditions x_0 ($x_0 = 0.1$, $x_0 = 1$, $x_0 = 5$, and $x_0 = 10$) are reported in Figure 8.6. The analysis indicates that the settling time is almost the same for any initial conditions. This is due to the fact that a novel nonlinear optimal control law that possesses unique features, as discussed earlier, is designed. However, the control bounds will increase the settling time. ☐

Example 8.13
The differential equations that describe a rigid-body mechanical systems are

$$\frac{dx_1}{dt} = ax_1 + bu, \quad \frac{dx_2}{dt} = x_1,$$

where x_1 and x_2 are the velocity and displacement.

The performance integrands are designed utilizing $\omega(x) = \tanh(x)$. Using the identity matrices $Q=I$ and $P=I$, functional (8.70) is expressed as

$$J = \int_{t_0}^{t_f} \frac{1}{2} \left\{ [\tanh x_1 \ \ \tanh x_2] \begin{bmatrix} 1 & 0 \\ 0 & 1 \end{bmatrix} \begin{bmatrix} \tanh x_1 \\ \tanh x_2 \end{bmatrix} + [\dot{x}_1 \text{sech}^2 x_1 \ \ \dot{x}_2 \text{sech}^2 x_2] \begin{bmatrix} 1 & 0 \\ 0 & 1 \end{bmatrix} \begin{bmatrix} \dot{x}_1 \text{sech}^2 x_1 \\ \dot{x}_2 \text{sech}^2 x_2 \end{bmatrix} \right\} dt$$

$$= \int_{t_0}^{t_f} \frac{1}{2} \left[\tanh^2 x_1 + \tanh^2 x_2 + \text{sech}^4 x_1 (a^2 x_1^2 + 2abx_1 u + b^2 u^2) + x_1^2 \text{sech}^4 x_2 \right] dt.$$

The first-order necessary condition for optimality (8.5) results in the following control law

$$u = -\frac{a}{b} x_1 - \frac{1}{b \text{sech}^4 x_1} \frac{\partial V}{\partial x_1}.$$

The closed-loop system is described as

$$\frac{dx_1}{dt} = -\frac{1}{\text{sech}^4 x_1} \frac{\partial V}{\partial x_1},$$

$$\frac{dx_2}{dt} = x_1.$$

One can examine the stability for this system. We use a positive-definite Lyapunov function

$$V_L = \frac{1}{2} x_1^2 + \frac{1}{2} x_2^2.$$

To derive an explicit expression for u, the quadratic return function

$$V = \frac{1}{2} k_{11} x_1^2 + k_{12} x_1 x_2 + \frac{1}{2} k_{22} x_2^2$$

is used. For the closed-loop system the total derivative is

$$\frac{dV_L}{dt} = -\frac{1}{\text{sech}^4 x_1} \left(\frac{\partial V}{\partial x_1} \right)^2 + \frac{\partial V}{\partial x_2} x_1 = -\frac{1}{\text{sech}^4 x_1} (k_{11} x_1 + k_{12} x_2)^2 + x_1 x_2.$$

Hence,

$$dV_L/dt$$

is negative-definite. It will be reported that the solution of the Hamilton-Jacobi equation results in $k_{11} = 1$ and $k_{12} = 0.25$. The total derivative of the Lyapunov function is illustrated in Figure 8.7. The following MATLAB statement is used to perform calculations and plotting

```
x=linspace(-1,1,25); y=x; [X,Y]=meshgrid(x,y); k11=1 ; k12=0.25 ;
dV=X.*Y-((k11*X+k12*Y).^2)./sech(X).^4; surf(x,y,dV);
xlabel('x _ 1','FontSize',14); ylabel('x _ 2','FontSize',14);
zlabel('dV _ L/dt','FontSize',14);
title('Lyapunov Function Total Derivative, dV _ L/ dt','FontSize',14);
```

Lyapunov Function Total Derivative, dV$_L$/ dt

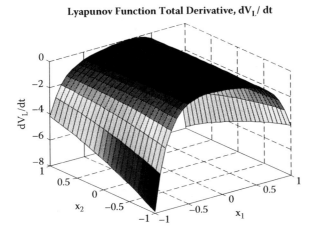

FIGURE 8.7
Total derivative of the Lyapunov function.

We solve the differential equation

$$-\frac{\partial V}{\partial t} = \frac{1}{2}\tanh^2 x_1 + \frac{1}{2}\tanh^2 x_2 - \frac{1}{2\text{sech}^4 x_1}\left(\frac{\partial V}{\partial x_1}\right)^2 + \frac{1}{2}x_1^2\text{sech}^4 x_2 + \frac{\partial V}{\partial x_2}x_1$$

by approximating its solution by the nonquadratic return function

$$V = \frac{1}{2}k_{11}x_1^2 + k_{12}x_1x_2 + \frac{1}{2}k_{22}x_2^2 + \frac{1}{4}k_{41}x_1^4 + \frac{1}{4}k_{42}x_2^4.$$

Having found the coefficients of $V(x)$, we have

$$u = -\frac{a}{b}x_1 - \frac{1}{b\text{sech}^4 x_1}\left(x_1 + 0.25x_2 + 0.5x_1^3\right).$$

The Simulink model is reported in Figure 8.8. Although there are no a and b coefficients in the resulting closed-loop system with the control law designed, we used these constant letting $a = 1$ and $b = 1$. The transient dynamics of the closed-loop system for the initial conditions

$$\begin{bmatrix} x_{10} \\ x_{20} \end{bmatrix} = \begin{bmatrix} 0.1 \\ -0.1 \end{bmatrix} \quad \text{and} \quad \begin{bmatrix} x_{10} \\ x_{20} \end{bmatrix} = \begin{bmatrix} 1 \\ -1 \end{bmatrix}$$

are reported in Figure 8.9. The plotting statement is

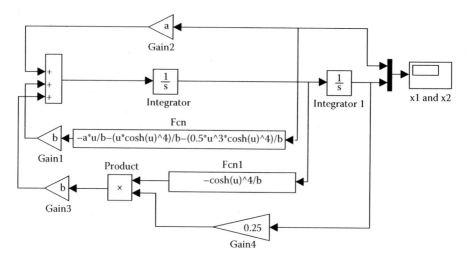

FIGURE 8.8
Simulink diagram to simulate the closed-loop system (ch8 _ 01.mdl).

```
plot(x(:,1),x(:,2),x(:,1),x(:,3));
xlabel('Time (seconds)','FontSize',14);
title('Transient Dynamics, x _ 1 and x _ 2','FontSize',14);
```

We conclude that the control law designed ensures optimal system evolution with respect to the functional minimized.   □

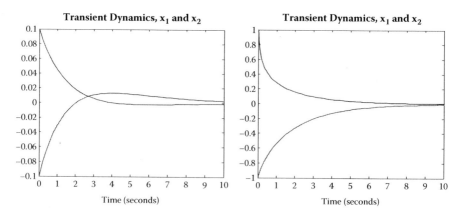

FIGURE 8.9
Transient dynamics in the closed-loop system.

8.9 Lyapunov Stability Theory in Analysis and Control of Electromechanical Systems

The electromechanical systems dynamics is described by nonlinear state-space differential equations, whereas the output equation is $y = H(x)$. The control bounds are $u_{min} \le u \le u_{max}$.

We examine stability of time-varying nonlinear dynamic systems described by

$$\dot{x}(t) = F(t, x), \quad x(t_0) = x_0, \quad t \ge 0. \tag{8.85}$$

The following theorem is formulated applying the results of the Lyapunov stability theory.

THEOREM 8.1

Consider the system described by nonlinear differential equations (8.85). If there exists a positive-definite scalar function

$$V(t,x), \quad V(\cdot):\mathbb{R}_{\ge 0} \times \mathbb{R}^n \to \mathbb{R}_{\ge 0},$$

called the Lyapunov function, with continuous first-order partial derivatives with respect to t and x

$$\frac{dV}{dt} = \frac{\partial V}{\partial t} + \left(\frac{\partial V}{\partial x}\right)^T \frac{dx}{dt} = \frac{\partial V}{\partial t} + \left(\frac{\partial V}{\partial x}\right)^T F(t, x),$$

then

- The equilibrium state of (8.85) is stable if the total derivative of the positive-definite function $V(t,x) > 0$ is

$$\frac{dV(t, x)}{dt} \le 0;$$

- The equilibrium state of (8.85) is uniformly stable if the total derivative of the positive-definite decreasing function $V(t,x) > 0$ is

$$\frac{dV(t, x)}{dt} \le 0;$$

- The equilibrium state of (8.85) is uniformly asymptotically stable in the large if the total derivative of $V(t,x) > 0$ is negative definite; that is,

$$\frac{dV(t, x)}{dt} < 0;$$

- The equilibrium state of (8.85) is exponentially stable in the large if the exist the K_∞-functions $\rho_1(\cdot)$ and $\rho_2(\cdot)$, and K-function $\rho_3(\cdot)$ such that

$$\rho_1(\|x\|) \le V(t, x) \le \rho_2(\|x\|) \quad \text{and} \quad \frac{dV(x)}{dt} \le -\rho_3(\|x\|).$$

□

Example 8.14

Consider a system that is described by two nonlinear time-invariant differential equations

$$\dot{x}_1(t) = x_1 - x_1^5 - x_1^3 x_2^4,$$

$$\dot{x}_2(t) = x_2 - x_2^9, \quad t \ge 0.$$

These differential equations describe the uncontrolled or controlled dynamics. For example, consider

$$\dot{x}_1(t) = x_1 + u_1, \quad \dot{x}_2(t) = x_2 + u_2,$$

where

$$u_1 = -x_1^5 - x_1^3 x_2^4 \quad \text{and} \quad u_2 = -x_2^9.$$

A scalar positive-definite function is expressed in the quadratic form as

$$V(x_1, x_2) = \frac{1}{2}\left(x_1^2 + x_2^2\right).$$

The total derivative is found to be

$$\frac{dV(x_1, x_2)}{dt} = \left(\frac{\partial V}{\partial x}\right)^T \frac{dx}{dt} = \left(\frac{\partial V}{\partial x}\right)^T F(x) = \frac{\partial V}{\partial x_1}\left(x_1 - x_1^5 - x_1^3 x_2^4\right)$$

$$+ \frac{\partial V}{\partial x_2}\left(x_2 - x_2^9\right) = x_1^2 - x_1^6 - x_1^4 x_2^4 + x_2^2 - x_2^{10}$$

The total derivative of a positive-definite $V(x_1, x_2) > 0$ is

$$\frac{dV(x_1, x_2)}{dt} < 0.$$

Hence, dV/dt is negative definite. Therefore, the equilibrium state of the system is uniformly asymptotically stable.

□

Example 8.15

Consider time-varying nonlinear differential equations

$$\dot{x}_1(t) = -x_1 + x_2^3,$$

$$\dot{x}_2(t) = -e^{-10t} x_1 x_2^2 - 5x_2 - x_2^3, \quad t \ge 0.$$

A scalar positive-definite function $V(t,x_1,x_2) > 0$ is chosen in the quadratic form as

$$V(t, x_1, x_2) = \frac{1}{2}\left(x_1^2 + e^{10t} x_2^2\right).$$

The total derivative is

$$\frac{dV(t, x_1, x_2)}{dt} = \frac{\partial V}{\partial t} + \frac{\partial V}{\partial x_1}\left(-x_1 + x_2^3\right) + \frac{\partial V}{\partial x_2}\left(-e^{-10t} x_1 x_2^2 - 5x_2 - x_2^3\right) = -x_1^2 - e^{10t} x_2^4.$$

The total derivative is negative definite,

$$\frac{dV(x_1, x_2)}{dt} < 0.$$

Using Theorem 8.1, one concludes that the equilibrium state is uniformly asymptotically stable. □

Example 8.16

The system dynamics is described by the differential equations

$$\dot{x}_1(t) = -x_1 + x_2,$$

$$\dot{x}_2(t) = -x_1 - x_2 - x_2\left|x_2\right|, \quad t \geq 0.$$

The positive-definite scalar Lyapunov candidate is chosen in the following form

$$V(x_1, x_2) = \frac{1}{2}\left(x_1^2 + x_2^2\right).$$

Thus, $V(x_1,x_2) > 0$. The total derivative is

$$\frac{dV(x_1, x_2)}{dt} = x_1\dot{x}_1 + x_2\dot{x}_2 = -x_1^2 - x_2^2(1 + |x_2|).$$

Therefore,

$$\frac{dV(x_1, x_2)}{dt} < 0.$$

Hence, the equilibrium state of the system is uniformly asymptotically stable, and the quadratic function

$$V(x_1, x_2) = \frac{1}{2}\left(x_1^2 + x_2^2\right)$$

is the Lyapunov function. □

Example 8.17　Stability of Permanent-Magnet Synchronous Motors

Consider a permanent-magnet synchronous motor in the rotor reference frame, which was studied in Example 8.8. The mathematical model, assuming that $T_L = 0$, is

$$\frac{di_{qs}^r}{dt} = -\frac{r_s}{L_{ls} + \frac{3}{2}\bar{L}_m}i_{qs}^r - \frac{\psi_m}{L_{ls} + \frac{3}{2}\bar{L}_m}\omega_r - i_{ds}^r\omega_r + \frac{1}{L_{ls} + \frac{3}{2}\bar{L}_m}u_{qs}^r,$$

$$\frac{di_{ds}^r}{dt} = -\frac{r_s}{L_{ls} + \frac{3}{2}\bar{L}_m}i_{ds}^r + i_{qs}^r\omega_r + \frac{1}{L_{ls} + \frac{3}{2}\bar{L}_m}u_{ds}^r,$$

$$\frac{d\omega_r}{dt} = \frac{3P^2\psi_m}{8J}i_{qs}^r - \frac{B_m}{J}\omega_r.$$

For an open-loop system

$$u_{qs}^r = 0 \quad \text{and} \quad u_{ds}^r = 0,$$

whereas for a closed-loop system, one has

$$u_{qs}^r \neq 0 \quad \text{and} \quad u_{ds}^r = 0.$$

For example,

$$u_{qs}^r = -k_\omega \omega_r.$$

For an open-loop drive

$$u_{qs}^r = 0 \quad \text{and} \quad u_{ds}^r = 0.$$

Hence, we study

$$\frac{di_{qs}^r}{dt} = -\frac{r_s}{L_{ls} + \frac{3}{2}\bar{L}_m}i_{qs}^r - \frac{\psi_m}{L_{ls} + \frac{3}{2}\bar{L}_m}\omega_r - i_{ds}^r\omega_r,$$

$$\frac{di_{ds}^r}{dt} = -\frac{r_s}{L_{ls} + \frac{3}{2}\bar{L}_m}i_{ds}^r + i_{qs}^r\omega_r,$$

$$\frac{d\omega_r}{dt} = \frac{3P^2\psi_m}{8J}i_{qs}^r - \frac{B_m}{J}\omega_r.$$

Using the quadratic positive-definite Lyapunov function

$$V\left(i_{qs}^r, i_{ds}^r, \omega_r\right) = \frac{1}{2}\left(i_{qs}^{r\,2} + i_{ds}^{r\,2} + \omega_r^2\right),$$

the expression for the total derivative is

$$\frac{dV(i_{qs}^r, i_{ds}^r, \omega_r)}{dt} = -\frac{r_s}{L_{ss}}\left(i_{qs}^{r\,2} + i_{ds}^{r\,2}\right) - \frac{B_m}{J}\omega_r^2 - \frac{8J\psi_m - 3P^2 L_{ss}\psi_m}{8JL_{ss}}i_{qs}^r\omega_r.$$

Thus,

$$\frac{dV(i_{qs}^r, i_{ds}^r, \omega_r)}{dt} < 0.$$

One concludes that the equilibrium state of an open-loop drive is uniformly asymptotically stable.

Consider the closed-loop system. To guarantee the balanced operation, we define the control law assigning the qd voltage components to be

$$u_{qs}^r = -k_\omega \omega_r \quad \text{and} \quad u_{ds}^r = 0.$$

The following differential equations result

$$\frac{di_{qs}^r}{dt} = -\frac{r_s}{L_{ls} + \frac{3}{2}\bar{L}_m} i_{qs}^r - \frac{\psi_m}{L_{ls} + \frac{3}{2}\bar{L}_m} \omega_r - i_{ds}^r \omega_r - \frac{1}{L_{ls} + \frac{3}{2}\bar{L}_m} k_\omega \omega_r,$$

$$\frac{di_{ds}^r}{dt} = -\frac{r_s}{L_{ls} + \frac{3}{2}\bar{L}_m} i_{ds}^r + i_{qs}^r \omega_r,$$

$$\frac{d\omega_r}{dt} = \frac{3P^2\psi_m}{8J} i_{qs}^r - \frac{B_m}{J}\omega_r.$$

Using the quadratic positive-definite Lyapunov function

$$V\left(i_{qs}^r, i_{ds}^r, \omega_r\right) = \frac{1}{2}\left(i_{qs}^{r\,2} + i_{ds}^{r\,2} + \omega_r^2\right),$$

we obtain

$$\frac{dV\left(i_{qs}^r, i_{ds}^r, \omega_r\right)}{dt} = -\frac{r_s}{L_{ss}}\left(i_{qs}^{r\,2} + i_{ds}^{r\,2}\right) - \frac{B_m}{J}\omega_r^2 - \frac{8J(\psi_m + k_\omega) - 3P^2 L_{ss}\psi_m}{8JL_{ss}} i_{qs}^r \omega_r.$$

Hence,

$$V\left(i_{qs}^r, i_{ds}^r, \omega_r\right) > 0 \quad \text{and} \quad \frac{dV(i_{qs}^r, i_{ds}^r, \omega_r)}{dt} < 0.$$

Therefore, the conditions for asymptotic stability are guaranteed. The rate of decreasing of

$$\frac{dV(i_{qs}^r, i_{ds}^r, \omega_r)}{dt}$$

affects the drive dynamics. The derived expression for

$$\frac{dV(i_{qs}^r, i_{ds}^r, \omega_r)}{dt}$$

illustrates the role of the proportional feedback gain k_ω. □

It has been shown that dynamic systems can be controlled to attain the desired transient dynamics, stability margins, and so on. Let us study how to solve the motion control problem with the ultimate goal to synthesize tracking control laws applying Lyapunov's stability theory. Using the reference (command) vector $r(t)$ and the system output $y(t)$, the tracking error (which ideally approaches to zero) is $e(t) = Nr(t) - y(t)$.

The Lyapunov theory is applied to derive the *admissible* control laws. That is, the *admissible* bounded control law should be designed as a continuous function within the constrained control set

$$U = \{u \in \mathbb{R}^m : u_{\min} \le u \le u_{\max}, u_{\min} < 0, u_{\max} > 0\} \subset \mathbb{R}^m.$$

Unbounded and bounded control laws can be designed applying Lyapunov stability theory. The control laws affect the system dynamics and change the total derivative of the Lyapunov function $V(t,x,e)$. For $V(t,x,e) > 0$, one can derive u such that $dV(t,x,e)/dt<0$. The feedback gains are found by solving nonlinear matrix inequality assigning the negative value or rate for

$$dV(t,x,e)/dt.$$

THEOREM 8.2

Consider the closed-loop electromechanical systems (8.3) with a control law (8.1) under the references $r \in R$ and disturbances $d \in D$. For the closed-loop system (8.3)–(8.1),

1. Solutions are uniformly ultimately bounded;
2. Equilibrium is exponentially stable in the convex and compact state evolution set $X(X_0,U,R,D) \subset \mathbb{R}^n$;
3. Tracking is ensured and disturbance attenuation is guaranteed in the state-error evolution set $XE(X_0,E_0,U,R,D) \subset \mathbb{R}^n \times \mathbb{R}^b$,

if there exists a continuous differentiable function $V(t,x,e)$,

$$V(\cdot):\mathbb{R}_{\ge 0} \times \mathbb{R}^n \times \mathbb{R}^b \to \mathbb{R}_{\ge 0}$$

in XE such that for all $x \in X$, $e \in E$, $u \in U$, $r \in R$ and $d \in D$ on $[t_0, \infty)$

i. $\rho_1 \|x\| + \rho_2 \|e\| \le V(t, x, e) \le \rho_3 \|x\| + \rho_4 \|e\|,$

ii. $\dfrac{dV(t, x, e)}{dt} \le -\rho_5 \|x\| - \rho_6 \|e\|$ holds.

Here,

$$\rho_1(\cdot):\mathbb{R}_{\ge 0} \to \mathbb{R}_{\ge 0}, \quad \rho_2(\cdot):\mathbb{R}_{\ge 0} \to \mathbb{R}_{\ge 0}, \quad \rho_3(\cdot):\mathbb{R}_{\ge 0} \to \mathbb{R}_{\ge 0}, \quad \text{and} \quad \rho_4(\cdot):\mathbb{R}_{\ge 0} \to \mathbb{R}_{\ge 0}$$

are the K_∞-functions;

$$\rho_5(\cdot):\mathbb{R}_{\ge 0} \to \mathbb{R}_{\ge 0} \quad \text{and} \quad \rho_6(\cdot):\mathbb{R}_{\ge 0} \to \mathbb{R}_{\ge 0}$$

arc the K-functions. $\qquad\qquad\Box$

The quadratic and nonquadratic Lyapunov candidates are applied. Using the system dynamics, the total derivative of the Lyapunov candidate $V(t,x,e)$ is obtained. The inequality

$$\frac{dV(t,x,e)}{dt} \le -\rho_5 \|x\| - \rho_6 \|e\|$$

may be solved to find the feedback coefficients. The Lyapunov candidate functions should be designed. For example, the nonquadratic scalar Lyapunov function $V(x,e)$, $V(\cdot): \mathbb{R}^n \times \mathbb{R}^b \to \mathbb{R}_{\ge 0}$ is

$$V(x,e) = \sum_{i=1}^{\eta} \frac{1}{2i}(x^i)^T K_{xi} x^i + \sum_{i=1}^{\varsigma} \frac{1}{2i}(e^i)^T K_{ei} e^i,$$

$$K_{xi} \in \mathbb{R}^{n \times n}, K_{ei} \in \mathbb{R}^{b \times b}, \quad \eta = 1,2,3,\dots \quad \text{and} \quad \varsigma = 1,2,3,\dots.$$

Using the matrix-functions

$$K_{xi}(\cdot): \mathbb{R}_{\ge 0} \to \mathbb{R}^{n \times n} \quad \text{and} \quad K_{ei}(\cdot): \mathbb{R}_{\ge 0} \to \mathbb{R}^{b \times b},$$

the time-varying Lyapunov function

$$V(t,x,e), \ V(\cdot): \mathbb{R}_{\ge 0} \times \mathbb{R}^n \times \mathbb{R}^b \to \mathbb{R}_{\ge 0}$$

can be given as

$$V(t,x,e) = \sum_{2i}^{\eta} \frac{1}{2i}(x^i)^T K_{xi}(t) x^i + \sum_{i=1}^{\varsigma} \frac{1}{2i}(e^i)^T K_{ei}(t) e^i, \ K_{xi}(\cdot): \mathbb{R}_{\ge 0} \to \mathbb{R}^{n \times n} \text{ and }$$

$$K_{ei}(\cdot): \mathbb{R}_{\ge 0} \to \mathbb{R}^{b \times b}.$$

The scalar Lyapunov function $V(t,x,e)$, $V(\cdot): \mathbb{R}_{\ge 0} \times \mathbb{R}^n \times \mathbb{R}^b \to \mathbb{R}_{\ge 0}$ can be expressed as

$$V(t,x,e) = \sum_{i=1}^{\eta} \frac{1}{2i}(x^i)^T K_{xi}(t) x^i + \sum_{i=1}^{\lambda} \frac{1}{2i}(x^i)^T K_{xei}(t) e^i + \sum_{i=1}^{\varsigma} \frac{1}{2i}(e^i)^T K_{ei}(t) e^i,$$

$$\lambda = 1,2,3,\dots, \quad K_{xei}(\cdot): \mathbb{R}_{\ge 0} \to \mathbb{R}^{n \times b}.$$

The results in analytic design of control laws reported in previous sections can be applied. From (8.46) or (8.64), one may derive unconstrained control laws $u = f(t,e,x_m)$ or constrained control laws $u = \phi(t,e,x_m)$ using the directly measurable (or observable) x_m and e. For example, using (8.46), we have

$$v_m = -G_z^{-1} B_z^T(z) \frac{\partial V_m(x,e)}{\partial z}$$

where $V_m(x,e)$ is designed to ensure $u = f(e,x_m)$. Thus, departing from (8.45)

$$u(t) = K_{F1}\begin{bmatrix} x(t) \\ e(t) \end{bmatrix} + \int K_{F2}\begin{bmatrix} x(\tau) \\ e(\tau) \end{bmatrix} d\tau,$$

using the concept reported, one has a linear proportional-integral tracking control law

$$u(t) = K_{mF1}\begin{bmatrix} x_m(t) \\ e(t) \end{bmatrix} + \int K_{mF2}\begin{bmatrix} x_m(\tau) \\ e(\tau) \end{bmatrix} d\tau.$$

This control law utilizes measurable $x_m(t)$ and $e(t)$. To study the closed-loop system stability the closed-loop dynamics is examined using the Lyapunov function which is not necessarily $V_m(\cdot):\mathbb{R}^n \times \mathbb{R}^b \to \mathbb{R}_{\geq 0}$. One can design constrained control functions with nonlinear feedback mappings using $x_m(t)$ and $e(t)$.

Example 8.18

Consider an electric drive with a permanent-magnet DC motor and a *step-down* converter, which was covered in Section 3.2.2. Using the Kirchhoff law and the *averaging* concept, we have the following nonlinear state-space model

$$\begin{bmatrix} \dfrac{du_a}{dt} \\ \dfrac{di_L}{dt} \\ \dfrac{di_a}{dt} \\ \dfrac{d\omega_r}{dt} \end{bmatrix} = \begin{bmatrix} 0 & \dfrac{1}{C} & -\dfrac{1}{C} & 0 \\ -\dfrac{1}{L} & -\dfrac{r_L + r_s}{L} & \dfrac{r_c}{L} & 0 \\ \dfrac{1}{L_a} & \dfrac{r_c}{L_a} & -\dfrac{r_a + r_c}{L_a} & -\dfrac{k_a}{L_a} \\ 0 & 0 & \dfrac{k_a}{J} & -\dfrac{B_m}{J} \end{bmatrix} \begin{bmatrix} u_a \\ i_L \\ i_a \\ \omega_r \end{bmatrix}$$

$$+ \begin{bmatrix} 0 \\ \dfrac{V_d}{Lu_{t\max}} - \dfrac{r_s}{Lu_{t\max}} i_L \\ 0 \\ 0 \end{bmatrix} u_c - \begin{bmatrix} 0 \\ 0 \\ 0 \\ \dfrac{1}{J} \end{bmatrix} T_L, \quad u_c \in [0 \quad 10] \text{ V.}$$

The positive-definite Lyapunov function is

$$V(x,e) = \frac{1}{2}[u_a \quad i_L \quad i_a \quad \omega_r]K_{x1}\begin{bmatrix} u_a \\ i_L \\ i_a \\ \omega_r \end{bmatrix} + \frac{1}{2}k_{e1}e^2 + \frac{1}{4}k_{e2}e^4,$$

where $K_{x1} = I \in \mathbb{R}^{4\times4}$.

The measured state variable is only the angular velocity $\omega_r(t)$. The tracking error, $e(t) = r(t) - \omega_r(t)$ is also available. A bounded control law is synthesized as

$$u = \mathrm{sat}_0^{+10}\left(k_{p1}e + k_{p2}e^3 + k_{i1}\int edt + k_{i2}\int e^3dt - k_{14}\omega_r\right).$$

The criteria (i) and (ii) of Theorem 8.2, imposed on the Lyapunov pair, are guaranteed to satisfy the stability conditions. The positive-definite nonquadratic Lyapunov function was used. The feedback gains are found by solving the inequality

$$\frac{dV(e,x)}{dt} \le -\frac{1}{2}\|x\|^2 - \frac{1}{2}\|e\|^2 - \frac{1}{4}\|e\|^4$$

yielding

$$k_{p1} = 5.4, \quad k_{p2} = 1.8, \quad k_{i1} = 4.9, \quad k_{i2} = 2, \quad \text{and} \quad k_{14} = 0.085. \qquad \square$$

Example 8.19

We study the eight-layered lead magnesium niobate actuator (3 mm diameter, 0.25 mm thickness). A set of differential equations to model the actuator dynamics is

$$\frac{dF}{dt} = -8500F + 14Fu + 450u, \quad \frac{dv}{dt} = 1000F - 100000v - 2500v^3 - 2750x, \quad \frac{dx}{dt} = v.$$

The control is bounded, and $-100 \le u \le 100$ [V]. The error is the difference between the reference and actuator linear displacements. That is, $e(t) = r(t) - y(t)$, where $y(t) = x(t)$.

A bounded control law is synthesized using nonlinear error feedback as

$$u = \mathrm{sat}_{-100}^{+100}\left(950e + 26e^3 + 45\int edt + 8.4\int e^3dt\right).$$

The feedback gains are found by solving the inequality

$$\frac{dV(e,x)}{dt} \le -\|e\|^2 - \|e\|^4 - \|x\|^2,$$

where

$$V(x,e) = \frac{1}{2}[F \ v \ x]K_{x1}\begin{bmatrix} F \\ v \\ x \end{bmatrix} + \frac{1}{2}k_{e1}e^2 + \frac{1}{4}k_{e2}e^4.$$

The criteria (i) and (ii) imposed on the Lyapunov pair are satisfied because

$$V(x,e) > 0 \quad \text{and} \quad \frac{dV(x,e)}{dt} \le 0.$$

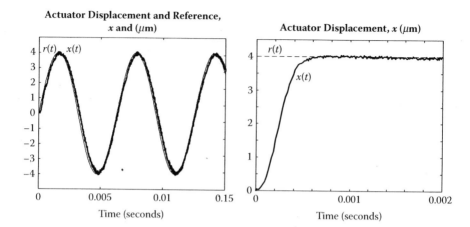

FIGURE 8.10

Transient output dynamics if $r(t) = 4 \times 10^{-6}\sin 1000t$ and $r(t) = \text{const} = 4 \times 10^{-6}$m.

Hence, the bounded control law guarantees stability and ensures tracking. The controller is experimentally tested. Figure 8.10 illustrates the transient dynamics for $x(t)$ if the reference signal (desired position) is assigned to be

$$r(t) = 4 \times 10^{-6}\sin 1000t \quad \text{and} \quad r(t) = \text{const} = 4 \times 10^{-6}\,\text{m}.$$

We conclude that the stability is guaranteed, desired performance is achieved, and the output precisely follows the reference $r(t)$.

□

8.10 Control of Linear Discrete-Time Electromechanical Systems Using the Hamilton-Jacobi Theory

8.10.1 Linear Discrete-Time Systems

Consider a discrete-time system described by the state-space difference equation

$$x_{n+1} = A_n x_n + B_n u_n, \quad n \ge 0. \tag{8.86}$$

The notations for the discrete-time case is similar for those used in continuous-time design. Different performance indexes are applied to optimize the closed-loop system dynamics. For example, the quadratic performance index to be minimized is

$$J = \sum_{n=0}^{N-1} \left[x_n^T Q_n x_n + u_n^T G_n u_n \right], \quad Q_n \ge 0, \quad G_n > 0. \tag{8.87}$$

Using the Hamilton-Jacobi theory, our goal is to find the control law that guarantees that the value of the performance index is minimum or maximum. For linear dynamic systems (8.86) and quadratic performance index (8.87), the solution of the Hamilton-Jacobi-Bellman equation

$$V(x_n) = \min_{u_n} \left[x_n^T Q_n x_n + u_n^T G_n u_n + V(x_{n+1}) \right] \qquad (8.88)$$

is satisfied by the quadratic return function

$$V(x_n) = x_n^T K_n x_n. \qquad (8.89)$$

From (8.88) and using (8.89), we have

$$V(x_n) = \min_{u_n} \left[x_n^T Q_n x_n + u_n^T G_n u_n + (A_n x_n + B_n u_n)^T K_{n+1} (A_n x_n + B_n u_n) \right]$$

$$= \min_{u_n} \left[x_n^T Q_n x_n + u_n^T G_n u_n + x_n^T A_n^T K_{n+1} A_n x_n + x_n^T A_n^T K_{n+1} B_n u_n \right.$$

$$\left. + u_n^T B_n^T K_{n+1} A_n x_n + u_n^T B_n^T K_{n+1} B_n u_n \right]. \qquad (8.90)$$

The first-order necessary condition for optimality (8.5) gives

$$u_n^T G_n + x_n^T A_n^T K_{n+1} B_n + u_n^T B_n^T K_{n+1} B_n = 0. \qquad (8.91)$$

From (8.91), the digital control law is

$$u_n = -\left(G_n + B_n^T K_{n+1} B_n \right)^{-1} B_n^T K_{n+1} A_n x_n. \qquad (8.92)$$

The second-order necessary condition for optimality (8.8) is guaranteed because

$$\frac{\partial^2 H(x_n, u_n, V(x_{n+1}))}{\partial u_n \times \partial u_n^T} = \frac{\partial^2 \left(u_n^T G_n u_n + u_n^T B_n^T K_{n+1} B_n u_n \right)}{\partial u_n \times \partial u_n^T}$$

$$= 2G_n + 2B_n^T K_{n+1} B_n > 0, \quad K_{n+1} > 0.$$

Using the control law derived as (8.92), from 8.90, one finds

$$x_n^T K_n x_n = x_n^T Q_n x_n + x_n^T A_n^T K_{n+1} A_n x_n$$

$$- x_n^T A_n^T K_{n+1} B_n \left(G_n + B_n^T K_{n+1} B_n \right)^{-1} B_n K_{n+1} A_n x. \qquad (8.93)$$

From (8.93), the difference equation to find the unknown symmetric matrix of the quadratic return function is

$$K_n = Q_n + A_n^T K_{n+1} A_n - A_n^T K_{n+1} B_n \left(G_n + B_n^T K_{n+1} B_n \right)^{-1} B_n K_{n+1} A_n. \qquad (8.94)$$

If in performance index (8.87) $N = \infty$, we have

$$J = \sum_{n=0}^{\infty} \left[x_n^T Q_n x_n + u_n^T G_n u_n \right], \quad Q_n \geq 0, \quad G_n > 0. \qquad (8.95)$$

The optimal control law is

$$u_n = -\left(G_n + B_n^T K_n B_n \right)^{-1} B_n^T K_n A_n x_n, \qquad (8.96)$$

where the unknown symmetric matrix K_n is found by solving the non-linear equation

$$-K_n + Q_n + A_n^T K_n A_n - A_n^T K_n B_n (G_n + B_n^T K_n B_n)^{-1} B_n K_n A_n = 0, \quad K_n = K_n^T. \qquad (8.97)$$

Matrix K_n is positive-definite, and the MATLAB dlqr command is used to find the matrix K_n, feedback matrix $(G_n + B_n^T K_n B_n)^{-1} B_n^T K_n A_n$ and eigenvalues. This command description is

```
>> help dlqr
DLQR Linear-quadratic regulator design for discrete-time systems.
   [K,S,E] = DLQR(A,B,Q,R,N) calculates the optimal gain matrix K
     such that the state-feedback law u[n] = -Kx[n] minimizes the
     cost function
              J = Sum {x'Qx + u'Ru + 2*x'Nu}
     subject to the state dynamics x[n+1] = Ax[n] + Bu[n].
   The matrix N is set to zero when omitted. Also returned are the
     Riccati equation solution S and the closed-loop eigenvalues E:
                                    -1
       A'SA - S - (A'SB+N)(R+B'SB) (B'SA+N') + Q = 0, E = EIG(A-B*K).
   See also dlqry, lqrd, lqgreg, and dare.
```

The closed-loop systems (8.86) with (8.96) is expressed as

$$x_{n+1} = \left[A_n - B_n \left(G_n + B_n^T K_{n+1} B_n \right)^{-1} B_n^T K_{n+1} A_n \right] x_n. \qquad (8.98)$$

Example 8.20

For the second-order discrete-time system

$$x_{n+1} = \begin{bmatrix} x_{1n+1} \\ x_{2n+1} \end{bmatrix} = A_n x_n + B_n u_n = \begin{bmatrix} 1 & 2 \\ 0 & 3 \end{bmatrix} \begin{bmatrix} x_{1n} \\ x_{2n} \end{bmatrix} + \begin{bmatrix} 4 & 5 \\ 6 & 7 \end{bmatrix} \begin{bmatrix} u_{1n} \\ u_{2n} \end{bmatrix},$$

we find the digital control law by minimizing the performance index

$$J = \sum_{n=0}^{\infty} \left[x_n^T Q_n x_n + u_n^T G_n u_n \right]$$

$$= \sum_{n=0}^{\infty} \left[\begin{bmatrix} x_{1n} & x_{2n} \end{bmatrix} \begin{bmatrix} 10 & 0 \\ 0 & 10 \end{bmatrix} \begin{bmatrix} x_{1n} \\ x_{2n} \end{bmatrix} + \begin{bmatrix} u_{1n} & u_{2n} \end{bmatrix} \begin{bmatrix} 5 & 0 \\ 0 & 5 \end{bmatrix} \begin{bmatrix} u_{1n} \\ u_{2n} \end{bmatrix} \right]$$

$$= \sum_{n=0}^{\infty} \left(10x_{1n}^2 + 10x_{2n}^2 + 5u_{1n}^2 + 5u_{2n}^2 \right).$$

We upload the matrices and use the `dlqr` command. In particular,

```
A=[1 2;0 3];B=[4 5;6 7];
Q=10*eye(size(A)); G=5*eye(size(B));
[Kfeedback,Kn,Eigenvalues]=dlqr(A,B,Q,G)
```

The following numerical results are found

```
Kfeedback =
        -2.8456e-001      2.3091e-001
         3.3224e-001      2.2197e-001
Kn =
         1.9963e+001     -7.1140e-001
        -7.1140e-001      1.0577e+001
Eigenvalues =
         5.2192e-001
         1.5903e-002
```

Having calculated

$$K_n = \begin{bmatrix} 20 & -0.71 \\ -0.71 & 10.6 \end{bmatrix},$$

the control law is

$$u_n = -\left(G_n + B_n^T K_{n+1} B_n \right)^{-1} B_n^T K_n A_n x_n = - \begin{bmatrix} -0.285 & 0.231 \\ 0.332 & 0.222 \end{bmatrix} \begin{bmatrix} x_{1n} \\ x_{2n} \end{bmatrix},$$

$$u_{1n} = 0.285x_{1n} - 0.231x_{2n}, \quad u_{2n} = -0.332x_{1n} - 0.222x_{2n}.$$

The system is stable because the eigenvalues are within the unit circle. The eigenvalues calculated are 0.522 and 0.0159. ☐

Example 8.21
For the third-order system

$$x_{n+1} = \begin{bmatrix} x_{1n+1} \\ x_{2n+1} \\ x_{3n+1} \end{bmatrix} = A_n x_n + B_n u_n = \begin{bmatrix} 1 & 1 & 2 \\ 3 & 3 & 4 \\ 5 & 5 & 6 \end{bmatrix} \begin{bmatrix} x_{1n} \\ x_{2n} \\ x_{2n} \end{bmatrix} + \begin{bmatrix} 10 \\ 20 \\ 30 \end{bmatrix} u_n,$$

we will find a control law by minimizing the quadratic performance index

$$J = \sum_{n=0}^{\infty} \left[x_n^T Q_n x_n + u_n^T G_n u_n \right]$$

$$= \sum_{n=0}^{\infty} \left[\begin{bmatrix} x_{1n} & x_{2n} & x_{3n} \end{bmatrix} \begin{bmatrix} 1 & 0 & 0 \\ 0 & 10 & 0 \\ 0 & 0 & 100 \end{bmatrix} \begin{bmatrix} x_{1n} \\ x_{2n} \\ x_{3n} \end{bmatrix} + 1000 u_n^2 \right]$$

$$= \sum_{n=0}^{\infty} \left(x_{1n}^2 + 10 x_{2n}^2 + 100 x_{3n}^2 + 1000 u_n^2 \right).$$

Our goal is also to simulate the closed-loop system. The matrices of the output equation

$$y_n = C_n x_n + H_n u_n,$$

are $C_n = [1\ 1\ 1]$ and $H_n = [0]$.

We upload the system matrices by typing in the Command Window

```
A=[1 1 2; 3 3 4; 5 5 6]; B=[10; 20; 30];
```

The weighting matrices Q_n and G_n are uploaded as

```
Q=eye(size(A)); Q(2,2)=10; Q(3,3)=100; G=1000;
```

The feedback coefficients and eigenvalues are found by using the dlqr solver. In particular,

```
[Kfeedback,Kn,Eigenvalues]=dlqr(A,B,Q,G)
```

The following results are displayed in the Command Window

```
Kfeedback =
        1.5536e-001        1.5536e-001        1.9913e-001
Kn =
        4.8425e+001        4.7425e+001        3.1073e+001
        4.7425e+001        5.7425e+001        3.1073e+001
        3.1073e+001        3.1073e+001        1.3983e+002
Eigenvalues =
       -6.7355e-001
        3.8806e-002
        1.0119e-015
```

Thus,

$$K_n = \begin{bmatrix} 48.4 & 47.4 & 31.1 \\ 47.4 & 57.4 & 31.1 \\ 31.1 & 31.1 & 140 \end{bmatrix},$$

and the control law is

$$u_n = -0.155 x_{1n} - 0.155 x_{2n} - 0.2 x_{3n}.$$

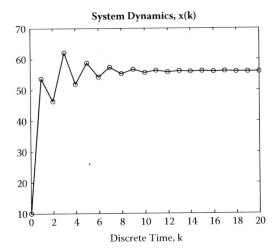

FIGURE 8.11
Output dynamics of the closed-loop system.

The dynamics of the closed-loop system, which is stable because the eigenvalues are within the unit circle, is simulated using the `filter` command. Having derived the closed-loop system dynamics as

$$x_{n+1} = \left[A_n - B_n \left(G_n + B_n^T K_{n+1} B_n \right)^{-1} B_n^T K_n A_n \right] x_n,$$

one finds the numerator and denominator of the transfer function in the z-domain. The MATLAB statement is

```
A _ closed=A-B*Kfeedback;C=[1 1 1]; H=[0];
[num,den]=ss2tf(A _ closed,B,C,H);
x0=[10 0 -10];
k=0:1:20; u=1*[ones(1,21)];
x=filter(num,den,u,x0);
plot(k,x,'-',k,x,'o');
title('System Dynamics, x(k)','FontSize',14);
xlabel('Discrete Time, k','FontSize',14);
```

The simulation results are illustrated in Figure 8.11.

The stabilization problem was solved, and the closed-loop system is stable. The tracking control problem must be approached and solved to guarantee that the system output follows the reference. □

8.10.2 Constrained Optimization of Discrete-Time Electromechanical Systems

Because of the constraints imposed on controls $u_{n\,min} \le u_n \le u_{n\,max}$, the designer must synthesize bounded control laws. In this section, the constrained optimization problem for multivariable discrete-time systems is covered.

Our goal is to design the constrained digital control laws using the Hamilton-Jacobi theory. We study discrete-time systems with bounded control. The system is described by the difference equation

$$x_{n+1} = A_n x_n + B_n u_n, \quad x_{n0} \in X_0, \quad u_n \in U, \quad n \geq 0 \tag{8.99}$$

mapping the control constraints by continuous, integrable, one-to-one bounded functions ϕ, $\phi \in U$, for which the inverse function ϕ^{-1} exist.

The nonquadratic performance index to be minimized is

$$J = \sum_{n=0}^{N-1} \left[x_n^T Q_n x_n + 2 \int (\phi^{-1}(u_n))^T G_n du_n - u_n^T B_n^T K_{n+1} B_n u_n \right]. \tag{8.100}$$

Performance indexes must be positive-definite. Hence, to attain the positive-definiteness,

$$\left[x_n^T Q_n x_n + 2 \int (\phi^{-1}(u_n))^T G_n du_n \right] > u_n^T B_n^T K_{n+1} B_n u_n$$

for all $x_n \in X$ and $u_n \in U$. \tag{8.101}

The Hamilton-Jacobi-Bellman recursive equation is

$$V(x_n) = \min_{u_n \in U} \left[x_n^T Q_n x_n + 2 \int (\phi^{-1}(u_n))^T G_n du_n - u_n^T B_n^T K_{n+1} B_n u_n + V(x_{n+1}) \right]. \tag{8.102}$$

Using the quadratic return function $V(x_n) = x_n^T K_n x_n$, from (8.102), one finds

$$x_n^T K_n x_n = \min_{u_n \in U} \left[x_n^T Q_n x_n + 2 \int (\phi^{-1}(u_n))^T G_n du_n - u_n^T B_n^T K_{n+1} B_n u_n \right.$$

$$\left. + (A_n x_n + B_n u_n)^T K_{n+1} (A_n x_n + B_n u_n) \right]. \tag{8.103}$$

Applying the first-order necessary condition for optimality, the bounded control law yields as

$$u_n = -\phi \left(G_n^{-1} B_n^T K_{n+1} A_n x_n \right), \quad u_n \in U. \tag{8.104}$$

It is evident that

$$\frac{\partial^2 (2 \int (\phi^{-1}(u_n))^T G_n du_n)}{\partial u_n \times \partial u_n^T}$$

is positive-definite because one-to-one integrable functions ϕ and ϕ^{-1} lie in the first and third quadrants and weighting matrix G_n is positive-definite.

We conclude that the second-order necessary condition for optimality is satisfied.

From (8.103) and (8.104), we have

$$
x_n^T K_n x_n = x_n^T Q_n x_n + 2 \int x_n^T A_n^T K_{n+1} B_n d\left(\phi\left(G_n^{-1} B_n^T K_{n+1} A_n x_n\right)\right)
$$

$$
+ x_n^T A_n^T K_{n+1} A_n x_n - 2 x_n^T A_n^T K_{n+1} B_n \phi\left(G_n^{-1} B_n^T K_{n+1} A_n x_n\right), \quad (8.105)
$$

where

$$
2 \int x_n^T A_n^T K_{n+1} B_n d\left(\phi\left(G_n^{-1} B_n^T K_{n+1} A_n x_n\right)\right) = 2 x_n^T A_n^T K_{n+1} B_n \phi\left(G_n^{-1} B_n^T K_{n+1} A_n x_n\right)
$$

$$
- 2 \int \left(\phi\left(G_n^{-1} B_n^T K_{n+1} A_n x_n\right)\right)^T d\left(B_n^T K_{n+1} A_n x_n\right).
$$

$$(8.106)$$

Using (8.105), one concludes that the unknown matrix K_{n+1} is found by solving

$$
x_n^T K_n x_n = x_n^T Q_n x_n + x_n^T A_n^T K_{n+1} A_n x_n - 2 \int \left(\phi\left(G_n^{-1} B_n^T K_{n+1} A_n x_n\right)\right)^T d\left(B_n^T K_{n+1} A_n x_n\right).
$$

$$(8.107)$$

Describing the control bounds imposed by using the continuous integrable one-to-one bounded functions $\phi \in U$, one finds the expression for

$$
2 \int \left(\phi\left(G_n^{-1} B_n^T K_{n+1} A_n x_n\right)\right)^T d\left(B_n^T K_{n+1} A_n x_n\right).
$$

For example, using the tanh function to describe the saturation-type constraints, one obtains

$$
\int \tanh z \, dz = \log \cosh z \quad \text{and} \quad \int \tanh^g z \, dz = -\frac{\tanh^{g-1} z}{g-1} + \int \tanh^{g-2} z \, dz, \quad g \neq 1.
$$

Matrix K_{n+1} is found by solving the recursive equation (8.107), and the feedback gains result.

Minimizing the nonquadratic performance index (8.100) for $N = \infty$, we obtain the bounded control law as

$$
u_n = -\phi\left(G_n^{-1} B_n^T K_n A_n x_n\right), \quad u_n \in U. \quad (8.108)
$$

The *admissibility* concept, which is based on the Lyapunov stability theory, is applied to verify the stability of the resulting closed-loop system in the operating envelope for all $x_n \in X$ and $u_n \in U$. This problem is of a particular importance for open-loop unstable systems. The resulting closed-loop system (8.99)–(8.108) evolves in X. A subset of the admissible domain of stability $S \subset \mathbb{R}^n$ is found by using the Lyapunov stability theory as

$$S = \{x_n \in R^n : x_{n0} \in X_0, u_n \in U \,|\, V(0) = 0, V(x_n) > 0, \Delta V(x_n) < 0, \forall x_n \in X(X_0, U)\}$$

The region of attraction can be studied, and S is an *invariant* domain. The quadratic Lyapunov function

$$V(x_n) = x_n^T K_n x_n,$$

is positive-definite if $K_n > 0$. Hence, the first difference, as given by

$$\Delta V(x_n) = V(x_{n+1}) - V(x_n) = x_n^T A_n^T K_{n+1} A_n x_n - 2x_n^T A_n^T K_{n+1} B_n \phi \left(G_n^{-1} B_n^T K_{n+1} A_n x_n \right)$$

$$+ \phi \left(G_n^{-1} B_n^T K_{n+1} A_n x_n \right)^T B_n^T K_{n+1} B_n \phi \left(G_n^{-1} B_n^T K_{n+1} A_n x_n \right) - x_n^T K_n x_n,$$

$$(8.109)$$

must be negative-definite for all $x_n \in X$ to ensure the stability. The evolution of the closed-loop systems depends on the initial conditions, constraints, references, disturbances, and so on. For the initial conditions within the operating envelope and control designed, we have the evolution set $X(X_0, U)$. The *sufficiency* analysis of stability is performed studying sets $S \subset \mathbb{R}^n$ and $X(X_0, U) \subset \mathbb{R}^n$. Stability is guaranteed if $X \subseteq S$.

The constrained optimization problem must be solved for nonlinear systems, which are modeled by nonlinear difference equations. Nonlinear discrete-time electromechanical systems are described as

$$x_{n+1} = F(x_n) + B(x_n)u_n, \quad x_{n0} \in X_0, \ u_n \in U, \ u_{n\,min} \leq u_n \leq u_{n\,max}, \ n \geq 0. \quad (8.110)$$

The Hamilton-Jacobi theory is applied to design bounded control laws using nonquadratic performance indexes. To design a nonlinear *admissible* control law $u_n \in U$, we describe the imposed control bounds by a continuous integrable one-to-one bounded function $\phi \in U$. The nonquadratic performance index is

$$J = \sum_{n=0}^{N-1} \left[x_n^T Q_n x_n - u_n^T B^T(x_n) K_{n+1} B(x_n) u_n + 2 \int (\phi^{-1}(u_n))^T G_n du_n \right]. \quad (8.111)$$

The integrand

$$2 \int \left(\phi^{-1}(u_n) \right)^T G_n du_n$$

is positive-definite because the integrable one-to-one function ϕ lies in the first and third quadrants, integrable function ϕ^{-1} exists, and $G_n > 0$. The positive definiteness of the performance index is guaranteed if

$$\left(x_n^T Q_n x_n + 2\int (\phi^{-1}(u_n))^T G_n du_n \right) > u_n^T B^T(x_n)K_{n+1}B(x_n)u_n$$

$$\text{for all} \quad x_n \in X \quad \text{and} \quad u_n \in U. \tag{8.112}$$

The positive definiteness of the performance index in $x_n \in X$ and $u_n \in U$ can be studied as positive-definite symmetric matrix K_{n+1} is found. The inequality (8.112) is ensured by using Q_n and G_n.

The first- and second-order necessary conditions for optimality are applied. For the quadratic return function (8.89), we have

$$V(x_{n+1}) = x_{n+1}^T K_{n+1} x_{n+1} = (F(x_n)+B(x_n)u_n)^T K_{n+1}(F(x_n)+B(x_n)u_n). \tag{8.113}$$

Therefore,

$$x_n^T K_n x_n = \min_{u_n \in U}\Big[x_n^T Q_n x_n - u_n^T B^T(x_n)K_{n+1}B(x_n)u_n + 2\int (\phi^{-1}(u_n))^T G_n du_n$$

$$+ (F(x_n)+B(x_n)u_n)^T K_{n+1}(F(x_n)+B(x_n)u_n)\Big]. \tag{8.114}$$

Using the first-order necessary condition for optimality, a bounded control law is

$$u_n = -\phi\big(G_n^{-1}B^T(x_n)K_{n+1}F(x_n)\big), \quad u_n \in U. \tag{8.115}$$

The second-order necessary condition for optimality is satisfied because

$$\frac{\partial^2 (2\int(\phi^{-1}(u_n))^T G_n du_n)}{\partial u_n \times \partial u_n^T} > 0.$$

From (8.114), using the bounded control law (8.115), we have the following recursive equation

$$x_n^T K_n x_n = x_n^T Q_n x_n + 2\int F^T(x_n)K_{n+1}B(x_n)d\big(\phi\big(G_n^{-1}B^T(x_n)K_{n+1}F(x_n)\big)\big)$$

$$+ F^T(x_n)K_{n+1}F(x_n) - 2F^T(x_n)K_{n+1}B(x_n)\phi\big(G_n^{-1}B^T(x_n)K_{n+1}F(x_n)\big). \tag{8.116}$$

The integration by parts gives

$$2\int F^T(x_n)K_{n+1}B(x_n)d\big(\phi\big(G_n^{-1}B^T(x_n)K_{n+1}F(x_n)\big)\big)$$

$$= 2F^T(x_n)K_{n+1}B(x_n)\,\phi\big(G_n^{-1}B^T(x_n)K_{n+1}F(x_n)\big)$$

$$-2\int \big(\phi\big(G_n^{-1}B^T(x_n)K_{n+1}F(x_n)\big)\big)^T d\big(B^T(x_n)K_{n+1}F(x_n)\big).$$

The equation to find the unknown symmetric matrix K_{n+1} is

$$x_n^T K_n x_n = x_n^T Q_n x_n + F^T(x_n) K_{n+1} F(x_n)$$

$$- 2 \int \left(\phi \left(G_n^{-1} B^T(x_n) K_{n+1} F(x_n) \right) \right)^T d(B^T(x_n) K_{n+1} F(x_n)). \quad (8.117)$$

By describing the control bounds imposed by the continuous integrable one-to-one bounded functions $\phi \in U$, one finds and approximates

$$2 \int \left(\phi \left(G_n^{-1} B^T(x_n) K_{n+1} F(x_n) \right) \right)^T d \left(B^T(x_n) K_{n+1} F(x_n) \right).$$

Hence, equation (8.117) can be solved.

The *admissibility* concept is applied to verify the stability of the resulting closed-loop system. The closed-loop system evolves in $X \subset \mathbb{R}^n$, and

$$\left\{ x_{n+1} = F(x_n) - B(x_n) \phi \left(G_n^{-1} B^T(x_n) K_{n+1} F(x_n) \right), \quad x_{n0} \in X_0 \right\} \in X(X_0, U).$$

Using the Lyapunov stability theory, the domain of stability $S \subset \mathbb{R}^n$ is found by applying the sufficient conditions under which the discrete-time closed-loop system (8.110)–(8.115) is stable. The positive-definite quadratic function (8.89) is used. To guarantee the stability, the first difference

$$\Delta V(x_n) = V(x_{n+1}) - V(x_n) = F^T(x_n) K_{n+1} F(x_n)$$

$$- 2 F^T(x_n) K_{n+1} B(x_n) \phi(G_n^{-1} B^T(x_n) K_{n+1} F(x_n))$$

$$+ \phi(G_n^{-1} B^T(x_n) K_{n+1} F(x_n))^T B^T(x_n) K_{n+1} B(x_n) \phi(G_n^{-1} B^T(x_n) K_{n+1} F(x_n))$$

$$- x_n^T K_n x_n. \quad (8.118)$$

must be negative-definite for all $x_n \in X$ and $u_n \in U$. Define the set

$$S = \{ x_n \in \mathbb{R}^n : x_{n0} \in X_0, u \in U \mid V(0) = 0, \ V(x_n) > 0, \ \Delta V(x_n) < 0, \ \forall x \in X(X_0, U) \}$$

The stability analysis is performed by studying S and $X(X_0, U)$. The constrained optimization problem is solved via the bounded admissible control law (8.115) and the stability is guaranteed if $X \subseteq S$.

8.10.3 Tracking Control of Discrete-Time Systems

We study systems modeled by the following difference equation in the state-space form

$$x_{n+1}^{system} = A_n x_n^{system} + B_n u_n, \quad x_{n0}^{system} \in X_0, \quad u_n \in U, \quad n \geq 0. \quad (8.119)$$

The output equation is

$$y_n = H_n x_n^{system}.$$

The *exogeneous* system is given as

$$x_n^{ref} = x_{n-1}^{ref} + r_n - y_n. \tag{8.120}$$

Using (8.119) and (8.120), one finds

$$x_{n+1}^{ref} = x_n^{ref} + r_{n+1} - y_{n+1} = x_n^{ref} + r_{n+1} - H_n \left(A_n x_n^{system} + B_n u_n \right). \tag{8.121}$$

Hence,

$$x_{n+1} = \begin{bmatrix} x_{n+1}^{system} \\ x_{n+1}^{ref} \end{bmatrix} = \begin{bmatrix} A_n & 0 \\ -H_n A_n & I_n \end{bmatrix} x_n + \begin{bmatrix} B_n \\ -H_n B_n \end{bmatrix} u_n + \begin{bmatrix} 0 \\ I_n \end{bmatrix} r_{n+1}, \quad x_n = \begin{bmatrix} x_n^{system} \\ x_n^{ref} \end{bmatrix}. \tag{8.122}$$

To synthesize the bounded control law, we minimize the nonquadratic performance index

$$J = \sum_{n=0}^{N-1} \left[x_n^T Q_n x_n + 2 \int (\phi^{-1}(u_n))^T G_n du_n - u_n^T \begin{bmatrix} B_n \\ -H_n B_n \end{bmatrix}^T K_{n+1} \begin{bmatrix} B_n \\ -H_n B_n \end{bmatrix} u_n \right]. \tag{8.123}$$

Using the quadratic return function (8.89), from the Hamilton-Jacobi equation

$$x_n^T K_n x_n = \min_{u_n \in U} \left[x_n^T Q_n x_n + 2 \int (\phi^{-1}(u_n))^T G_n du_n - u_n^T \begin{bmatrix} B_n \\ -H_n B_n \end{bmatrix}^T K_{n+1} \begin{bmatrix} B_n \\ -H_n B_n \end{bmatrix} u_n \right.$$

$$\left. + \left(\begin{bmatrix} A_n & 0 \\ -H_n A_n & I_n \end{bmatrix} x_n + \begin{bmatrix} B_n \\ -H_n B_n \end{bmatrix} u_n \right)^T K_{n+1} \left(\begin{bmatrix} A_n & 0 \\ -H_n A_n & I_n \end{bmatrix} x_n + \begin{bmatrix} B_n \\ -H_n B_n \end{bmatrix} u_n \right) \right] \tag{8.124}$$

using the first-order necessary condition for optimality, one obtains the following bounded tracking control law

$$u_n = -\phi \left(G_n^{-1} \begin{bmatrix} B_n \\ -H_n B_n \end{bmatrix}^T K_{n+1} \begin{bmatrix} A_n & 0 \\ -H_n A_n & I_n \end{bmatrix} x_n \right), \quad u_n \in U. \tag{8.125}$$

The unknown matrix K_{n+1} is found by solving the equation

$$x_n^T K_n x_n = x_n^T Q_n x_n + x_n^T \begin{bmatrix} A_n & 0 \\ -H_n A_n & I_n \end{bmatrix}^T K_{n+1} \begin{bmatrix} A_n & 0 \\ -H_n A_n & I_n \end{bmatrix} x_n$$

$$-2\int \left(\phi \left(G_n^{-1} \begin{bmatrix} B_n \\ -H_n B_n \end{bmatrix}^T K_{n+1} \begin{bmatrix} A_n & 0 \\ -H_n A_n & I_n \end{bmatrix} x_n \right) \right)^T d\left(\begin{bmatrix} B_n \\ -H_n B_n \end{bmatrix}^T K_{n+1} \begin{bmatrix} A_n & 0 \\ -H_n A_n & I_n \end{bmatrix} x_n \right).$$

$$(8.126)$$

The tracking control problem can be solved for linear and nonlinear discrete-time systems applying the *state transformation* concept reported in Section 8.4. One recalls that the proportional-integral control law with the state feedback results. The problem, however, is to implement the control law designed using analog and digital controller hardware. Not all state variables can be directly measured or observed. Therefore, the *minimal complexity* control laws can be considered as a viable solution. Among various practical solutions, the PID-centered controllers usually are prioritized. For nonlinear systems these controllers can be designed by using the Lyapunov theory. The Hamilton-Jacobi method also can be used to derive control laws. The designer also attempts to minimize the sampling period, which affects the dynamics of electromechanical systems. The design of control laws must be integrated with the hardware used ensuring a coherent codesign.

Homework Problems

Problem 8.1

For the second-order system

$$\frac{dx_1}{dt} = -x_1 + x_2, \quad \frac{dx_2}{dt} = x_1 + u,$$

find a control law that minimizes the quadratic functional

$$J = \frac{1}{2}\int_0^\infty \left(x_1^2 + 2x_2^2 + 3u^2 \right) dt$$

applying MATLAB. Study the closed-loop system stability.

Problem 8.2

Using the Lyapunov stability theory, study stability of the system that is described by the differential equations

$$\dot{x}_1(t) = -x_1 + 10x_2,$$
$$\dot{x}_2(t) = -10x_1 - x_2^7, \quad t \geq 0.$$

Problem 8.3

Let the electromechanical system be described by the differential equations

$$x_1(t) = -x_1 + 10x_2,$$
$$x_2(t) = x_1 + u.$$

Derive (synthesize) the control law that will stabilize this system. Using the Lyapunov stability theory, prove the stability of the closed-loop system.

References

1. M. Athans and P.L. Falb, *Optimal Control: An Introduction to the Theory and its Applications.* McGraw-Hill Book Company, New York, 1966.
2. R.C. Dorf and R.H. Bishop, *Modern Control Systems*, Addison-Wesley Publishing Company, Reading, MA, 1995.
3. J.F. Franklin, J.D. Powell, and A. Emami-Naeini, *Feedback Control of Dynamic Systems*, Addison-Wesley Publishing Company, Reading, MA, 1994.
4. H.K. Khalil, *Nonlinear Systems*, Prentice-Hall, Inc., NJ, 1996.
5. B.C. Kuo, *Automatic Control Systems*, Prentice Hall, Englewood Cliffs, NJ, 1995.
6. W.S. Levine (Editor), *Control Handbook*, CRC Press, FL, 1996.
7. S.E. Lyshevski, *Control Systems Theory with Engineering Applications*, Birkhauser, Boston, MA, 2000.
8. S.E. Lyshevski, *MEMS and NEMS: Systems, Devices, and Structures*, CRC Press, Boca Raton, FL, 2005.
9. K. Ogata, *Discrete-Time Control Systems*, Prentice-Hall, Upper Saddle River, NJ, 1995.
10. K. Ogata, *Modern Control Engineering*, Prentice-Hall, Upper Saddle River, NJ, 1997.
11. C. L. Phillips and R.D. Harbor, *Feedback Control Systems*, Prentice Hall, Englewood Cliffs, NJ, 1996.

Index